The 'PIPING GUIDE' .... ■ Discusses in detail the design and drafting of piping systems

- ■ Describes pipe, piping components most commonly used, valves, and equipment

- ■ Presents charts, tables, and examples for daily reference

- ■ Provides a design reference for companies and consultants

- ■ Supplements existing company standards, information, and methods

- ■ Serves as an instructional aid

PART I - TEXT: explains ....■ Techniques of piping design

- ■ Assembling of piping from components, and methods for connecting to equipment

- ■ Office organization, and methods to translate concepts into finished designs from which plants are built

- ■ Terms and abbreviations concerned with piping

PART II - TABLES: provide .....■ Frequently needed data and information, arranged for quick reference

- ■ Factors for establishing widths of pipeways

- ■ Spacing between pipes, with and without flanges, and for 'jumpovers' and 'rununders'

- ■ Principal dimensions and weights for pipe fittings, flanges, valves, structural steel, etc.

- ■ Conversion for customary and metric units

- ■ Direct-reading metric conversion tables for dimensions

and ....■ A metric supplement with principal dimensional data in millimeters

For PART II, turn to the back cover

Dust cover photographs by Johnny Hamilton. Dust cover design by Johnny Hamilton with lots of help from and collaboration with Margaret Hamilton, Mary Westheimer and Sandy Aska. We appreciate Mr. David Sherwood writing the piece about how *The Piping Guide* came into being.

the Piping Guide

or the Design and Drafting of

ndustrial Piping Systems

David R. Sherwood
Member, American Society of Mechanical Engineers
Member, Institution of Production Engineers

Dennis J. Whistance BS, MS

Copyright 1973, David R. Sherwood and Dennis J. Whistance
Second Edition, Copyright 1991, Syentek Books Company, Inc.
Copyright Transferred in 2008 to Construction Trades Press, LLC

Printed in the United States of America

Published and Distributed by:
Construction Trades Press, LLC
2265 Southeast Blvd
Clinton, North Carolina 28328

Hard Cover   ISBN 978-0-9624197-6-8
Soft Cover   ISBN 978-0-9624197-7-5

PART I

CONTENTS:

Sections, figures, charts and tables in **Part I** are referred to numerically, and are located by the margin index. Charts and tables in **Part II** are identified by letter.

The text refers to standards and codes, using designations such as ANSI B31.1, ASTM A-53, ISA S5.1, etc. Full titles of these standards and codes will be found in tables 7.3 thru 7.14.

FOR TERMS NOT EXPLAINED IN THE TEXT, REFER TO THE INDEX.

ABBREVIATIONS ARE GIVEN IN CHAPTER 8.

PIPING: Uses, and Plant Construction

USES OF PIPING   1.1

Piping is used for industrial (process), marine, transportation, civil engineering, and for 'commercial' (plumbing) purposes.

This book is primarily concerned with industrial piping for processing and service systems. *Process piping* is used to transport fluids between storage tanks and processing units. *Service piping* is used to convey steam, air, water, etc., for processing. Piping here defined as 'service' piping is sometimes referred to as 'utility' piping, but, in the Guide, the term 'utility piping' is reserved for major lines supplying water, fuel gases, and fuel oil (that is, for commodities usually purchased from utilities companies and bulk suppliers).

Marine piping for ships is often extensive. Much of it is fabricated from welded and screwed carbon-steel piping, using pipe and fittings described in this book.

Transportation piping is normally large-diameter piping used to convey liquids, slurries and gases, sometimes over hundreds of miles. Crude oils, petroleum products, water, and solid materials such as coal (carried by water) are transported thru pipelines. Different liquids can be transported consecutively in the same pipeline, and branching arrangements are used to divert flows to different destinations.

Civil piping is used to distribute public utilities (water, fuel gases), and to collect rainwater, sewage, and industrial waste waters. Most piping of this type is placed underground.

Plumbing (commercial piping) is piping installed in commercial buildings, schools, hospitals, residences, etc., for distributing water and fuel gases, for collecting waste water, and for other purposes.

COMMISSIONING, DESIGNING,   1.2
& BUILDING A PLANT

When a manufacturer decides to build a new plant, or to expand an existing one, the manufacturer will either employ an engineering company to undertake design and construction, or, if the company's own engineering department is large enough, they will do the design work, manage the project, and employ one or more contractors to do the construction work.

In either procedure, the manufacturer supplies information concerning the purposes of buildings, processes, production rates, design criteria for specific requirements, details of existing plant, and site surveys, if any.

Chart 1.1 shows the principals involved, and the flow of information and material.

SCHEMATIC FOR PLANT CONSTRUCTION   **CHART 1.1**

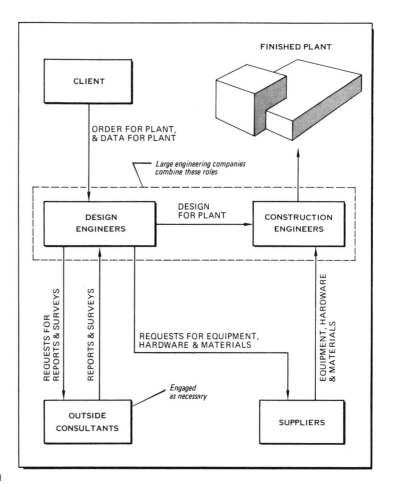

CHART 1.1

The designing and building of an industrial plant is a complex undertaking. Except for the larger industrial concerns, who may maintain their own design staffs, the design and construction of plants and related facilities is usually undertaken by specialist companies.

The Guide describes in 4.1 the organization and responsibilities of design engineering, with special reference to the duties of individuals engaged in the development of piping designs for plants.

PIPE, FITTINGS, FLANGES, REINFORCEMENTS, In-line Equipment and Support Equipment

PROCESS PIPE   2.1

PIPE & TUBE   2.1.1

Tubular products are termed 'tube' or 'pipe'. Tube is customarily specified by its outside diameter and wall thickness, expressed either in BWG (Birmingham wire gage) or in thousandths of an inch. Pipe is customarily identified by 'nominal pipe size', with wall thickness defined by 'schedule number', 'API designation', or 'weight', as explained in 2.1.3. Non-standard pipe is specified by nominal size with wall thickness stated.

The principal uses for tube are in heat exchangers, instrument lines, and small interconnections on equipment such as compressors, boilers, and refrigerators.

SIZES & LENGTHS COMMONLY USED FOR STEEL PIPE   2.1.2

ANSI standard B36.10M establishes wall thicknesses for pipe ranging from 1/8 to 80-inch nominal diameter ('nominal pipe size'). Pipe sizes normally stocked include: 1/2, 3/4, 1, 1¼, 1½, 2, 2½, 3, 3½, 4, 5, 6, 8, 10, 12, 14, 16, 18, 20 and 24. Sizes 1¼, 2½, 3½, and 5 inch are seldom used (unusual sizes are sometimes required for connecting to equipment, but piping is normally run in the next larger stock size after connection has been made). 1/8, 1/4, 3/8 and 1/2-inch pipe is usually restricted to instrument lines or to service and other lines which have to mate with equipment. 1/2-inch pipe is extensively used for steam tracing and for auxiliary piping at pumps, etc.

Straight pipe is supplied in 'random' lengths (17 to 25 ft), and sometimes 'double random' lengths (38 to 48 ft), if preferred. The ends of these lengths are normally either plain (PE), beveled for welding (BE), or threaded and supplied with one coupling per length ('threaded and coupled', or 'T&C'). If pipe is ordered 'T&C', the rating of the coupling is specified—see chart 2.3. Other types of ends, such as grooved for special couplings, can be obtained to order.

DIAMETERS & WALL THICKNESSES OF PIPE   2.1.3

The size of all pipe is identified by the nominal pipe size, abbreviated 'NPS', which is seldom equal to the true bore (internal diameter) of the pipe—the difference in some instances is large. NPS 14 and larger pipe has outside diameter equal to the nominal pipe size.

Pipe in the various sizes is made in several wall thicknesses for each size, which have been established by three different sources:—

(1)   The American National Standards Institute, thru 'schedule numbers'

(2)   The American Society of Mechanical Engineers and the American Society for Testing and Materials, thru the designations 'STD' (standard), 'XS' (extra-strong), and 'XXS' (double-extra-strong), drawn from dimensions established by manufacturers. *In the Guide, these designations are termed 'manufacturers' weights'*

(3)   The American Petroleum Institute, through its standard 5L, for 'Line pipe'. Dimensions in this standard have no references for individual sizes and wall thicknesses

'Manufacturers' weights' (second source) were intended, as long ago as 1939, to be superseded by schedule numbers. However, demand for these wall thicknesses has caused their manufacture to continue. Certain fittings are available only in manufacturers' weights.

Pipe dimensions from the second and third sources are incorporated in American National Standard B36.10M. Tables P-1 list dimensions for welded and seamless steel pipe in this standard, and give derived data.

IRON PIPE SIZES were initially established for wrought-iron pipe, with wall thicknesses designated by the terms 'standard (weight)', 'extra-strong', and 'double-extra-strong'. Before the schedule number scheme for steel pipe was first published by the American Standards Association in 1935, the iron pipe sizes were modified for steel pipe by slightly decreasing the wall thicknesses (leaving the outside diameters constant) so that the weights per foot (lb/ft) equalled the iron pipe weights.

Wrought-iron pipe (no longer made) has been completely supplanted by steel pipe, but schedule numbers, intended to supplant iron pipe designations did not. Users continued to specify pipe in iron pipe terms, and as the mills responded, these terms are included in ANSI standard B36.10M for steel pipe. Schedule numbers were introduced to establish pipe wall thicknesses by formula, but as wall thicknesses in common use continued to depart from those proposed by the scheme, schedule numbers now identify wall thicknesses of pipe in the different nominal sizes as ANSI B36.10M states "as a convenient designation system for use in ordering".

STAINLESS-STEEL SIZES American National Standard B36.19 established a range of thin-walled sizes for stainless-steel pipe, indentified by schedules 5S and 10S.

MATERIALS FOR PIPE   2.1.4

STEEL PIPE Normally refers to carbon-steel pipe. Seam-welded steel pipe is made from plate. Seamless pipe is made using dies. Common finishes are 'black' ('plain' or 'mill' finish) and galvanized.

Correctly selected steel pipe offers the strength and durability required for the application, and the ductility and machinability required to join it and form it into piping ('spools' -- see 5.2.9). The selected pipe must withstand the conditions of use, especially pressure, temperature and corrosion conditions. These requirements are met by selecting pipe made to an appropriate standard; in almost all instances an ASTM or API standard (see 2.1.3 and table 7.5).

The most-used steel pipe for process lines, and for welding, bending, and coiling, is made to ASTM A-53 or ASTM A-106, principally in wall thicknesses defined by schedules 40, 80, and manufacturers' weights, STD and XS. Both ASTM A-53 and ASTM A-106 pipe is fabricated seamless or seamed, by electrical resistance welding, in Grades A and B. Grades B have the higher tensile strength. Three grades of A-106 are available—Grades A, B, and C, in order of increasing tensile strength.

The most widely stocked pipe is to ASTM A-120 which covers welded and seamless pipe for normal use in steam, water, and gas (including air) service. ASTM A-120 is not intended for bending, coiling or high temperature service. It is not specified for hydrocarbon process lines.

In the oil and natural gas industries, steel pipe used to convey oil and gas is manufactured to the American Petroleum Institute's standard API 5L, which applies tighter control of composition and more testing than ASTM-120.

Steel specifications in other countries may correspond with USA specifications. Some corresponding european standards for carbon steels and stainless steels are listed in table 2.1.

IRON pipe is made from cast-iron and ductile-iron. The principal uses are for water, gas, and sewage lines.

OTHER METALS & ALLOYS Pipe or tube made from copper, lead, nickel, brass, aluminum and various stainless steels can be readily obtained. These materials are relatively expensive and are selected usually either because of their particular corrosion resistance to the process chemical, their good heat transfer, or for their tensile strength at high temperatures. Copper and copper alloys are traditional for instrument lines, food processing, and heat transfer equipment, but stainless steels are increasingly being used for these purposes.

PLASTICS Pipe made from plastics may be used to convey actively corrosive fluids, and is especially useful for handling corrosive or hazardous gases and dilute mineral acids. Plastics are employed in three ways: as all-plastic pipe, as 'filled' plastic materials (glass-fiber-reinforced, carbon-filled, etc.) and as lining or coating materials. Plastic pipe is made from polypropylene, polyethylene (PE), polybutylene (PB), polyvinyl chloride (PVC), acrylonitrilebutadiene-styrene (ABS), cellulose acetate-butyrate (CAB), polyolefins, and polyesters. Pipe made from polyester and epoxy resins is frequently glassfiber-reinforced ('FRP') and commercial products of this type have good resistance to wear and chemical attack.

COMPARABLE USA & EUROPEAN SPECIFICATIONS FOR STEEL PIPE   TABLE 2.1

	USA	UK	W. GERMANY	SWEDEN
CARBON-STEEL PIPE	**ASTM A53** Grade A SMLS Grade B SMLS	**BS 3601** HFS 22 & CDS 22 HFS 27 & CDS 27	**DIN 1629** St 35 St 45	SIS 1233-05 SIS 1434-05
	ASTM A53 Grade A ERW Grade B ERW	**BS 3601** ERW 22 ERW 27	**DIN 1626** Blatt 3 St 34-2 ERW Blatt 3 St 37-2 ERW	
	ASTM A53 FBW	**BS 3601** BW 22	**DIN 1626** Blatt 3 St 34-2 FBW	
	ASTM A106 Grade A Grade B Grade C	**BS 3602** HFS 23 HFS 27 HFS 35	**DIN 17175*** St 35-8 St 45-8	SIS 1234-05 SIS 1435-05
	ASTM A134	**BS 3601** EFW	**DIN 1626** Blatt 2 EFW	
	ASTM A135 Grade A Grade B	**BS 3601** ERW 22 ERW 27	**DIN 1626** Blatt 3 St 34-2 ERW Blatt 3 St 37-2 ERW	SIS 1233-06 SIS 1434-06
	ASTM A139 Grade A Grade B	**BS 3601** EFW 22 EFW 27	**DIN 1626** Blatt 2 St 37 Blatt 2 St 42	
	ASTM A155 **Class 2** C 45 C 50 C 55 KC 55 KC 60 KC 65 KC 70	**BS 3602** EFW 28 EFW 28S	**DIN 1626, Blatt 3, with certification C** St 34-2 St 37-2 St 42-2 St 42-2 * St 42-2 * St 52-3 St 52-3	
	API 5L Grade A SMLS Grade B SMLS	**BS 3601** HFS 22 & CDS 22 HFS 27 & CDS 27	**DIN 1629** St 35 St 45	SIS 1233-05 SIS 1434-05
	API 5L Grade A ERW Grade B ERW	**BS 3601** ERW 22 ERW 27 †	**DIN 1625** Blatt 3 St 34-2 ERW Blatt 4 St 37-2 ERW	SIS 1233-06 SIS 1434-06 †
	API 5L Grade A EFW Grade B EFW	**BS 3601** **Double-welded** EFW 22 EFW 27 †	**DIN 1626** Blatt 3 St 34-2 FW Blatt 4 St 37-2 FW	
	API 5L FBW	**BS 3601** BW 22	**DIN 1626** Blatt 3 St 34-2 FBW	
	*Specify "Si-killed" †Specify API 5L Grade B testing procedures for these steels			
STAINLESS-STEEL PIPE	**ASTM A312** TP 304 TP 304H TP 304L TP 310 TP 316	**BS 3605** Grade 801 Grade 811 Grade 801L Grade 805 Grade 845	**WSN Designation:** 4301 X 5 CrNi 18 9 4306 X 2 CrNi 18 9 4841 X 15 CrNiSi 25 20 4401/ X 5 CrNiMo 18 10 4436	SIS 2333-02 SIS 2352-02 SIS 2361-02 SIS 2343-02
	TP 316H TP 316L TP 317 TP 321 TP 321H TP 347 TP 347H	Grade 855 Grade 845L Grade 846 Grade 822 Ti Grade 832 Ti Grade 822 Nb Grade 832 Nb	 4404 X 2 CrNiMo 18 10 4541 X 10 CrNiTi 18 9 4550 X 10 CrNiNb 18 9	SIS 2353-02 SIS 2337-02 SIS 2338-02

The American National Standards Institute has introduced several schedules for pipe made from various plastics. These ANSI standards and others for plastic pipe are listed in table 7.5.

GLASS All-glass piping is used for its chemical resistance, cleanliness and transparency. Glass pipe is not subject to 'crazing' often found in glass-lined pipe and vessels subject to repeated thermal stresses. Pipe, fittings, and hardware are available both for process piping and for drainage. Corning Glass Works offers a Pyrex 'Conical' system for process lines in 1, 1½, 2, 3, 4 and 6-inch sizes (ID) with 450 F as the maximum operating temperature, and pressure ranges 0—65 PSIA (1 in. thru 3 in.), 0—50 PSIA (4 in.) and 0—35 PSIA (6 in.). Glass cocks, strainers and thermowells are available. Pipe fittings and equipment are joined by flange assemblies which bear on the thickened conical ends of pipe lengths and fittings. Corning also offers a Pyrex Acid-Waste Drainline system in 1½, 2, 3, 4 and 6-inch sizes (ID) with beaded ends joined by Teflon-gasketed nylon compression couplings. Both Corning systems are made from the same borosilicate glass.

LININGS & COATINGS Lining or coating carbon-steel pipe with a material able to withstand chemical attack permits its use to carry corrosive fluids. Lengths of lined pipe and fittings are joined by flanges, and elbows, tees, etc., are available already flanged. Linings (rubber, for example) can be applied after fabricating the piping, but pipe is often pre-lined, and manufacturers give instructions for making joints. Linings of various rubbers, plastics, metals and vitreous (glassy) materials are available. Polyvinyl chloride, polypropylene and copolymers are the most common coating materials. Carbon-steel pipe zinc-coated by immersion into molten zinc (hot-dip galvanized) is used for conveying drinking water, instrument air and various other fluids. Rubber lining is often used to handle abrasive fluids.

TEMPERATURE & PRESSURE LIMITS   2.1.5

Carbon steels lose strength at high temperatures. Electric-resistance-welded pipe is not considered satisfactory for service above 750 F, and furnace-butt-welded orpe above about 650 F. For higher temperatures, pipe made from stainless steels or other alloys should be considered.

Pressure ratings for steel pipe at different temperatures are calculated according to the ANSI B31 Code for Pressure Piping (detailed in table 7.2). ANSI B31 gives stress/temperature values for the various steels from which pipe is fabricated.

METHODS FOR JOINING PIPE   2.2

The joints used for most carbon-steel and stainless-steel pipe are:

BUTT-WELDED	SEE 2.3
SOCKET-WELDED	SEE 2.4
SCREWED	SEE 2.5
BOLTED FLANGE	SEE 2.3.1, 2.4.1 & 2.5.1
BOLTED QUICK COUPLINGS	SEE 2.8.2

WELDED & SCREWED JOINTS   2.2.1

Lines NPS 2 and larger are usually butt-welded, this being the most economic leakproof way of joining larger-diameter piping. Usually such lines are subcontracted to a piping fabricator for prefabrication in sections termed 'spools', then transported to the site. Lines NPS 1½ and smaller are usually either screwed or socket-welded, and are normally field-run by the piping contractor from drawings. Field-run and shop-fabricated piping are discussed in 5.2.9.

SOCKET-WELDED JOINTS   2.2.2

Like screwed piping, socket welding is used for lines of smaller sizes, but has the advantage that absence of leaking is assured: this is a valuable factor when flammable, toxic, or radioactive fluids are being conveyed—the use of socket-welded joints is not restricted to such fluids, however.

BOLTED-FLANGE JOINTS   2.2.3

Flanges are expensive and for the most part are used to mate with flanged vessels, equipment, valves, and for process lines which may require periodic cleaning.

Flanged joints are made by bolting together two flanges with a gasket between them to provide a seal. Refer to 2.6 for standard forged-steel flanges and gaskets.

FITTINGS   2.2.4

Fittings permit a change in direction of piping, a change in diameter of pipe, or a branch to be made from the main run of pipe. They are formed from plate or pipe, machined from forged blanks, cast, or molded from plastics.

Chart 2.1 shows the ratings of butt-welding fittings used with pipe of various schedule numbers and manufacturers' weights. For dimensions of butt-welding fittings and flanges, see tables D-1 thru D-6, and tables F-1 thru F-7. Drafting symbols are given in charts 5.3 thru 5.5.

Threaded fittings have Pressure Class designations of: 2000, 3000 and 6000. Socket-welding fittings have Pressure Class designations of: 3000, 6000 and 9000. How these Pressure Class designations relate to schedule numbers and manufacturers' weights for pipe is shown in table 2.2.

CORRELATION OF CLASS OF THREADED   **TABLE 2.2**
& SOCKET-WELDING FITTINGS
WITH SCHEDULES/WEIGHTS OF PIPE

	PIPE DESIGNATION SCH/MFR'S			
Pressure Class	2000	3000	6000	9000
Threaded fittings	80/XS	160	XXS	
Socketed fittings		80/XS	160	XXS

Sections 2.1.3 thru 2.2.4 have shown that there is a wide variety of differently-rated pipe, fittings and materials from which to make a choice. Charts 2.1 thru 2.3 show how various weights of pipe, fittings and valves can be combined in a piping system.

COMPONENTS FOR BUTT-WELDED PIPING SYSTEMS   2.3

WHERE USED: For most process, utility and service piping

ADVANTAGE OF JOINT: Most practicable way of joining larger pipes and fittings which offers reliable, leakproof joints

DISADVANTAGE OF JOINT: Intruding weld metal may affect flow

HOW JOINT IS MADE: The end of the pipe is beveled as shown in chart 2.1. Fittings are similarly beveled by the manufacturer. The two parts are aligned, properly gapped, tack welded, and then a continuous weld is made to complete the joint

Chart 2.1 shows the ratings of pipe, fittings and valves that are commonly combined or may be used together. It is a guide only, and not a substitute for a project specification.

FITTINGS, BENDS, MITERS & FLANGES FOR BUTT-WELDED SYSTEMS   2.3.1

Refer to tables D, F and W-1 for dimensions and weights of fittings and flanges.

ELBOWS or 'ELLS' make 90- or 45-degree changes in direction of the run of pipe. The elbows normally used are 'long radius' (LR) with centerline radius of curvature equal to 1½ times the nominal pipe size for NPS 3/4 and larger sizes. 'Short radius' (SR) elbows with centerline radius of curvature equal to the nominal pipe size are also available. 90-degree LR elbows with a straight extension at one end ('long tangent') are still available in STD weight, if required.

REDUCING ELBOW makes a 90-degree change in direction with change in line size. Reducing elbows have centerline radius of curvature 1½ times the nominal size of the pipe to be attached to the larger end.

RETURN changes direction of flow thru 180 degrees, and is used to construct heating coils, vents on tanks, etc.

BENDS are made from straight pipe. Common bending radii are 3 and 5 times the pipe size (3R and 5R bends, where R = nominal pipe size—nominal diameter, *not* radius). 3R bends are available from stock. Larger radius bends can be custom made, preferably by hot bending. Only seamless or electric-resistance-welded pipe is suitable for bending.

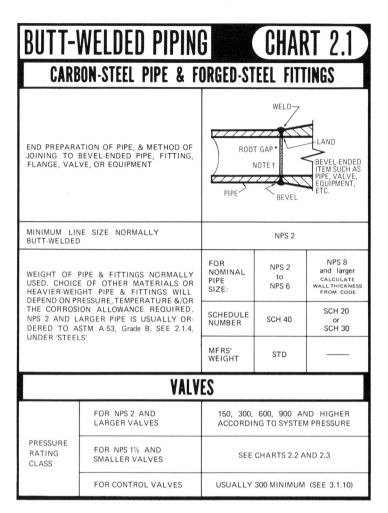

BUTT-WELDED PIPING   CHART 2.1

CARBON-STEEL PIPE & FORGED-STEEL FITTINGS

END PREPARATION OF PIPE, & METHOD OF JOINING TO BEVEL-ENDED PIPE, FITTING, FLANGE, VALVE, OR EQUIPMENT			
MINIMUM LINE SIZE NORMALLY BUTT-WELDED	NPS 2		

WEIGHT OF PIPE & FITTINGS NORMALLY USED. CHOICE OF OTHER MATERIALS OR HEAVIER-WEIGHT PIPE & FITTINGS WILL DEPEND ON PRESSURE, TEMPERATURE &/OR THE CORROSION ALLOWANCE REQUIRED. NPS 2 AND LARGER PIPE IS USUALLY ORDERED TO ASTM A-53, Grade B. SEE 2.1.4, UNDER 'STEELS'	FOR NOMINAL PIPE SIZE:	NPS 2 to NPS 6	NPS 8 and larger CALCULATE WALL THICKNESS FROM CODE
	SCHEDULE NUMBER	SCH 40	SCH 20 or SCH 30
	MFRS' WEIGHT	STD	——

VALVES

PRESSURE RATING CLASS	FOR NPS 2 AND LARGER VALVES	150, 300, 600, 900 AND HIGHER ACCORDING TO SYSTEM PRESSURE
	FOR NPS 1½ AND SMALLER VALVES	SEE CHARTS 2.2 AND 2.3
	FOR CONTROL VALVES	USUALLY 300 MINIMUM (SEE 3.1.10)

*See 5.3.5 under 'Dimensioning spools'
†A 'backing ring'—sometimes termed a 'chill ring'—may be inserted between any butt-welding joint prior to welding. Preventing weld spatter and spikes ('icicles') of weld metal from forming inside the pipe during welding, the ring also serves as an alignment aid. Normally used for severe service, but should be considered for process fluids such as fibrous suspensions, where weld icicles could result in material collecting at joints and choking lines. See 2.11

BACKING RING   **FIGURE 2.1**

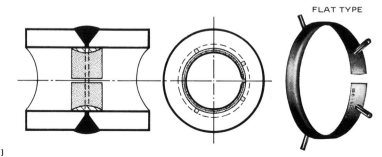

FLAT TYPE

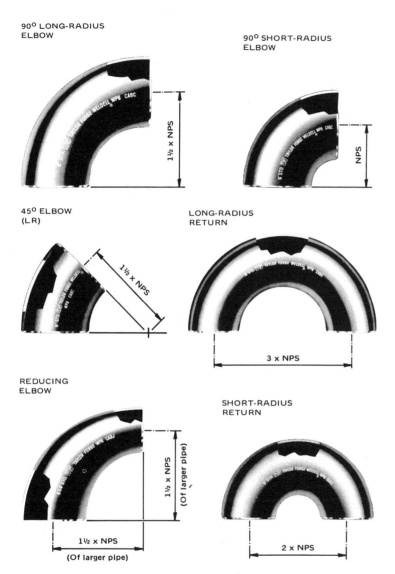

90° LONG-RADIUS ELBOW

1½ x NPS

90° SHORT-RADIUS ELBOW

NPS

45° ELBOW (LR)

1½ x NPS

LONG-RADIUS RETURN

3 x NPS

REDUCING ELBOW

1½ x NPS (Of larger pipe)

1½ x NPS (Of larger pipe)

SHORT-RADIUS RETURN

2 x NPS

REDUCER (or INCREASER) joins a larger pipe to a smaller one. The two available types, concentric and eccentric, are shown. The eccentric reducer is used when it is necessary to keep either the top or the bottom of the line level—offset equals ½ x (larger ID minus smaller ID).

REDUCERS

FIGURE 2.3

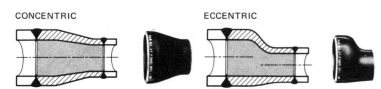

CONCENTRIC   ECCENTRIC

SWAGE is employed to connect butt-welded piping to smaller screwed or socket-welded piping. In butt-welded lines, used as an alternative to the reducer when greater reductions in line size are required. Regular swages in concentric or eccentric form give abrupt change of line size, as do reducers. The 'venturi' swage allows smoother flow. Refer to table 2.3 for specifying swages for joining to socket-welding items, and to table 2.4 for specifying swages for joining to screwed piping. For offset, see 'Reducer'.

SWAGES, or SWAGED NIPPLES

FIGURE 2.4

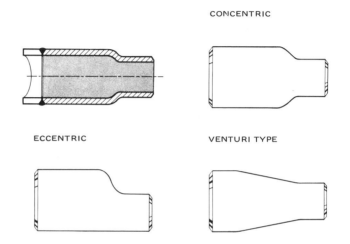

CONCENTRIC

ECCENTRIC

VENTURI TYPE

MITERED ELBOWS are fabricated as required from pipe—they are not fittings. The use of miters to make changes in direction is practically restricted to low-pressure lines 10-inch and larger if the pressure drop is unimportant; for these uses regular elbows would be costlier. A 2-piece, 90-degree miter has four to six times the hydraulic resistance of the corresponding regular long-radius elbow, and should be used with caution. A 3-piece 90-degree miter has about double the resistance to flow of the regular long-radius elbow—refer to table F-10. Constructions for 3-, 4-, and 5-piece miters are shown in tables M-2.

3-PIECE & 2-PIECE MITERS

FIGURE 2.5

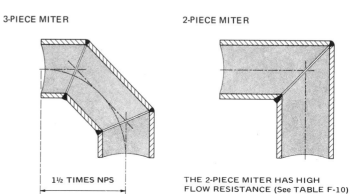

3-PIECE MITER   2-PIECE MITER

1½ TIMES NPS

THE 2-PIECE MITER HAS HIGH FLOW RESISTANCE (See TABLE F-10)

The following five flange types are used for butt-welded lines. The different flange facings available are discussed in 2.6.

WELDING-NECK FLANGE, REGULAR & LONG *Regular welding-neck flanges are used with butt-welding fittings.* Long welding-neck flanges are primarily used for vessel and equipment nozzles, rarely for pipe. Suitable where extreme temperature, shear, impact and vibratory stresses apply. Regularity of the bore is maintained. Refer to tables F for bore diameters of these flanges.

WELDING-NECK FLANGE   **FIGURE 2.6**

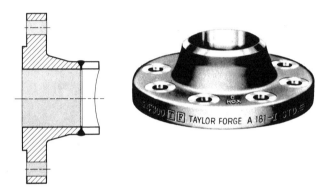

SLIP-ON FLANGE is properly used to flange pipe. Slip-on flanges can be used with long-tangent elbows, reducers, and swages (not usual practice). The internal weld is slightly more subject to corrosion than the butt weld. The flange has poor resistance to shock and vibration. It introduces irregularity in the bore. It is cheaper to buy than the welding-neck flange, but is costlier to assemble. It is easier to align than the welding-neck flange. Calculated strengths under internal pressure are about one third that of the corresponding welding-neck flanges. The pipe or fitting is set back from the face of the flange a distance equal to the wall thickness —0″ + 1/16″.

SLIP-ON FLANGE   **FIGURE 2.7**

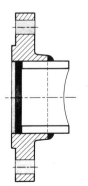

REDUCING FLANGE   Suitable for changing line size, but should not be used if abrupt transition would create undesirable turbulence, as at pump connections. Available to order in welding-neck and eccentric types, and usually from stock in slip-on type. Specify by nominal pipe sizes, stating the size of the larger pipe first. Example: a slip-on reducing flange to connect a NPS 4 pipe to a Class 150 NPS 6 line-size flange is specified:

RED FLG NPS 6 x 4 Class 150 SO

For a welding-neck reducing flange, correct bore is obtained by giving the pipe schedule number or manufacturers' weight of the pipe to be welded on.

REDUCING SLIP-ON FLANGE   **FIGURE 2.8**

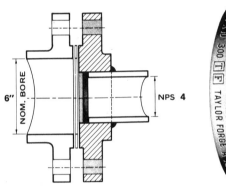

EXPANDER FLANGE   Application as for welding-neck flange—see above. Increases pipe size to first or second larger size. Alternative to using reducer and welding-neck flange. Useful for connecting to valves, compressors and pumps. Pressure ratings and dimensions are in accord with ANSI B16.5.

EXPANDER (or INCREASER) FLANGE   **FIGURE 2.9**

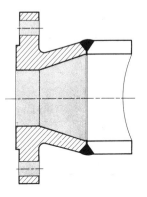

LAP-JOINT, or 'VAN STONE', FLANGE Economical if costly pipe such as stainless steel is used, as the flange can be of carbon steel and only the lap-joint stub end need be of the line material. A stub end must be used in a lap joint, and the cost of the two items must be considered. If both stub and flange are of the same material they will be more expensive than a welding-neck flange. Useful where alignment of bolt holes is difficult, as with spools to be attached to flanged nozzles of vessels.

LAP-JOINT FLANGE (with Stub-end)   **FIGURE 2.10**

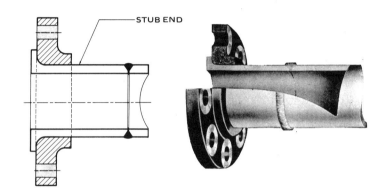

STUB END

BUTT-WELDING FITTINGS FOR BRANCHING   **2.3.2**
FROM BUTT-WELDED SYSTEMS

STUB-IN Term for a branch pipe welded directly into the side of the main pipe run—it is not a fitting. This is the commonest and least expensive method of welding a full-size or reducing branch for pipe 2-inch and larger. A stub-in can be reinforced by means set out in 2.11.

STUB-IN   **FIGURE 2.11**

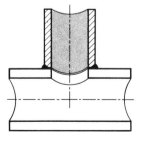

BUTT-WELDING TEES, STRAIGHT or REDUCING, are employed to make 90-degree branches from the main run of pipe. Straight tees, with branch the same size as the run, are readily available. Reducing tees have branch smaller than the run. Bullhead tees have branch larger than the run, and are very seldom used but can be made to special order. None of these tees requires reinforcement. Reducing tees are ordered as follows:—

SPECIFYING SIZE OF BUTT-WELDING REDUCING TEES

HOW TO SPECIFY TEES:	RUN INLET	RUN OUTLET	BRANCH	EXAMPLE
REDUCING ON BRANCH	6"	6"	4"	RED TEE 6 x 6 x 4

BUTT-WELDING TEES   **FIGURE 2.12**

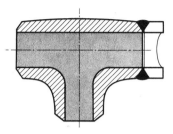

STRAIGHT BUTT-WELDING TEE   REDUCING BUTT-WELDING TEE

The next four branching fittings are made by Bonney Forge.
These fittings offer an alternate means of connecting into the main run, and do not require reinforcement. They are preshaped to the curvature of the run pipe.

WELDOLET makes a 90-degree branch, full-size or reducing, on straight pipe. Closer manifolding is possible than with tees. Flat-based weldolets are available for connecting to pipe caps and vessel heads.

WELDOLET   **FIGURE 2.13**

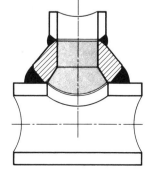   

BUTT-WELDING ELBOLET makes a reducing tangent branch on long-radius and short-radius elbows.

ELBOLET
FIGURE 2.14

BUTT-WELDING LATROLET
FIGURE 2.15

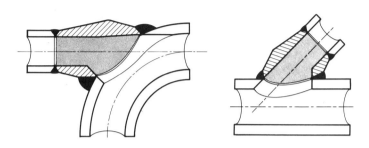

BUTT-WELDING LATROLET makes a 45-degree reducing branch on straight pipe.

SWEEPOLET makes a 90-degree reducing branch from the main run of pipe. Primarily developed for high-yield pipe used in oil and gas transmission lines. Provides good flow pattern, and optimum stress distribution.

SWEEPOLET   FIGURE 2.16

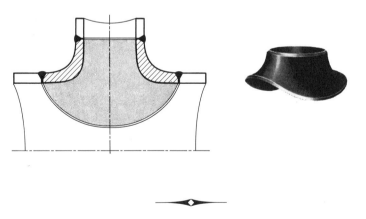

The next three fittings are usually used for special designs:

CROSS, STRAIGHT or REDUCING Straight crosses are usually stock items. Reducing crosses may not be readily available. For economy, availability and to minimize the number of items in inventory, it is preferred to use tees, etc., and not crosses, except where space is restricted, as in marine piping or 're-vamp' work. Reinforcement is not needed.

BUTT-WELDING CROSS   FIGURE 2.17

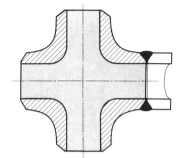

LATERAL, STRAIGHT or REDUCING, permits odd-angled entry into the pipe run where low resistance to flow is important. Straight laterals with branch bore equal to run bore are available in STD and XS weights. Reducing laterals and laterals at angles other than 45 degrees are usually available only to special order. Reinforcement is required where it is necessary to restore the strength of the joint to the full strength of the pipe. Reducing laterals are ordered similarly to butt-welding tees, except that the angle between branch and run is also stated.

LATERAL   FIGURE 2.18

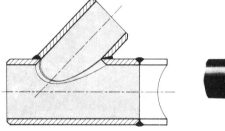

SHAPED NIPPLE Now rarely used, but can be obtained from stock in 90- and 45-degree angles, and in any size and angle, including offset, to special order. The run is field-cut, using the nipple as template. Needs reinforcement if it is necessary to bring the strength of the joint up to the full strength of the pipe.

SHAPED NIPPLE   FIGURE 2.19

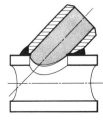

[10]

CAP is used to seal the end of pipe. (See figure 2.20(a).)

FLAT CLOSURES Flat plates are normally cut especially from platestock by the fabricator or erector. (See figure 2.20 (b) and (c).)

THREE WELDED CLOSURES   FIGURE 2.20

(a) BUTT-WELDING CAP   (b) FLAT CLOSURE (c) FLAT CLOSURE

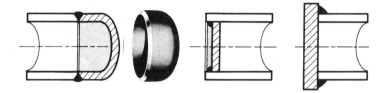

ELLIPSOIDAL, or DISHED, HEADS are used to close pipes of large diameter, and are similar to those used for constructing vessels.

COMPONENTS FOR SOCKET-WELDED PIPING SYSTEMS   2.4

WHERE USED: For lines conveying flammable, toxic, or expensive material, where no leakage can be permitted. For steam: 300 to 600 PSI, and sometimes 150 PSI steam. For corrosive conditions, see Index under 'Corrosion'

ADVANTAGES OF JOINT:
(1) Easier alignment on small lines than butt welding. Tack welding is unnecessary
(2) No weld metal can enter bore
(3) Joint will not leak, when properly made

DISADVANTAGES OF JOINT:
(1) The 1/16-inch recess in joint (see chart 2.2) pockets liquid
(2) Use not permitted by ANSI B31.1 - 1989 if severe vibration or crevice corrosion is anticipated

HOW JOINT IS MADE: The end of the pipe is finished flat, as shown in chart 2.2. It is located in the fitting, valve, flange, etc., and a continuous fillet weld is made around the circumference

Chart 2.2 shows the ratings of pipe, fittings and valves that are commonly combined, or may be used together. The chart is a guide only, and not a substitute for a project specification.

SOCKET-WELDED PIPING   CHART 2.2

CARBON-STEEL PIPE & FORGED-STEEL FITTINGS

END PREPARATION OF PIPE, AND METHOD OF JOINING TO FITTING, FLANGE, VALVE, OR EQUIPMENT				
WELD — PLAIN END — PIPE —	1/16" EXPANSION GAP *	BORE	SOCKET-ENDED ITEM SUCH AS COUPLING, EQUIPMENT, VALVE, Etc.	

MAXIMUM LINE SIZE NORMALLY SOCKET WELDED		NPS 1½ (NPS 2½ IN MARINE PIPING)		
AVAILABILITY OF FORGED-STEEL SOCKET-WELDING FITTINGS		NPS 1/8 to NPS 4		

WEIGHTS OF PIPE AND PRESSURE CLASSES OF FITTINGS WHICH ARE COMPATIBLE	PIPE	SCHEDULE NUMBER	SCH 80	SCH 160	—
		MFRS' WEIGHT	XS	—	XXS
	FITTINGS	FITTING CLASS	3000	6000	9000
		FITTING BORED TO:	SCH 40	SCH 160	XXS

MOST COMMON COMBINATION: CHOICE OF MATERIAL OR HEAVIER-WEIGHT PIPE AND FITTING WILL DEPEND ON PRESSURE, TEMPERATURE AND/OR CORROSION ALLOWANCE REQUIRED. PIPE NPS 1½ AND SMALLER IS USUALLY ORDERED TO ASTM SPECIFICATION A-106 Grade B. REFER TO 2.1.4, UNDER 'STEELS'

VALVES

MINIMUM PRESSURE (RATING) CLASS	CONTROL VALVES (USUALLY FLANGED)	USUALLY 300 (SEE 3.1.10)
	VALVES OTHER THAN CONTROL VALVES	600 (ANSI) 800 (API)

* ANSI B16.11 recommends a 1/16th-inch gap to prevent weld from cracking under thermal stress

† Socket-ended fittings are now only made in classes 3000 6000 and 9000 (ANSI B16.11)

CHART 2.2

Dimensions of fittings and flanges are given in tables D-8 and F-1 thru F-6.

FULL-COUPLING (termed 'COUPLING) joins pipe to pipe, or to a nipple, swage, etc.

FULL-COUPLING   **FIGURE 2.21**

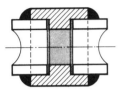

REDUCER joins two different diameters of pipe.

REDUCER   **FIGURE 2.22**

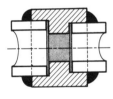

REDUCER INSERT A reducing fitting used for connecting a small pipe to a larger fitting. Socket-ended reducer inserts can be made in any reduction by boring standard forged blanks.

SOCKET-WELDING REDUCING INSERTS   **FIGURE 2.23**

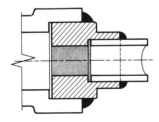

SOCKET-ENDED
FITTING, FLANGE,
OR EQUIPMENT

THREE FORMS
OF REDUCER
INSERT:

UNION is used primarily for maintenance and installation purposes. This is a screwed joint designed for use with socket-welded piping systems. See explanation in 2.5.1 of uses given under 'threaded union'. Union should be screwed tight before the ends are welded, to minimize warping of the seat.

SOCKET-WELDING UNION   **FIGURE 2.24**

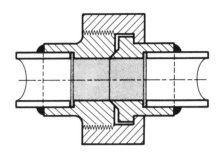

SWAGED NIPPLES According to type, these allow joining: (1) Socket-ended items of different sizes—this type of swaged nipple has both ends plain (PBE) for insertion into socket ends. (2) A socket-ended item to a larger butt-welding pipe or fitting—this type of swaged nipple has the larger end beveled (BLE) and the smaller end plain (PSE) for insertion into a socket-ended item. A swaged nipple is also referred to as a 'swage' (pronounced 'swedge') abbreviated on drawings as 'SWG' or 'SWG NIPP'. When ordering a swage, state the weight designations of the pipes to be joined. For example, NPS 2 (SCH 40) x NPS 1 (SCH 80). Examples of the different end terminations that may be specified are as follows:-

**SPECIFYING SIZE & END FINISH
OF SOCKET-WELDING SWAGES**   **TABLE 2.3**

SWAGE FOR JOINING——		EXAMPLE NOTE ON DRAWING
LARGER	to SMALLER	
SW ITEM	SW ITEM	SWG 1½ x 1 PBE
BW FITTING or PIPE	SW ITEM	SWG 2 x 1 BLE—PSE
ABBREVIATIONS:	SW = Socket welding BW = Butt welding PBE = Plain both ends PLE = Plain large end PSE = Plain small end BLE = Bevel large end	

SWAGE (PBE)   **FIGURE 2.25**

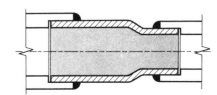

ELBOWS make 90- or 45-degree changes of direction in the run of pipe.

SOCKET-WELDING ELBOWS FIGURE 2.26

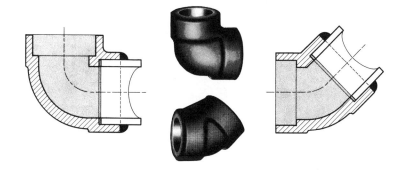

SOCKET--WELDING FLANGE Regular type is available from stock. Reducing type is available to order. For example, a reducing flange to connect a NPS 1 pipe to a Class 150 NPS 1½ line-size flange is specified:

RED FLG NPS 1½ x 1 Class 150 SW

SOCKET-WELDING FLANGE FIGURE 2.27

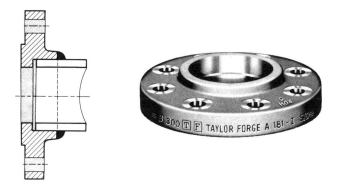

FITTINGS FOR BRANCHING FROM SOCKET-WELDED SYSTEMS 2.4.2

BRANCH FROM SOCKET-WELDED RUN

TEE, STRAIGHT or REDUCING, makes 90-degree branch from the main run of pipe. Reducing tees are custom-fabricated by boring standard forged blanks.

SPECIFYING SIZE OF SOCKET-WELDING TEES

HOW TO SPECIFY TEES:	RUN INLET	RUN OUTLET	BRANCH	EXAMPLE
REDUCING ON BRANCH	$1\frac{1}{2}"$	$1\frac{1}{2}"$	$1"$	RED TEE 1½ x 1½ x 1
REDUCING ON RUN (SPECIAL APPLICATIONS ONLY)	$1\frac{1}{2}"$	$1"$	$1\frac{1}{2}"$	RED TEE 1½ x 1 x 1½

SOCKET-WELDING TEE FIGURE 2.28

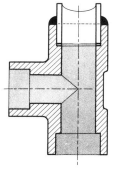

LATERAL makes full-size 45-degree branch from the main run of pipe.

SOCKET-WELDING LATERAL FIGURE 2.29

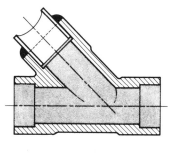

CROSS Remarks for butt-welding cross apply—see 2.3.2. Reducing crosses are custom-fabricated by boring standard forged blanks.

SOCKET-WELDING CROSS FIGURE 2.30

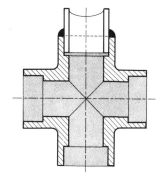

TABLE 2.3

HALF-COUPLING   The full-coupling is not used for branching or for vessel connections, as the half-coupling is the same length and is stronger. The half-coupling permits 90-degree entry into a larger pipe or vessel wall. The sockolet is more practicable as shaping is necessary with the coupling.

SOCKET-WELDING HALF-COUPLING   FIGURE 2.31

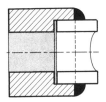

The next four fittings are made by Bonney Forge and offer an alternate method of entering the main pipe run. They have the advantage that the beveled welding ends are shaped to the curvature of the run pipe. Reinforcement for the butt-welded piping or vessel is not required.

SOCKOLET   makes a 90-degree branch, full-size or reducing, on straight pipe. Flat-based sockolets are available for branch connections on pipe caps and and vessel heads.

SOCKOLET   FIGURE 2.32

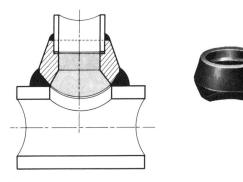

SOCKET-WELDING ELBOLET makes a reducing tangent branch on long-radius and short-radius elbows.

SOCKET-WELDING ELBOLET   FIGURE 2.33

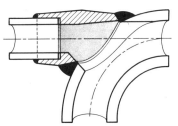

SOCKET-WELDING LATROLET   makes a 45-degree reducing branch on straight pipe.

SOCKET-WELDING LATROLET   FIGURE 2.34

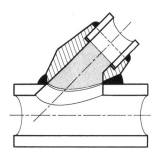

NIPOLET   A variant of the sockolet, having integral plain nipple. Primarily developed for small valved connections—see figure 6.47.

NIPOLET   FIGURE 2.35

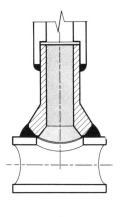

STUB-IN   See comments in 2.3.2. Not preferred for lines under 2-inch due to risk of weld metal entering line and restricting flow.

CLOSURE   2.4.4

SOCKET-WELDING CAP seals plain-ended pipe.

SOCKET-WELDING CAP   FIGURE 2.36

WHERE USED: For lines conveying services, and for smaller process piping

ADVANTAGES:
(1) Easily made from pipe and fittings on site
(2) Minimizes fire hazard when installing piping in areas where flammable gases or liquids are present

DISADVANTAGES:
(1)* Use not permitted by ANSI B31.1-1989, if severe erosion, crevice corrosion, shock, or vibration is anticipated, nor at temperatures over 925 F. (Also see footnote table F-9)
(2) Possible leakage of joint
(3)* Seal welding may be required—see footnote to chart 2.3
(4) Strength of the pipe is reduced, as forming the screwthread reduces the wall thickness

*These remarks apply to systems using forged-steel fittings.

FITTINGS & FLANGES FOR SCREWED SYSTEMS   2.5.1

Screwed piping is piping assembled from threaded pipe and fittings.

Threaded malleable-iron and cast-iron fittings are extensively used for plumbing in buildings. In industrial applications, Class 150 and 300 galvanized malleable-iron fittings and similarly rated valves are used for drinking water and air lines. Dimensions of malleable-iron fittings are given in table D-11.

In process piping, forged-steel fittings are preferred over cast-iron and malleable-iron fittings (although their pressure/temperature ratings may be suitable), for their greater mechanical strength. To simplify material specifications, drafting, checking, purchasing and warehousing, the overall economics are in favor of utilizing as few different types of threaded fittings as possible. Dimensions of forged-steel threaded fittings are given in table D-9.

FULL-COUPLING (termed 'COUPLING') joins pipe or items with threaded ends.

FULL-COUPLING   FIGURE 2.37

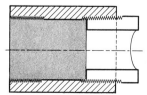

Chart 2.3 shows the ratings of pipe, fittings and valves that are commonly combined, or may be used together. The chart is a guide only, and not a substitute for a project specification.

SCREWED PIPING   CHART 2.3
CARBON-STEEL PIPE & FORGED-STEEL FITTINGS

END PREPARATION OF PIPE, AND METHOD OF JOINING TO FITTING, FLANGE, VALVE OR EQUIPMENT			

THREAD ENGAGEMENT / PIPE / NPT / ITEM SUCH AS VALVE, COUPLING, EQUIPMENT, ETC. / OPTIONAL SEAL WELD *

MAXIMUM LINE SIZE NORMALLY THREADED		NPS 1½		
AVAILABILITY OF FORGED-STEEL THREADED FITTINGS		NPS 1/8 to NPS 4		

WEIGHTS OF PIPE AND PRESSURE CLASSES OF FITTINGS WHICH ARE COMPATIBLE	PIPE	SCHEDULE NUMBER	SCH 40	SCH 80	—
		MFRS' WEIGHT	STD	XS	XXS
	FITTING CLASS		2000	3000	6000

MOST COMMON COMBINATION: THE MINIMUM CLASS FOR FITTINGS PREFERRED IN MOST INSTANCES FOR MECHANICAL STRENGTH IS 3000. CHOICE OF MATERIAL OR HEAVIER-WEIGHT PIPE & FITTING WILL DEPEND ON PRESSURE, TEMPERATURE AND /OR CORROSION ALLOWANCE REQUIRED. PIPE NPS 1½ AND SMALLER IS USUALLY ORDERED TO ASTM SPECIFICATION A-106 Grade B. REFER TO 2.1.4, UNDER 'STEELS'

VALVES

MINIMUM PRESSURE (RATING) CLASS	CONTROL VALVES (USUALLY FLANGED)	USUALLY 300 (SEE 3.1.10)
	VALVES OTHER THAN CONTROL VALVES	600 (ANSI) 800 (API)

* ANSI B31.1.0 states that seal welding shall not be considered to contribute to the strength of the joint

SEAL WELDING APPLICATIONS

On-plot: On all screwed connections within battery limits, with the exception of piping carrying air or other inert gas, and water
Off-plot: On screwed lines for hydrocarbon service and for lines conveying dangerous, toxic, corrosive or valuable fluids

CHART 2.3

FIGURES 2.31–2.37

REDUCING COUPLING, or REDUCER, joins threaded pipes of different sizes. Can be made in any reduction by boring and tapping standard forged blanks.

REDUCING COUPLING   FIGURE 2.38

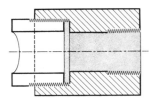

NIPPLES join unions, valves, strainers, fittings, etc. Basically a short length of pipe either fully threaded (close nipple) or threaded both ends (TBE), or plain one end and threaded one end (POE—TOE). Available in various lengths -refer to table D-11. Nipples can be obtained with a Victaulic groove at one end.

NIPPLES FOR THREADED ITEMS   FIGURE 2.39

(a) CLOSE NIPPLE   (b) LONG or SHORT NIPPLE (TBE)   (c) NIPPLE (POE—TOE)

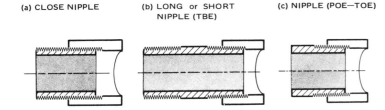

(d) TANK NIPPLE

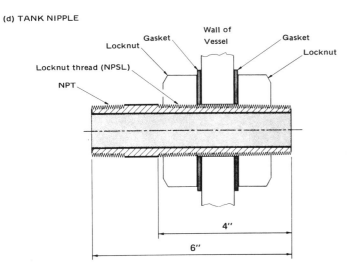

TANK NIPPLE is used for making a screwed connection to a non-pressure vessel or tank in low-pressure service. Overall length is usually 6 inches with a standard taper pipe thread at each end. On one end only, the taper pipe thread runs into a ANSI lock-nut thread.

UNION makes a joint which permits easy installation, removal or replacement of lengths of pipe, valves or vessels in screwed piping systems. Examples: to remove a valve it must have at least one adjacent union, and to remove piping from a vessel with threaded connections, each outlet from the vessel should have one union between valve and vessel. Ground-faced joints are preferred, although other facings are available.

THREADED UNION   FIGURE 2.40

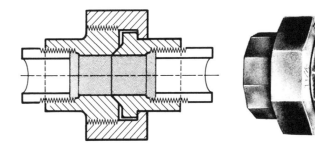

PIPE-TO-TUBE CONNECTOR For joining threaded pipe to tube. Figure 2.41 shows a connector fitted to specially-flared tube. Other types are available.

PIPE-TO-TUBE CONNECTOR   FIGURE 2.41

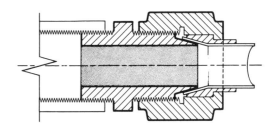

HEXAGON BUSHING A reducing fitting used for connecting a smaller pipe into a larger threaded fitting or nozzle. Has many applications to instrument connections. Reducing fittings can be made in any reduction by boring and tapping standard forged blanks. Normally not used for high-pressure service.

HEXAGON BUSHING   FIGURE 2.42

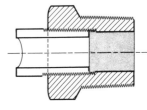

SWAGED NIPPLE This is a reducing fitting, used for joining larger diameter to smaller diameter pipe. Also referred to as a 'swage (pronounced 'swedge') and abbreviated as 'SWG' or 'SWG NIPP' on drawings. When ordering a swage, state the weight designations of the pipes to be joined: for example, NPS 2 (SCH 40) x NPS 1 (SCH 80). A swage may be used for joining: (1) Screwed piping to screwed piping. (2) Screwed piping to butt-welded piping. (3) Butt-welded piping to a threaded nozzle on equipment. It is necessary to specify on the piping drawing the terminations required.

SPECIFYING SIZE & END FINISH OF THREADED SWAGES   **TABLE 2.4**

SWAGE FOR JOINING——— LARGER to SMALLER		EXAMPLE NOTE ON DRAWING
THRD ITEM	THRD ITEM	SWG 1½ x 1 TBE
BW ITEM or PIPE	THRD ITEM	SWG 2 x 1 BLE—TSE
THRD ITEM*	BW ITEM*	SWG 3 x 2 TLE—BSE
ABBREVIATIONS:	BW = Butt welding THRD = Threaded TBE = Threaded both ends TSE = Threaded small end	TLE = Threaded large end TOE = Threaded one end BLE = Beveled large end BSE = Beveled small end

* A larger threaded item is seldom joined to a smaller buttwelding item. However, the connection of a buttwelded line to a threaded nozzle on a vessel is an example.

SWAGED NIPPLES, TBE and BLE—TSE   **FIGURE 2.43**

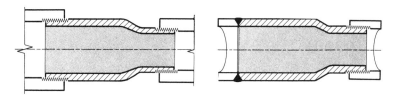

ELBOWS make 90- or 45-degree changes in direction of the run of pipe. Street elbows having a integral nipple at one end (see table D-11), are available

THREADED ELBOWS, 45 and 90 DEGREE   **FIGURE 2.44**

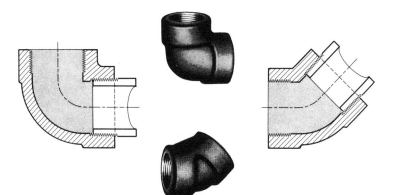

THREADED FLANGES are used to connect threaded pipe to flanged items. Regular and reducing types are available from stock. For example, a reducing flange to connect a NPS 1 pipe to a Class 150 NPS 1½ line-size flange is specified:

RED FLG NPS 1½ x 1 Class 150 THRD

THREADED FLANGE   **FIGURE 2.45**

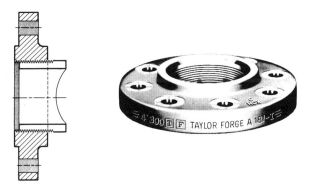

FITTINGS FOR BRANCHING FROM SCREWED SYSTEMS   2.5.2

BRANCH FROM SCREWED MAIN RUN

TEE, STRAIGHT or REDUCING, makes a 90-degree branch from the run of pipe. Reducing tees are made by boring and tapping standard forged blanks.

SPECIFYING SIZE OF THREADED REDUCING TEES

HOW TO SPECIFY TEES:	RUN INLET	RUN OUTLET	BRANCH	EXAMPLE
REDUCING ON BRANCH	1½"	1½"	1"	RED TEE 1½ x 1½ x 1
REDUCING ON RUN (SPECIAL APPLICATIONS ONLY)	1½"	1"	1½"	RED TEE 1½ x 1 x 1½

THREADED TEES, STRAIGHT and REDUCING   **FIGURE 2.46**

STRAIGHT TEE   REDUCING TEE

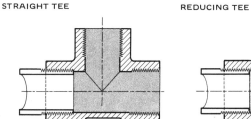

LATERAL makes full-size 45-degree branch from the main run of pipe.

THREADED LATERAL   FIGURE 2.47

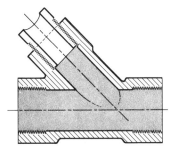

CROSS   Remarks for butt-welding cross apply — see 2.3.2.   Reducing crosses are made by boring and tapping standard forged blanks.

THREADED CROSS   FIGURE 2.48

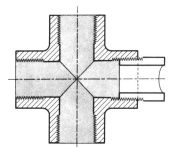

FITTINGS FOR SCREWED BRANCH   2.5.3
FROM VESSEL OR BUTT-WELDED MAIN RUN

HALF-COUPLING can be used to make 90-degree threaded connections to pipes for instruments, or for vessel nozzles. Welding heat may cause embrittlement of the threads of this short fitting. Requires shaping.

THREADED HALF-COUPLING & FULL-COUPLING   FIGURE 2.49

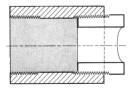

FULL-COUPLING   Superior to half-coupling. Also requires shaping for connecting to pipe.

TANK NIPPLE See 2.5.1, figure 2.39(d).

The next four fittings for branching are made by Bonny Forge. These fittings offer a means of joining **screwed** piping to a **welded** run, and for making instrument connections. The advantages are that the welding end does not require reinforcement and that the ends are shaped to the curvature of the run pipe.

THREDOLET makes a 90-degree branch, full or reducing, on straight pipe. Flat-based thredolets are available for branch connections on pipe caps and vessel heads.

THREDOLET   FIGURE 2.50

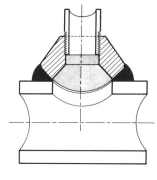

THREADED ELBOLET makes reducing tangent branch on long-radius and short radius elbows.

THREADED ELBOLET   FIGURE 2.51

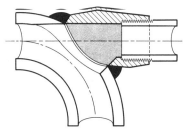

THREADED LATROLET makes a 45-degree reducing branch on a straight pipe.

THREADED LATROLET   FIGURE 2.52

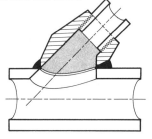

THREADED NIPOLET A variant of the thredolet with integral threaded nipple. Primarily developed for small valved connections—see figure 6.47.

THREADED NIPOLET   **FIGURE 2.53**

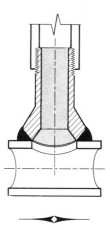

STUB-IN See comments in 2.3.2. Not preferred for branching from pipe smaller than NPS 2 as weld metal may restrict flow.

CLOSURES   2.5.4

CAP seals the threaded end of pipe.

THREADED CAP   **FIGURE 2.54**

BARSTOCK PLUG seals the threaded end of a fitting. Also termed 'round-head plug'.

BARSTOCK PLUG (IN TEE)   **FIGURE 2.55**

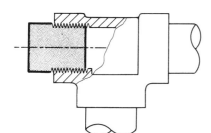

PIPE THREADS   2.5.5

Standard ANSI/ASME B1.20.1 defines general purpose pipe threads: tapered and straight threads for pipe (and fittings, etc.). For the same nominal pipe size, the number of threads per inch is the same for straight and tapered threads. Most pipe joints are made using the tapered thread form.

Tapered and straight threads will mate. Taper/taper and taper/straight (both types) joints are self sealing with the use of pipe dope (a compound spread on the threads which lubricates and seals the joint on assembly), or plastic tape (Teflon). Tape is wrapped around the external thread before the joint is assembled. A straight/straight screwed joint requires locknuts and gaskets to ensure sealing - see fig. 2.39 (d).

Standard ANSI B1.20.3 defines 'dryseal' threads. Dryseal threads seal against line pressure without the use of pipe dope or tape. The seal is obtained by using a modified thread form of sharp crest and flat root. This causes inter-ference (metal-to-metal contact) between the engaged threads, and prevents leakage through the spiral cavity of mating threads.

Symbols used for specifying threads:

    N = American National Standard Thread Form, P = Pipe, T = Taper,
    C = Coupling, F = Fuel & Oil, H = Hose coupling, I = Intermediate,
    L = Locknut, M = Mechancal, R= Railing fittings, S = Straight

ANSI B1.20.1: PIPE THREADS, GENERAL PURPOSE

Taper Pipe Thread	NPT
- Rigid mechanical joint for Railings	NPTR
Straight Pipe Thread:	
- Internal, in Pipe Couplings	NPSC
- Free-fitting, Mechanical Joints for Fixtures	NPSM
- Loose-fitting, Mechanical Joints with Locknuts	NPSL
- Loose-fitting, Mechanical Joints for Hose Couplings	NPSH

ANSI B1.20.3: DRYSEAL PIPE THREADS

Taper Pipe Thread:	
- Dryseal Standard	NPTF
- Dryseal SAE Short (NPTF type, shortened by one thread)	PTF-SAE SHORT
Straight Pipe Thread (internal only):	
- Dryseal, Fuel (for use in soft/ductile materials)	NPSF
- Dryseal, Intermediate (for use in hard/brittle materials)	NPSI

(NPTF is the only type that ensures sealing against line pressure. If there is no objection to its use, pipe dope may be used with all threads to improve sealing, and lessen galling of the threads.)

Specify pipe threads by : NPS - Threads per inch - Thread type

      Example: 3 - 8 NPT

FLANGE FACINGS & FINISHES   2.6.1

Many facings for flanges are offered by flange manufacturers, including various 'tongue and groove' types which must be used in pairs. However, only four types of facing are widely used, and these are shown in figure 2.56.

The raised face is used for about 80% of all flanges. The ring-joint facing, employed with either an oval-section or octagon-section gasket, is used mainly in the petrochemical industry.

THE MOST-USED FLANGE FACINGS   **FIGURE 2.56**

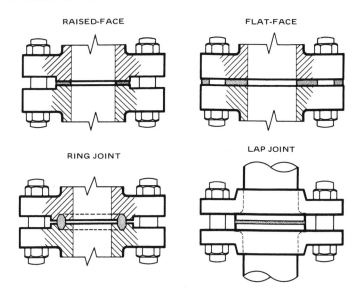

The **RAISED FACE** is 1/16-inch high for Classes 150 and 300 flanges, and 1/4-inch high for all other classes. Class 250 cast-iron flanges and flanged fittings also have the 1/16-inch raised face.

> *Suppliers' catalogs give 'length thru hub' dimensions which include the 0.06-inch raised face on flanges in Classes 150 and 300, but exclude the 0.25-inch raised face on flanges in Classes 400 thru 2500. Tables F include the raised face for all flange Classes.*

FLAT FACE Most common uses are for mating with non-steel flanges on bodies of pumps, etc. and for mating with Class 125 cast-iron valves and fittings. Flat-faced flanges are used with a gasket whose outer diameter equals that of the flange — this reduces the danger of cracking a cast-iron, bronze or plastic flange when the assembly is tightened.

RING-JOINT FACING is a more expensive facing, and considered the most efficient for high-temperature and high-pressure service. Both flanges of a pair are alike. The ring-joint facing is not prone to damage in handling as the surfaces in contact with the gasket are recessed. Use of facings of this type may increase as hollow metal O-rings gain acceptance for process chemical seals.

LAP-JOINT FLANGE is shaped to accommodate the stub end. The combination of flange and stub end presents similar geometry to the raised-face flange and can be used where severe bending stresses will not occur. Advantages of this flange are stated in 2.3.1.

The term 'finish' refers to the type of surface produced by machining the flange face which contacts the gasket. Two principal types of finish are produced, the 'serrated' and 'smooth'.

Forged-steel flanges with raised-face are usually machined to give a 'serrated-concentric' groove, or a 'serrated-spiral' groove finish to the raised-face of the flange. The serrated-spiral finish is the more common and may be termed the 'stock' or 'standard finish' available from suppliers.

The pitch of the groove and the surface finish vary depending on the size and class of the flange. For raised-face steel flanges, the pitch varies from 24 to 40 per inch. It is made using a cutting tool having a minimum radius at the tip of 0.06-inch. The maximum roughness of surface finish is 125-500 microinches.

'Smooth' finish is usually specially-ordered, and is available in two qualities. (1) A fine machined finish leaving no definite tool marks. (2) A 'mirror-finish', primarily intended for use without gaskets.

BOLT HOLES IN FLANGES   2.6.2

Bolt holes in flanges are equally spaced. Specifying the number of holes, diameter of the bolt circle and hole size sets the bolting configuration. Number of bolt holes per flange is given in tables F.

Flanges are positioned so that bolts straddle vertical and horizontal center-lines. This is the normal position of bolt holes on all flanged items.

BOLTS FOR FLANGES   2.6.3

Two types of bolting are available: the studbolt using two nuts, and the machine bolt using one nut. Both boltings are illustrated in figure 2.57. Studbolt thread lengths and diameters are given in tables F.

Studbolts have largely displaced regular bolts for bolting flanged piping joints. Three advantages of using studbolts are:

(1)   The studbolt is more easily removed if corroded

(2)   Confusion with other bolts at the site is avoided

(3)   Studbolts in the less frequently used sizes and materials can be readily made from round stock

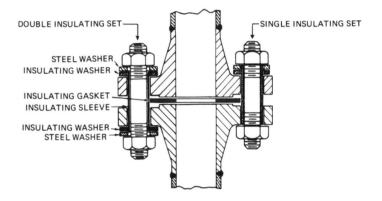

SQUARE-HEAD
MACHINE BOLT

STUDBOLT

HEX NUT   HEX NUT   HEX NUT

UNIFIED INCH SCREW THREADS (UN AND UNR THREAD FORM) UNR indicates rounded root contour, and applies to external threads only. Flat, or rounded root is optional with the UN thread. There are four Unified Screw Threads: Unified Coarse (UNC/ UNCR), Unified Fine (UNF/UNFR), Unified Extra-fine (UNEF/UNEFR) and Unified Selected (UNS/UNSR), with three classes of fit: 1A, 2A and 3A for external threads; 1B, 2B, and 3B for internal threads. (Class 3 has the least clearance.) The standard is ANSI B1.1. which incorporates a metric translation.

UNC (Class 2 medium fit bolt and nut) is used for bolts and studbolts in piping, and specified in the following order:

     Diameter - Threads per inch - Thread - Class of fit.

     Example: BOLT: ½ - 13 UNC 2A
                     NUT: ½ - 13 UNC 2B

GASKETS   2.6.4

Gaskets are used to make a fluid-resistant seal between two surfaces. The common gasket patterns for pipe flanges are the full-face and ring types, for use with flat-faced and raised-face flanges respectively. Refer to figure 2.56. Widely-used materials for gaskets are compressed asbestos (1/16-inch thick) and asbestos-filled metal ('spiral-wound', 0.175-inch thick). The filled-metal gasket is especially useful if maintenance requires repeated uncoupling of flanges, as the gasket separates cleanly and is often reusable.

Choice of gasket is decided by:

(1)   Temperature, pressure and corrosive nature of the conveyed fluid
(2)   Whether maintenance or operation requires repeated uncoupling
(3)   Code/environmental requirements that may apply
(4)   Cost

Garlock Incorporated's publication 'Engineered gasketing products' provides information on the suitability of gasket materials for different applications. Tables 2.5 gives some characteristics of gaskets, to aid selection.

It may be required that adjacent parts of a line are electrically insulated from one another, and this may be effected by inserting a flanged joint fitted with an insulating gasket set between the parts. A gasket electrically insulates the flange faces, and sleeves and washers insulate the bolts from one or both flanges, as illustrated in figure 2.58.

GASKET CHARACTERISTICS   **TABLE 2.5**

GASKET MATERIAL	EXAMPLE USE	MAXIMUM TEMPERATURE (Deg F)	MAXIMUM 'TP' FACTOR Temperature x Pressure (Deg F x PSI)	AVAILABLE THICKNESS (INCHES)
Synthetic rubbers	Water, Air	250	15,000	1/32,1/16,3/32,1/8,1/4
Vegetable fiber	Oil	250	40,000	1/64,1/32,1/16,3/32,1/8
Synthetic rubbers with cloth insert ('CI')	Water, Air	250	125,000	1/32,1/16,3/32,1/8,1/4
Solid Teflon	Chemicals	500	150,000	1/32,1/16,3/32,1/8
Compressed asbestos	Most	750	250,000	1/64,1/32,1/16,1/8
Carbon steel	High-pressure fluids	750	1,600,000	For ring-joint gaskets, refer to part II
Stainless steel	High-pressure &/or corrosive fluids	1200	3,000,000	
Spiral-wound: SS/Teflon CS/Asbestos SS/Asbestos SS/Ceramic	Chemicals Most Corrosive Hot gases	500 750 1200 1900	} 250,000+	Most-used thickness for spiral-wound gaskets is 0.175. Alternative gasket thickness: 0.125.

INSULATING GASKET SET   **FIGURE 2.58**

DOUBLE INSULATING SET   SINGLE INSULATING SET

STEEL WASHER
INSULATING WASHER

INSULATING GASKET
INSULATING SLEEVE

INSULATING WASHER
STEEL WASHER

TEMPORARY CLOSURES FOR LINES   2.7

IN-LINE CLOSURES   2.7.1

A completely leak-proof means of stopping flow in lines is necessary in piping systems when: (1) A change in process material to flow in the line is to be made and cross-contamination is to be avoided. (2) Periodic maintenance is to be carried out, and a hazard would be presented by flammable and/or toxic material passing a valve.

The valves described in 3.1 may not offer complete security against leakage, and one of the following methods of temporary closure can be used: Line-blind valve, line blind (including special types-for use with ring-joint flanges), spectacle plate (so-called from its shape), 'double block and bleed', and blind flanges replacing a removable spool. The last three closures are illustrated in figures 2.59 thru 2.61.

FIGURES 2.56–2.58

TABLE 2.5

SPECTACLE PLATE & LINE BLIND

FIGURE 2.59

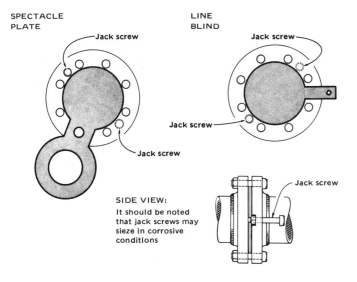

SPECTACLE PLATE

LINE BLIND

Jack screw

Jack screw

Jack screw

Jack screw

Jack screw

Jack screw

SIDE VIEW:
It should be noted that jack screws may sieze in corrosive conditions

DOUBLE-BLOCK-AND-BLEED

FIGURE 2.60

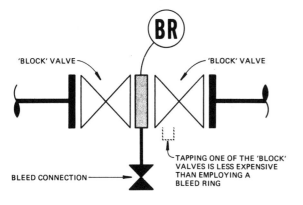

BR

'BLOCK' VALVE

'BLOCK' VALVE

TAPPING ONE OF THE 'BLOCK' VALVES IS LESS EXPENSIVE THAN EMPLOYING A BLEED RING

BLEED CONNECTION

REMOVABLE SPOOL

FIGURE 2.61

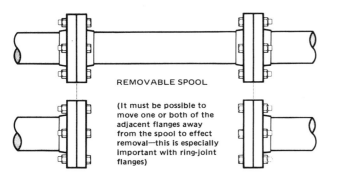

REMOVABLE SPOOL

(It must be possible to move one or both of the adjacent flanges away from the spool to effect removal—this is especially important with ring-joint flanges)

If a line is to be temporarily closed down with double-block-and-bleed, both valves are closed, and the fluid between drawn off with the bleed valve. The bleed valve is then left open to show whether the other valves are tightly shut.

Figure 2.60 shows the bleed ring connected to a bleed valve—see 3.1.11. The use of a tapped valve rather than a bleed ring should be considered, as it is a more economic arrangement, and usually can be specified merely by adding a suffix to the valve ordering number.

A line-blind valve is not illustrated as construction varies. This type of valve incorporates a spectacle plate sandwiched between two flanges which may be expanded or tightened (by some easy means), allowing the spectacle plate to be reversed. Constant-length line-blind valves are also available, made to ANSI dimensions for run length.

Table 2.6 compares the advantages of the four in-line temporary closures:

IN-LINE CLOSURES   TABLE 2.6

CLOSURE / CRITERION	LINE BLIND VALVE	SPECTACLE PLATE, or LINE BLIND	DOUBLE BLOCK, & BLEED	REMOVABLE SPOOL
RELATIVE OVERALL COST	LEAST EXPENSIVE	MEDIUM EXPENSE, DEPENDING ON FREQUENCY OF CHANGEOVER		MOST EXPENSIVE
MANHOURS FOR DOUBLE CHANGEOVER	NEGLIGIBLE	1 to 3	NEGLIGIBLE	2 to 6
INITIAL COST	FAIRLY HIGH	LOW	VERY HIGH	HIGH
CERTAINTY OF SHUT-OFF	COMPLETE	COMPLETE	DOUBTFUL	COMPLETE
VISUAL INDICATION?	YES	YES	YES, BUT SUSPECT	YES
WHO OPERATES?	PLANT OPERATOR	PIPEFITTER	PLANT OPERATOR	PIPEFITTER

CLOSURES FOR PIPE ENDS & VESSEL OPENINGS   2.7.2

Temporary bolted closures include blind flanges using flat gaskets or ring joints, T-bolt closures, welded-on closures with hinged doors — including the boltless manhole cover (Robert Jenkins, England) and closures primarily intended for vessels, such as the Lanape range (Bonney Forge) which may also be used with pipe of large diameter. The blind flange is mostly used with a view to future expansion of the piping system, or for cleaning, inspection, etc. Hinged closures are often installed on vessels; infrequently on pipe.

QUICK CONNECTORS & COUPLINGS   2.8

QUICK CONNECTORS   2.8.1

Two forms of connector specifically designed for temporary use are:
(1) Lever type with double lever clamping, such as Evertite 'Standard' and Victaulic 'Snap Joint'. (2) Screw type with captive nut — 'hose connector'.

Typical use is for connecting temporarily to tank cars, trucks or process vessels. Inter-trades agreements permit plant operators to attach and uncouple these boltless connectors. Certain temporary connectors have built-in valves. Evertite manufactures a double shut-off connector for liquids, and Schrader a valved connector for air lines.

BOLTED QUICK-COUPLINGS   2.8.2

Connections of this type may be suitable for either permanent or temporary use, depending on the joint and gasket, and service conditions. Piping can be built rapidly with them, and they are especially useful for making repairs to lines, for constructing short-run process installations such as pilot plants, and for process modification.

COUPLINGS FOR GROOVED COMPONENTS & PIPE

Couplings of this type are manufactured by the Victaulic Company of America for use with steel, cast-iron, FRP or plastic pipe, either having grooved ends, or with Victaulic collars welded or cemented to the pipe ends.

The following special fittings with grooved ends are available: elbow, tee (all types), lateral, cross, reducer, nipple, and cap. Groove-ended valves and valve adaptors are also available. Advantages: (1) Quick fitting and removal. (2) Joint can take up some deflection and expansion. (3) Suitable for many uses, with correct gaskets.

The manufacturer states that the biggest uses are for permanent plant air, water (drinking, service, process, waste) and lubricant lines.

COMPRESSION SLEEVE COUPLINGS are extensively used for air, water, oil and gas. Well-known manufacturers include Victaulic, Dresser and Smith-Blair. Advantages: (1) Quick fitting and removal. (2) Joint may take up some deflection and expansion. (3) End preparation of pipe is not needed.

VICTAULIC COMPRESSION SLEEVE COUPLING   FIGURE 2.62

EXPANSION JOINTS & FLEXIBLE PIPING   2.9

EXPANSION JOINTS   2.9.1

Figures 2.63 thru 2.66 show methods of accommodating movement in piping due to temperature changes, if such movement cannot be taken up by:

(1) Re-routing or re-spacing the line. (2) Expansion loops—see figure 6.1. (3) Calculated placement of anchors. (4) Cold springing—see 6.1. Bellows-type expansion joints of the type shown in figure 2.63 are also used to absorb vibration.

SIMPLE BELLOWS   FIGURE 2.63

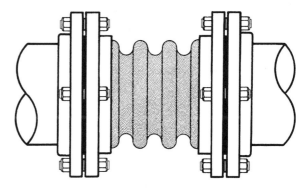

ARTICULATED BELLOWS   FIGURE 2.64

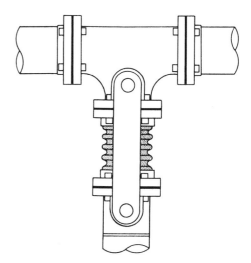

ARTICULATED TWIN-BELLOWS ASSEMBLY   FIGURE 2.65

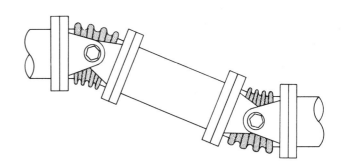

TABLE 2.6

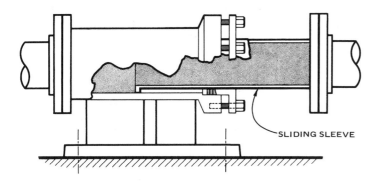

SLIDING SLEEVE

FLEXIBLE PIPING   2.9.2

For filling and emptying railcars, tankers, etc., thru rigid pipe, it is necessary to design articulated piping, using 'swiveling' joints, or 'ball' joints (the latter is a 'universal' joint). Flexible hose has many uses especially where there is a need for temporary connections, or where vibration or movement occurs. Chemical-resistant and/or armored hoses are available in regular or jacketed forms (see figure 6.39).

SEPARATORS, STRAINERS, SCREENS & DRIPLEGS   2.10

COLLECTING UNWANTED MATERIAL FROM THE FLOW   2.10.1

Devices are included in process and service lines to separate and collect undesirable solid or liquid material. Pipe scale, loose weld metal, unreacted or decomposed process material, precipitates, lubricants, oils, or water may harm either equipment or the process.

Common forms of line-installed separator are illustrated in figures 2.67 and 2.68. Other more elaborate separators mentioned in 3.3.3 are available, but these fall more into the category of process equipment, normally selected by the process engineer.

Air and some other gases in liquid-bearing lines are normally self-collecting at piping high points and at the remote ends of headers, and are vented by discharge valves — see 3.1.9.

SEPARATORS   2.10.2

These permanent devices are used to collect droplets from a gaseous stream, for example, to collect oil droplets from compressed air, or condensate droplets from wet steam. Figure 2.67 shows a separator in which droplets in the stream collect in chevroned grooves in the barrier and drain to the small well. Collected liquid is discharged via a trap—see 3.1.9 and 6.10.7.

STRAINERS   2.10.3

Inserted in lines immediately upstream of sensitive equipment, strainers collect solid particles in the approximate size range 0.02–0.5 inch, which can be separated by passing the fluid bearing them thru the strainer's screen. Typical locations for strainers are before a control valve, pump, turbine, or traps on steam systems. 20-mesh strainers are used for steam, water, and heavy or medium oils. 40-mesh is suitable for steam, air, other gases, and light oils.

The commonest strainer is the illustrated wye type where the screen is cylindric and retains the particles within. This type of strainer is easily dismantled. Some strainers can be fitted with a valve to facilitate blowing out collected material without shutting the line down—see figure 6.9, for example. Jacketed strainers are available.

SEPARATOR   FIGURE 2.67

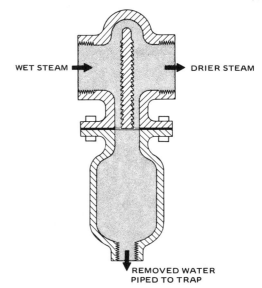

WET STEAM →   → DRIER STEAM

REMOVED WATER
PIPED TO TRAP

STRAINER   FIGURE 2.68

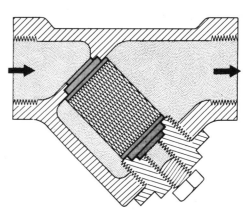

SCREENS 2.10.4

Simple temporary strainers made from perforated sheet metal and/or wire mesh are used for startup operations on the suction side of pumps and compressors, especially where there is a long run of piping before the unit that may contain weld spatter or material inadvertently left in the pipe. After startup, the screen usually is removed.

It may be necessary to arrange for a small removable spool to accommodate the screen. It is important that the flow in suction lines should not be restricted. Cone-shaped screens are therefor preferred, with cylindric types as second choice. Flat screens are better reserved for low-suction heads.

SCREEN BETWEEN FLANGES   **FIGURE 2.69**

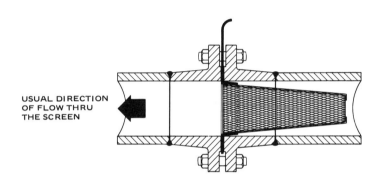

USUAL DIRECTION
OF FLOW THRU
THE SCREEN

DRIPLEG CONSTRUCTION   **FIGURE 2.70**

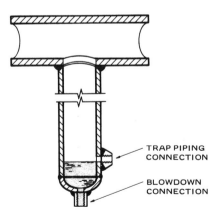

TRAP PIPING
CONNECTION

BLOWDOWN
CONNECTION

DRIPLEGS 2.10.5

Often made from pipe and fittings, the dripleg is an inexpensive means of collecting condensate. Figure 2.70 shows a dripleg fitted to a horizontal pipe. Removal of condensate from steam lines is discussed in 6.10. Recommended sizes for driplegs are given in table 6.10.

REINFORCEMENTS 2.11

BRANCH CONNECTIONS

'Reinforcement' is the addition of extra metal at a branch connection made from a pipe or vessel wall. The added metal compensates for the structural weakening due to the hole.

Stub-ins may be reinforced with regular or wraparound saddles, as shown in figure 2.71. Rings made from platestock are used to reinforce branches made with welded laterals and butt-welded connections to vessels. Small welded connections may be reinforced by adding extra weld metal to the joint.

Reinforcing pieces are usually provided with a small hole to vent gases produced by welding; these gases would otherwise be trapped. A vent hole also serves to indicate any leakage from the joint.

STRAIGHT PIPE

If a butt weld joining two sections of straight pipe is subject to unusual external stress, it may be reinforced by the addition of a 'sleeve' (formed from two units, each resembling the lower member in figure 2.71 (b)).

The code applicable to the piping should be consulted for reinforcement requirements. Backing rings are not considered to be reinforcements—see the footnote to chart 2.1.

REINFORCING SADDLES   **FIGURE 2.71**

(a) REGULAR SADDLE

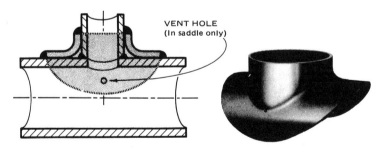

VENT HOLE
(In saddle only)

(b) WRAPAROUND SADDLE

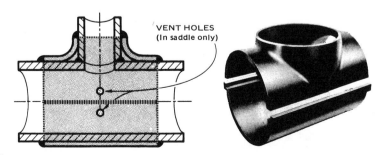

VENT HOLES
(In saddle only)

FIGURES
2.66–2.71

PIPE SUPPORTS

FIGURE 2.72A

HANGERS

SUPPORTS

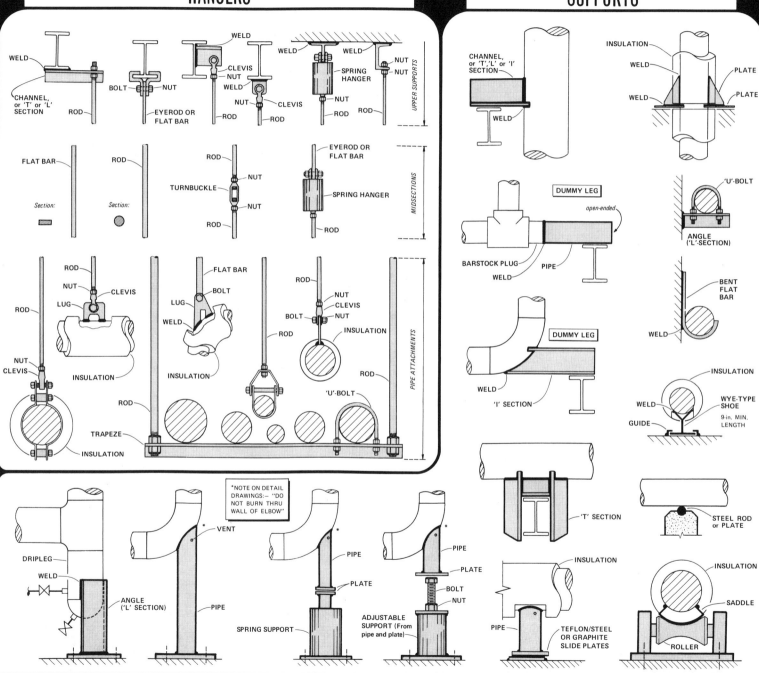

*NOTE ON DETAIL DRAWINGS:— "DO NOT BURN THRU WALL OF ELBOW"

SUPPORTING PIPE CLOSE TO STRUCTURAL STEEL

(COURTESY STEEL CITY DIVISION, MIDLAND-ROSS CORP)

 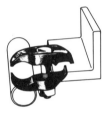

'KINDORF SYSTEM'

(COURTESY UNISTRUT CORPORATION)

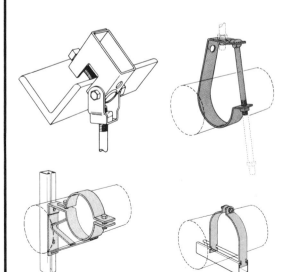

SPRING HANGERS

(COURTESY VOKES-BERGEN-GENSPRING LTD)

1. CONSTANT LOAD TYPE

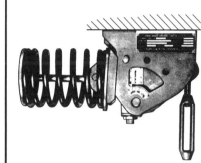

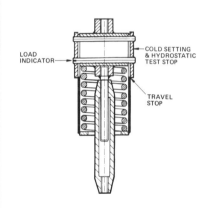

LOAD INDICATOR

COLD SETTING & HYDROSTATIC TEST STOP

TRAVEL STOP

SPRING SUPPORT

(COURTESY VOKES-BERGEN-GENSPRING LTD)

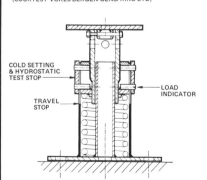

COLD SETTING & HYDROSTATIC TEST STOP

LOAD INDICATOR

TRAVEL STOP

SUPPORTS ALLOWING FREE MOVEMENT OF PIPE

(COURTESY STEEL CITY DIVISION, MIDLAND-ROSS CORP)

(COURTESY UNION CARBIDE)

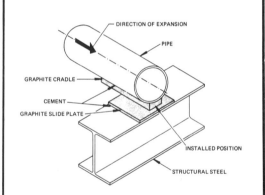

DIRECTION OF EXPANSION

PIPE

GRAPHITE CRADLE

CEMENT

GRAPHITE SLIDE PLATE

INSTALLED POSITION

STRUCTURAL STEEL

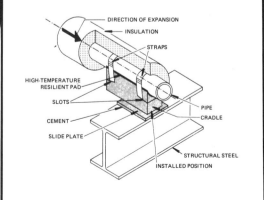

DIRECTION OF EXPANSION

INSULATION

STRAPS

HIGH-TEMPERATURE RESILIENT PAD

SLOTS

CEMENT

SLIDE PLATE

PIPE

CRADLE

STRUCTURAL STEEL

INSTALLED POSITION

FIGURES
2.72A&B

Symbols for drafting various types of support are shown in chart 5.7. For designing support systems, see 6.2.

PIPE SUPPORTS   2.12.1

Pipe supports should be as simple as conditions allow. Stock items are used where practicable, especially for piping held from above. To support piping from below, supports are usually made to suit from platestock, pipe, and pieces of structural steel.

A selection of available hardware for supporting is illustrated in figures 2.72A and B.

TERMS FOR SUPPORTS   2.12.2

SUPPORT The weight of piping is usually carried on supports made from structural steel, or steel and concrete. (The term 'support' is also used in reference to hangers.)

HANGER Device which suspends piping (usually a single line) from structural steel, concrete or wood. Hangers are usually adjustable for height.

ANCHOR A rigid support which prevents transmission of movement (thermal, vibratory, etc.) along piping. Construction may be from steel plate, brackets, flanges, rods, etc. Attachment of an anchor to pipe should preferably encircle the pipe and be welded all around as this gives a better distribution of stress in the pipe wall.

TIE An arrangement of one or more rods, bars, etc., to restrain movement of piping.

DUMMY LEG An extension piece (of pipe or rolled steel section) welded to an elbow in order to support the line—see figure 2.72A and table 6.3.

The following hardware is used where mechanical and/or thermal movement is a problem:

GUIDE A means of allowing a pipe to move along its length, but not sideways.

SHOE A metal piece attached to the underside of a pipe which rests on supporting steel. Primarily used to reduce wear from sliding for lines subject to movement. Permits insulation to be applied to pipe.

SADDLE A welded attachment for pipe requiring insulation, and subject to longitudinal or rolling movement (resulting from temperature changes other than climatic). Saddles may be used with guides as shown in 6.2.8.

SLIDE PLATE A slide plate support is illustrated in figure 2.72A. Figure 2.72B shows applications of 'Ucar' graphite slide plates which are offered by Union Carbide Inc. The two plates used in a support are made from or faced with a material of low friction able to withstand mechanical stress and temperature changes. Plates are often made from graphite blocks. Steel plates with a teflon facing are available and may be welded to steel.

Spring hangers or supports allow variations in the length of pipe due to changes in temperature, and are often used for vertical lines. Refer to 6.2.5 figure 6.16. There are two types of spring hanger or support:

'CONSTANT LOAD' HANGER This device consists of a coil spring and lever mechanism in a housing. Movement of the piping, within limits, will not change the spring force holding up the piping; thus, no additional forces will be introduced to the piping system.

'VARIABLE SPRING' HANGER, and SUPPORT These devices consist of a coil spring in a housing. The weight of the piping rests on the spring in compression. The spring permits a limited amount of thermal movement. A variable spring hanger holding up a vertical line will reduce its lifting force as the line expands toward it. A variable spring support would increase its lifting force as the line expands toward it. Both place a load on the piping system. Where this is undesirable, a constant-load hanger can be used instead.

———◆———

HYDRAULIC DAMPENER, SHOCK, SNUBBER, or SWAY SUPPRESSOR One end of the unit is attached to piping and the other to structural steel or concrete. The unit expands or contracts to absorb slow movement of piping, but is rigid to rapid movement.

SWAY BRACE, or SWAY ARRESTOR, is essentially a helical spring in a housing which is fitted between piping and a rigid structure. Its function is to buffer vibration and sway.

WELDING TO PIPE   2.12.3

If the applicable code permits, lugs may be welded to pipe. Figure 2.72A illustrates some common arrangements using welded lugs, rolled steel sections and pipe, for:—

(1)   Fixing hangers to structural steel, etc.
(2)   Attaching to pipe
(3)   Supporting pipe

Welding supports to prelined pipe will usually spoil the lining, and therefor lugs, etc., must be welded to pipe and fittings before the lining is applied. Welding of supports and lugs to pipes and vessels to be stress-relieved should be done before heat treatment.

VALVES, PUMPS, COMPRESSORS, and Types of Process Equipment

VALVES   3.1

FUNCTIONS OF VALVES   3.1.1

Table 3.1 gives a basis for classifying valves according to function:

USES OF VALVES   TABLE 3.1

VALVE ACTION	EXPLANATION	SEE SECTION:
ON/OFF	STOPPING OR STARTING FLOW	3.1.4 and 3.1.6
REGULATING	VARYING THE RATE OF FLOW	3.1.5, 3.1.6 and 3.1.10
CHECKING	PERMITTING FLOW IN ONE DIRECTION ONLY	3.1.7
SWITCHING	SWITCHING FLOW ALONG DIFFERENT ROUTES	3.1.8
DISCHARGING	DISCHARGING FLUID FROM A SYSTEM	3.1.9

Types of valve suitable for on/off and regulating functions are listed in chart 3.2. The suitability of a valve for a required purpose depends on its construction, discussed in 3.1.3.

PARTS OF VALVES   3.1.2

Valve manufacturers' catalogs offer a seemingly endless variety of constructions. Classification is possible, however, by considering the basic parts that make up a valve:

(1)   The 'disc' and 'seat' that directly affect the flow

(2)   The 'stem' that moves the disc — in some valves, fluid under pressure does the work of a stem

(3)   The 'body' and 'bonnet' that house the stem

(4)   The 'operator' that moves the stem (or pressurizes fluid for squeeze valves, etc.)

Figures 3.1 thru 3.3 show three common types of valve with their parts labeled.

DISC, SEAT, & PORT

Chart 3.1 illustrates various types of disc and port arrangements, and mechanisms used for stopping or regulating flow. The moving part directly affecting the flow is termed the 'disc' regardless of its shape, and the non-moving part it bears on is termed the 'seat'. The 'port' is the maximum internal opening for flow (that is, when the valve is fully open). Discs may be actuated by the conveyed fluid or be moved by a stem having a linear, rotary or helical movement. The stem can be moved manually or be driven hydraulically, pneumatically or electrically, under remote or automatic control, or mechanically by weighted lever, spring, etc.

The size of a valve is determined by the size of its ends which connect to the pipe, etc. The port size may be smaller.

STEM

There are two categories of screwed stem: The rising stem shown in figures 3.1 and 3.2, and the non-rising stem shown in figure 3.3.

Rising stem (gate and globe) valves are made either with 'inside screw' (IS) or 'outside screw' (OS). The OS type has a yoke on the bonnet and the assembly is referred to as 'outside screw and yoke', abbreviated to 'OS&Y'. The handwheel can either rise with the stem, or the stem can rise thru the handwheel.

TABLE
3.1

BASIC VALVE MECHANISMS
FLUID CONTROL ELEMENTS (DISCS)

CHART 3.1

IN THESE SCHEMATIC DIAGRAMS, THE DISC IS SHOWN WHITE, THE SEAT IN SOLID COLOR, & THE CONVEYED FLUID SHADED.

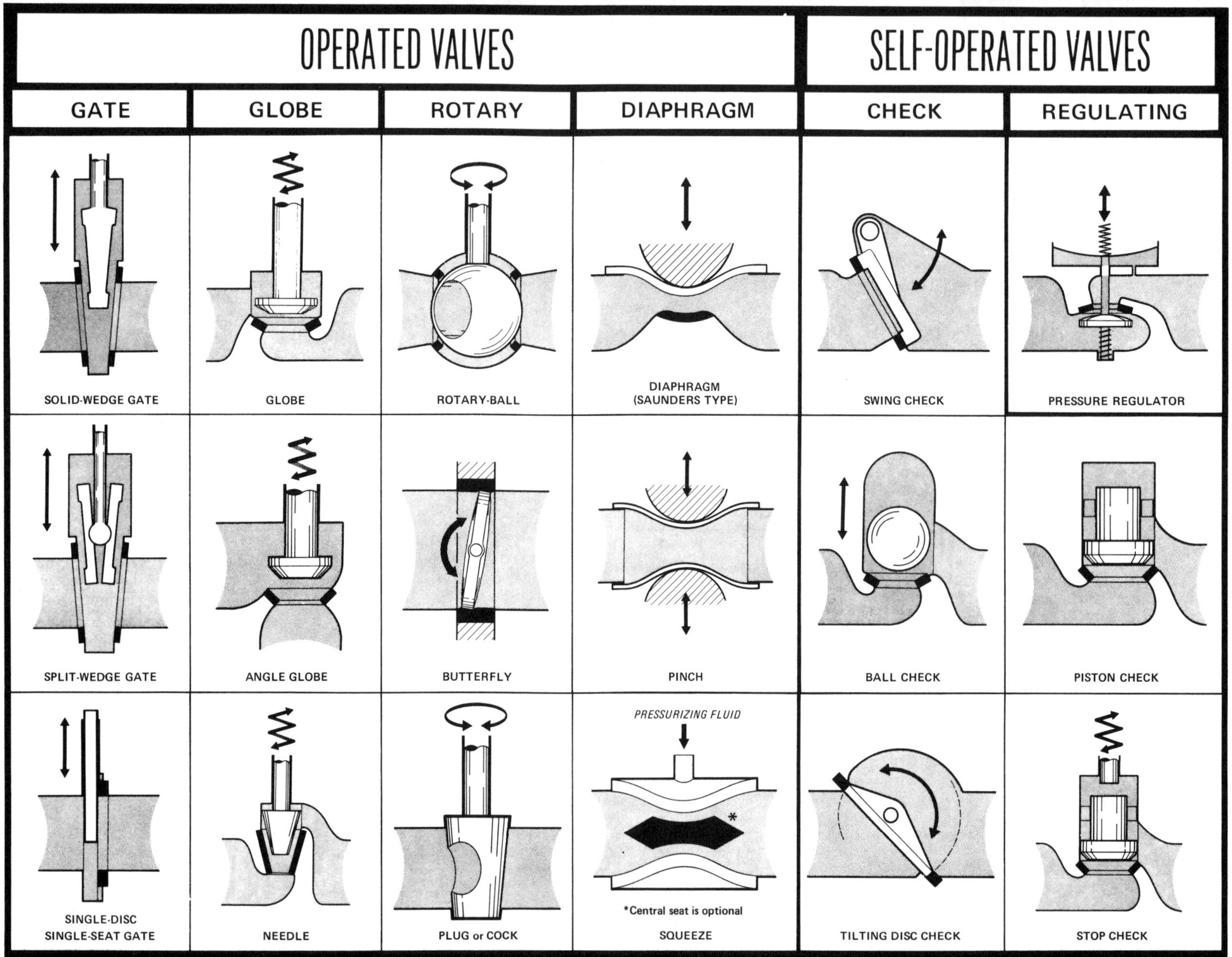

OPERATED VALVES

SELF-OPERATED VALVES

GATE	GLOBE	ROTARY	DIAPHRAGM	CHECK	REGULATING
SOLID-WEDGE GATE	GLOBE	ROTARY-BALL	DIAPHRAGM (SAUNDERS TYPE)	SWING CHECK	PRESSURE REGULATOR
SPLIT-WEDGE GATE	ANGLE GLOBE	BUTTERFLY	PINCH	BALL CHECK	PISTON CHECK
SINGLE-DISC SINGLE-SEAT GATE	NEEDLE	PLUG or COCK	SQUEEZE	TILTING DISC CHECK	STOP CHECK

PRESSURIZING FLUID

*Central seat is optional

[30]

Non-rising stem valves are of the gate type. The handwheel and stem are in the same position whether the valve is open or closed. The screw is inside the bonnet and in contact with the conveyed fluid.

A 'floor stand' is a stem extension for use with both types of stem, where it is necessary to operate a valve thru a floor or platform. Alternately, rods fitted with universal joints may be used to bring a valve handwheel within an operator's reach.

Depending on the size of the required valve and availabilities, selection of stem type can be based on:

(1) Whether it is undesirable for the conveyed fluid to be in contact with the threaded bearing surfaces

(2) Whether an exposed screw is liable to be damaged by abrasive atmospheric dust

(3) Whether it is necessary to see if the valve is open or closed

In addition to the preceding types of stem used with gate and globe valves, most other valves have a simple rotary stem. Rotary-ball, plug and butterfly valves have a rotary stem which is moved by a permanent lever, or tool applied to a square boss at the end of the stem.

BONNET

There are three basic types of attachment for valve bonnets: screwed (including union), bolted, and breechlock.

A screwed bonnet may occasionally stick and turn when a valve is opened. Although sticking is less of a problem with the union type bonnet, valves with screwed bonnets are best reserved for services presenting no hazard to personnel. Union bonnets are more suitable for small valves requiring frequent dismantling than the simple screwed type.

The bolted bonnet has largely displaced screwed and union bonnet valves in hydrocarbon applications. A U-bolt or clamp-type bonnet is offered on some small gate valves for moderate pressures, to facilitate frequent cleaning and inspection.

The 'pressure seal' is a variation of the bolted bonnet used for high-pressure valves, usually combined with OS&Y construction. It makes use of line pressure to tighten and seal an internal metal ring or gasket against the body.

The breechlock is a heavier infrequently-used and more expensive construction, also for high-pressure use, and involves seal-welding of the bonnet with the body.

FIGURE 3.1

GATE VALVE (OS&Y, bolted bonnet, rising stem)

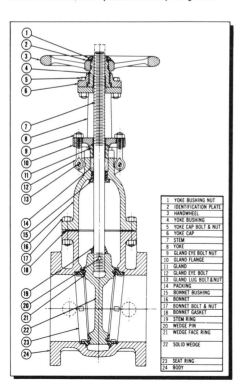

1	YOKE BUSHING NUT
2	IDENTIFICATION PLATE
3	HANDWHEEL
4	YOKE BUSHING
5	YOKE CAP BOLT & NUT
6	YOKE CAP
7	STEM
8	YOKE
9	GLAND EYE BOLT NUT
10	GLAND FLANGE
11	GLAND
12	GLAND EYE BOLT
13	GLAND LUG BOLT & NUT
14	PACKING
15	BONNET BUSHING
16	BONNET
17	BONNET BOLT & NUT
18	BONNET GASKET
19	STEM RING
20	WEDGE PIN
21	WEDGE FACE RING
22	SOLID WEDGE
23	SEAT RING
24	BODY

FIGURE 3.2

GLOBE VALVE (OS&Y, bolted bonnet, rising stem)

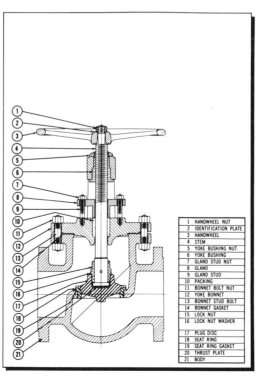

1	HANDWHEEL NUT
2	IDENTIFICATION PLATE
3	HANDWHEEL
4	STEM
5	YOKE BUSHING NUT
6	YOKE BUSHING
7	GLAND STUD NUT
8	GLAND
9	GLAND STUD
10	PACKING
11	BONNET BOLT NUT
12	YOKE BONNET
13	BONNET STUD BOLT
14	BONNET GASKET
15	LOCK NUT
16	LOCK NUT WASHER
17	PLUG DISC
18	SEAT RING
19	SEAT RING GASKET
20	THRUST PLATE
21	BODY

FIGURE 3.3

GATE VALVE (IS, bolted bonnet, non-rising stem)

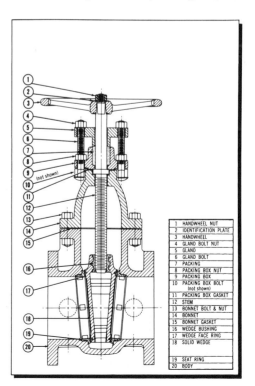

1	HANDWHEEL NUT
2	IDENTIFICATION PLATE
3	HANDWHEEL
4	GLAND BOLT NUT
5	GLAND
6	GLAND BOLT
7	PACKING
8	PACKING BOX NUT
9	PACKING BOX
10	PACKING BOX BOLT (not shown)
11	PACKING BOX GASKET
12	STEM
13	BONNET BOLT & NUT
14	BONNET
15	BONNET GASKET
16	WEDGE BUSHING
17	WEDGE FACE RING
18	SOLID WEDGE
19	SEAT RING
20	BODY

CHART 3.1

FIGURES 3.1–3.3

A critical factor for valves used for process chemicals is the lubrication of the stem. Care has to be taken in the selection of packing, gland design, and choice and application of lubricant. As an option the bonnet may include a 'lantern ring' which serves two purposes — either to act as a collection point to drain off any hazardous seepages, or as a point where lubricant can be injected.

LANTERN RING

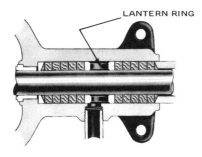

LANTERN RING

BODY

Selection of material to fabricate the interior of the valve body is important with a valve used for process chemicals. There is often a choice with regard to the body and trim, and some valves may be obtained with the entire interior of the body lined with corrosion-resistant material.

Valves are connected to pipe, fittings or vessels by their body ends, which may be flanged, screwed, butt- or socket-welding, or finished for hose, Victaulic coupling, etc. Jacketed valves are also available—see 6.8.2.

SEAL

In most stem-operated valves, whether the stem has rotary or lineal movement, packing or seals are used between stem and bonnet (or body). If high vacuum or corrosive, flammable or toxic fluid is to be handled, the disc or stem may be sealed by a metal bellows, or by a flexible diaphragm (the latter is termed 'packless' construction). A gasket is used as a seal between a bolted bonnet and valve body.

BELLOWS-SEAL VALVE

'PACKLESS' VALVE

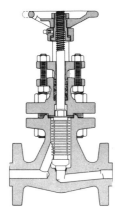

Flanged valves use gaskets to seal against the line flanges. Butterfly valves may extend the resilient seat to also serve as line gaskets. The pressure-seal bonnet joint utilizes the pressure of the conveyed fluids to tighten the seal — see 'Pressure seal' under 'Bonnet', this section.

MANUAL OPERATORS

HANDLEVER is used to actuate the stems of small butterfly and rotary-ball valves, and small cocks. Wrench operation is used for cocks and small plug valves.

HANDLEVERS ON SMALL VALVES

COCK

WRENCH

WRENCH USED AS OPERATOR ON COCK

HANDWHEEL is the most common means for rotating the stem on the majority of popular smaller valves such as the gate, globe and diaphragm types. Additional operating torque for gate and globe valves is offered by 'hammerblow' or 'impact' handwheels which may be substituted for normal handwheels if easier operation is needed but where gearing is unnecessary.

HAMMER-BLOW HANDWHEEL

HAMMER ACTION IS PROVIDED BY TWO LUGS CAST ON UNDER-SIDE OF HANDWHEEL, WHICH HIT ANVIL PROJECTING BETWEEN

CHAIN operator is used where a handwheel would be out of reach. The stem is fitted with a chainwheel or wrench (for lever-operated valves) and the loop of the chain is brought within 3 ft of working floor level. Universal-type chainwheels which attach to the regular handwheel have been blamed for accidents: in corrosive atmospheres where an infrequently-operated valve has stuck, the attaching bolts have been known to fail. This problem does not arise with the chainwheel that replaces the regular valve handwheel.

GEAR operator is used to reduce the operating torque. For manual operation, consists of a handwheel-operated gear train actuating the valve stem. As a guide, gear operators should be considered for valves of the following sizes and classes: 125, 150, and 300, 14-inch and larger; 400 and 600, 8-inch and larger; 900 and 1500, 6-inch and larger; 2500, 4-inch and larger.

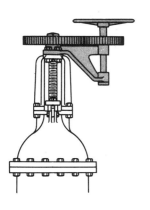

SPUR-GEAR OPERATOR

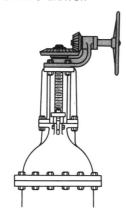

BEVEL-GEAR OPERATOR

POWERED OPERATORS

Electric, pneumatic or hydraulic operation is used: (1) Where a valve is remote from the main working area. (2) If the required frequency of operation would need unreasonable human effort. (3) If rapid opening and/or closing of a valve is required.

ELECTRIC MOTOR The valve stem is moved by the electric motor, thru reducing gears.

SOLENOID may be used with fast-acting check valves, and with on/off valves in light-duty instrumentation applications.

ELECTRIC MOTOR OPERATOR   **PNEUMATIC OPERATOR**

PNEUMATIC & HYDRAULIC OPERATORS may be used where flammable vapor is likely to be present. They take the following forms: (1) Cylinder with double-acting piston driven by air, water, oil, or other liquid which usually actuates the stem directly. (2) Air motor which actuates the stem thru

gearing—these motors are commonly piston-and-cylinder radial types. (3) A double-acting vane with limited rotary movement in a sector casing, actuating the stem directly. (4) Squeeze type (refer to 'Squeeze valve').

QUICK-ACTING OPERATORS FOR NON-ROTARY VALVES
(Manually-operated valves)

Quick-acting operators are used with gate and globe valves. Two stem movements are employed:—

(1)   Rotating stem, rotated by a lever
(2)   Sliding stem, in which the stem is raised and lowered by lever

QUICK-ACTING LEVERS ON VALVES

(1) Rotating stem on globe valve   (2) Sliding stem on gate valve

Steam and air whistles are examples of the use of sliding-stem quick-acting operators with globe valves.

SELECTING ON/OFF & REGULATING VALVES   3.1.3

The suitability of a valve for a particular service is decided by its materials of construction in relation to the conveyed fluid as well as its mechanical design. Referring to the descriptions in 3.1.2, the steps in selection are to choose: (1) Material(s) of construction. (2) The disc type. (3) Stem type. (4) Means of operating the stem — the 'operator'. (5) Bonnet type. (6) Body ends — welding, flanged, etc. (7) Delivery time. (8) Price. (9) Warranty of performance for severe conditions.

Chart 3.2 is a guide to valve selection, and indicates valves which may be chosen for a given service. The chart should be read from left to right. First, ascertain whether a liquid, gas or powder is to be handled by the valve. Next, consider the nature of the fluid—whether it is foodstuffs or drugs to be handled hygienically, chemicals that are corrosive, or whether the fluid is substantially neutral or non-corrosive.

Next consider the function of the valve — simple open-or-closed operation ('on/off'), or regulating for control or for dosing. These factors decided, the chart will then indicate types of valves which should perform satisfactorily in the required service.

If the publication is available, reference should also be made to the Crane Company's 'Choosing the right valve'.

VALVE SELECTION GUIDE — CHART 3.2

CONVEYED FLUID	NATURE OF FLUID See Note (2) in Key	VALVE FUNCTION	TYPE OF DISC	SPECIAL FEATURES [.....] denotes Limitation. (.....) denotes Option.
LIQUID	NEUTRAL (WATER, OIL, Etc.)	ON/OFF	GATE ROTARY BALL PLUG DIAPHRAGM BUTTERFLY PLUG GATE	NONE NONE NONE [For oil: No natural rubber] NONE NONE
		REGULATING	GLOBE BUTTERFLY PLUG GATE DIAPHRAGM NEEDLE	NONE NONE NONE [For oil: No natural rubber] NONE, [Small flows only]
	CORROSIVE (ALKALINE, ACID, Etc.)	ON/OFF	GATE PLUG GATE ROTARY BALL PLUG DIAPHRAGM BUTTERFLY	ANTI-CORROSIVE*,(OS&Y),(Bellows seal) ANTI-CORROSIVE*,(OS&Y) ANTI-CORROSIVE*,(Lined) ANTI-CORROSIVE*,(Lubricated),(Lined) ANTI-CORROSIVE*,(Lined) ANTI-CORROSIVE*,(Lined)
		REGULATING	GLOBE DIAPHRAGM BUTTERFLY PLUG GATE	ANTI-CORR.*,(OS&Y),(Diaphragm or Bellows Seal) ANTI-CORROSIVE*,(Lined) ANTI-CORROSIVE*,(Lined) ANTI-CORROSIVE*,(OS&Y)
	HYGIENIC (BEVERAGES, FOOD and DRUGS)	ON/OFF	BUTTERFLY DIAPHRAGM	SPECIAL DISC†, WHITE SEAT † SANITARY LINING, WHITE DIAPHRAGM †
		REGULATING	BUTTERFLY DIAPHRAGM SQUEEZE PINCH	SPECIAL DISC†, WHITE SEAT † SANITARY LINING, WHITE DIAPHRAGM † WHITE FLEXIBLE TUBE† WHITE FLEXIBLE TUBE†
	SLURRY	ON/OFF	ROTARY BALL BUTTERFLY DIAPHRAGM PLUG PINCH SQUEEZE	ABRASION-RESISTANT LINING ABRASION-RESIST. DISC, RESILIENT SEAT ABRASION-RESISTANT LINING LUBRICATED, (Lined) NONE CENTRAL SEAT
		REGULATING	BUTTERFLY DIAPHRAGM SQUEEZE PINCH GATE	ABRASION-RESIST. DISC, RESILIENT SEAT LINED* NONE NONE SINGLE SEAT, NOTCHED DISC
	FIBROUS SUSPENSIONS	ON/OFF & REGULATING	GATE DIAPHRAGM SQUEEZE PINCH	SINGLE SEAT, KNIFE-EDGED DISC, NOTCHED DISC NONE NONE NONE
GAS	NEUTRAL (AIR, STEAM, Etc.)	ON/OFF	GATE GLOBE ROTARY BALL PLUG DIAPHRAGM	NONE (Composition Disc),(Plug-Type Disc) NONE NONE, [Unsuitable for steam service] NONE, [Unsuitable for steam service]
		REGULATING	GLOBE NEEDLE BUTTERFLY DIAPHRAGM GATE	NONE NONE, [Small flows only] NONE NONE, [Unsuitable for steam service] SINGLE SEAT
	CORROSIVE (ACID VAPORS, CHLORINE, Etc.)	ON/OFF	BUTTERFLY ROTARY BALL DIAPHRAGM PLUG	ANTI-CORROSIVE* ANTI-CORROSIVE* ANTI-CORROSIVE* ANTI-CORROSIVE*
		REGULATING	BUTTERFLY GLOBE NEEDLE DIAPHRAGM	ANTI-CORROSIVE* ANTI-CORROSIVE*, (OS&Y) ANTI-CORROSIVE*, [Small flows only] ANTI-CORROSIVE*
	VACUUM	ON/OFF	GATE GLOBE ROTARY BALL BUTTERFLY	BELLOWS SEAL DIAPHRAGM or BELLOWS SEAL NONE RESILIENT SEAT
SOLID	ABRASIVE POWDER (SILICA, Etc.)	ON/OFF & REGULATING	PINCH SQUEEZE SPIRAL SOCK	NONE (CENTRAL SEAT) NONE
	LUBRICATING POWDER (GRAPHITE, TALC, Etc.)	ON/OFF & REGULATING	PINCH GATE SQUEEZE SPIRAL SOCK	NONE SINGLE SEAT (CENTRAL SEAT) NONE

* Suitability of materials of construction with respect to the great variety of fluids encountered is a complex topic. A good general reference is the current edition of the Chemical Engineer's Handbook

† The disc should be smooth, without bolts and recesses, in a sanitary material such as stainless steel, or fully coated with 'white' plastic or rubber material. 'White' means that the material does not contain a filler which is toxic or can discolor the product.

KEY TO VALVE SELECTION GUIDE CHART 3.2

(1) Determine type of conveyed fluid—liquid, gas slurry, or powder

(2) Determine nature of fluid:
- Substantially neutral—not noticeably acid or alkaline, such as various oils, drinking water, nitrogen, gas, air,etc.
- Corrosive—markedly acid, alkaline, or otherwise chemically reactive
- 'Hygienic'—materials for the food, drug, cosmetic or other industries
- Slurry—suspension of solid particles in a liquid can have an abrasive effect on valves, etc. Non-abrasive slurries such as wood-pulp slurries can choke valve mechanisms

(3) Determine operation:
- 'On/off'—fully open or fully closed
- Regulating—including close regulation (throttling)

(4) Look into other factors affecting choice:
- Pressure and temperature of conveyed fluid
- Method of operating stem—consider closing time
- Cost
- Availability
- Special installation problems—such as welding valves into lines. Welding heat will sometimes distort the body and affect the sealing of small valves.

In industrial piping, on/off control of flow is most commonly effected with gate valves. Most types of gate valve are unsuitable for regulating: erosion of the seat and disc occurs in the throttling position due to vibration of the disc ("chattering"). With some fluids, it may be desirable to use globe valves for on/off service, as they offer tighter closure. However, as the principal function of globe valves is regulation, they are described in 3.1.5.

SOLID WEDGE GATE VALVE has either a solid or flexible wedge disc. In addition to on/off service, these valves can be used for regulating, usually in sizes 6-inch and larger, but will chatter unless disc is fully guided throughout travel. Suitable for most fluids including steam, water, oil, air and gas. The flexible wedge was developed to overcome sticking on cooling in high-temperature service, and to minimize operating torque. The flexible wedge is not illustrated—it can be likened to two wheels set on a very short axle.

SOLID WEDGE GATE VALVE

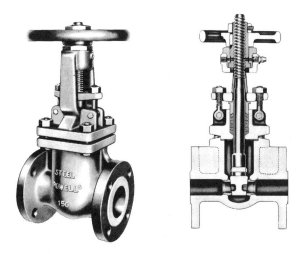

DOUBLE-DISC PARALLEL-SEATS GATE VALVE has two parallel discs which are forced, on closure, against parallel seats by a 'spreader'. Used for liquids and gases at normal temperatures. Unsuitable for regulation. To prevent jamming, installation is usually vertical with handwheel up.

DOUBLE-DISC (SPLIT-WEDGE) WEDGE GATE VALVE Discs wedge against inclined seats without use of a spreader. Remarks for double-disc parallel seats gate valve apply, but smaller valves are made for steam service. Often, construction allows the discs to rotate, distributing wear.

SINGLE-DISC SINGLE-SEAT GATE VALVE, or SLIDE VALVE, is used for handling paper pulp slurry and other fibrous suspensions, and for low-pressure gases. Will not function properly with inflow on the seat side. Suitable for regulating flow if tight closure is not required.

SINGLE-DISC PARALLEL-SEATS GATE VALVE Unlike the single-seat slide valve, this valve affords closure with flow in either direction. Stresses on stem and bonnet are lower than with wedge-gate valves. Primarily used for liquid hydrocarbons and gases.

SINGLE-DISC PARALLEL-SEATS GATE VALVE   **PLUG GATE VALVE**

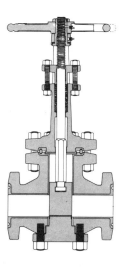 

CHART 3.2

PLUG GATE VALVE This valve has a round tapered disc which moves up and down. Suitable for throttling and full-flow use, but only available in the smaller sizes.

PLUG VALVE Mechanism is shown in chart 3.1, but the disc may be cylindric as well as tapered. Advantages are compactness, and rotary 90-degree stem movement. The tapered plug tends to jam and requires a high operating torque: this is overcome to some extent by the use of a low-friction (teflon, etc.) seat, or by lubrication (with the drawback that the conveyed fluid is contaminated). The friction problem is also met by mechanisms raising the disc from the seat before rotating it, or by using the 'eccentric' design (see rotary-ball valve). Principal uses are for water, oils, slurries, and gases.

LINE-BLIND VALVE This is a positive shutoff device which basically consists of a flanged assembly sandwiching a spectacle-plate or blind. This valve is described and compared with other closures in 2.7.1.

VALVES MAINLY FOR REGULATING SERVICE 3.1.5

GLOBE VALVE, STRAIGHT & ANGLE TYPE These are the valves most used for regulating. For line sizes over 6-inch, choice of a valve for flow control tends to go to suitable gate or butterfly valves. For more satisfactory service, the direction of flow thru valve recommended by manufacturers is from stem to seat, to assist closure and to prevent the disc chattering against the seat in the throttling position. Flow should be from seat to stemside (1) if there is a hazard presented by the disc detaching from the stem thus closing the valve, or (2) if a composition disc is used, as this direction of flow then gives less wear.

ANGLE VALVE This is a globe valve with body ends at right angles, saving the use of a 90-degree elbow. However, the angles of piping are often subject to higher stresses than straight runs, which must be considered with this type of valve.

GLOBE VALVES

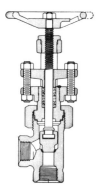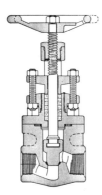

REGULAR-DISC GLOBE VALVE Unsuitable for close regulation as disc and seat have narrow (almost line) contact.

PLUG-TYPE DISC GLOBE VALVE Used for severe regulating service with gritty liquids, such as boiler feedwater, and for blow-off service. Less subject to wear under close regulation than the regular-seated valve.

WYE-BODY GLOBE VALVE has in-line ports and stem emerging at about 45 degrees; hence the 'Y'. Preferred for erosive fluids due to smoother flow pattern.

WYE-BODY GLOBE VALVE (Incorporating composition disc)

COMPOSITION-DISC GLOBE VALVE Suitable for coarse regulation and tight shutoff. Replaceable composition-disc construction is similar to that of a faucet. Grit will imbed in the soft disc preventing seat damage and ensuring good closure. Close regulating will rapidly damage the seat.

DOUBLE-DISC GLOBE VALVE features two discs bearing on separate seats spaced apart on a single shaft, which frees the operator from stresses set up by the conveyed fluid pressing into the valve. Principle is used on control valves and pressure regulators for steam and other gases. Tight shutoff is not ensured.

NEEDLE VALVE is a small valve used for flow control and for dosing liquids and gases. Resistance to flow is precisely controlled by a relatively large seat area and the adjustment afforded by fine threading of the stem.

NEEDLE VALVE

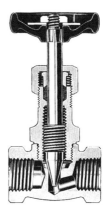

SQUEEZE VALVE is well-suited to regulating the flow of difficult liquids, slurries and powders. Maximum closure is about 80%, which limits the range of regulation, unless the variation of this type of valve with a central core (seat) is used, offering full closure.

PINCH VALVE Also suited to regulating flow of difficult liquids, slurries and powders. Complete closure is possible but tends to rapidly wear the flexible tube, unless of special design.

VALVES FOR BOTH REGULATING & ON/OFF SERVICE   3.1.6

ROTARY-BALL VALVE Advantages are low operating torque, availability in large sizes, compactness, rotary 90-degree stem movement, and 'in-line' replaceability of all wearing parts in some designs. Possible disadvantages are that fluid is trapped within the body (and within the disc on closure), and that compensation for wear is effected only by resilient material behind the seats: the latter problem is avoided in the single-seat 'eccentric' version, which has the ball slightly offset so that it presses into the seat, on closure. Principal uses are for water, oils, slurries, gases and vacuum. Valve is available with a ball having a shaped port for regulation.

ROTARY-BALL VALVE

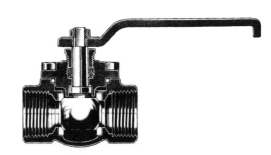

BUTTERFLY VALVE offers the advantages of rotary stem movement (90 degrees or less), compactness, and absence of pocketing. It is available in all sizes, and can be produced in chemical-resistant and hygienic forms. The valves are used for gases, liquids, slurries, powders and vacuum. The usual resilient plastic seat has a temperature limitation, but tight closure at high temperatures is available with a version having a metal ring seal around the disc. If the valve is flanged, it may be held between flanges of any type. Slip-on and screwed flanges do not form a proper seal with some wafer forms of the valve, in which the resilient seat is extended to serve also as line gaskets.

BUTTERFLY VALVE
(Wafer type)

VALVES FOR CHECKING BACKFLOW   3.1.7

All valves in this category are designed to permit flow of liquid or gas in one direction and close if flow reverses.

SWING CHECK VALVE The regular swing check valve is not suitable if there is frequent flow reversal as pounding and wearing of disc occurs. For gritty liquids a composition disc is advisable to reduce damage to the seat. May be mounted vertically with flow upward, or horizontally. Vertically-mounted valve has a tendency to remain open if the stream velocity changes slowly. An optional lever and outside weight may be offered either to assist closing or to counterbalance the disc in part, and allow opening by low-pressure fluid.

SWING CHECK VALVES

Outside Lever & Weight
for swing check valve

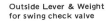

TILTING-DISC VALVE Suitable where frequent flow reversal occurs. Valve closes rapidly with better closure and less slamming than the swing check valve, which it somewhat resembles. It has higher pressure drop with large

flow velocities and lower-pressure drop with small velocities than a comparable swing-check valve. May be installed vertically with flow upward, or horizontally. Disc movement can be controlled by an integral dashpot or snubber.

LIFT-CHECK VALVE resembles the piston-check valve. The disc is guided, but the dashpot feature is absent. Spring-loaded types can operate at any orientation, but unsprung valves have to be arranged so that the disc will close by gravity. Composition-disc valves are available for gritty liquids.

PISTON-CHECK VALVE Suitable where frequent change of direction of flow occurs as these valves are much less subject to pounding with pulsating flow due to the integral dash-pot. Spring-loaded types can operate at any orientation. Unsprung valves have to be orientated for gravity closure. Not suitable for gritty liquids.

STOP CHECK VALVE

PISTON-CHECK VALVE

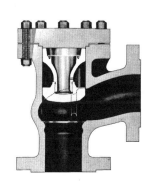

STOP-CHECK VALVE Principal example of use is in steam generation by multiple boilers, where a valve is inserted between each boiler and the main steam header. Basically, a check valve that optionally can be kept closed automatically or manually.

BALL-CHECK VALVE is suitable for most services. The valve can handle gases, vapors and liquids, including those forming gummy deposits. The ball seats by gravity and/or back pressure, and is free to rotate, which distributes wear and aids in keeping contacting surfaces clean.

WAFER CHECK VALVE effects closure by two semicircular 'doors', both hinged to a central post in a ring-shaped body which is installed between flanges. Frequently used for non-fouling liquids, as it is compact and of relatively low cost. A single disc type is also available.

FOOT VALVE Typical use is to maintain a head of water on the suction side of a sump pump. The valve is basically a lift-check valve with a strainer integrated.

VALVES FOR SWITCHING FLOW   3.1.8

MULTIPORT VALVE Used largely on hydraulic and pneumatic control circuits and sometimes used directly in process piping, these valves have rotary-ball or plug-type discs with one or more ports arranged to switch flow.

DIVERTING VALVE Two types of 'diverting' valve are made. Both switch flow from a line into one of two outlets. One type is of wye pattern with a hinged disc at the junction which closes one of the two outlets, and is used to handle powders and other solids. The second type handles liquid only, and has no moving parts—flow is switched by two pneumatic control lines. It is available in sizes to 6-inch.

VALVES FOR DISCHARGING   3.1.9

These valves allow removal of fluid from within a piping system either to atmosphere, to a drain, or to another piping system or vessel at a lower pressure. Operation is often automatic. Relief and safety valves, steam traps, and rupture discs are included in this section. Pressure-relieving valves are usually spring loaded, as those worked by lever and weight can be easily rendered inoperative by personnel. The first three valves are operated by system pressure, and are usually mounted directly onto the piping or vessel to be protected, in a vertical, upright position. Refer to the governing code for the application of these valves, including the need for an external lifting device (handlever, etc.).

SAFETY VALVE A rapid-opening (popping action) full-flow valve for air and other gases.

RELIEF VALVE Intended to relieve excess pressure in liquids, in situations where full-flow discharge is not required, when release of a small volume of liquid would rapidly lower pressure. Mounting is shown in figure 6.4.

SAFETY VALVE   **RELIEF VALVE**

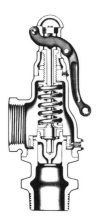 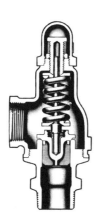

SAFETY-RELIEF VALVE Relieves excess pressure of either gas or liquid which may suddenly develop a vapor phase due to rapid and uncontrolled heating from chemical reaction in liquid-laden vessels. Refer to figure 6.4.

BALL FLOAT VALVE These automatic valves are used: (1) As air traps to remove water from air systems. (2) To remove air from liquid systems and act as vacuum breakers or breather valves. (3) To control liquid level in tanks. They are not intended to remove condensate.

BALL FLOAT VALVE
(For first use above)   **BLOWOFF VALVE**

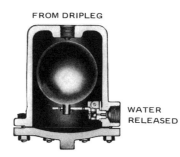 

BLOWOFF VALVE A variety of globe valve conforming with boiler code requirements and especially designed for boiler blowoff service. Sometimes suitable also for blowdown service. Wye-pattern and angle types often used. Used to remove air and other gases from boilers, etc. Manually-operated.

FLUSH-BOTTOM TANK VALVE Usually a globe type, designed to minimize pocketing, primarily for conveniently discharging liquid from the low point of a tank.

FLUSH-BOTTOM TANK VALVE (GLOBE TYPE)

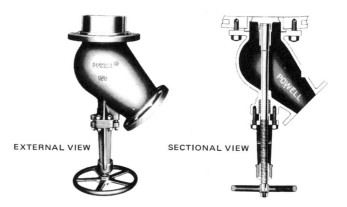

EXTERNAL VIEW   SECTIONAL VIEW

RUPTURE DISC A safety device designed to burst at a certain excess pressure and rapidly discharge gas or liquid from a system. Usually made in the form of a replaceable metal disc held between flanges. Disc may also be of graphite or, for lowest bursting pressures, plastic film.

SAMPLING VALVE A valve, usually of needle or globe pattern, placed in a branch line for the purpose of drawing off samples of process material thru the branch. Sampling from very high pressure lines is best done thru a double valved collecting vessel. A cooling arrangement may be needed for sampling from high-temperature lines.

TRAP An automatic valve for: (1) Discharging condensate, air and gases from steam lines without releasing steam. (2) Discharging water from air lines without releasing air—see 'Ball float valve', this section.

INVERTED-BUCKET TRAP

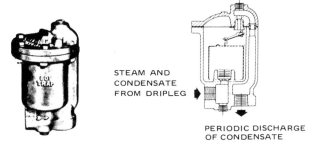

STEAM AND
CONDENSATE
FROM DRIPLEG

PERIODIC DISCHARGE
OF CONDENSATE

CONTROL VALVES & PRESSURE REGULATORS   3.1.10

CONTROL VALVES

Control valves automatically regulate pressure and/or flow rate, and are available for any pressure. If different plant systems operate up to, and at pressure/temperature combinations that require Class 300 valves, sometimes (where the design permits), all control valves chosen will be Class 300 for interchangeability. However, if none of the systems exceeds the ratings for Class 150 valves, this is not necessary. The control valve is usually chosen to be smaller than line size to avoid throttling and consequent rapid wear of the seat.

Globe-pattern valves are normally used for control, and their ends are usually flanged for ease of maintenance. The disc is moved by a hydraulic, pneumatic, electrical, or mechanical operator.

Figure 3.4 shows schematically how a control valve can be used to control rate of flow in a line. Flow rate is related to the pressure drop across the 'sensing element' (an orifice plate in this instance—see 6.7.5). The 'controller' receives the pressure signals, compares them with the pressure drop for the desired flow and, if the actual flow is different, adjusts the control valve to increase or decrease the flow.

Comparable arrangements to figure 3.4 can be devised to control any of numerous process variables—temperature, pressure, level and flow rate are the most common controlled variables.

Control valves may be self-operating, and not require the addition of a controller, sensing element, etc. Pressure regulators are a common example of this type of valve, and chart 3.1 shows the principles of operation of a pressure regulator.

PRESSURE REGULATOR Control valve of globe type which adjusts downstream pressure of liquid or gas (including steam or vapors) to a lower desired value ('set pressure').

BACK-PRESSURE REGULATOR Control valve used to maintain upstream pressure in a system.

SCHEMATIC FOR A CONTROL VALVE ARRANGEMENT   **FIGURE 3.4**

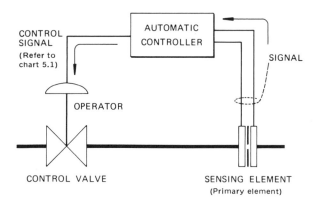

CONTROL
SIGNAL
(Refer to
chart 5.1)

AUTOMATIC
CONTROLLER

SIGNAL

OPERATOR

CONTROL VALVE

SENSING ELEMENT
(Primary element)

UNCLASSIFIED VALVES & TERMS   3.1.11

With few exceptions, the following are not special valve types different from those previously discussed, but are terms used to describe valves by service or function.

BARSTOCK VALVE Any valve having a body machined from solid metal (barstock). Usually needle or globe type.

BIBB A small valve with turned-down end, like a faucet.

BLEED VALVE Small valve provided for drawing off fluid.

BLOCK VALVE An on/off valve, nearly always a gate valve, placed in lines at battery limits.

BLOWDOWN VALVE Usually refers to a plug-type disc globe valve used for removing sludge and sedimentary matter from the bottom of boiler drums, vessels, driplegs, etc.

BREATHER VALVE A special self-acting valve installed on storage tanks, etc., to release vapor or gas on slight increase of internal pressure (in the region of ½ to 3 ounces per square inch).

BYPASS VALVE Any valve placed in a bypass arranged around another valve or equipment—see 6.1.3 under 'If there is no P&ID....' and figures 6.6 thru 6.11.

DIAPHRAGM VALVE Examples of true diaphragm valves, where the diaphragm closes off the flow, are shown in chart 3.1. These forms of diaphragm valve are popular for regulating the flow of slurries and corrosive fluids and for vacuum. The term 'diaphragm valve' is also applied to valves which have a diaphragm seal between stem and body, but these are better referred to as 'diaphragm seal' or 'packless' valves—see 3.1.2, under 'Seal'.

DRAIN VALVE A valve used for the purpose of draining liquids from a line or vessel. Selection of a drain valve, and the method of attachment, is influenced by the undesirability of pocketing the material being drained—this is important with slurries and liquids which are subject to: (1) Solidification on cooling or polymerization. (2) Decomposition.

DRIP VALVE A drain valve fitted to the bottom of a dripleg to permit blowdown.

FIGURE
3.4

FLAP VALVE A non-return valve having a hinged disc or rubber or leather flap, used for low-pressure lines.

HEADER VALVE An isolating valve installed in a branch where it joins a header.

HOSE VALVE A gate or globe valve having one of its ends externally threaded to one of the hose thread standards in use in the USA. These valves are used for vehicular and firewater connections.

ISOLATING VALVE An on/off valve isolating a piece of equipment or a process from piping.

KNIFE-EDGE VALVE A single-disc single-seat gate valve (slide gate) with a knife-edged disc.

MIXING VALVE regulates the proportions of two inflows to produce a controlled outflow.

NON-RETURN VALVE Any type of stop-check valve—see 3.1.7.

PAPER-STOCK VALVE A single-disc single-seat gate valve (slide gate) with knife-edged or notched disc used to regulate flow of paper slurry or other fibrous slurry.

PRIMARY VALVE See 'Root valve', this section.

REGULATING VALVE Any valve used to adjust flow.

ROOT VALVE (1) A valve used to isolate a pressure element or instrument from a line or vessel. (2) A valve placed at the beginning of a branch from a header.

SAMPLING VALVE Small valve provided for drawing off fluid. See 3.1.9.

SHUTOFF VALVE An on/off valve placed in lines to or from equipment, for the purpose of stopping and starting flow.

SLURRY VALVE A knife-edge valve used to control flow of non-abrasive slurries.

SPIRAL-SOCK VALVE A valve used to control flow of powders by means of a twistable fabric tube or sock.

STOP VALVE An on/off valve, usually a globe valve.

THROTTLING VALVE Any valve used to closely regulate flow in the just-open position.

VACUUM BREAKER A special self-acting valve, or any valve suitable for vacuum service, operated manually or automatically, installed to admit gas (usually atmospheric air) into a vacuum or low-pressure space. Such valves are installed on high points of piping or vessels to permit draining, and sometimes to prevent siphoning.

UNLOADING VALVE See 3.2.2, under 'Unloading', and figure 6.23.

QUICK-ACTING VALVE Any on/off valve rapidly operable, either by manual lever, spring, or by piston, solenoid or lever with heat-fusible link releasing a weight which in falling operates the valve. Quick-acting valves are desirable in lines conveying flammable liquids. Unsuitable for water or for liquid service in general without a cushioning device (hydraulic accumulator, 'pulsation pot' or 'standpipe') to protect piping from shock. See 3.1.2, under 'Quick-acting operators for non-rotary valves'.

PUMPS   **3.2.1**

DRIVERS

Electric motors are the most frequently used drivers. Larger pumps may be driven by steam-, gas-, or diesel-engines, or by turbines.

'HEADS' (PRESSURES) IN PUMP PIPING   **FIGURE 3.5**

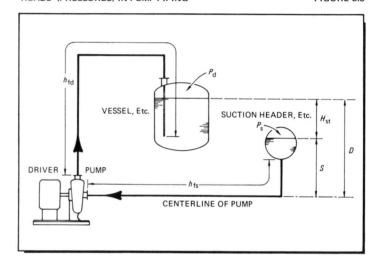

NOTES

The total head, H, which must be provided by the pump in the arrangement shown, is:—

$$H = h_d - h_s = H_{st} + (h_{fd} + h_{fs}) + (P_d - P_s)$$

Heads may be expressed either all in absolute units or all in gage units, but not in mixed units. The various head terms in this equation are, with reference to the illustration:—

h_d = total discharge head

h_s = total suction head

H_{st} = static head (differential) = $D - S$

h_{fd} = friction head loss in discharge piping, including exit loss (as liquid discharges into vessel, etc.) and loss at increaser located at pump outlet*

h_{fs} = friction head loss in suction piping, including entrance loss (as liquid enters line from header, etc.) and loss at reducer located at pump inlet*

P_d = pressure head above liquid level in discharge vessel or header

P_s = pressure head above liquid level in suction header or vessel

NET POSITIVE SUCTION HEAD (NPSH)

'NPSH' is defined by:— $S - h_{fs} + P_s - P_{vp}$, where

P_{vp} = vapor pressure of liquid at temperature of liquid at suction header, etc. Vapor pressures are given in absolute units

---

*Table F-10 gives entrance loss, exit loss, flow resistance of reducers and swages, etc., expressed in equivalent lengths of pipe.

PUMP SELECTION GUIDE — CHART 3.3

	I. IMPELLOR			II. CHAMBER-CRANK TRAIN		III. CHAMBER-WHEEL TRAIN			IV. RECIPROCATING		V. MISCELLANEOUS	
CLASS OF MECHANISM												
BASIC PUMP TYPE	CENTRIFUGAL	PROPELLOR	TURBINE	VANE	NUTATOR	SPURGEAR	BEHRENS	SCREW	PISTON	DIAPHRAGM	MOYNO	PERISTALTIC
OTHER RELATED TYPES OF PUMP	VOLUTE DIFFUSER		AXIAL-FLOW TURBINE	CAM & PISTON, SHUTTLE-BLOCK, SWINGING VANE	NUTATING DISC	GEAR, STAR AND CRESCENT		TRIPLE-SCREW	SWASH-PLATE, RADIAL, RAM		'SINGLE SCREW'	
BASIC FORM OF MECHANISM; SHOWN SCHEMATICALLY (FLOW IS FROM LEFT TO RIGHT)												
FLOW RATE AT CONSTANT DRIVE SPEED	UNIFORM IF TOTAL HEAD UNCHANGED			SOME VARIATION	UNIFORM AT CONSTANT DRIVE SPEED				PULSATING UNDER ALL CONDITIONS		UNIFORM	NEARLY UNIFORM
DISCHARGE PRESSURE	LOW TO MEDIUM			LOW TO HIGH	LOW TO MEDIUM	MEDIUM	LOW TO HIGH	MEDIUM	LOW TO HIGH	LOW TO HIGH	LOW TO MEDIUM	LOW

TYPICAL FLUIDS HANDLED WITH APPROPRIATE CONSTRUCTION

	CENTRIFUGAL	PROPELLOR	TURBINE	VANE	NUTATOR	SPURGEAR	BEHRENS	SCREW	PISTON	DIAPHRAGM	MOYNO	PERISTALTIC
CLEAN LIQUIDS	●	●	●	●	●	●	●	●	●	●	●	●
OILS	●	●	●	●	●	●	●	●	●	●	●	●
VISCOUS LIQUIDS	●	●	×	●	●	●	●	●	●	●	●	●
SLURRIES	●	●	×	×	×	×	×	×	×	●	●	●
EMULSIONS	●	●	×	●	●	●	●	●	●	●	●	●
PASTES	×	×	×	×	×	×	●	●	×	×	●	●
LUMPS	×	×	×	●	×	×	●	×	×	×	●	×
POWDERS	×	×	×	×	×	×	×	×	×	×	●	×

● = SUITABLE MECHANISM; ✕ = MECHANISM EITHER UNSUITABLE OR NOT PREFERRED

TYPES OF PUMP

A pump is a device for moving a fluid from one place to another thru pipes or channels. Chart 3.3, a selection guide for pumps, puts various types of pump used industrially into five catagories, based on operating principle. In common reference, the terms centrifugal, rotary, screw, and reciprocating are used. Chart 3.3 is not comprehensive: pumps utilizing other principles are in use. *About nine out of ten pumps used in industry are of the centrifugal type.*

The following information is given to enable an estimate to be made of required total head, pump size, capacity, and horsepower for planning purposes. Data in the Guide permit estimating pump requirements *for water systems.*

PUMP 'TOTAL HEAD'

A pump imparts energy to the pumped liquid. This energy is able to raise the liquid to a height, or 'head'. The 'total head' of a pump (in ft) is the energy (in ft-lb) imparted by the pump to each pound of liquid. In piped systems, part of the total head is used to overcome friction in the piping, which results in a pressure drop (or 'headloss').

For a centrifugal pump, the same total head can be imparted to all liquids of comparable viscosity, and is independent of the liquid's density: the required driving power increases with density. Figure 3.3 relates the total head provided by the pump to the headlosses in the pumped system.

PRESSURE & 'HEAD'

In US customary units, pressure (p) in PSI is related to head (h) in ft: p [PSI] = $(d)(h)/(144)$ = $(S.G.)(h)/(2.31)$, where d is liquid density in lb/ft³, and S.G. is specific gravity. Atmospheric pressure at sea level is equal to 14.7 PSIA, the pressure generated by a 34-ft height of water.

VELOCITY HEAD

Usually the liquid being pumped is stationary before entering the suction piping, and some power is absorbed in accelerating it to the suction line velocity. This causes a small 'velocity head' loss (usually about 1 ft) and may be found from table 3.2, which is applicable to liquid of any density, if the velocity head is read as feet of the liquid concerned.

VELOCITY & VELOCITY HEAD — TABLE 3.2

VELOCITY (Ft/sec)	4	5	6	7	8	9	10	12	15
VELOCITY HEAD (Ft.)	0.25	0.39	0.56	0.76	0.99	1.26	1.55	2.24	3.50

Flow rate, liquid velocity and cross-sectional area (at right angles to flow) are related by the formulas:

Flow rate in cubic feet per second = $(v)(a)/(144)$

Flow rate in US gallons per minute = $(3.1169)(v)(a)$

where: v = liquid velocity in feet per second

a = cross-sectional area in square inches (table P-1)

POWER CALCULATIONS

If S.G. = specific gravity of the pumped liquid, H = total head in feet of the pumped liquid, and p = pressure drop in PSI, then:

Hydraulic horsepower = $\dfrac{(GPM)(H)(S.G.)}{3960} = \dfrac{(GPM)(p)}{1714}$

CHART 3.3 / FIGURE 3.5 / TABLE 3.2

The mechanical efficiency, *e*, of a pump is defined as the hydraulic horsepower (power transferred to the pumped liquid) divided by the brake horsepower (power applied to the driving shaft of the pump).

If the pump is driven by an electric motor which has a mechanical efficiency e_m, the electricity demand is:

$$\text{Kilowatt (KW)} = \frac{(GPM)(H)(S.G.)}{(5310)(e)(e_m)} = \frac{(GPM)(p)}{(2299)(e)(e_m)}$$

Often, estimates of brake horsepower, electricity demand, etc., must be made without proper knowledge of the efficiencies. To obtain estimates, the mechanical efficiency of a centrifugal pump may be assumed to be 60%, and that of an electric motor 80%.

COMPRESSORS, BLOWERS & FANS 3.2.2

REFERENCES

'Compressed air and gas data'. Editor Gibbs C.W. (Ingersoll-Rand)
'Air receivers'. Section 1910.169 of the Code of Federal Regulations; CFR Occupational Safety and Health Administration (OSHA)

Compressors are used to supply high-pressure air for plant use, to pressurize refrigerant vapors for cooling systems, to liquefy gases, etc. They are rated by their maximum output pressure and the number of cubic feet per minute of a gas handled at a specified speed or power, stated at 'standard conditions', 60 F and 14.7 PSIA (not at compressed volume). 60 F is accepted as standard temperature by the gas industry.

The term 'compressor' is usually reserved for machines developing high pressures in closed systems, and the terms 'blower' and 'fan' for machines working at low pressures in open-ended systems.

COMPRESSOR PRESSURE RANGES   TABLE 3.3

MACHINE	DISCHARGE PRESSURE RANGE
COMPRESSOR	15 thru 20,000 PSIG, and higher
BLOWER	1 thru 15 PSIG
FAN	Up to 1 PSIG (about 30 in. water)

COMPRESSING IN STAGES

Gases (including air) can be compressed in one or more operations termed 'stages'. Each stage can handle a practicable increase in pressure—before temperature increase due to the compression necessitates cooling the gas. Cooling between stages is effected by passing the gas thru an intercooler. Staging permits high pressures, and lower discharge temperatures, with reduced stresses on the compressor.

TYPES OF COMPRESSOR

RECIPROCATING COMPRESSOR Air or other gas is pressurized in cylinders by reciprocating pistons. If the compressor is lubricated, the outflow may be contaminated by oil. If an oil-free outflow is required, the pistons may be fitted with graphite or teflon piston rings. Flow is pulsating.

ROTARY SCREW COMPRESSOR Air or other gas enters pockets formed between mating rotors and a casing wall. The pockets rotate away from the inlet, taking the gas toward the discharge end. The rotors do not touch each other or the casing wall. Outflow is uncontaminated in the 'dry type' of machine, in which power is applied to both rotors thru external timing gears. In the 'wet type', power is applied to one rotor, and both rotors are separated by an oil film, which contaminates the discharge. Flow is uniform.

ROTARY VANE COMPRESSOR resembles the rotary vane pump shown in chart 3.3. Variation in the volume enclosed by adjacent vanes as they rotate produces compression. Ample lubrication is required, which may introduce contamination. Flow is uniform.

ROTARY LOBE COMPRESSOR consists of two synchronized lobed rotors turning within a casing, in the same way as the pump shown in chart 3.3 (under 'spurgear' type). The rotors do not touch each other or the casing. No lubrication is used within the casing, and the outflow is not contaminated. Flow is uniform. This machine is often referred to as a 'blower'.

DYNAMIC COMPRESSORS resemble gas turbines acting in reverse. Both axial-flow machines and centrifugal machines (with radial flow) are available. Centrifugal compressors commonly have either one or two stages. Axial compressors have at least two stages, but seldom more than 16 stages. The outflow is not contaminated. Flow is uniform.

LIQUID RING COMPRESSOR This type of compressor consists of a single multi-bladed rotor which turns within a casing of approximately elliptic cross section. A controlled volume of liquid in the casing is thrown to the casing wall with rotation of the vanes. This liquid serves both to compress and to seal. Inlet and outlet ports located in the hub communicate with the pockets formed between the vanes and the liquid ring. These compressors have special advantages: wet gases and liquid carryover including hydrocarbons which are troublesome with other compressors are easily handled. Additional cooling is seldom required. Condensible vapor can be recovered by using liquid similar to that in the ring. Flow is uniform.

EQUIPMENT FOR COMPRESSORS

INTERCOOLER A heat exchanger used for cooling compressed gas between stages. Air must not be cooled below the dew point (at the higher pressure) as moisture will interfere with lubrication and cause wear in the next stage.

AFTERCOOLER A heat exchanger used for cooling gas after compression is completed. If air is being compressed, chilling permits removal of much of the moisture.

DAMPENER or SNUBBER; VOLUME BOTTLE or SURGE DRUM Reciprocating compressors create pulsations in the air or gas which may cause the discharge and/or suction piping to resonate and damage the compressor or its valves. A dampener, or snubber, is a baffled vessel which smooths pulsations in flow. A volume bottle or surge drum has the same purpose, but lacks baffles. These devices are not normally part of the compressor package, and are often bought separately (with the compressor maker's recommendations). Large compressors may require an arrangement of 'choke tubes' (restrictions) and 'bottles' (vessels), conforming to a theoretical design and located near the compressor's outlet, upstream of the aftercooler.

The location of the following four items of equipment is shown in figure 6.23:

SEPARATOR (normally used only with air compressors) A water separator is often provided following the aftercooler, and, sometimes, also at the intake to a compressor having a long suction line, if water is likely to collect in the line. Each separator is provided with a drain to allow continuous removal of water.

RECEIVER Refer to 'Discharge (supply) lines' and 'Storing compressed air', this section.

SILENCER is used to suppress objectionable sound which may radiate from an air intake.

FILTER is provided in the suction line to an air compressor to collect particulate matter.

The following information is given as a guide for engineering purposes

LINE SIZES FOR AIR SUCTION & DISTRIBUTION

SUCTION LINE Suction lines and manifolds should be large enough to prevent excessive noise and starvation of the air supply. If the first compression stage is reciprocating, the suction line should allow a 10 to 23 ft/sec flow: if a single-stage reciprocating compressor is used, the intake flow should not be faster than 20 ft/sec. Dynamic compressors can operate with faster intake velocities, but 40 ft/sec is suggested as a maximum. The inlet reducer for a dynamic compressor should be placed close to the inlet nozzle.

DISCHARGE (SUPPLY) LINES are sized for 150 to 175% of average flow, depending on the number of outlets in use at any time. The pressure loss in a branch should be limited to 3 PSI. The pressure drop in a hose should not exceed 5 PSI. The pressure drop in distribution piping, from the compressor to the most remote part of the system, should not be greater than 5 PSI (not including hoses).

These suggested pressure drops may be used to select line sizes with the aid of table 3.5. From the required SCFM flow in the line to be sized, find the next higher flow in the table. Multiply the allowed pressure drop (PSI) in the line by 100 and divide by the length of the line in feet to obtain the PSI drop per 100 ft—find the next lower figure to this in the table, and read required line size.

Equipment drawing air at a high rate for a short period is best served by a receiver close to the point of maximum use—lines can then be sized on average demand. A minimum receiver size of double the SCF used in intermittent demand should limit the pressure drop at the end of the period of use to about 20% in the worst instances and keep it under 10% in most others.

COMPRESSOR CHARACTERISTICS TABLE 3.4

COMPRESSOR TYPE	MAXIMUM OUTPUT PRESSURE (PSIG)	CONTAM-INANT IN OUTPUT	INFLOW (CFM/HP)	ECONOMIC RANGE (Inflow CFM)
			DATA FOR 100 PSIG OUTFLOW	
RECIPROCATING Lubicated Non-lubricated	35,000 700	OIL NONE	4 to 7	10,000
DYNAMIC Centrifugal Axial	4,000 90	NONE NONE	4 4½	500 to 110,000 5,000 to 13,000,000
ROTARY VANE	125	OIL	4	150 to 6,000
ROTARY LOBE	30	NONE		50,000
ROTARY SCREW NON-LUBED/LUBED	125	NONE/ OIL	4	30 to 150
LIQUID RING	75*	WATER or other	1.6 to 2.2	20 to 5,000

*Figure applies to a two-stage machine

FLOW OF COMPRESSED AIR:
PRESSURE DROPS OVER 100 Ft PIPE,
WITH AIR ENTERING AT 100 PSIG* TABLE 3.5
(Adapted from data published by Ingersoll-Rand)

FREE AIR INFLOW (SCFM)	NOMINAL PIPE SIZE (INCHES) — SCHEDULE 40 PIPE							
	¾	1	1½	2	2½	3	4	6
40	1.24	0.37						
70	3.77	1.05	0.12	*Pressure drop smaller than*				
90	6.00	1.69	0.19	*than 0.1 PSI per 100 ft*				
100	7.53	2.09	0.24					
400		32.2	3.59	0.98	0.41	0.13		
700			10.8	2.92	1.19	0.38	0.10	
900			17.9	4.78	1.97	0.62	0.15	
1,000			22.0	5.90	2.43	0.76	0.19	
4,000						11.9	2.90	0.35
7,000							8.77	1.06
9,000		*Pressure drop larger*					14.6	1.75
10,000		*than 35 PSI per 100 ft*					18.0	2.13
40,000								33.8

*Pressure drop varies inversely as absolute pressure of entering air.

POWER CONSUMPTION

The power consumption of the different compressor types is characteristic. Table 3.4 gives the horsepower needed at an output pressure of 100 PSIG. Power consumption per CFM rises with rising output pressure. Air cooling adds 3-5% to power consumption (including fan drive). 'FAD' power consumption figures for compressors of 'average' power consumption are given. 'FAD' denotes 'free air delivered corresponding to standard cubic ft per minute (SCFM) or liters per minute measured as set out in ASME PTC9, BS 1571 or DIN 1945.'

SPECIFIC POWER CONSUMPTION (FAD)

P S I G		50	75	100	125
HP per 100 CFM INFLOW	SINGLE-STAGE	14	18	22	24
	TWO-STAGE	13	16	18	21

COOLING-WATER REQUIREMENTS

Cooling-water demand is normally shown on the vendor's P&ID or data sheet. Most of the water demand is for the aftercooler (and intercooler, with a two-stage compressor). Jackets and lube oil may also require cooling. As a guide, 8 US gallons per hour are needed for each horsepower supplied to the compressor. If the final compression is 100 PSIG, the water demand will usually be about 2 US GPH per each SCFM inflow. These approximate demands are based on an 40 F temperature increase of the cooling water. Demand for cooling water increases slightly with relative humidity of the incoming air.

QUANTITIES OF MOISTURE CONDENSED FROM COMPRESSED AIR

The following calculation (taken from the referenced Atlas Copco manual) is for a two-stage compressor, and is based on moisture content given in the table below:

DATA:
Capacity of the compressor = 2225 SCFM
Temperature of the incoming air = 86 F
Relative humidity of the incoming air = 75%

Intercooler
$\begin{cases}\text{Outlet temperature = 86 F}\\ \text{Air pressure = 25.3 PSIG, or 40 PSIA}\\ \text{Water separation efficiency = 80\%}\end{cases}$

Aftercooler
$\begin{cases}\text{Outlet air temperature = 86 F}\\ \text{Air pressure = 100 PSIG, or 115 PSIA}\\ \text{Water separation efficiency = 90\%}\end{cases}$

CALCULATIONS:

(1) From the table, weight of water vapor in 2225 SCFM air at 86 F and 75% RH = (0.00189)(2225)(0.75) = 3.15 lb/min.

(2) Rate of removal of condensed water from intercooler, thru trap = (0.8)[3.15 − (0.00189)(2225)(14.7)/(40)] = 1.28 lb/min., or (1.28)(60)/(8.33) = 9.2 US GPH

(3) Rate of removal of condensed water from aftercooler, thru trap = (0.9)[3.15 − 1.28 − (0.00189)(2225)(14.7)/(115)] = 1.20 lb/min., or (1.20)(60)/(8.33) = 8.6 US GPH

(4) Total rate at which water is removed from both coolers = 9.2 + 8.6 = 17.8 US GPH

MOISTURE CONTENT OF AIR AT 100% RH

TEMPERATURE (Degrees F)	14	32	50	68	86	104	122
MOISTURE (10^{-4} lb/ft^3)	1.35	3.02	5.87	10.9	18.9	31.6	51.3

UNLOADING (POSITIVE-DISPLACEMENT COMPRESSORS)

'Unloading' is the removal of the compression load from the running compressor. Compressors are unloaded at startup and for short periods when demand for gas falls off. Damage to the compressor's drive motor can result if full compression duties are applied suddenly.

If the vendor does not provide means of unloading the compressor, a manual or automatic bypass line should be provided between suction and discharge (on the compressor's side of any isolating valves)—see figure 6.23.

Provision should be made so that the discharge pressure cannot rise above a value which would damage the compressor or its driver. Automatic unloading will ensure this, and the control actions are listed in table 3.6.

AUTOMATIC UNLOADING ACTIONS FOR COMPRESSORS   **TABLE 3.6**

COMPRESSOR	DISCHARGE PRESSURE	AUTOMATIC CONTROL ACTION
Not running	Low—reaches lower set value	Starts compressor unloaded, accelerates to normal speed, and brings on load
Running	High—reaches higher set value	Unloads compressor for a preset period
Idling	Low—reaches reload pressure before idling period is over	Reloads compressor
	Medium—idling period ends before reload pressure is reached	Switches off compressor

STORING COMPRESSED AIR

A limited amount of compressed air or other gas can be stored in receivers. One or more receivers provided in the compressor's discharge piping also serve to suppress surges (which can be due to demand, as well as supply) to assist cooling, and to collect moisture. Receivers storing air or other gas are classed as pressure vessels—refer to 6.5.1.

RECEIVER CONSTRUCTION Usual construction is a long vertical cylinder with dished heads, supported on a pad. Water will collect in the base, and therefor a valved drain must be provided for manual blowdown. Collected water may freeze in cold climates. Feeding the warm air or gas at the base of the receiver may prevent freezing, but the inlet must be designed so that it cannot be closed by water if it does freeze.

CAPACITY NEEDED A simple rule to decide the total receiver volume is to divide the compressor rating in SCFM by ten to get the volume in cubic feet for the receiver. For example, if the compressor is designed to take 5500 cubic feet per minute, a receiver volume of about 550 cubic feet is adequate. This rule is considered suitable for outflow pressures up to about 125 PSIG and where the continuously running compressor is unloaded by automatic valves—see 'Unloading' above. An extensive piping system for distributing compressed air or other gas may have a capacity sufficiently large in itself to serve as a receiver.

Process equipment is a term used to cover the many types of equipment used to perform one or more of these basic operations on the process material:

    (1)   **CHEMICAL REACTION**
    (2)   **MIXING**
    (3)   **SEPARATION**
    (4)   **CHANGE OF PARTICLE SIZE**
    (5)   **HEAT TRANSFER**

Equipment manufacturers give all information necessary for installation and piping.

This section is a quick reference to the function of some items of equipment used in process work. In table 3.7, the function of the equipment is expressed in terms of the phase (solid, liquid or gas) of the process materials mixed. Examples: (1) A blender can mix two powders, and its function is tabulated as "S+S". (2) An agitator can be used to stir a liquid into another liquid—this function is tabulated "L+L". Another large and varied group of equipment achieves separations, and a similar method of tabulating function is used in table 3.8.

CHEMICAL REACTION   3.3.1

Chemical reactions are carried out in a wide variety of specialized equipment, termed reactors, autoclaves, furnaces, etc. Reactions involving liquids, suspensions, and sometimes gases, are often performed in 'reaction vessels'. The vessel and its contents frequently have to be heated or cooled, and piping to a jacket or internal system of coils has to be arranged. If reaction takes place under pressure, the vessel may need to comply with the ASME Boiler and Pressure Vessel Code. Refer also to 6.5.1, under 'Pressure vessels', and to the standards listed in table 7.10.

MIXING   3.3.2

A variety of equipment is made for mixing operations. The principal types of equipment are listed in table 3.7:

MIXING EQUIPMENT   **TABLE 3.7**

EQUIPMENT	PHASES MIXED
AGITATOR	S + L, L + L
BLENDER (TUMBLER TYPE)	S + S, S + L
EDUCTOR	L + L, L + G, G + G
MIXER (RIBBON, SCROLL, OR OTHER TYPE)	S + S, S + L
PROPORTIONING PUMP	L + L
PROPORTIONING VALVE	L + L
(G = GAS, L = LIQUID, S = SOLID)	

SEPARATION   3.3.3

Equipment for separation is even more varied. Equipment separating solids on the basis of particle size or specific gravity alone are in general termed classifiers. The broader range of separation equipment separates phases (solid, liquid, gas) and some of the types used are listed in the table below:

SEPARATION EQUIPMENT   **TABLE 3.8**

EQUIPMENT	FEED MATERIAL	RETAINED MATERIAL	OUTFLOW MATERIAL
CENTRIFUGE	S + L	S	L
CONTINUOUS CENTRIFUGE	L(1) + L(2)	None	L(1), L(2), †
CYCLONE	S + G	None	G, S †
DEAERATOR	L + G	L	G
DEFOAMER	L + G	L	G
DISTILLATION COLUMN	L(1) + L(2)	L(1)	L(2) *
DRYER	S + L	S	L *
DRY SCREEN	S(1) + S(2)	S(1)	S(2)
EVAPORATOR	L + S / L(1) + L(2)	L + S / L(1)	L * / L(2) *
FILTER PRESS	S + L	S	L
FLOTATION TANK	S + L	S	L
FRACTIONATION COLUMN	L(1) + L(2) + L(3) + etc.	None	L(1), L(2), L(3), etc. †
SCRUBBER	S + G	S	G
SETTLING TANK	S + L	S	L
STRIPPER	L(1) + L(2)	L(1)	L(2)

†Separate flows   *Removed as vapor
(G = GAS, L = LIQUID, S = SOLID, S(1), S(2), L(1), L(2), etc. = DIFFERENT SOLIDS OR LIQUIDS)

CHANGE OF PARTICLE SIZE   3.3.4

Reduction of particle size is a common operation, and can be termed 'attrition'. Equipment used includes crushers, rod-, ball- and hammer-mills, and—to achieve the finest reductions—energy mills, which run on compressed air. Emulsions ('creams' or 'milks'), which are liquid-in-liquid dispersions, are stabilized by homogenizers, typically used on milk to reduce the size of the fat globules and thus prevent cream from separating.

Occasionally, particle or lump size of the product is increased. Equipment for agglomerating, pelletizing, etc., is used. Examples: tablets, sugar cubes, powdered beverage and food products.

PROCESS HEAT TRANSFER   3.3.5

Adding and removing heat is a significant part of chemical processing. Heating or cooling of process material is accomplished with heat exchangers, jacketed vessels, or other heat transfer equipment. The project and piping groups specify the duty and mechanical arrangement, but the detail design is normally left to the manufacturer.

The term 'heat exchanger' in chemical processing refers to an unfired vessel exchanging heat between two fluids which are kept separated. The commonest form of heat exchanger is the 'shell-and-tube' exchanger, consisting of a bundle of tubes held inside a 'shell' (the vessel part). One fluid passes inside the tubes, the other thru the space between the tubes and shell. Exchanged heat has to flow thru the tube walls. Refer to 6.8 ('Keeping process material at the right temperature') and to 6.6 for piping shell-and-tube heat exchangers.

Heat exchange with process material can take place in a variety of other equipment, such as condensers, evaporators, heaters, chillers, etc.

MULTIFUNCTION EQUIPMENT   3.3.6

Sometimes, items of equipment are designed to perform more than one of the functions listed at the beginning of 3.3.

Mixing and heating (or cooling) may be simultaneously carried out in mixers having blades provided with internal channels to carry hot (or cold) fluid.

Separation and attrition may be achieved in a single mill, designed to output particles of the required degree of fineness and recycle and regrind particles which are still too coarse.

ORGANIZATION OF WORK : Job Responsibilities, Drawing-Office Equipment and Procedures

THE PIPING GROUP   4.1

Plant design is divided into several areas, each the responsibility of a 'design group'. Chart 4.1(a) shows the main groups of people cooperating on the plant design, and the types of drawings for which they are responsible. Other groups, involved with instrumentation, stress analysis, pipesupport, etc., contribute to the design at appropriate stages.

The personnel responsible for the piping design may be part of an engineering department's mechanical design group, or they may function as a separate section or department. For simplicity, this design group is referred to as the 'piping group', and its relationship with the organization and basic activities are indicated in chart 4.1(a).

Chart 4.1(c) shows the structure of a design group.

RESPONSIBILITIES OF THE PIPING GROUP   4.1.1

The piping group produces designs in the form of drawings and model(s), showing equipment and piping.

The following are provided by the piping group as its contribution to the plant design:—

(1)   AN EQUIPMENT ARRANGEMENT DRAWING, USUALLY TERMED THE 'PLOT PLAN'

(2)   PIPING DESIGN (DRAWINGS OR MODEL)

(3)   PIPING DETAILS FOR FABRICATION AND CONSTRUCTION

(4)   REQUISITIONS FOR PURCHASE OF PIPING MATERIEL

JOB FUNCTIONS   4.1.2

On joining a design office it is important that the new member should know what line of authority exists. This is especially important when information is required and it saves the wrong people from being interrupted. Chart 4.2 shows two typical lines of authority. (Different companies will have different set-ups and job titles.)

JOB		FUNCTIONS
DESIGN SUPERVISOR	(1)	RESPONSIBLE FOR ALL PERSONNEL IN GROUPS INCLUDING HIRING
	(2)	COORDINATING WITH OTHER GROUPS (AND THE CLIENT)
	(3)	OVERALL PLANNING AND SUPERVISING THE GROUP'S WORK
	(4)	LIAISON WITH PROJECT ENGINEER(S)
GROUP LEADER	(1)	SUPERVISING DESIGN & DRAFTING IN AREA(S) ALLOCATED BY DESIGN SUPERVISOR
NOTE: On small projects, may also assume Design Supervisor's duties	(2)	ASSIGNING WORK TO DESIGNERS & DRAFTERS
	(3)	RESPONSIBLE FOR PLOT PLANS, PLANT DESIGNS & PRESENTATION & COMPLETENESS OF FINISHED DRAWINGS
	(4)	COORDINATES MECHANICAL, STRUCTURAL, ELECTRICAL, AND CIVIL DETAILS FROM OTHER GROUPS
	(5)	CHECKING & MARKING VENDORS' DRAWINGS
	(6)	OBTAINING INFORMATION FOR MEMBERS OF THE GROUP
	(7)	ESTABLISHING THE NUMBER OF DRAWINGS REQUIRED FOR EACH JOB (DRAWING CONTROL OR REGISTER)—SEE INDEX
	(8)	ASSIGNING TITLES FOR EACH DRAWING AND MAINTAINING UP-TO-DATE DRAWING CONTROL OR REGISTER OF DRAWINGS, CHARTS, GRAPHS, AND SKETCHES FOR EACH CURRENT PROJECT
	(9)	ESTABLISHING A DESIGN GROUP FILING SYSTEM FOR ALL INCOMING & OUTGOING PAPERWORK
	(10)	KEEPING A CURRENT SCHEDULE AND RECORD OF HOURS WORKED
	(11)	REQUISITIONING VIA PURCHASING DEPARTMENT ALL PIPING MATERIALS
CHECKER	(1)	CHECKING DESIGNERS' AND DRAFTERS' DESIGNS AND DETAILS FOR DIMENSIONAL ACCURACY AND CONFORMITY WITH SPECIFICATIONS, P&ID's, VENDORS' DRAWINGS, ETC.
	(2)	IF AGREED WITH THE DESIGNER &/OR GROUP LEADER, MAY MAKE IMPROVEMENTS AND ALTERATIONS TO THE DESIGN
DESIGNER	(1)	PRODUCING STUDIES AND LAYOUTS OF EQUIPMENT AND PIPING WHICH MUST BE ECONOMIC, SAFE, OPERABLE AND EASILY MAINTAINED
	(2)	MAKING ANY NECESSARY ADDITIONAL CALCULATIONS FOR THE DESIGN
	(3)	SUPERVISING DRAFTERS
DRAFTER		MINIMUM RESPONSIBILITIES ARE:—
	(1)	PRODUCING DETAILED DRAWINGS FROM DESIGNERS' OR GROUP LEADERS' STUDIES OR SKETCHES
	(2)	SECONDARY DESIGN WORK
	(3)	FAMILIARIZATION WITH THE RECORDS, FILES, INFORMATION SHEETS AND COMPANY PRACTICES RELATING TO THE PROJECT

OFFICE ORGANIZATION

CHART 4.1

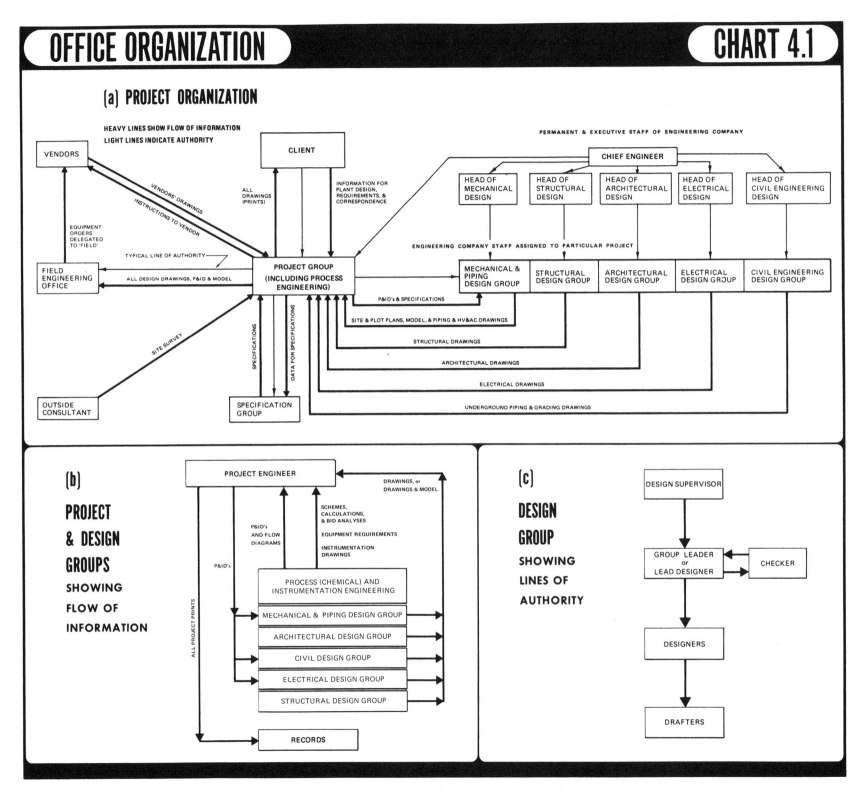

(a) PROJECT ORGANIZATION

HEAVY LINES SHOW FLOW OF INFORMATION
LIGHT LINES INDICATE AUTHORITY

PERMANENT & EXECUTIVE STAFF OF ENGINEERING COMPANY

VENDORS

CLIENT

CHIEF ENGINEER

HEAD OF MECHANICAL DESIGN

HEAD OF STRUCTURAL DESIGN

HEAD OF ARCHITECTURAL DESIGN

HEAD OF ELECTRICAL DESIGN

HEAD OF CIVIL ENGINEERING DESIGN

VENDORS' DRAWINGS
INSTRUCTIONS TO VENDOR

ALL DRAWINGS (PRINTS)

INFORMATION FOR PLANT DESIGN, REQUIREMENTS, & CORRESPONDENCE

EQUIPMENT ORDERS DELEGATED TO 'FIELD'

TYPICAL LINE OF AUTHORITY

ENGINEERING COMPANY STAFF ASSIGNED TO PARTICULAR PROJECT

FIELD ENGINEERING OFFICE

ALL DESIGN DRAWINGS, P&ID & MODEL

PROJECT GROUP (INCLUDING PROCESS ENGINEERING)

MECHANICAL & PIPING DESIGN GROUP

STRUCTURAL DESIGN GROUP

ARCHITECTURAL DESIGN GROUP

ELECTRICAL DESIGN GROUP

CIVIL ENGINEERING DESIGN GROUP

P&ID's & SPECIFICATIONS

SITE & PLOT PLANS, MODEL, & PIPING & HV&AC DRAWINGS

STRUCTURAL DRAWINGS

ARCHITECTURAL DRAWINGS

ELECTRICAL DRAWINGS

UNDERGROUND PIPING & GRADING DRAWINGS

SITE SURVEY

SPECIFICATIONS

DATA FOR SPECIFICATIONS

OUTSIDE CONSULTANT

SPECIFICATION GROUP

(b)

PROJECT & DESIGN GROUPS

SHOWING FLOW OF INFORMATION

PROJECT ENGINEER

DRAWINGS, or DRAWINGS & MODEL

P&ID's AND FLOW DIAGRAMS

SCHEMES, CALCULATIONS, & BID ANALYSES

EQUIPMENT REQUIREMENTS

INSTRUMENTATION DRAWINGS

P&ID's

ALL PROJECT PRINTS

PROCESS (CHEMICAL) AND INSTRUMENTATION ENGINEERING

MECHANICAL & PIPING DESIGN GROUP

ARCHITECTURAL DESIGN GROUP

CIVIL DESIGN GROUP

ELECTRICAL DESIGN GROUP

STRUCTURAL DESIGN GROUP

RECORDS

(c)

DESIGN GROUP

SHOWING LINES OF AUTHORITY

DESIGN SUPERVISOR

GROUP LEADER or LEAD DESIGNER

CHECKER

DESIGNERS

DRAFTERS

DESIGN INFORMATION TO PIPING GROUP   4.2

The following information is required by the piping group:—

FROM THE PROJECT GROUP

(1) 'JOB SCOPE' DOCUMENT, WHICH DEFINES PROCEDURES TO BE USED IN PREPARING DESIGN SKETCHES AND DIAGRAMS

(2) PIPING & INSTRUMENTATION DIAGRAM (P&ID—SEE 5.2.4)

(3) LIST OF MAJOR EQUIPMENT (EQUIPMENT INDEX), SPECIAL EQUIPMENT AND MATERIALS OF FABRICATION

(4) LINE DESIGNATION SHEETS OR TABLES, INCLUDING ASSIGNATION OF LINE NUMBERS—SEE 4.2.3 AND 5.2.5

(5) SPECIFICATIONS FOR MATERIALS USED IN PIPING SYSTEMS—SEE 4.2.1

(6) SCHEDULE OF COMPLETION DATES (UPDATED ON FED-BACK INFORMATION)

(7) CONTROLS (METHODS OF WORKING,ETC.) TO BE ADOPTED FOR EXPEDITING THE JOB

FROM OTHER GROUPS

(8) DRAWINGS—SEE 5.2.7

FROM SUPPLIERS

(9) VENDORS' PRINTS—SEE 5.2.7

SPECIFICATIONS   4.2.1

These consist of separate specifications for plant layout, piping materials, supporting, fabrication, insulation, welding, erection, painting and testing. The piping designer is mostly concerned with plant layout and materiel specifications, which detail the design requirements and materials for pipe, flanges, fittings, valves, etc., to be used for the particular project.

The piping materials specification usually has an index to the various services or processes. The part of the specification dealing with a particular service can be identified from the piping drawing line number or P&ID line number—see 5.2.4 under 'Flow lines'. All piping specifications must be strictly adhered to as they are compiled from information supplied by the project group. Although the fittings, etc., described in the Guide are those most frequently used, they will not necessarily be seen in every piping specification.

On some projects (such as 'revamp' work) where there is no specification, the designer may be responsible for selecting materials and hardware, and it is important to give sufficient information to specify the hardware in all essential details. Non-standard items are often listed by the item number and/or model specification for ordering taken from the catalog of the particular manufacturer.

LIST OF EQUIPMENT, or EQUIPMENT INDEX   4.2.2

This shows, for each item of equipment, the equipment number, equipment title, and status—that is whether the item has been approved, ordered, and whether certified vendor's prints have been received.

LINE DESIGNATION SHEETS, or TABLES   4.2.3

These sheets contain tabulated data showing nominal pipe size, material specification, design and operating conditions. Line numbers are assigned in sequence of flow, and a separate sheet is prepared for each conveyed fluid —see 5.2.5.

DRAWING CONTROL (REGISTER)   4.2.4

A drawing number relates the drawing to the project, and may be coded to show such information as project (or 'job') number, area of plant, and originating group (which may be indicated 'M' for mechanical, etc.). Figure 5.15 shows a number identifying part of a piping system.

The drawing control shows the drawing number, title, and progress toward completion. The status of revision and issues is shown—see 5.4.3. The drawing control is kept up-to-date by the group leader.

DESIGN GROUP—TWO TYPICAL LINES OF AUTHORITY   CHART 4.2

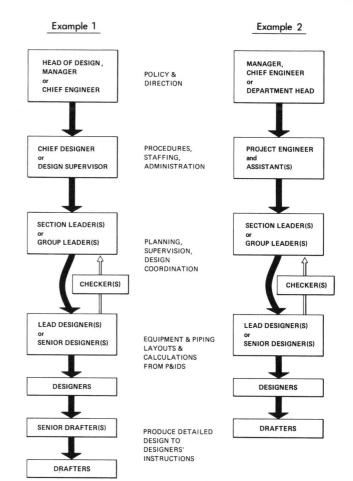

There are two types of drawings to file—those produced by the group and those received by the group. The former are filed in numerical order under plant or unit number in the drawing office on a 'stick file' or in a drawer— see 4.4.10. The filing of the latter, 'foreign', prints is often poorly done, causing time to be wasted and information to be lost. These prints are commonly filed by equipment index number, placing all information connected with that item of equipment in the one file.

A suggested method for filing these incoming prints is illustrated in chart 4.3, which cross-references process, function, or area with the group originating the drawing, and with associated vessels, equipment, etc. All correspondence between the project and design groups, client, vendors, and field would be filed under 'zero', as shown.

MATERIALS & TOOLS FOR THE DRAFTING ROOM   4.4

PAPER   4.4.1

Vellum paper and mylar film are used for drawings. Drawing sheets must be translucent to the light used in copying machines. Mylar with a coated drawing surface is more expensive than vellum, but is preferable where durability and dimensional stability are important. Sheets can be supplied printed with border and title block and with a 'fade-out' ruled grid on the reverse side. 'Isometric' sheets with fade-out 30-degree grid are available for drawing isos.

ANSI 14.1 defines the following flat drawing-sheet sizes (in inches): (A) 8½x11, (B) 11x17, (C) 17x22, (D) 22x34, (E) 34x44.

International drawing sheet sizes of approximately the same dimensions are defined (in inches) as: (A4) 8.27x11.69, (A3) 11.69x16.54, (A2) 16.54x23.39, (A1) 23.39x33.11, (A0) 33.11x46.81.

PAPERS FOR COPYING MACHINES Photosensitive paper is used for making prints for checking, issuing and filing purposes. 'Sepia' photocopying paper (Ozalid Company, etc.) gives brown positive prints which may be amended with pencil or ink, and the revision used as an original for photocopying in a diazo machine. Sepias may also be used to give a faint background print for drawing other work over, such as ducting or pipe supports. The quality of sepia prints is not good. Positive photocopies of superior quality are made on clear plastic film, which may have either continuous emulsion to give heavy copies, or screened emulsion to yield faint background prints (emulsion should preferably be water-removable).

LEADS & PENCILS   4.4.2

Pencil leads used in the drawing office are available in the following grades, beginning with the softest : B (used for shading), HB (usually used for writing only), F (usually softest grade used for drafting), H (grade most often used for drafting), 2H (used for drawing thinner lines such as dimension lines), 3H and 4H (used for faint lines for layout or background). Softer penciling is prone

UNIT OR AREA	CORRESPONDENCE	MECHANICAL	PIPING	ELECTRICAL	CIVIL	STRUCTURAL	ARCHITECTURAL	INSTRUMENTATION	PUMPS	VESSELS	TANKS	EXCHANGERS	CONDENSERS	SEPARATORS	CONVEYOR
	.0	.1	.2	.3	.4	.5	.6	.7	.8	.9	.10	.11	.12	.13	.14	.15	.16
1 COMPRESSED AIR															
2 COOLING WATER															
3 FLARESTACK															
4 FUEL OIL															
5 SOLVENTS															
6 STEAM SYSTEM															
7 VENTILATION – OFFICES															
8 VENTILATION – PROCESS AREA															
9															
10															

Paperwork classified according to a system of this type may be located in a filing cabinet fitted with numbered dividers as shown :—

STANDARD DIVIDERS FOR FILING CABINET

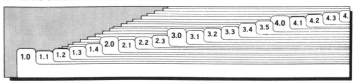

to smearing on handling, Grades harder than 3H tend to cut paper making lines difficult to erase. Conventional leads are 2 mm in diameter and require frequent repointing. 0.5 mm and 0.3 mm leads speed work, as they need no repointing. Conventional leads are not suitable for use on plastic films as they smear and are difficult to erase. 'Film' leads and pencils are available in the same sizes as conventional leads, and in different grades of hardness.

Clutch pencils (lead holders) suitable for use with either type of the smaller diameter leads have a push-button advance.

SCALES   4.4.3

The architect's scale is used for piping drawings, and is divided into fractions of an inch to one foot—for example, 3/8 inch per foot. The engineer's scale is used to draw site plans, etc., and is divided into one inch per stated number of feet, such as 1 inch per 30 feet.

ERASERS & ERASING SHIELDS 4.4.4

Several types of eraser and erasing methods are available—use of each is given in table 4.1: Rubber in various hardnesses from pure gum rubber (artgum) for soft pencilling and cleaning lead smears, to hard rubber for hard pencilling and ink; 'plastic' is cleaner to use, as it has less tendency to absorb graphite; 'magic rub' for erasing pencil from plastic films. Most types of eraser are available for use with electric erasing machines.

An erasing shield is a thin metal plate with holes of various shapes and sizes so that parts of the drawing not to be erased may be protected.

ERASING GUIDE   TABLE 4.1

MEDIUM MATERIAL	SOFT PENCIL	HARD PENCIL	INDIAN INK	PHOTOGRAPHIC BACKGROUND
TRACING PAPER, or LINEN	SRE, or artgum	HRE, or SRE	IHRE	————
SEPIA (OZALID), or PHOTOCOPY PAPER (PHOTOSTAT)	SRE	HRE, or SRE	Blade, or IHRE	Bleach *
PLASTIC FILM	Wet PE	Wet PE	Wet PE, or Blade	Wet PE, or Bleach*

KEY: E = eraser. SR = soft rubber, HR = hard rubber, I = ink, P = plastic.
* Chemical bleach for removing black photographic silver deposit

CLEANING POWDER 4.4.5

Fine rubber granules are supplied in 'salt-shaker' drums. Sprinkled on a drawing, these granules reduce smearing of pencil lines during working. The use of cleaning powder is especially helpful when using a teesquare. The powder is brushed off after use.

LETTERING AIDS 4.4.6

Title blocks, notes, and subtitles on drawings or sections should be in capitals. Capitals, either upright or sloped, are preferred. Pencilled lettering is normally used. Where ink work is required on drawings for photography, charts, reports, etc., ink stylus pens (Technos, Rapidograph, etc.) are available for stencil lettering (and for line drawing in place of ruling pens). The Leroy equipment is also used for inked lettering. Skeleton lettering templates are used for lettering section keys. The parallel line spacer is a small, inexpensive tool useful for ruling guide lines for lettering.

As alternatives to hand-inked lettering, machines such as Kroy which print onto adhesive-backed transparent film which is later positioned on the drawing. Adhesive or transferable letters and numbers are available in sheets, and special patterns and panels can be supplied to order for title blocks or detailing, symbolism, abbreviations, special notes, etc. Printed adhesive tapes

are limited in application, but are useful for making drawings for photographic reproduction, such as panel boards, charts, and special reports—see 4.4.13, under 'Photographic layouts'.

TEMPLATES 4.4.7

Templates having circular and rectangular openings are common. Orthogonal and isometric drafting templates are available for making process piping drawings and flow diagrams. These piping templates give the outlines for ANSI valves, flanges, fittings and pipe diameters to 3/8 inch per foot, or 1/4-inch per foot.

MACHINES 4.4.8

The first two machines are usually used in drawing offices in place of the slower teesquare:

DRAFTING MACHINE allows parallel movement of a pair of rules set at right angles. The rules are set on a protractor, and their angle on the board may be altered. The protractor usually has 15-degree clickstops and vernier scale.

PARALLEL RULE, or SLIDER, permits drawing of long horizontal lines only, and is used with a fixed or adjustable triangle.

PLANIMETER A portable machine for measuring areas. When set to the scale of the drawing, the planimeter will measure areas of any shape.

PANTOGRAPH System of articulated rods permitting reduction or enlargement of a drawing by hand. Application is limited.

LIGHT BOX 4.4.9

A light box has a translucent glass or plastic working surface fitted underneath with electric lights. The drawing to be traced is placed on the illuminated surface.

FILING METHODS 4.4.10

Original drawings are best filed flat in shallow drawers. Prints filed in the drawing office are usually retained on a 'stick', which is a clamp for holding several sheets. Sticks are housed in a special rack or cabinet.

Original drawings will eventually create a storage problem, as it is inadvisable to scrap them. If these drawings are not sent to an archive, after a period of about three years they are photographed to a reduced scale for filing, and only the film is retained. Equipment is available for reading such films, or large photographic prints can be made.

CHART
4.3

TABLE
4.1

COPYING PROCESSES   4.4.11

'Diazo' or 'dyeline' processes reproduce to the same scale as the original drawing as a positive copy or print. Bruning and Ozalid machines are often employed. The drawing that is to be copied must be on tracing paper, linen or film, and the copy is made on light-sensitive papers or films. The older reversed-tone 'blue-print' is no longer in use.

SCALED PLANT MODELS   4.4.12

Plant models are often used in designing large installations involving much piping. When design of the plant is completed, the model is sent to the site as the basis of construction in the place of orthographic drawings. Some engineering companies strongly advocate their use, which necessitates maintaining a model shop and retaining trained personnel. Scaled model piping components are available in a wide range of sizes. The following color coding may be used on models:—

PIPING	YELLOW, RED or BLUE
EQUIPMENT	GREY
INSTRUMENTS	ORANGE
ELECTRICAL	GREEN

ADVANTAGES

- Available routes for piping are easily seen
- Interferences are easily avoided
- Piping plan and elevation drawings can be eliminated; only the model, plot plan, P&ID's, and piping fabrication drawings (isos) are required
- The model can be photographed — see 4.4.13.
- Provides a superior visual aid for conferences, for construction crews and for training plant personnel

DISADVANTAGES

- Duplication of the model is expensive
- The model is not easily portable and is liable to damage during transportation
- Changes are not recorded in the model itself

PHOTOGRAPHIC AIDS   4.4.13

'DRAWINGS' FROM THE MODEL

The lack of portability of a scaled plant model can be partially overcome by photographing it. To do this it must be designed so that it can be taken apart easily. Photographs can be made to correspond closely to the regular plan, elevation and isometric projections by photographing the model from 40 ft or more away with long focal length lenses—'vanishing points' (converging lines) in the picture are effectively eliminated.

The negative is projected through a contact screen and a print made on 'reproducible' film. Dimensions, notes, etc., are added to the reproducible film which can be printed by a diazo process—see 4.4.11. These prints are used as working drawings, and distributed to those needing information.

REVAMP WORK FOR EXISTING PLANTS

A Polaroid (or video) camera can be used to supply views of the plant and unrecorded changes. Filed drawings of a plant do not always include alterations, or deviation from original design.

Photographs of sections of a plant can be combined with drawings to facilitate installation of new equipment, or to make further changes to the existing plant. To do this, photographs are taken of the required views, using a camera fitted with a wide-angle lens (to obtain a wider view).

The negatives obtained are printed onto screened positive films which are attached to the back of a clear plastic drawing sheet. Alterations to the piping system are then drawn on the front face of this sheet, linking the photographs as desired. Reproductions of the composite drawing are made in the usual way by diazo process.

Alternately, positives may be marked directly for minor changes or instructions to the field.

PHOTOGRAPHIC LAYOUTS

The following technique produces equipment layout 'drawings', and is especially useful for areas where method study or investigational reports are required.

First, equipment outlines are produced to scale on photographic film, either in the regular way or by xerography. Next, a drawing-sized sheet of clear film is laid on a white backing sheet having a correctly-scaled grid marked on it.

The building outline and other features can be put onto the film using the variety of printed transparent tapes and decals available. The pieces of film with equipment outlines may then be positioned with clear tape, and any other parts of the 'drawing' completed. Alterations to the layout may be rapidly made with this technique, which photographs well for reports, and allows prints to be made in the usual ways for marking and comment. The film layout should be covered with an acetate or other protective sheet before insertion in a copying machine.

REDUCTION BY PHOTOGRAPHY

It is frequently required to include reproductions of diagrams and drawings in reports, etc. Photographic reduction to less than half-size (on lengths) is not recommended because normal-sized printing and details may not be legible. A graphic scale should be included on drawings to be reduced — see chart 5.8.

DRAFTING: PROCESS AND PIPING DRAWINGS
including Drawing Symbols, Showing Dimensions, Showing Instrumentation, and Bills of Materiel

PIPING SYMBOLS   5.1

SHOWING PIPE & JOINTS   5.1.1

Hand-drawn piping layouts depict pipe by single lines for clarity and economy. Pipe and flanges are sometimes drawn partially 'double line' to display clearances. Computer drawn layouts can show piping in plan, elevational and isometric views in single line, or (without additional effort or expense) in double line. Double line representation is best reserved for three-dimensional views, such as isos.

In double-line drawing, valves are shown by the symbols in chart 5.6 (refer to the panel 'Drafting valves'). Double-line representation is not used for entire piping arrangements, as it is very time-consuming, difficult to read, and not justified technically.

DOUBLE-LINE PRESENTATION   SINGLE-LINE PRESENTATION

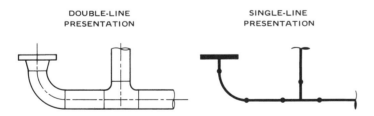

In presenting piping 'single line' on piping drawings, only the centerline of the pipe is drawn, using a solid line (see chart 5.1), and the line size is written. Flanges are shown as thick lines drawn to the scaled outside diameter of the flange. Valves are shown by special symbols drawn to scale. Pumps are shown by drawing the pads on which they rest, and their nozzles: figure 6.21 illustrates this simplified presentation. Equipment and vessels are shown by drawing their nozzles, outlines, and supporting pads.

If there is a piping specification, it is not necessary to indicate welded or screwed joints, except to remove ambiguities—for example, to differentiate between a tee and a stub-in. In most current practice, the symbols for screwed joints and socket welds are normally omitted, although butt welds are often shown.

The ways of showing joints set out in the standard ANSI Y32.2.3 are not typical of current industrial practice. The standard's symbol for a butt-weld as shown in table 5.1 is commonly used to indicate a butt-weld to be made 'in the field' (field weld).

SHOWING NON-FLANGED JOINTS AT ELBOWS   **TABLE 5.1**

	BUTT WELD	SOCKET WELD	SCREWED JOINT
SIMPLIFIED PRACTICE *			
CONVENTIONAL PRACTICE			
ANSI Y32.2.3 (Not current practice)			

*The joint symbol may be omitted if the type of joint is determined by a piping specification. It is usually preferred to use the dot weld symbol to make the type of construction clear: for example, to distinguish between a tee and a stub-in.

TABLE 5.1

[53]

LINE SYMBOLS WHICH MAY BE USED ON ALL DRAWINGS 5.1.2

Chart 5.1 shows commonly accepted ways of drawing various lines. Many other line symbols have been devised but most of these are not readily recognized, and it is better to state in words the function of special lines, particularly on process flow diagrams and P&ID's. The designer or draftsman should use his current employer's symbols.

MISCELLANEOUS SYMBOLS FOR PIPING DRAWINGS 5.1.6

Symbols that are shown in a similar way in all systems are collected in chart 5.7.

GENERAL ENGINEERING SYMBOLS 5.1.7

Chart 5.8 gives some symbols, signs, etc., which are used generally and are likely to be found or needed on piping drawings.

VALVE & EQUIPMENT SYMBOLS FOR P&ID's & PROCESS FLOW DIAGRAMS 5.1.3

Practice in showing equipment is not uniform. Chart 5.2 is based on ANSI Y32.11, and applies to P&ID's and process flow diagrams.

REPRESENTING PIPING ON PIPING DRAWINGS 5.1.4

Charts 5.3-6 show symbols used in butt-welded, screwed and socket-welded systems. The various aspects of the fitting, valve, etc., are given. These symbols are based on conventional practice rather than the ANSI standard Z32.2.3, titled 'Graphic symbols for pipe fittings, valves and piping'.

REPRESENTING VALVES ON PIPING DRAWINGS 5.1.5

Chart 5.6 shows ways of denoting valves, including stems, handwheels and other operators. The symbols are based on ANSI Z32.2.3, but more valve types are covered and the presentation is up-dated. Valve handwheels should to be drawn to scale with valve stem shown fully extended.

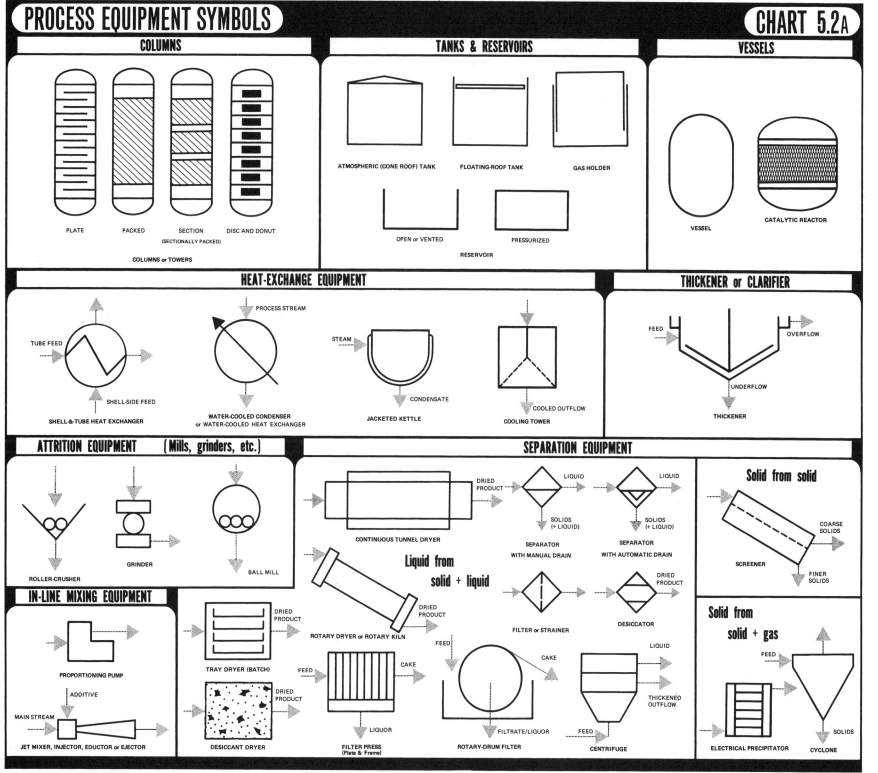

PROCESS EQUIPMENT SYMBOLS

CHART 5.2A

5

COLUMNS

PLATE PACKED SECTION (SECTIONALLY PACKED) DISC AND DONUT

COLUMNS or TOWERS

TANKS & RESERVOIRS

ATMOSPHERIC (CONE ROOF) TANK FLOATING-ROOF TANK GAS HOLDER

OPEN or VENTED PRESSURIZED

RESERVOIR

VESSELS

VESSEL CATALYTIC REACTOR

HEAT-EXCHANGE EQUIPMENT

PROCESS STREAM

TUBE FEED

SHELL-SIDE FEED

SHELL-&-TUBE HEAT EXCHANGER

WATER-COOLED CONDENSER or WATER-COOLED HEAT EXCHANGER

STEAM

CONDENSATE

JACKETED KETTLE

COOLED OUTFLOW

COOLING TOWER

THICKENER or CLARIFIER

FEED OVERFLOW

UNDERFLOW

THICKENER

ATTRITION EQUIPMENT (Mills, grinders, etc.)

ROLLER-CRUSHER

GRINDER

BALL MILL

SEPARATION EQUIPMENT

DRIED PRODUCT

CONTINUOUS TUNNEL DRYER

LIQUID

SOLIDS (+ LIQUID)

SEPARATOR WITH MANUAL DRAIN

LIQUID

SOLIDS (+ LIQUID)

SEPARATOR WITH AUTOMATIC DRAIN

Liquid from solid + liquid

ROTARY DRYER or ROTARY KILN

DRIED PRODUCT

FILTER or STRAINER

DRIED PRODUCT

DESICCATOR

Solid from solid

COARSE SOLIDS

SCREENER

FINER SOLIDS

Solid from solid + gas

IN-LINE MIXING EQUIPMENT

PROPORTIONING PUMP

ADDITIVE

MAIN STREAM

JET MIXER, INJECTOR, EDUCTOR or EJECTOR

DRIED PRODUCT

TRAY DRYER (BATCH)

DRIED PRODUCT

DESICCANT DRYER

FEED CAKE

LIQUOR

FILTER PRESS (Plate & Frame)

FEED

CAKE

FILTRATE/LIQUOR

ROTARY-DRUM FILTER

LIQUID

THICKENED OUTFLOW

FEED

CENTRIFUGE

FEED

ELECTRICAL PRECIPITATOR

SOLIDS

CYCLONE

[55]

PROCESS EQUIPMENT SYMBOLS

CHART 5.2B

VALVES

NORMALLY OPEN   NORMALLY CLOSED   CONTROL

VALVES (GENERAL)

Special types of valve may be indicated by the symbols given in chart 5.6

VALVE OPERATORS (ISA S5.1)

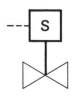

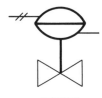

| ROTARY MOTOR (Electric Signal) | SINGLE-ACTING CYLINDER | DOUBLE-ACTING CYLINDER | SOLENOID | DIAPHRAGM (Pressure-Balanced) |

PUMPS, COMPRESSOR, BLOWER, & FAN

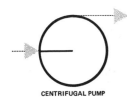

CENTRIFUGAL PUMP

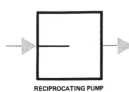

RECIPROCATING PUMP

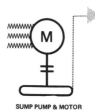

SUMP PUMP & MOTOR

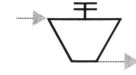

BLOWER or FAN

TURBINE COMPRESSOR

ROTARY PUMP

RECEIVER

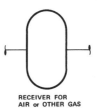

RECEIVER FOR AIR or OTHER GAS

DRAIN

VISIBLE DRAIN

ACCUMULATORS

| GENERAL SYMBOL | SPRING-LOADED TYPE | GAS-CHARGED TYPE | WEIGHTED TYPE |

THESE SYMBOLS CAN BE USED FOR HYDRAULIC OR PNEUMATIC ACCUMULATORS, USED TO SMOOTH THE PULSATING OUTFLOW FROM PUMPS AND COMPRESSORS, OR TO ACT AS RESERVOIRS FOR VARIABLE DEMAND.

DRIVERS

Drive Coupling (TYP)

2-PHASE ELECTRIC MOTOR

3-PHASE ELECTRIC MOTOR

ENGINE DRIVER

STEAM OR AIR

STEAM- or AIR-PISTON DRIVER

STEAM OR AIR

TURBINE DRIVER

CONVEYORS

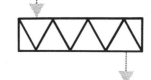

SCREW CONVEYOR

ROLLER CONVEYOR

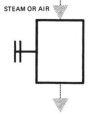

BELTS or SHAKERS

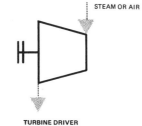

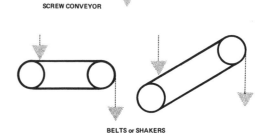

BUCKET or FLIGHT CONVEYOR

SYMBOLS FOR BUTT-WELDED SYSTEMS

CHART 5.3

5

NOTE — IN CHARTS 5.3 THRU 5.5, THE SYMBOL IS SHOWN IN HEAVY LINE. LIGHTER LINES SHOW CONNECTED PIPE, AND ARE NOT A PART OF THE SYMBOL.

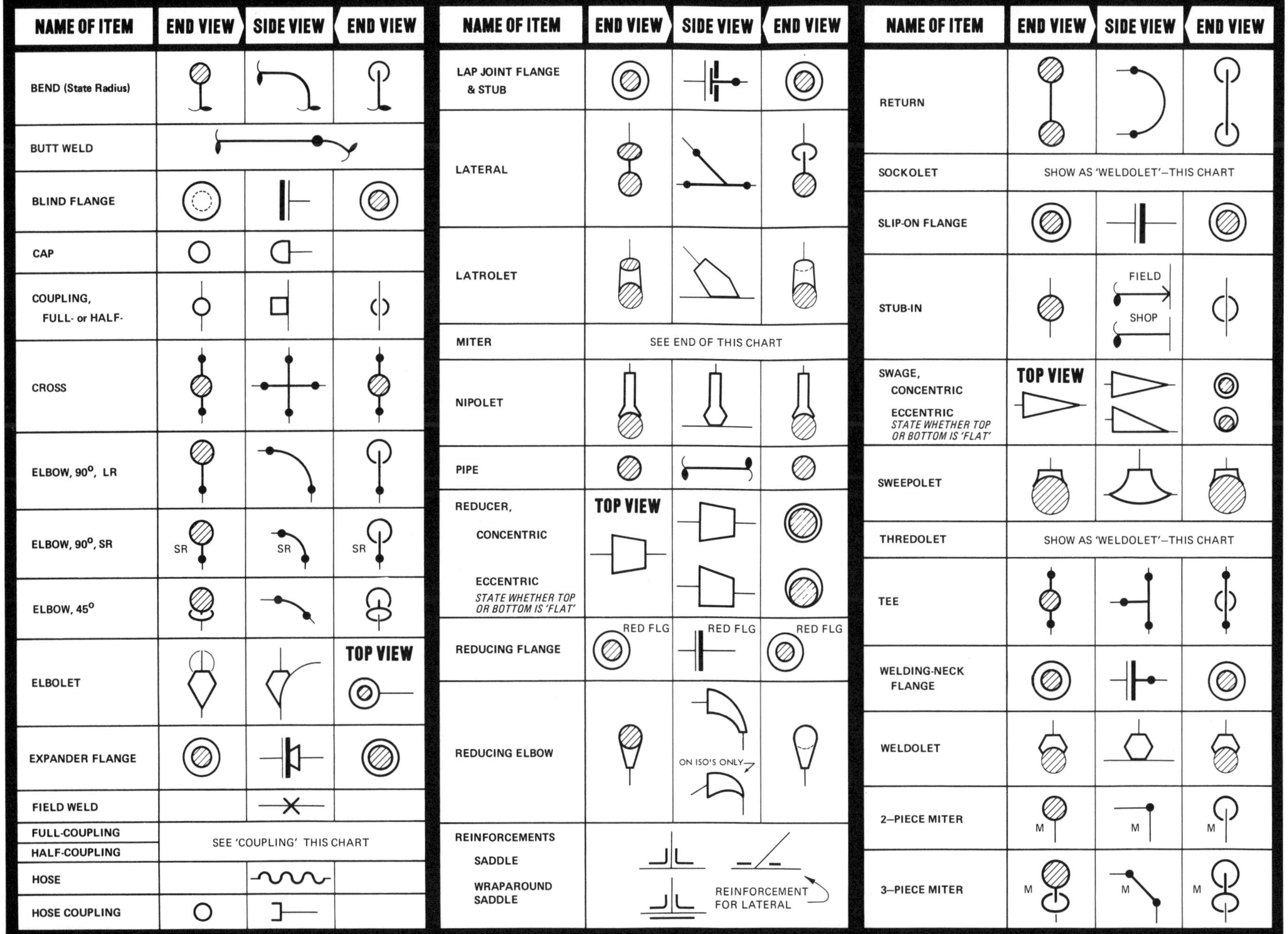

CHARTS
5.2B & 5.3

NAME OF ITEM	END VIEW	SIDE VIEW	END VIEW
BEND (State Radius)			
BUTT WELD			
BLIND FLANGE			
CAP			
COUPLING, FULL- or HALF-			
CROSS			
ELBOW, 90°, LR			
ELBOW, 90°, SR	SR	SR	SR
ELBOW, 45°			
ELBOLET			TOP VIEW
EXPANDER FLANGE			
FIELD WELD			
FULL-COUPLING / HALF-COUPLING	SEE 'COUPLING' THIS CHART		
HOSE			
HOSE COUPLING			

NAME OF ITEM	END VIEW	SIDE VIEW	END VIEW
LAP JOINT FLANGE & STUB			
LATERAL			
LATROLET			
MITER	SEE END OF THIS CHART		
NIPOLET			
PIPE			
REDUCER, CONCENTRIC	TOP VIEW		
ECCENTRIC STATE WHETHER TOP OR BOTTOM IS 'FLAT'			
REDUCING FLANGE	RED FLG	RED FLG	RED FLG
REDUCING ELBOW		ON ISO'S ONLY	
REINFORCEMENTS SADDLE			
WRAPAROUND SADDLE		REINFORCEMENT FOR LATERAL	

NAME OF ITEM	END VIEW	SIDE VIEW	END VIEW
RETURN			
SOCKOLET	SHOW AS 'WELDOLET'—THIS CHART		
SLIP-ON FLANGE			
STUB-IN		FIELD / SHOP	
SWAGE, CONCENTRIC	TOP VIEW		
ECCENTRIC STATE WHETHER TOP OR BOTTOM IS 'FLAT'			
SWEEPOLET			
THREDOLET	SHOW AS 'WELDOLET'—THIS CHART		
TEE			
WELDING-NECK FLANGE			
WELDOLET			
2–PIECE MITER	M	M	M
3–PIECE MITER	M	M	M

SYMBOLS FOR SCREWED SYSTEMS — CHART 5.4

NAME OF ITEM	END VIEW	SIDE VIEW	END VIEW
CAP	○	⊏	
COUPLING, FULL- & HALF-	SHOW FOR BRANCH CONNECTIONS ONLY— SEE 'COUPLING' IN CHART 5.3		
CROSS	⊕	⊞	⊕
ELBOW, 90°	⊕	⌐	⊕
ELBOW, 45°	⊕	⌐	⊕
FLANGE	◎	┼	◎
HOSE		∿	
HOSE CONNECTION	○	⊐	
PIPE	⊘	⊷	⊘
PLUG		⊏	
REDUCER	⊘	▷	⊘
RETURN *Only malleable-iron and cast-iron returns are available. For forged-steel systems, combine forged-steel elbows.*	⊘⊘	⊃	⊘⊘
SEAL WELD	SHOW BY NOTING 'SEAL WELD'		
SWAGE, CONCENTRIC	**TOP VIEW** ▷	▷	◎
ECCENTRIC *STATE WHETHER TOP OR BOTTOM IS 'FLAT'*	▷	▷	◎
TEE, STRAIGHT or REDUCING	⊕	┬	⊕
THREDOLET	SHOW AS 'WELDOLET'—CHART 5.3		
UNION		⫛	

SYMBOLS FOR SOCKET-WELDED SYSTEMS — CHART 5.5

NAME OF ITEM	END VIEW	SIDE VIEW	END VIEW
CAP	○	⊏	
COUPLING, FULL- & HALF-	SHOW FOR BRANCH CONNECTIONS ONLY— SEE 'COUPLING' IN CHART 5.3		
CROSS	⊕	⊞	⊕
ELBOLET	SEE 'ELBOLET'—CHART 5.3		
ELBOW, 90°	⊕	⌐	⊕
ELBOW, 45°	⊕	⌐	⊕
FLANGE	◎	┼	◎
HOSE		∿	
PIPE	⊘	⊷	⊘
REDUCER,	◎	▷	◎
RETURN	NO SOCKET-WELDING FORGED-STEEL FITTING IS AVAILABLE. IF A 180-DEGREE RETURN IS REQUIRED, IT MAY BE MADE USING A BUTT-WELDING RETURN, OR TWO SOCKET-WELDING ELBOWS WITH NIPPLE BETWEEN.		
SOCKOLET	SHOW AS 'WELDOLET'—CHART 5.3		
SWAGE, CONCENTRIC	**TOP VIEW** ▷	▷	◎
ECCENTRIC *STATE WHETHER TOP OR BOTTOM IS 'FLAT'*	▷	▷	◎
TEE, STRAIGHT or REDUCING	⊕	┬	⊕
UNION		⫛	

DRAFTING VALVES

CHART 5.6 GIVES THE BASIC SYMBOLS FOR VALVES. THESE BASIC SYMBOLS ARE USED OR ADAPTED AS FOLLOWS:

P & I D's

USE THE RELEVANT VALVE SYMBOL TO SHOW THE TYPE OF VALVE. DRAW MOST SYMBOLS 1/4-in. LONG. MANUAL OPERATORS ARE NOT SHOWN.

PIPING DRAWINGS

OPERATOR IS SHOWN IF IMPORTANT

(1) SCREWED VALVES
USE THE BASIC VALVE SYMBOL. DRAW THE LENGTH OF THE VALVE TO SCALE.

(2) SOCKET-ENDED VALVES
IF THE PROJECT HAS A PIPING SPECIFICATION, USE THE BASIC VALVE SYMBOL. IF NOT, SHOW SOCKET ENDS TO THE VALVES:

VALVE WITH:	Sockets both ends	Socket one end, other end plain
SYMBOL EXAMPLE	⊢⋈⊣	⊣⋈⊢

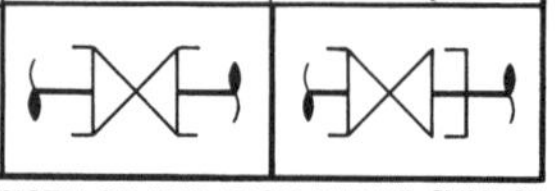

DRAW THE LENGTH OF THE BASIC VALVE SYMBOL TO SCALE OVER SOCKET ENDS.

(3) FLANGED VALVES
USE THE BASIC VALVE SYMBOL, WITH OPERATOR, AND SHOW MATING FLANGES AS DETAILED BELOW:

SINGLE-LINE	DOUBLE-LINE
1. Drawing the symbol	

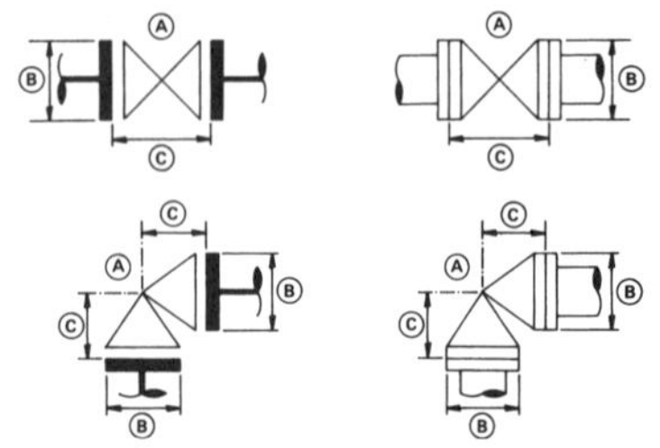

(A) Show the basic valve symbol between flanges.
(B) Draw flange OD to scale.
(C) Draw these lengths scaled to the flange-face-to-flange-face or center-to flange-face dimensions for the valve.

2. Dimensioning nonstandard valves

Refer to 5.3.3, under 'Dimensioning to valves'

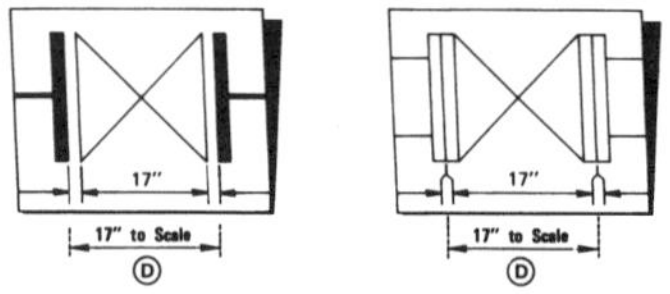

(D) Draw this length to scale (overall length of valve without gaskets) but place arrowheads on the drawing as shown. This convention ensures that:
[1] The line will be made to the correct length.
[2] The fabricator will be reminded to allow for gaskets.

TYPE OF VALVE	SIDE VIEW	TOP VIEW
ANGLE GLOBE		
BALL, ROTARY		
BUTTERFLY		
CHECK (SWING) *Position of dot here shows flow from left to right*		
COCK	SEE 'PLUG VALVE'	
CONTROL		
DIAPHRAGM		
FLUSH-BOTTOM TANK VALVE		
GATE		
GLOBE		

TYPE OF VALVE	SIDE VIEW	TOP VIEW
(a) LINE-BLIND VALVE (Using spectacle plate) (b) LINE BLIND (Shown between flanges)	(a)	(b)
NEEDLE		
PINCH	USE 'SQUEEZE VALVE' SYMBOL	
PLUG		
'QUICK OPENING'		
RELIEF		
SAFETY		
SAFETY-RELIEF		
STOP CHECK		
SQUEEZE		
TRAP		

TYPE OF VALVE	SIDE VIEW	TOP VIEW
VACUUM BREAKER (or Breather)		
WYE-PATTERN GLOBE		
3-WAY		
4-WAY		

OPERATOR	SIDE VIEW	END VIEW	TOP VIEW
SPUR GEAR			
BEVEL GEAR			
CHAIN WHEEL			
CHAIN WRENCH			

THIS CHART GIVES THE BASIC VALVE SYMBOL WHICH IS USED ON P&ID's AND FLOW DIAGRAMS.
ADAPTATION OF THE SYMBOLS TO PIPING DRAWINGS IS EXPLAINED ON THE FACING PAGE <

CHARTS
5.4–5.6

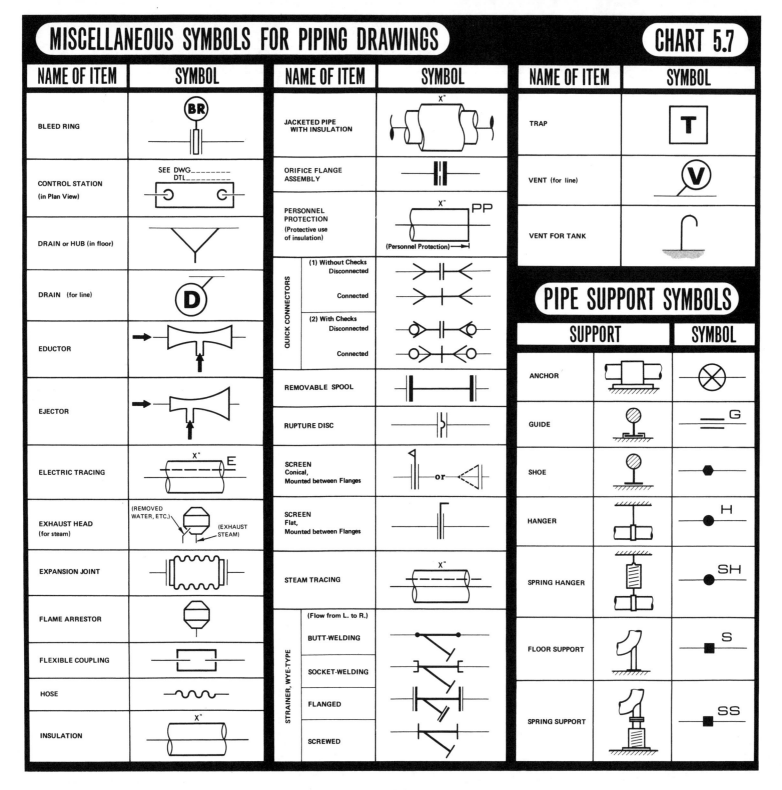

MISCELLANEOUS SYMBOLS FOR PIPING DRAWINGS

CHART 5.7

NAME OF ITEM	SYMBOL
BLEED RING	BR
CONTROL STATION (in Plan View)	SEE DWG_____ DTL_____
DRAIN or HUB (in floor)	
DRAIN (for line)	D
EDUCTOR	
EJECTOR	
ELECTRIC TRACING	X" E
EXHAUST HEAD (for steam)	(REMOVED WATER, ETC.) (EXHAUST STEAM)
EXPANSION JOINT	
FLAME ARRESTOR	
FLEXIBLE COUPLING	
HOSE	
INSULATION	X"

NAME OF ITEM	SYMBOL
JACKETED PIPE WITH INSULATION	X"
ORIFICE FLANGE ASSEMBLY	
PERSONNEL PROTECTION (Protective use of insulation)	X" PP (Personnel Protection)
QUICK CONNECTORS (1) Without Checks — Disconnected	
(1) Without Checks — Connected	
(2) With Checks — Disconnected	
(2) With Checks — Connected	
REMOVABLE SPOOL	
RUPTURE DISC	
SCREEN Conical, Mounted between Flanges	or
SCREEN Flat, Mounted between Flanges	
STEAM TRACING	X"
STRAINER, WYE-TYPE (Flow from L. to R.) BUTT-WELDING	
STRAINER, WYE-TYPE SOCKET-WELDING	
STRAINER, WYE-TYPE FLANGED	
STRAINER, WYE-TYPE SCREWED	

NAME OF ITEM	SYMBOL
TRAP	T
VENT (for line)	V
VENT FOR TANK	

PIPE SUPPORT SYMBOLS

SUPPORT		SYMBOL
ANCHOR		⊗
GUIDE		G
SHOE		
HANGER		H
SPRING HANGER		SH
FLOOR SUPPORT		S
SPRING SUPPORT		SS

SYMBOL	DESCRIPTION	SYMBOL	DESCRIPTION
(1) (2) N N	NORTH ARROWS. (1) FOR PLANS AND ELEVATIONS (2) FOR ISOMETRIC DRAWINGS	ADJACENT TO AREA ON FRONT OF SHEET — **HOLD** FOR ..... — STATE REASON FOR 'HOLD' — ENCIRCLE AREA IN QUESTION AND THE 'HOLD' MARKING ON REAR OF SHEET	'CONSTRUCTION HOLD' MARKING. IF SUFFICIENT INFORMATION IS NOT AVAILABLE TO FINALIZE PART OF THE DESIGN, THE 'HOLD' MARKING IS USED TO INSTRUCT THE CONTRACTOR TO AWAIT A LATER REVISION OF THE DRAWING BEFORE STARTING THE WORK IN QUESTION
10 0 10 20 30	GRAPHIC SCALE REQUIRED ON DRAWINGS LIKELY TO BE CHANGED IN SIZE PHOTOGRAPHICALLY FOR REPORTS, etc.	PLACE TRIANGLE ADJACENT TO REVISED AREA ON FRONT OF SHEET — 4 — ENCIRCLE AREA OF CHANGE INCLUDING REVISION TRIANGLE ON REAR OF SHEET	REVISION TRIANGLE. THE LATEST REVISION NUMBER OF THE DRAWING IS SHOWN WITHIN THE TRIANGLE WHICH IS ENCIRCLED ON THE REAR OF THE SHEET. ALL REVISION TRIANGLES REMAIN ON THE DRAWING, BUT ENCIRCLING OF THE PREVIOUS TRIANGLE IS ERASED
SYMBOL LOCATING AXES OF REFERENCE	SYMBOL LOCATING AXES OF REFERENCE: INTERSECTION OF ORDINATES (COORDINATE POINT)	(1) ⊠ (2) ▭ or ◯	OPENINGS. (1) OPENING WHICH MAY BE COVERED. (ARCH. AND H&V DRAWINGS) (2) HOLE. (ARCH.)
A A DWG NO or A A DWG NO.	TYPICAL SECTION INDICATORS. LETTERS 'I' AND 'O' SHOULD NOT BE USED TO AVOID CONFUSION WITH NUMERALS '1' AND '0'. IF MORE THAN 24 SECTIONS ARE NEEDED, USE COMBINATIONS OF LETTERS AND NUMERALS. SHOW NUMBER OF THE DRAWING ON WHICH SECTION WILL APPEAR	(1) ∟ (2) ⊏ (3) I	STRUCTURAL STEEL SECTIONS: (1) ANGLE. (2) CHANNEL. (3) I-BEAM
₵	CENTERLINE SYMBOL	(railing elevation symbols)	ELEVATION SYMBOLS FOR RAILING
Dimension	DIMENSION LINE SYMBOL USED TO SHOW A DIMENSION NOT TO SCALE	(1) (2) (3)	DISCONTINUED VIEWS: (1) PIPE, ROUND SHAFT, etc. (2) SLAB, SQUARE BAR, etc. (3) VESSEL, EQUIPMENT, etc. (Also used to terminate drawing)
✳	'FITTING MAKEUP' SYMBOL (NOT PREFERRED — SEE 5.3.3, UNDER 'FITTING MAKEUP')	(screwthread) or (screwthread)	SCREWTHREAD SYMBOLS
TYPE OF INSTRUMENT — PROCESS VARIABLE — FG 8 — Upper line: FUNCTIONAL IDENTIFICATION — 'LOOP' NUMBER — Lower line: 'LOOP' IDENTIFICATION	INSTRUMENT BALLOON, USUALLY DRAWN 7/16-INCH DIAMETER ON P&ID's AND PIPING DRAWINGS (TO 3/8 IN. PER FT SCALE)	— · — · — · — · — · —	CHAIN SYMBOL

SHADINGS

THESE SHADINGS ARE USED FOR SHOWING MATERIALS AND SECTIONS OF SOLIDS

GRADE or EARTH	SOLID MATERIAL (and pipe cross section)	STEEL	CONCRETE	BRICK & STONE MASONRY	WOOD	CHECKER PLATE (Use 30° lines)	GRATING
(crosshatch)	(diagonal)	(diagonal)	(stipple)	(diagonal)	(wood grain)	(crosshatch)	(grid)

CHART 5.9

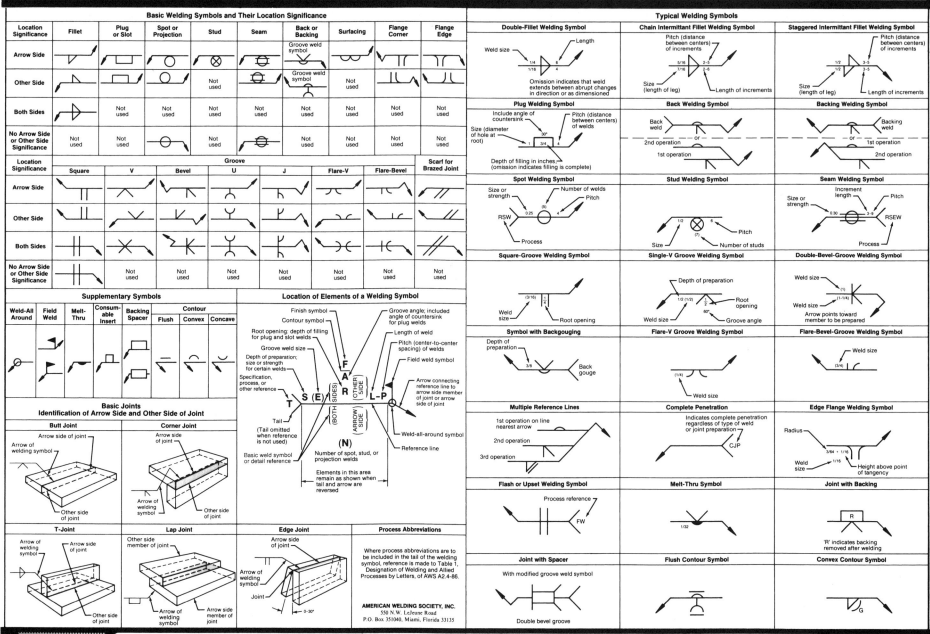

Standard welding symbols are published by the American Welding Society. These symbols should be used as necessary on details of attachments, vessels, piping supports, etc. The practice of writing on drawings instructions such as 'TO BE WELDED THROUGHOUT', or 'TO BE COMPLETELY WELDED' transfers the design responsibility for all attachments and connections from the designer to the welder, which the Society considers to be a dangerous and uneconomic practice.

The 'welding symbol' devised by the American Welding Society has eight elements. Not all of these elements are necessarily needed by piping designers. The assembled welding symbol which gives the welder all the necessary instruction, and locations of its elements, is shown in chart 5.9. The elements are:

- **REFERENCE LINE**
- **ARROW**
- **BASIC WELD SYMBOLS**
- **DIMENSIONS & OTHER DATA**
- **SUPPLEMENTARY SYMBOLS**
- **FINISH SYMBOLS**
- **TAIL**
- **SPECIFICATIONS, PROCESS or OTHER REFERENCE**

The following is a quick guide to the scheme. Full details will be found in the current revision of 'Standard Welding Symbols' available from the American Welding Society.

ASSEMBLING THE WELDING SYMBOL

Reference line and arrow: The symbol begins with a reference line and arrow pointing to the joint where the weld is to be made. The reference line has two 'sides': 'other side' (above the line) and 'arrow side' (below the line)—refer to the following examples and to chart 5.9.

BASIC WELDING ARROW   **FIGURE 5.1**

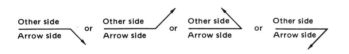

BASIC WELDING SYMBOLS

(a) The weld symbol

FILLET	BACK, or BACKING	PLUG or SLOT	SPOT, or PROJECTION	SEAM	EDGE FLANGE	CORNER FLANGE

(b) The groove symbol

SQUARE	'V'	BEVEL	'U'	'J'	FLARE-'V'	FLARE-BEVEL

EXAMPLE USE OF THE FILLET WELD SYMBOL

If a continuous fillet weld is needed, like this:

the fillet weld symbol is placed on the 'arrow side' of the reference line, thus:

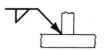

If the weld is required on the far side from the arrow, thus:

the weld symbol is shown on the 'other side' of the reference line:

If a continuous fillet weld is needed on both sides of the joint,

the fillet weld symbol is placed on both sides of the reference line:

EXAMPLE USE OF THE BEVEL GROOVE SYMBOL

If a bevel groove is required, like this:

The 'groove' symbol for a bevel is shown, with the fillet weld symbol, and a break is made in the arrow toward the member to be beveled, thus:

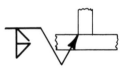

Only the bevel and 'J' groove symbols require a break in the arrow —see chart 5.9.

DIMENSIONING THE WELD CROSS SECTION

Suppose the weld is required to be 1/4 inch in size, and the bevel is to be 3/16 inch deep:

These dimensions are shown to the left of the weld symbol:

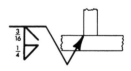

Alternatively, the bevel can be expressed in degrees of arc:

and be indicated thus on the symbol:

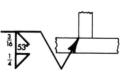

If a root gap is required, thus:

the symbol is:

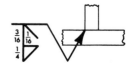

CHART | **5.9**

FIGURE | **5.1**

DIMENSIONING THE LENGTH OF THE WELD

Going back to the fillet weld joint without a bevel, if the weld needs to be 1/4-inch in size and 6 inches long, like this:

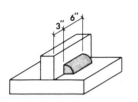

the weld symbol may be drawn:

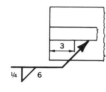

alternately:

If a series of 6-inch long welds is required with 6-inch gaps between them (that is, the pitch of the welds is 12 inches), thus:

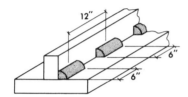

the symbol is:

alternately:

If these welds are required staggered on both sides—

like this:

the symbol is:

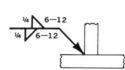

SUPPLEMENTARY SYMBOLS

These symbols give instructions for making the weld and define the required countour:

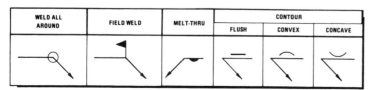

WELD ALL AROUND	FIELD WELD	MELT-THRU	CONTOUR		
			FLUSH	CONVEX	CONCAVE

Going back to the example of a simple fillet weld, if the weld is required all around a member,

like this:

or like this:

it is shown in this way:

If this same 'all around' weld has to be made in the field, it is shown thus:

The contour of the weld is shown by a contour symbol on the weld symbol:

FLUSH CONTOUR   **CONVEX CONTOUR**   **CONCAVE CONTOUR**

like this: like this: or:

The method of finishing the weld contour is indicated by adding a finish notation letter, thus,

where M = machining, G = grinding, and C = chipping.

FULL WELDING SYMBOL

Occasionally it is necessary to give other instructions in the welding symbol. The symbol can be elaborated for this as shown in 'Location of elements of a welding symbol' in chart 5.9.

Chart 5.9, reproduced by permission of the American Welding Society, summarizes and amplifies the explanations of this section.

All information for constructing piping systems is contained in drawings, apart from the specifications, and the possible use of a model and photographs.

> **THE MAIN PURPOSE OF A DRAWING IS TO COMMUNICATE INFORMATION IN A SIMPLE AND EXPLICIT WAY.**

PROCESS & PIPING DRAWINGS GROW FROM THE SCHEMATIC DIAGRAM 5.2.1

To design process piping, three types of drawing are developed in sequence from the schematic diagram (or 'schematic') prepared by the process engineer.

These three types of drawing are, in order of development:—

(1)   FLOW DIAGRAM (PROCESS, or SERVICE)

(2)   PIPING AND INSTRUMENTATION DIAGRAM, or 'P&ID'

(3)   PIPING DRAWING

EXAMPLE DIAGRAMS

Figure 5.2 shows a simple example of a 'schematic'. A solvent recovery system is used as an example. Based on the schematic diagram of figure 5.2, a developed process flow diagram is shown in figure 5.3. From this flow diagram, the P&ID (figure 5.4) is evolved.

As far as practicable, the flow of material(s) should be from left to right. Incoming flows should be arrowed and described down the left-hand edge of the drawing, and exiting flows arrowed and described at the right of the drawing, without intruding into the space over the title block.

Information normally included on the process drawings is detailed in sections 5.2.2 thru 5.2.4. Flow diagrams and P&ID's each have their own functions and should show only that information relevant to their functions, as set out in 5.2.3 and 5.2.4. Extraneous information such as piping, structural and mechanical notes should not be included, unless essential to the process.

SECURITY

A real or supposed need for industrial or national security may restrict information appearing on drawings. Instead of naming chemicals, indeterminate or traditional terms such as 'sweet water', 'brine', 'leach acid', 'chemical B', may be used. Data important to the reactions such as temperatures, pressures and flow rates may be withheld. Sometimes certain key drawings are locked away when not in use.

SCHEMATIC DIAGRAM 5.2.2

Commonly referred to as a 'schematic', this diagram shows paths of flow by single lines, and operations or process equipment are represented by simple figures such as rectangles and circles. Notes on the process will often be included.

The diagram is not to scale, but relationships between equipment and piping with regard to the process are shown. The desired spatial arrangement of equipment and piping may be broadly indicated. Usually, the schematic is not used after the initial planning stage, but serves to develop the process flow diagram which then becomes the primary reference.

FLOW DIAGRAM 5.2.3

This is an unscaled drawing describing the process. It is also referred to as a 'flow sheet'.

It should state the materials to be conveyed by the piping, conveyors, etc., and specify their rates of flow and other information such as temperature and pressure, where of interest. This information may be 'flagged' (on lines) within the diagram or be tabulated on a separate panel—such a panel is shown at the bottom left of figure 5.3.

LAYOUT OF THE FLOW DIAGRAM

Whether a flow diagram is to be in elevation or plan view should depend on how the P&ID is to be presented. To easily relate the two drawings, both should be presented in the same view. Elevations are suitable for simple systems arranged vertically. Installations covering large horizontal areas are best shown in plan view.

Normally, a separate flow diagram is prepared for each plant process. If a single sheet would be too crowded, two or more sheets may be used. For simple processes, more than one may be shown on a sheet. Process lines should have the rate and direction of flow, and other required data, noted. Main process flows should preferably be shown going from the left of the sheet to the right. Line sizes are normally not shown on a flow diagram. Critical internal parts of vessels and other items essential to the process should be indicated.

All factors considered, it is advisable to write equipment titles *either near the top or near the bottom of the sheet*, either directly above or below the equipment symbol. Sometimes it may be directed that all pumps be drawn at a common level near the bottom of the sheet, although this practice may lead to a complex-looking drawing. Particularly with flow diagrams, simplicity in presentation is of prime importance.

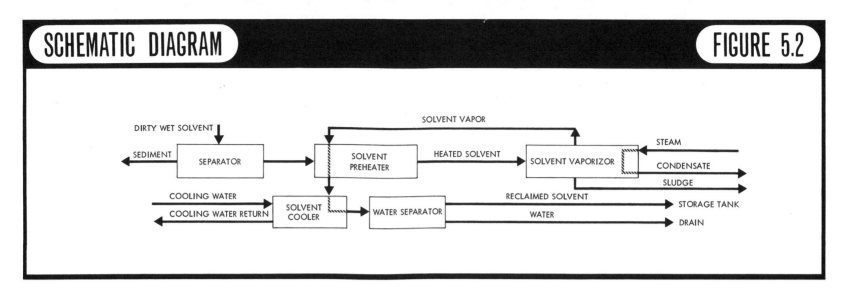

SCHEMATIC DIAGRAM

FIGURE 5.2

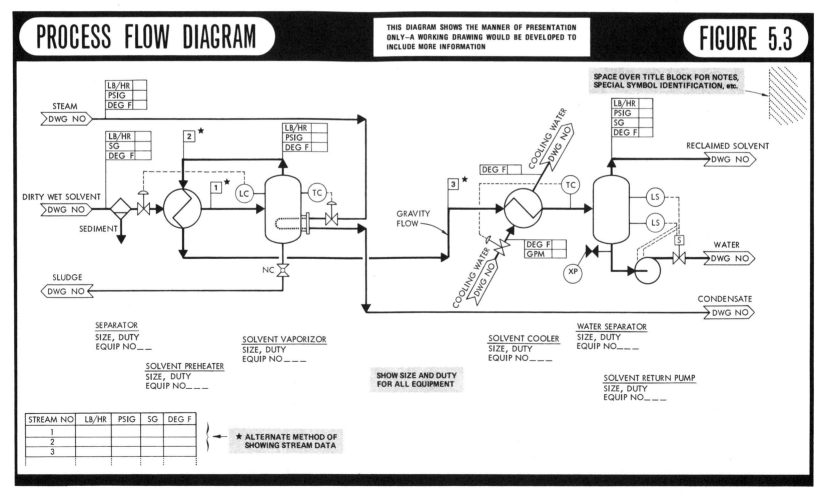

PROCESS FLOW DIAGRAM

THIS DIAGRAM SHOWS THE MANNER OF PRESENTATION ONLY—A WORKING DRAWING WOULD BE DEVELOPED TO INCLUDE MORE INFORMATION

FIGURE 5.3

SPACE OVER TITLE BLOCK FOR NOTES, SPECIAL SYMBOL IDENTIFICATION, etc.

SEPARATOR
SIZE, DUTY
EQUIP NO __

SOLVENT PREHEATER
SIZE, DUTY
EQUIP NO___

SOLVENT VAPORIZOR
SIZE, DUTY
EQUIP NO ___

SHOW SIZE AND DUTY
FOR ALL EQUIPMENT

SOLVENT COOLER
SIZE, DUTY
EQUIP NO ___

WATER SEPARATOR
SIZE, DUTY
EQUIP NO___

SOLVENT RETURN PUMP
SIZE, DUTY
EQUIP NO___

STREAM NO	LB/HR	PSIG	SG	DEG F
1				
2				
3				

★ ALTERNATE METHOD OF
SHOWING STREAM DATA

FLOW LINES

Directions of flow within the diagram are shown by solid arrowheads. The use of arrowheads at all junctions and corners aids the rapid reading of the diagram. The number of crossings can be minimized by good arrangement. Suitable line thicknesses are shown at full size in chart 5.1. For photographic reduction, lines should be spaced not closer than 3/8 inch.

Process and service streams entering or leaving the flow diagram are shown by large hollow arrowheads, with the conveyed fluid written over and the continuation sheet number within the arrowhead, as in figure 5.3.

ARROWS ON FLOW DIAGRAMS

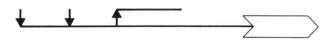

SHOWING VALVES ON THE FLOW DIAGRAM

Instrument-controlled and manual valves which are necessary to the process are shown. The following valves are shown if required by a governing code or regulation, or if they are essential to the process: isolating, bypassing, venting, draining, sampling, and valves used for purging, steamout, etc., for relieving excess pressure of gases or liquids (including rupture discs), breather valves and vacuum breakers.

SHOW ONLY SPECIAL FITTINGS

Piping fittings, strainers, and flame arrestors should not be shown unless of special importance to the process.

ESSENTIAL INSTRUMENTATION

Only instrumentation essential to process control should be shown. Simplified representation is suitable. For example, only instruments such as controllers and indicators need be shown: items not essential to the drawing (transmitters, for example) may be omitted.

EQUIPMENT DATA

Capacities of equipment should be shown. Equipment should be drawn schematically, using equipment symbols, and where feasible should be drawn in proportion to the actual sizes of the items. Equipment symbols should neither dominate the drawing, nor be too small for clear understanding.

STANDBY & PARALLELED EQUIPMENT

Standby equipment is not normally drawn. If identical units of equipment are provided for paralleled operation (that is, all units on stream), only one unit need normally be drawn. Paralleled or standby units should be indicated by noting the equipment number and the service function ('STANDBY' or 'PARALLEL OP').

It is advisable to draw equipment that is operated cyclically. For example, with filter presses operated in parallel, one may be shown on-stream, and the second press for alternate operation.

PROCESS DATA FOR EQUIPMENT

The basic process information required for designing and operating major items of equipment should be shown. This information is best placed immediately below the title of the equipment.

IDENTIFYING EQUIPMENT

Different types of equipment may be referred to by a classification letter (or letters). There is no generally accepted coding — each company has its own scheme if any standardization is made at all. Equipment classed under a certain letter is numbered in sequence from '1' upward. If a new installation is made in an existing plant, the method of numbering may follow previous practice for the plant.

Also, it is useful to divide the plant and open part of the site as necessary into areas, giving each a code number. An area number can be made the first part of an equipment number. For example, if a heat exchanger is the 53rd item of equipment listed under the classification letter 'E', located in area '1', (see 'Key plan' in 5.2.7) the exchanger's equipment number can be 1-E-53.

Each item of equipment should bear the same number on all drawings, diagrams and listings. Standby or identical equipment, if in the same service, may be identified by adding the letters, A, B, C, and so on, to the same equipment identification letter and number. For example, a heat exchanger and its standby may be designated 1-E-53A, and 1-E-53B.

SERVICES ON PROCESS FLOW DIAGRAMS

Systems for providing services should not be shown. However, the type of service, flow rates, temperatures and pressures should be noted at consumption rates corresponding to the material balance—usually shown by a 'flag' to the line—see figure 5.3.

DISPOSAL OF WASTES

The routes of disposal for all waste streams should be indicated. For example, arrows or drain symbols may be labelled with destination, such as 'chemical sewer' or 'drips recovery system'. In some instances the disposal or waste-treatment system may be detailed on one or more separate sheets. See 6.13 where 'effluent' is discussed.

MATERIAL BALANCE

The process material balance can be tabulated on separate 8½ x 11-inch sheets, or along the bottom of the process flow diagram.

This drawing is commonly referred to as the 'P&ID'. Its object is to indicate all process and service lines, instruments and controls, equipment, and data necessary for the design groups. The process flow diagram is the primary source of information for developing the P&ID. Symbols suitable for P&ID's are given in charts 5.1 thru 5.7.

The P&ID should define piping, equipment and instrumentation well enough for cost estimation and for subsequent design, construction, operation and modification of the process. Material balance data, flow rates, temperatures, pressures, etc., and piping fitting details are not shown, and purely mechanical piping details such as elbows, joints and unions are inappropriate to P&ID's.

INTERCONNECTING P&ID

This drawing shows process and service lines between buildings and units, etc., and serves to link the P&ID's for the individual processes, units or buildings. Like any P&ID, the drawing is not to scale. It resembles the layout of the site plan, which enables line sizes and branching points from headers to be established, and assists in planning pipeways.

P&ID LAYOUT

The layout of the P&ID should resemble as far as practicable that of the process flow diagram. The process relationship of equipment should corres-pond exactly. Often it is useful to draw equipment in proportion verti-cally, but to reduce horizontal dimensions to save space and allow room for flow lines between equipment. Crowding information is a common drafting fault — it is desirable to space generously, as, more often than not, revisions add information. On an elevational P&ID, a base line indicating grade or first-floor level can be shown. Critical elevations are noted.

For revision purposes, a P&ID is best made on a drawing sheet having a grid system—this is a sheet having letters along one border and numbers along the adjacent border. Thus, references such as 'A6', 'B5', etc., can be given to an area where a change has been made. (A grid system is applicable to P&ID's more complicated than the simple example of figure 5.4.)

DRAFTING GUIDELINES FOR P&ID's

- Suitable line thicknesses are shown at full size in chart 5.1
- Crossing lines must not touch—break lines going in one direction only. Break instrument lines crossing process and service lines
- Keep parallel lines at least 3/8 inch apart
- Preferably draw all valves the same size—1/4-inch long is suitable—as this retains legibility for photographic reduction. Instrument isolating valves and drain valves can be drawn smaller, if desired
- Draw instrument identification balloons 7/16th-inch diameter—see 5.5
- Draw trap symbols 3/8th-inch square

FLOW LINES ON P&ID's

All flow lines and interconnections should be shown on P&ID's. Every line should show direction of flow, and be labeled to show the area of project, conveyed fluid, line size, piping material or specification code number (company code), and number of the line. This information is shown in the 'line number'.

EXAMPLE LINE NUMBER: (74 | BZ | 6 | 412 | 23) may denote the 23rd line in area 74, a 6-inch pipe to company specification 412. 'BZ' identifies the conveyed fluid.

This type of full designation for a flow line need not be used, provided identification is adequate.

Piping drawings use the line numbering of the P&ID, and the following points apply to piping drawings as well as P&ID's.

- For a system of lines conveying the same fluid, allocate sequential numbers to lines, beginning with '1' *for each system*
- For a continuous line, retain the same number of line (such as 23 in the example) as the line goes thru valves, strainers, small filters, traps, venturis, orifice flanges and small equipment generally —unless the line changes in size
- Terminate the number of a line at a major item of equipment such as a tank, pressure vessel, mixer, or any equipment carrying an individual equipment number
- Allocate new numbers to branches

As with the process flow diagram, directions of flow within the drawing are shown by solid arrows placed at every junction, and all corners except where changes of direction occur closely together. Corners should be square. The number of crossings should be kept minimal by good arrangement.

Process and service streams entering or leaving the process are noted by hollow arrows with the name of the conveyed fluid written over the arrowhead and the continuation sheet number within it. No process flow data will normally be shown on a P&ID.

FLOW LINES ON P&ID's

NOTES FOR LINES

Special points for design and operating procedures are noted—such as lines which need to be sloped for gravity flow, lines which need careful cleaning before startup, etc.

P&ID SHOWS ALL EQUIPMENT & SPECIAL ITEMS

The P&ID should show all major equipment and information that is relevant to the process, such as equipment names, equipment numbers, the sizes, ratings, capacities, and/or duties of equipment, and instrumentation.

Standby and paralleled equipment is shown, including all connected lines. Equipment numbers and service functions ('STANDBY' or 'PARALLEL OP') are noted.

'Future' equipment, together with the equipment that will service it, is shown in broken outline, and labeled. Blind-flange terminations to accommodate future piping should be indicated on headers and branches. 'Future' additions are usually not anticipated beyond a 5-year period.

Pressure ratings for equipment are noted if the rating is different from the piping system. A 'typical' note may be used to describe multiple pieces of identical equipment in the same service, but all equipment numbers are written.

CLOSURES

Temporary closures for process operation or personnel protection are shown.

SEPARATORS, SCREENS & STRAINERS

These items should be shown upstream of equipment and processes needing protection, and are discussed in 2.10.

STEAM TRAPS ON THE P&ID

If the locations of traps are known they are indicated. For example, the trap required upstream of a pressure-reducing station feeding a steam turbine should be shown.

Steam traps on steam piping are not otherwise indicated, as these trap positions are determined when making the piping drawings. They can be added later to the P&ID if desired, after the piping drawings have been completed.

DRIPLEGS

Driplegs are not shown.

VENTS & DRAINS

Vents and drains on high and low points of lines respectively, to be used for hydrostatic testing, are not shown, as they are established on the piping arrangement drawings. Process vents and drains are shown.

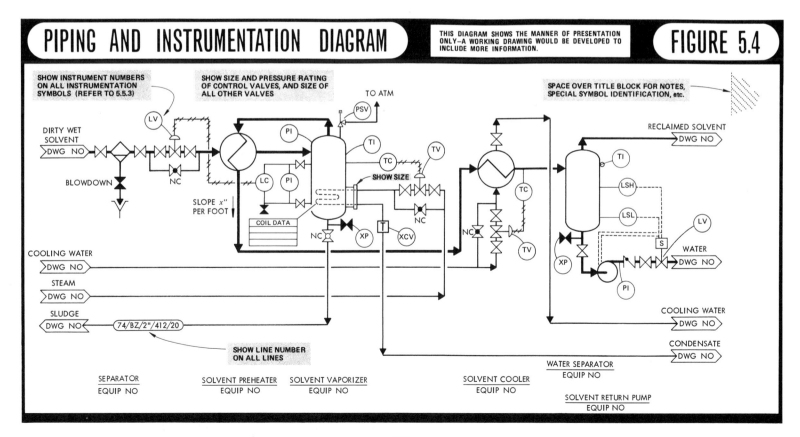

PIPING AND INSTRUMENTATION DIAGRAM — FIGURE 5.4

THIS DIAGRAM SHOWS THE MANNER OF PRESENTATION ONLY—A WORKING DRAWING WOULD BE DEVELOPED TO INCLUDE MORE INFORMATION.

FIGURE 5.4

VALVES ON THE P&ID

- Show and tag process and service valves with size and identifying number if applicable. Give pressure rating if different from line specification

- Indicate any valves that have to be locked open or locked closed

- Indicate powered operators

SHOWING INSTRUMENTATION ON THE P&ID

Signal-lead drafting symbols shown in chart 5.1 may be used, and the ISA scheme for designating instrumentation is described in 5.5. Details of instrument piping and conduit are usually shown on separate instrument installation drawings.

- Show all instrumentation on the P&ID, for and including these items: element or sensor, signal lead, orifice flange assembly, transmitter, controller, vacuum breaker, flame arrestor, level gage, sight glass, flow indicator, relief valve, rupture disc, safety valve. The last three items may be tagged with set pressure(s) also

- Indicate local- or board-mounting of instruments by the symbol—refer to the labeling scheme in 5.5.4

INSULATION & TRACING

Insulation on piping and equipment is shown, together with the thickness required. Tracing requirements are indicated. Refer to 6.8.

CONTROL STATIONS

Control stations are discussed in 6.1.4. Control valves are indicated by pressure rating, instrument identifying number and size—see figure 5.15, for example.

P&ID SHOWS HOW WASTES ARE HANDLED

Drains, funnels, relief valves and other equipment handling wastes are shown on the P&ID. If an extensive system or waste-treatment facility is involved, it should be shown on a separate P&ID. Wastes and effluents are discussed in 6.13.

SERVICE SYSTEMS MAY HAVE THEIR OWN P&ID

Process equipment may be provided with various services, such as steam for heating, water or refrigerant for cooling, or air for oxidizing. Plant or equipment providing these services is usually described on separate 'service P&ID's'. A service line such as a steam line entering a process P&ID is given a 'hollow arrow' line designation taken from the service P&ID. Returning service lines are designated in the same way. Refer to figure 5.4.

UTILITY STATIONS

Stations providing steam, compressed air, and water, are shown. Refer to 6.1.5.

LINE DESIGNATION SHEETS OR TABLES   5.2.5

These sheets are tabulated lists of lines and information about them. The numbers of the lines are usually listed at the right of the sheet. Other columns list line size, material of construction (using company's specification code, if there is one), conveyed fluid, pressure, temperature, flow rate, test pressure, insulation or jacketing (if required), and connected lines (which will usually be branches).

The sheets are compiled and kept up-to-date by the project group, taking all the information from the P&ID. Copies are supplied to the piping group for reference.

On small projects involving only a few lines line designation sheets may not be used. It is useful to add a note on the P&ID stating the numbers of the last line and last valve used.

VIEWS USED FOR PIPING DRAWINGS   5.2.6

Two types of view are used:

(1)   ORTHOGRAPHIC – PLANS AND ELEVATIONS

(2)   PICTORIAL – ISOMETRIC VIEW AND OBLIQUE PRESENTATION

Figure 5.5 shows how a building would appear in these different views.

PRESENTATIONS USED IN PIPING DRAWINGS   FIGURE 5.5

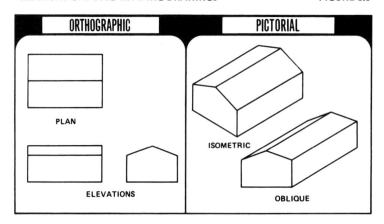

PLANS & ELEVATIONS

Plan views are more common than elevational views. Piping layout is developed in plan view, and elevational views and section details are added for clarity where necessary.

PICTORIAL VIEWS

In complex piping systems, where orthographic views may not easily illustrate the design, pictorial presentation can be used for clarity. In either isometric or oblique presentations, lines not horizontal or vertical on the drawing are usually drawn at 30 degrees to the horizontal.

Oblique presentation has the advantage that it can be distorted or expanded to show areas of a plant, etc. more clearly than an isometric view. It is not commonly used, but can be useful for diagramatic work.

Figure 5.6 illustrates how circular shapes viewed at different angles are approximated by means of a 35-degree ellipse template. Isometric templates for valves, etc., are available and neat drawings can be rapidly produced with them. Orthographic and isometric templates can be used to produce an oblique presentation.

PLAN, ELEVATION, ISOMETRIC & OBLIQUE PRESENTATIONS OF A PIPING SYSTEM

Figure 5.7 is used to show the presentations used in drafting. Isometric and oblique drawings both clearly show the piping arrangement, but the plan view fails to show the bypass loop and valve, and the supplementary elevation is needed.

PIPING DRAWINGS ARE BASED ON OTHER DRAWINGS   5.2.7

The purpose of piping drawings is to supply detailed information to enable a plant to be built. Prior to making piping drawings, the site plan and equipment arrangement drawings are prepared, and from these two drawings the plot plan is derived. These three drawings are used as the basis for developing the piping drawings.

SITE PLAN

The piping group produces a 'site plan' to a small scale (1 inch to 30 or 100 ft for example). It shows the whole site including the boundaries, roads, railroad spurs, pavement, buildings, process plant areas, large structures, storage areas, effluent ponds, waste disposal, shipping and loading areas. 'True' (geographic) and 'assumed' or 'plant' north are marked and their angular separation shown—see figure 5.11.

ISOMETRIC PRESENTATION OF CIRCULAR SECTIONS   **FIGURE 5.6**

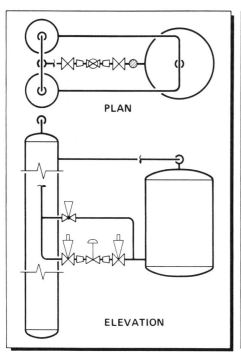

PIPING ARRANGEMENT IN DIFFERENT PRESENTATIONS

FIGURE 5.7

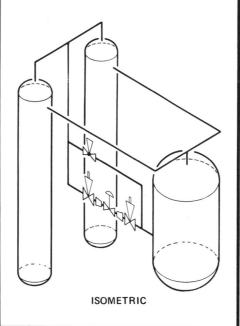

PLAN

ELEVATION

ISOMETRIC

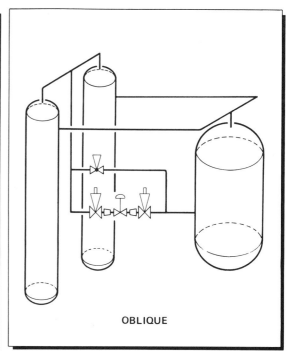

OBLIQUE

KEY PLAN

A 'key plan' is produced by adapting the site plan, dividing the area of the site into smaller areas identified by key letters or numbers. A small simplified inset of the key plan is added to plot plans, and may be added to piping and other drawings for reference purposes. The subject area of the particular drawing is hatched or shaded, as shown in figure 5.8.

DRAWING SHEET SHOWING KEY PLAN & MATCHLINE   **FIGURE 5.8**

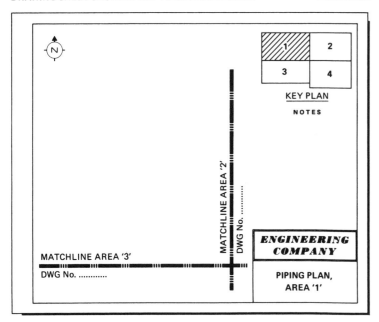

EQUIPMENT ARRANGEMENT DRAWING

Under project group supervision, the piping group usually makes several viable arrangements of equipment, seeking an optimal design that satisfies process requirements. Often, preliminary piping studies are necessary in order to establish equipment coordinates.

A design aid for positioning equipment is to cut out scaled outlines of equipment from stiff paper, which can be moved about on a plan view of the area involved. (If multiple units of the same type are to be used, xeroxing the equipment outlines is faster.) Another method which is useful for areas where method study or investigational reports are needed is described in 4.4.13 under 'Photographic layouts'.

PLOT PLAN

When the equipment arrangement drawings are approved, they are developed into 'plot plans' by the addition of dimensions and coordinates to locate all major items of equipment and structures.

North and east coordinates of the extremities of buildings, and centerlines of steelwork or other architectural constructions should be shown on the plot plan, preferably at the west and south ends of the installation. Both 'plant north' and true north should be shown—see figure 5.11.

Equipment coordinates are usually given to the centerlines. Coordinates for pumps are given to the centerline of the pump shaft and either to the face of the pump foundation, or to the centerline of the discharge port.

Up-dated copies of the above drawings are sent to the civil, structural and electrical or other groups involved in the design, to inform them of requirements as the design develops.

VESSEL DRAWINGS

When the equipment arrangement has been approved and the piping arrangement determined, small dimensioned drawings of process vessels are made (on sheets 8½ x 11 or 11 x 17 inches) in order to fix nozzles and their orientations, manholes, ladders, etc. These drawings are then sent to the vendor who makes the shop detail drawings, which are examined by the project engineer and sent to the piping group for checking and approval. Vessel drawings need not be to scale. (Figure 5.14 is an example vessel drawing.)

DRAWINGS FROM OTHER SOURCES

Piping drawings should be correlated with the following drawings from other design groups and from vendors. Points to be checked are listed:

Architectural drawings:
- Outlines of walls or sidings, indicating thickness
- Floor penetrations for stairways, lifts, elevators, ducts, drains, etc.
- Positions of doors and windows

Civil engineering drawings:
- Foundations, underground piping, drains, etc.

Structural-steel drawings:
- Positions of steel columns supporting next higher floor level
- Supporting structures such as overhead cranes, monorails, platforms or beams
- Wall bracing, where pipes may be taken thru walls

Heating, ventilating & air-conditioning (HVAC) drawings:
- Paths of ducting and rising ducts, fan room, plenums, space heaters, etc.

Electrical drawings:
- Positions of motor control centers, switchgear, junction boxes and control panels
- Major conduit or wiring runs (including buried runs)
- Positions of lights

Instrumentation drawings:
- Instrument panel and console locations

Vendors' drawings:
- Dimensions of equipment
- Positions of nozzles, flange type and pressure rating, instruments, etc.

Mechanical drawings:
- Positions and dimensions of mechanical equipment such as conveyors, chutes, etc.
- Piped services needed for mechanical equipment.

Process equipment and piping systems have priority. Drawings listed on the preceding page must be reviewed for compatibility with the developing piping design.

Pertinent background details (drawn faintly) from these drawings help to avoid interferences. Omission of such detail from the piping drawing often leads to the subsequent discovery that pipe has been routed thru a brace, stairway, doorway, foundation, duct, mechanical equipment, motor control center, fire-fighting equipment, etc.

Completed piping drawings will also show spool numbers, if this part of the job is not subcontracted — see 5.2.9. Electrical and instrument cables are not shown on piping drawings, but trays to hold the cables are indicated—for example, see figure 6.3, point (8).

It is not always possible for the piping drawing to follow exactly the logical arrangement of the P&ID. Sometimes lines must be routed with different junction sequence, and line numbers may be changed. During the preliminary piping studies, economies and practicable improvements may be found, and the P&ID may be modified to take these into account. However, it is not the piping designer's job to seek ways to change the P&ID.

SCALE

Piping is arranged in plan view, usually to 3/8 in./ft scale.

ALLOCATING SPACE ON THE SHEET

●   Obtain the drawing number and fill in the title block at the bottom right corner of the sheet

ALLOCATING SPACE ON A DRAWING SHEET   **FIGURE 5.9**

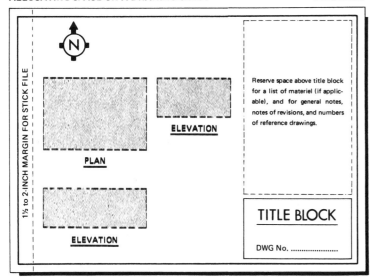

●   On non-standard sheets, leave a 1½- to 2-inch margin at the left edge of the sheet, to allow filing on a 'stick'. Standard drawing sheets usually have this margin

●   On drawings showing a plan view, place a north arrow at the top left corner of the sheet to indicate plant north—see figure 5.11

●   Do not draw in the area above the title block, as this space is allocated to the bill of materiel, or to general notes, brief descriptions of changes, and the titles and numbers of reference drawings

●   If plans and elevations are small enough to go on the same sheet, draw the plan at the upper left side of the sheet and elevations to the right and bottom of it, as shown in figure 5.9

BACKGROUND DETAIL

●   Show background detail as discussed in 5.2.8 under 'Piping drawings'. It is sometimes convenient to draw outlines on the reverse side of the drawing sheet

●   After background details have been determined, it is best to make a print on which nozzles on vessels, pumps, etc., to be piped can be marked in red pencil. Utility stations can also be established. This will indicate areas of major usage and the most convenient locations for the headers. Obviously, at times there will be a number of alternate routes offering comparable advantages

PROCESS & SERVICE LINES ON PIPING DRAWINGS

●   Take line numbers from the P&ID. Refer to 5.2.4 under 'Flow lines on P&ID's' for information on numbering lines. Include line numbers on all views, and arrowheads showing direction of flow

●   Draw all pipe 'single line' unless special instructions have been given for drawing 'double line'. Chart 5.1 gives line thicknesses (full size)

●   Line numbers are shown against lines, thus:

●   Take lines continued on another sheet to a matchline, and there code with line numbers only. Show the continuation sheet numbers on matchlines—see figure 5.8

●   Show where changes in line material specification occur. The change is usually indicated immediately downstream of a flange of a valve or equipment

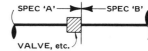

●   Show a definite break in a line crossing behind another line—see 'Rolled ell', under 'Plan view piping drawings', this section

- If pipe sleeves are required thru floors, indicate where they are needed and inform the group leader for transmitting this information to the group(s) concerned

- Indicate insulation, and show whether lines are electrically or steam traced—see chart 5.7

FITTINGS, FLANGES, VALVES & PUMPS ON PIPING DRAWINGS

- The following items should be labeled in one view only: tees and ells rolled at 45 degrees (see example, this page), short-radius ell, reducing ell, eccentric reducer and eccentric swage (note on plan views whether 'top flat' or 'bottom flat'), concentric reducer, concentric swage, non-standard or companion flange, reducing tee, special items of unusual material, of pressure rating different from that of the system, etc. Refer to charts 5.3, 5.4 and 5.5 for symbol usage

- Draw the outside diameters of flanges to scale

- Show valve identification number from P&ID

- Label control valves to show: size, pressure rating, dimension over flanges, and valve instrument number, from the P&ID—see figure 5.15

- Draw valve handwheels to scale with valve stem fully extended

- If a valve is chain-operated, note distance of chain from operating floor, which for safety should be approximately 3 ft

- For pumps, show outline of foundation and nozzles

DRIPLEGS & STEAM TRAPS

Driplegs are indicated on relevant piping drawing plan views. Unless identical, a separate detail is drawn for each dripleg. The trap is indicated on the dripleg piping by a symbol, and referred to a separate trap detail or data sheet. The trap detail drawing should show all necessary valves, strainers, unions, etc., required at the trap—see figures 6.43 and 6.44.

The piping shown on the dripleg details should indicate whether condensate is to be taken to a header for re-use, or run to waste. The design notes in 6.10.5 discuss dripleg details for steam lines in which condensate forms continuously. Refer to 6.10.9 also.

INSTRUMENTS & CONNECTIONS ON PIPING DRAWINGS

- Show location for each instrument connection with encircled instrument number taken from the P&ID. Refer to 5.5.3 and chart 6.2

- Show similar isolating valve arrangements on instrument connections as 'typical' detail, unless covered by standard company detail sheet

VENTS & DRAINS

Refer to 6.11 and figure 6.47.

PIPE SUPPORTS

Refer to 6.2.2, and chart 5.7. for symbols.

PLAN VIEW PIPING DRAWINGS

- Draw plan views for each floor of the plant. These views should show what the layout will look like between adjacent floors, viewed from above, or at the elevation thru which the plan view is cut

- If the plan view will not fit on one sheet, present it on two or more sheets, using matchlines to link the drawings. See figure 5.8

- Note the elevation below which a plan view is shown—for example, 'PLAN BELOW ELEVATION 15'—0" '. For clarity, both elevations can be stated: 'PLAN BETWEEN ELEVATIONS 30'—0" & 15'—0" '

- If a tee or elbow is 'rolled' at 45 degrees, note as shown in the view where the fitting is rolled out of the plane of the drawing sheet

'ROLLED' ELL

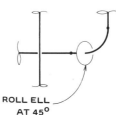

ROLL ELL AT 45°

'ROLLED' TEE

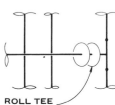

ROLL TEE AT 45°

- Figure 5.10 shows how lines can be broken to give sufficient information without drawing other views

- Indicate required field welds

ELEVATIONS (SECTIONS) & DETAILS

- Draw elevations and details to clarify complex piping or piping hidden in the plan view

- Do not draw detail that can be described by a note

- Show only as many sections as necessary. A section does not have to be a complete cross section of the plan

- Draw to a large scale any part needing fuller detail. Enlarged details are preferably drawn in available space on elevational drawings, and should be cross-referenced by the applicable detail and drawing number(s)

- Identify sections indicated on plan views by letters (see chart 5.8) and details by numbers. Letters I and O are not used as this can lead to confusion with numerals. If more than twentyfour sections are needed the letter identification can be broken down thus: A1—A1, A2—A2, B4—B4, ... and so on

- Do not section plan views looking toward the bottom of the drawing sheet

- Figure 5.10 shows how to break lines to give sufficient information whilst avoiding drawing another view or section

SHOWING 'HIDDEN' LINES ON PIPING DRAWINGS   **FIGURE 5.10**

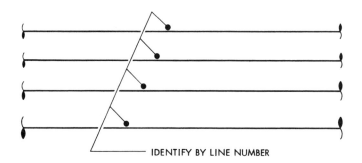

IDENTIFY BY LINE NUMBER

P L A N (or ELEVATION)

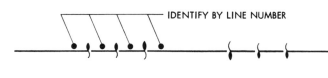

IDENTIFY BY LINE NUMBER

Corresponding E L E V A T I O N (or PLAN)

PIPING FABRICATION DRAWINGS—'ISOS' & 'SPOOLS'   5.2.9

The two most common methods for producing piping designs for a plant are by making either plan and elevation drawings, or by constructing a scaled model. For fabricating welded piping, plans and elevations are sent directly to a subcontractor, usually referred to as a 'shop fabricator'—if a model is used, isometric drawings (referred to as 'isos') are sent instead.

Isometric views are commonly used in prefabricating parts of butt-welded piping systems. Isos showing the piping to be prefabricated are sent to the shop fabricator. Figure 5.15 is an example of such an iso.

The prefabricated parts of the piping system are termed 'spools', described under 'Spools', this section. The piping group either produces isos showing the required spools, or marks the piping to be spooled on plans and elevations, depending on whether or not a model is used (as shown in chart 5.10). From these drawings, the subcontractor makes detail drawings termed 'spool sheets'. Figure 5.17 is an example spool sheet.

SPOOL FABRICATION   **CHART 5.10**

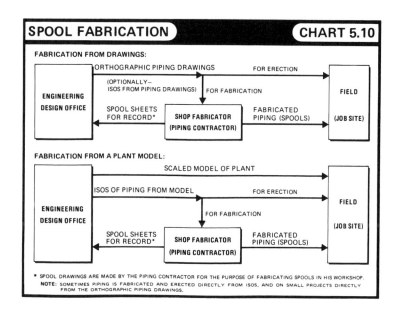

FABRICATION FROM DRAWINGS:

FABRICATION FROM A PLANT MODEL:

* SPOOL DRAWINGS ARE MADE BY THE PIPING CONTRACTOR FOR THE PURPOSE OF FABRICATING SPOOLS IN HIS WORKSHOP.
NOTE: SOMETIMES PIPING IS FABRICATED AND ERECTED DIRECTLY FROM ISOS, AND ON SMALL PROJECTS DIRECTLY FROM THE ORTHOGRAPHIC PIPING DRAWINGS.

ISOMETRIC DRAWINGS, or 'ISOS'

An iso usually shows a complete line from one piece of equipment to another—see figure 5.15. It gives all information necessary for fabrication and erection of piping.

Isos are usually drawn freehand, but the various runs of pipe, fittings and valves should be roughly in proportion for easy understanding. Any one line (that is, all the piping with the same line number) should be drawn on the minimum number of iso sheets. If continuation sheets are needed, break the line at natural breakpoints such as flanges (except orifice flanges), welds at fittings, or field welds required for installation.

Items and information to be shown on an iso include:

- North arrow (plant north)

- Dimensions and angles

- Reference number of plan drawing from which iso is made (unless model is used), line number, direction of flow, insulation and tracing

- Equipment numbers and locations of equipment (by centerlines)

- Identify all items by use of an understood symbol, and amplify by a description, as necessary

- Give details of any flanged nozzles on equipment to which piping has to be connected, if the flange is different from the specification for the connected piping

- Size and type of every valve

- Size, pressure rating and instrument number of control valves

- Number, location and orientation for each instrument connection

CHART 5.10

FIGURE 5.10

- Shop and field welds. Indicate limits of shop and field fabrication
- Iso sheet continuation numbers
- Unions required for installation and maintenance purposes
- On screwed and socket-welded assemblies, valve handwheel positions need not be shown
- Materials of construction
- Locations of vents, drains, and traps
- Locations of supports, identified by pipesupport number

The following information may also be given:

- Requirements for stress relieving, seal welding, pickling, lining, coating, or other special treatment of the line

Drawing style to be followed is shown in the example iso, figure 5.15, which displays some of the above points, and gives others as shaded notes. An iso may show more than one spool.

SPOOLS

A spool is an assembly of fittings, flanges and pipe that may be prefabricated. It does not include bolts, gaskets, valves or instruments. Straight mill-run lengths of pipe over 20 ft are usually not included in a spool, as such lengths may be welded in the system on erection (on the iso, this is indicated by noting the length, and stating 'BY FIELD').

The size of a spool is limited by the fabricator's available means of transportation, and a spool is usually contained within a space of dimensions 40 ft x 10 ft x 8 ft. The maximum permissible dimensions may be obtained from the fabricator.

FIELD-FABRICATED SPOOLS

Some States in the USA have a trades agreement that 2-inch and smaller carbon-steel piping must be fabricated at the site. This rule is sometimes extended to piping larger than 2-inch.

SHOP-FABRICATED SPOOLS

All alloy spools, and spools with 3 or more welds made from 3-inch (occasionally 4-inch) and larger carbon-steel pipe are normally 'shop-fabricated'. This is, fabricated in the shop fabricator's workshop, either at his plant or at the site. Spools with fewer welds are usually made in the field.

Large-diameter piping, being more difficult to handle, often necessitates the use of jigs and templates, and is more economically produced in a workshop.

SPOOL SHEETS

A spool sheet is an orthographic drawing of a spool made by the piping contractor either from plans and elevations, or from an iso—see chart 5.10.

Each spool sheet shows only one type of spool, and:—

(1) Instructs the welder for fabricating the spool

(2) Lists the cut lengths of pipe, fittings and flanges, etc. needed to make the spool

(3) Gives materials of construction, and any special treatment of the finished piping

(4) Indicates how many spools of the same type are required

NUMBERING ISOS, SPOOL SHEETS, & SPOOLS

Spool numbers are allocated by the piping group, and appear on all piping drawings. Various methods of numbering can be used as long as identification is easily made. A suggested method follows:—

Iso sheets can be identified by the line number of the section of line that is shown, followed by a sequential number. For example, the fourth iso sheet showing a spool to be part of a line numbered 74/BZ/6/412/23 could be identified: 74/BZ/6/412/23—4 .

Both the spool and the spool sheet can be identified by number or letter using the iso sheet number as a prefix. For example, the numbering of spool sheets relating to iso sheet 74/BZ/6/412/23—4 could be

	74/BZ/6/412/23—4—1,	74/BZ/6/412/23—4—2, ... etc.,
or	74/BZ/6/412/23—4—A,	74/BZ/6/412/23—4—B, ... etc.

The full line number need not be used if a shorter form would suffice for identification.

Spool numbers are also referred to as 'mark numbers'. They are shown on isos and on the following:—

(1) Spool sheets—as the sheet number
(2) The fabricated spool—so it can be related to drawings or isos
(3) Piping drawings—plans and elevations

DIMENSIONING   5.3

DIMENSIONING FROM REFERENCE POINTS   5.3.1

HORIZONTAL REFERENCE

When a proposed plant site is surveyed, a geographic reference point is utilized from which measurements to boundaries, roads, buildings, tanks, etc., can be made. The geographic reference point chosen is usually an officially-established one.

The lines of latitude and longitude which define the geographic reference point are not used, as a 'plant north' (see figure 5.11) is established, parallel to structural steelwork. The direction closest to true north is chosen for the 'plant north'.

The coordinates of the southwest corner of the plant in figure 5.11, as referred to 'plant north', are N 110.00 and E 200.00.

Sometimes coordinates such as those above may be written N 1+10 and E 2+00. The first coordinate is read as "one hundred plus 10 ft north" and the second as "two hundred plus zero ft east". This is a system used for traverse survey, and is more correctly applied to highways, railroads, etc.

Coordinates are used to locate tanks, vessels, major equipment and structural steel. In the open, these items are located directly with respect to a geographic reference point, but in buildings and structures, can be dimensioned from the building steel.

HORIZONTAL REFERENCE   **FIGURE 5.11**

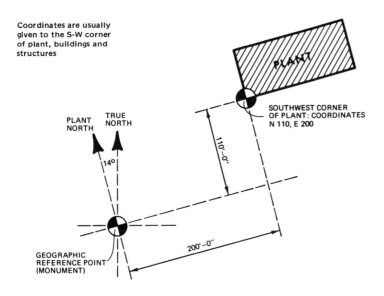

Coordinates are usually given to the S-W corner of plant, buildings and structures

The US Department of Commerce's Coast and Geodetic Survey has established a large number of references for latitude and longitude, and for elevations above sea level. These are termed 'geodetic control stations'.

Control stations for horizontal reference (latitude and longitude) are referred to as 'triangulation stations' or 'traverse stations', etc. Control stations for vertical reference are referred to as 'benchmarks'. Latitude and longitude have not been established for all benchmarks.

A geodetic control station is marked with a metal disc showing identity and date of establishment. To provide stable locations for the discs, they are set into tops of 'monuments', mounted in holes drilled in bedrock or large firmly-imbedded boulders, or affixed to a solid structure, such as a building, bridge, etc.

The geographic positions of these stations can be obtained from the Director, US Coast and Geodetic Survey, Rockville, Maryland 20852.

VERTICAL REFERENCE

Before any building or erecting begins , the site is leveled ('graded') with earth-moving equipment. The ground is made as flat as practicable, and after leveling is termed 'finished grade'.

The highest graded point is termed the 'high point of finished grade', (HPFG), and the horizontal plane passing thru it is made the vertical reference plane or 'datum' from which plant elevations are given. Figure 5.12 shows that this horizontal plane is given a 'false' or nominal elevation, usually 100 ft, and is not referred to mean sea level.

The 100 ft nominal elevation ensures that foundations, basements, buried pipes and tanks, etc., will have positive elevations. 'Minus' elevations, which would be a nuisance, are thus avoided.

Large plants may have several areas, each having its own high point of finished grade. Nominal grade elevation is measured from a benchmark, as illustrated in figure 5.12.

VERTICAL REFERENCE   **FIGURE 5.12**

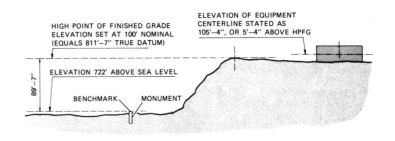

DIMENSIONING PIPING DRAWINGS   **5.3.2**

DRAWING DIMENSIONS—& TOLERANCES MAINTAINED IN ERECTED PIPING

On plot: Dimensions on piping drawings are normally maintained within the limits of plus or minus 1/16th inch. How this tolerance is met does not concern the designer. Any necessary allowances to ensure that dimensions are maintained are made by the fabricator and erector (contractor).

Off plot: Dimensions are maintained as closely as practicable by the erector.

WHICH DIMENSIONS SHOULD BE SHOWN?

Sufficient dimensions should be given for positioning equipment, for fabricating spools and for erecting piping. Duplication of dimensions in different views should be avoided, as this may easily lead to error if alterations are made.

Basically the dimensions to show are:

	TYPE OF DIMENSION		EXAMPLES
1	REFERENCE LINE* TO CENTERLINE		VESSELS PUMPS EQUIPMENT LINES
2	CENTERLINE TO CENTERLINE		LINES STANDARD VALVES
3	CENTERLINE TO FLANGE FACE †	NOZZLES ON	{ VESSELS PUMPS EQUIPMENT
4	FLANGE FACE TO FLANGE FACE†	NON-STANDARD	{ VALVES EQUIPMENT METERS INSTRUMENTS

* REFERENCE LINE CAN BE EITHER AN ORDINATE (LINE OF LATITUDE OR LONGITUDE) OR A CENTERLINE OF BUILDING STEEL

† IT IS NECESSARY TO SHOW THESE DIMENSIONS FOR ITEMS LACKING STANDARD DIMENSIONS (DEFINED BY ANY RECOGNIZED STANDARD)

Figure 5.13 illustrates the use of these types of dimensions.

PLAN VIEW DIMENSIONS

Plan views convey most of the dimensional information, and may also show dimensions for elevations in the absence of an elevational view or section.

EXAMPLE DIMENSIONS FOR PLAN VIEW   **FIGURE 5.13**

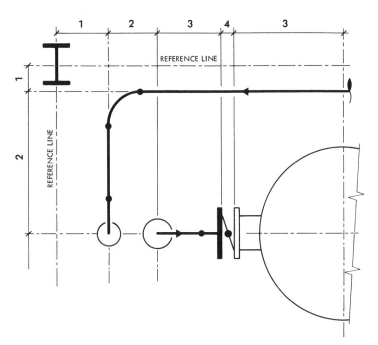

VERTICAL VIEW ELEVATIONS & DIMENSIONS

On piping drawings, elevations may be given as in table 5.2.

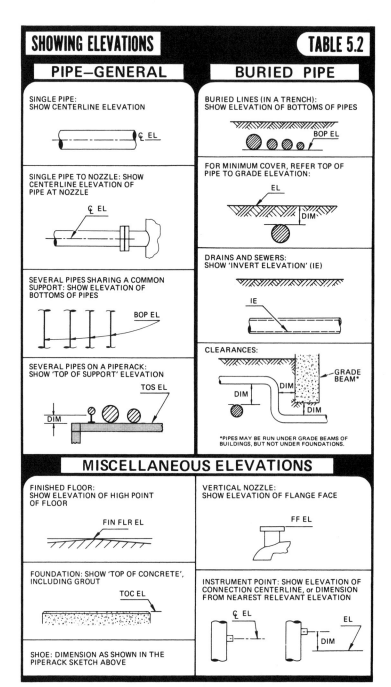

- Show all key dimensions, including elevations and coordinates

- Show dimensions outside of the drawn view unless unavoidable — do not clutter the picture

- Draw dimension lines unbroken with a fine line. Write the dimension just above a horizontal line. Write the dimension of a vertical line sideways, preferably at the left. It is usual to terminate the line with arrowheads, and these are preferable for isos. The oblique dashes shown are quicker and are suitable for plans and elevations, especially if the dimensions are cramped

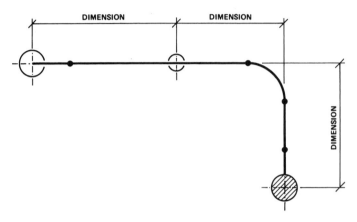

- If a series of dimensions is to be shown, string them together as shown in the sketch. (Do not dimension from a common reference line as in machine drawing.) Show the overall dimension of the string of dimensions if this dimension will be of repeated interest

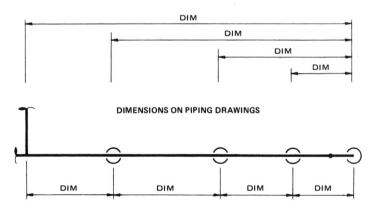

- Do not omit a *significant* dimension other than 'fitting makeup', even though it may be easily calculated — see 'fitting makeup', this section

- Most piping under 2-inch is screwed or socket-welded and assembled at the site (field run). Therefore, give only those dimensions necessary to route such piping clear of equipment, other obstructions, and thru walls, and to locate only those items whose safe positioning or accessability is important to the process

- Most lengths will be stated to the nearest sixteenth of an inch. Dimensions which cannot or need not be stated to this precision are shown with a plus-or-minus sign: 8'–7"±, 15'–3"±, etc.

- Dimensions under two feet are usually marked in inches, and those over two feet in feet and inches. Some companies prefer to mark all dimensions over one foot in feet and inches

- Attempt to round off non-critical dimensions to whole feet and inches. Reserve fractions of inches for dimensions requiring this precision

PLANS & ELEVATIONS—GENERAL DIMENSIONING POINTS

- Reserve horizontal dimensions for the plan view

- Underline all out-of-scale dimensions, or show as in chart 5.8

- If a certain piping arrangement is repeated on the same drawing, it is sufficient to dimension the piping in one instance and note the other appearances as 'TYP' (typical). This situation occurs where similar pumps are connected to a common header. For another example, see the pump base in figure 6.17

- Do not duplicate dimensions. Do not repeat them in different views

DIMENSIONING TO JOINTS

- Do not terminate dimensions at a welded or screwed joint

- Unless necessary, do not dimension to unions, in-line couplings or any other items that are not critical to construction or operation of the piping

- Where flanges meet it is usual to show a small gap between dimension lines to indicate the gasket. Gaskets should be covered in the piping specification, with gasket type and thickness stated. Refer to the panel 'Drafting valves', preceding chart 5.6.

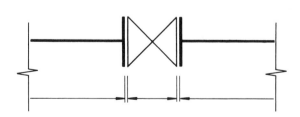

- As nearly all flanged joints have gaskets, a time-saving procedure is to note flanged joints without gaskets (for example, see 3.1.6 under 'Butterfly valve'). The fabricator and erector can be alerted to the need for gaskets elsewhere by a general note on all piping drawings:

"GASKETS AS SPECIFICATION EXCEPT AS NOTED"

FIGURE **5.13**

TABLE **5.2**

FITTING MAKEUP

If a number of items of standard dimensions are grouped together it is unnecessary to dimension each item, as the fabricator knows the sizes of standard fittings and equipment. It is necessary, however, to indicate that the overall dimension is 'fitting makeup' by the special cross symbol, or preferably by writing the overall dimension. Any non-standard item inserted between standard items should be dimensioned.

FITTING MAKEUP SYMBOL ✳

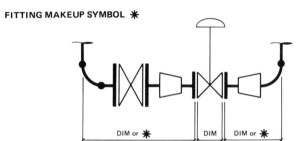

DIMENSIONING TO VALVES

● Locate flanged and welding-end valves with ANSI standard dimensions by dimensioning to their centers. Most gate and globe valves are standard—see table V-1

● Dimension non-standard flanged valves as shown in the panel opposite chart 5.6. Although a standard exists for control valves, face-to-face dimensions are usually given, as it is possible to obtain them in non-standard sizes

● Standard flanged check valves need not be dimensioned, but if location is important, dimension to the flange face(s)

● Non-flanged valves are dimensioned to their centers or stems

DIMENSIONING TO NOZZLES ON VESSELS & EQUIPMENT

● In plan view, a nozzle is dimensioned to its face from the centerline of the equipment it is on

● In elevation, a nozzle's centerline is either given its own elevation or is dimensioned from another reference. In the absence of an elevational view, nozzle elevations can be shown on the plan view

DIMENSIONING ISOS   5.3.4

In order to clearly show all dimensions, the best aspect of the piping must be determined. Freedom to extend lines and spread the piping without regard to scale is a great help in showing isometric dimensions. The basic dimensions set out in 5.3.2, 5.3.3, and the guidelines in 5.2.9 apply.

Figure 5.15 illustrates the main requirements of an isometric drawing, and includes a dimensioned offset. Figure 5.16 shows how other offsets are dimensioned.

● Dimension in the same way as plans and elevations
● Give sufficient dimensions for the fabricator to make the spool drawings —see figure 5.17

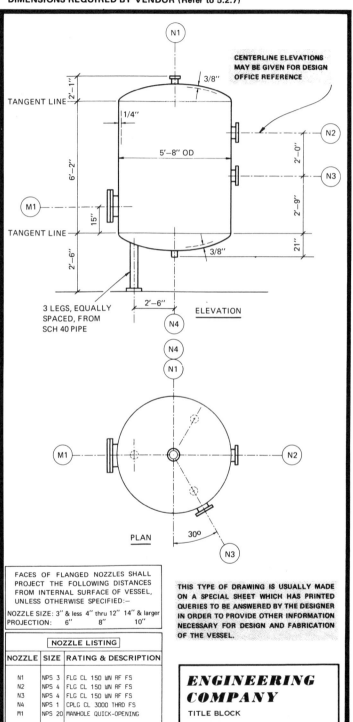

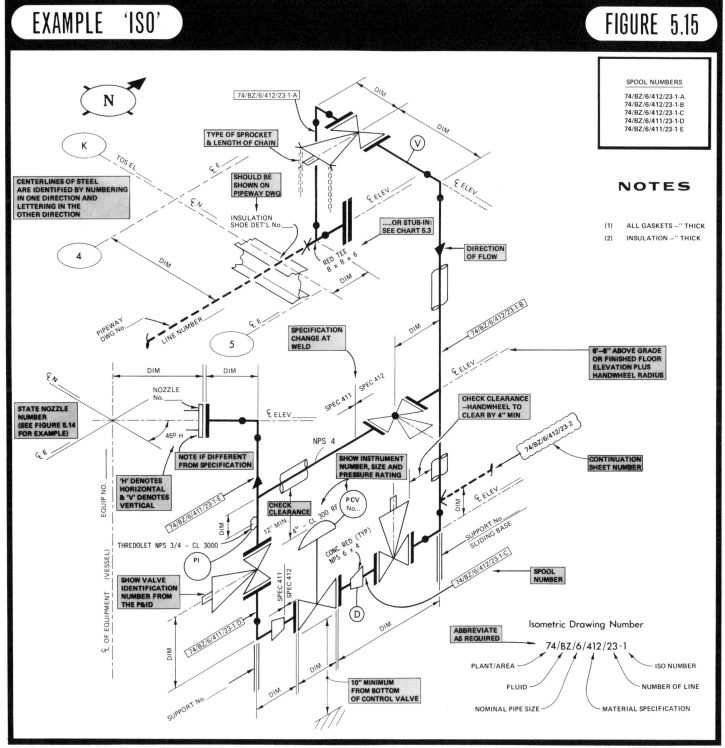

SPOOL NUMBERS

74/BZ/6/412/23-1-A
74/BZ/6/412/23-1-B
74/BZ/6/412/23-1-C
74/BZ/6/411/23-1-D
74/BZ/6/411/23-1-E

NOTES

(1)   ALL GASKETS –'' THICK
(2)   INSULATION –'' THICK

TYPE OF SPROCKET & LENGTH OF CHAIN

SHOULD BE SHOWN ON PIPEWAY DWG

CENTERLINES OF STEEL ARE IDENTIFIED BY NUMBERING IN ONE DIRECTION AND LETTERING IN THE OTHER DIRECTION

INSULATION SHOE DET'L No.

.....OR STUB-IN: SEE CHART 5.3

RED TEE 8 x 8 x 6

DIRECTION OF FLOW

PIPEWAY DWG No.

LINE NUMBER

SPECIFICATION CHANGE AT WELD

6'–6'' ABOVE GRADE OR FINISHED FLOOR ELEVATION PLUS HANDWHEEL RADIUS

NOZZLE No.

STATE NOZZLE NUMBER (SEE FIGURE 5.14 FOR EXAMPLE)

SPEC 411   SPEC 412

CHECK CLEARANCE —HANDWHEEL TO CLEAR BY 4'' MIN

45º H

NOTE IF DIFFERENT FROM SPECIFICATION

NPS 4

SHOW INSTRUMENT NUMBER, SIZE AND PRESSURE RATING

CONTINUATION SHEET NUMBER

'H' DENOTES HORIZONTAL & 'V' DENOTES VERTICAL

CHECK CLEARANCE

PCV No...

12'' MIN.   4'' – CL 300 RF

THREDOLET NPS 3/4 – CL 3000

PI

CONC RED (TYP) NPS 6 x 4

SUPPORT No. SLIDING BASE

SHOW VALVE IDENTIFICATION NUMBER FROM THE P&ID

SPEC 411   SPEC 412

SPOOL NUMBER

10'' MINIMUM FROM BOTTOM OF CONTROL VALVE

ABBREVIATE AS REQUIRED

Isometric Drawing Number

74/BZ/6/412/23-1

PLANT/AREA   ISO NUMBER

FLUID   NUMBER OF LINE

NOMINAL PIPE SIZE   MATERIAL SPECIFICATION

SUPPORT No.

EQUIP No.   (VESSEL)   ℄ OF EQUIPMENT

FIGURES 5.14 & 5.15

[81]

(Chart M-1 gives a formula for calculating the compound angle)

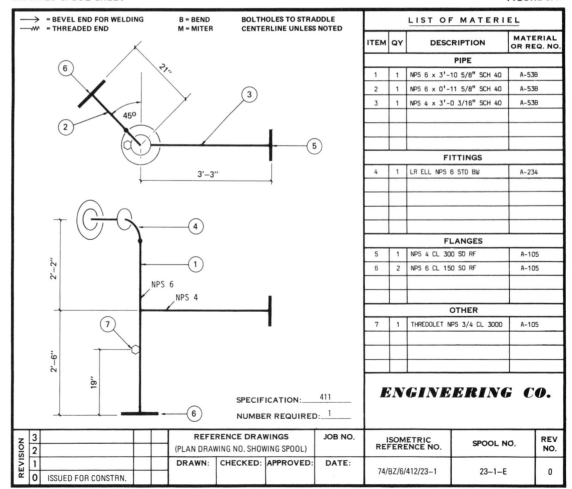

Allowance for weld spacing (root gap) is a shop set-up problem and should not be considered in making assembly drawings or detailed sketches. The Pipe Fabrication Institute recommends that an overall dimension is shown which is the sum of the nominal dimensions of the component parts.

A spool sheet deals with only one design of spool, and shows complete dimensional detail, lists material for making the spool, and specifies how many spools of that type are required. Figure 5.17 shows how a spool from figure 5.15 would be dimensioned.

EXAMPLE SPOOL SHEET   **FIGURE 5.17**

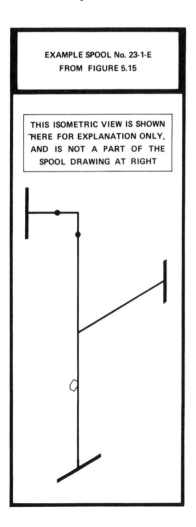

EXAMPLE SPOOL No. 23-1-E FROM FIGURE 5.15

THIS ISOMETRIC VIEW IS SHOWN HERE FOR EXPLANATION ONLY, AND IS NOT A PART OF THE SPOOL DRAWING AT RIGHT

→ = BEVEL END FOR WELDING
⋎⋎ = THREADED END

B = BEND
M = MITER

BOLTHOLES TO STRADDLE
CENTERLINE UNLESS NOTED

LIST OF MATERIEL			
ITEM	QY	DESCRIPTION	MATERIAL OR REQ. NO.
PIPE			
1	1	NPS 6 x 3'-10 5/8" SCH 40	A-53B
2	1	NPS 6 x 0'-11 5/8" SCH 40	A-53B
3	1	NPS 4 x 3'-0 3/16" SCH 40	A-53B
FITTINGS			
4	1	LR ELL NPS 6 STD BW	A-234
FLANGES			
5	1	NPS 4 CL 300 SO RF	A-105
6	2	NPS 6 CL 150 SO RF	A-105
OTHER			
7	1	THREDOLET NPS 3/4 CL 3000	A-105

ENGINEERING CO.

21"
45°
3'-3"
2'-2"
2'-6"
19"
NPS 6
NPS 4

SPECIFICATION: 411

NUMBER REQUIRED: 1

REVISION				REFERENCE DRAWINGS (PLAN DRAWING NO. SHOWING SPOOL)			JOB NO.	ISOMETRIC REFERENCE NO.	SPOOL NO.	REV NO.
	3									
	2									
	1			DRAWN:	CHECKED:	APPROVED:	DATE:	74/BZ/6/412/23-1	23-1-E	0
	0	ISSUED FOR CONSTRN.								

CHECKING & ISSUING DRAWINGS 5.4

RESPONSIBILITIES 5.4.1

P&ID's, process flow diagrams and line designation sheets are checked by engineers in the project group.

Except for spool drawings, all piping drawings are checked by the piping group.

Orthographic spool drawings produced by the piping fabricator are not usually checked by the piping group, except for 'critical' spools, such as spools for overseas shipment and intricate spools.

Usually an experienced designer within the piping group is given the task of checking. Some companies employ persons specifically as design checkers.

The checker's responsibilities are set out in 4.1.2.

CHECKING PIPING DRAWINGS 5.4.2

Prints of drawings are checked and corrected by marking with colored pencils. Areas to be corrected on the drawing are usually marked in red on the print. Correct areas and dimensions are usually marked in yellow.

Checked drawings to be changed should be returned to their originator whenever possible, for amendment. A new print is supplied to the checker with the original 'marked up' print for 'backchecking'.

ISSUING DRAWINGS 5.4.3

Areas of a drawing awaiting further information or decision are ringed clearly on the reverse side and labeled 'HOLD'—refer to chart 5.8. (A black, red, or yellow china marker is suitable for film with a slick finish on the reverse side.)

Changes or revisions are indicated on the fronts of the sheets by a small triangle in the area of the revision. The revision number is marked inside the triangle, noted above the title block (or in an allocated panel) with a description of the revision, required initials, and date. The revision number may be part of the drawing number, or it may follow the drawing number (preferred method—see figure 5.17). The drawing as first issued is numbered the 'zero' revision.

A drawing is issued in three stages. The first issue is 'FOR APPROVAL', by management or client. The second issue is 'FOR CONSTRUCTION BID', when vendors are invited to bid for equipment and work contracts. The third issue is 'FOR CONSTRUCTION' following awarding of all purchase orders and contracts. Drawings may be reissued at each stage if significant changes are made. Minor changes may be made after the third stage (by agreement on cost and extent of work) but major changes may involve all three stages of issue.

CHECKING PIPING DRAWINGS 5.4.4
(PLANS, ELEVATIONS, & ISOS)

Points to be checked on all piping drawings include the following:

- Title of drawing
- Number of issue, and revision number
- Orientation: North arrow against plot plan
- Inclusion of graphic scale (if drawing is to be photographically reduced)
- Equipment numbers and their appearance on piping drawings
- That correct identification appears on all lines in all views
- Line material specification changes
- Agreement with specifications and agreement with other drawings
- That the drawing includes reference number(s) and title(s) to any other relevant drawings
- That all dimensions are correct
- Agreement with certified vendors' drawings for dimensions, nozzle orientation, manholes and ladders
- That face-to-face dimensions and pressure ratings are shown for all non-standard flanged items
- Location and identification of instrument connections
- Provision of line vents, drains, traps, and tracing. Check that vents are at all high points and drains at all low points of lines for hydrostatic test. Driplegs should be indicated and detailed. Traps should be identified, and piping detailed
- The following items should be labeled in one view only: tees and ells rolled at 45 degrees (see example in 5.2.8), short-radius ell, reducing ell, eccentric reducer and eccentric swage (note on plan views whether 'top flat' or 'bottom flat'), concentric reducer, concentric swage, non-standard or companion flange, reducing tee, special items of unusual material, of pressure rating different from that of the system, etc. Refer to charts 5.3, 5.4 and 5.5 for symbol usage
- That insulation has been shown as required by the P&ID
- Pipe support locations with support numbers
- That all anchors, dummy legs and welded supports are shown
- That the stress group's requirements have been met
- That all field welds are shown
- Correctness of scale
- Coordinates of equipment against plot plan
- Piping arrangement against P&ID requirements
- Possible interferences
- Adequacy of clearances of piping from steelwork, doors, windows and braces, ductwork, equipment and major electric apparatus, including control consoles, cables from motor control centers (MCC's), and firefighting equipment. Check accessibility for operation and maintenance

FIGURES 5.16 & 5.17

- That floor and wall penetrations are shown correctly
- Accessibility for operation and maintenance, and that adequate manholes, hatches, covers, dropout and handling areas, etc. have been provided
- Foundation drawings with vendors' equipment requirements
- List of materiel, if any. Listed items should be identified once, either on the plan or the elevation drawings
- That section letters agree with the section markings on the plan view
- That drawings include necessary matchline information
- Appearance of necessary continuation sheet number(s)
- That spool numbers appear correctly
- Presence of all required signatures

This further point should be checked on isos:

- Agreement with model

These further points should be checked on spool sheets :

- That materiel is completely listed and described
- That the required number of spools of identical type is noted

INSTRUMENTATION (As shown on P&ID's)   5.5

This section briefly describes the purposes of instruments and explains how instrumentation may be read from P&ID's. Piping drawings will *also* show the connection (coupling, etc.) to line or vessel. However, piping drawings should show only instruments connected to (or located in) piping and vessels. The only purpose in adding instrumentation to a piping drawing is to identify the connection, orifice plate or equipment to be installed on or in the piping, and to correlate the piping drawing to the P&ID.

INSTRUMENT FUNCTION ONLY IS SHOWN   5.5.1

Instrumentation is shown on process diagrams and piping drawings by symbols. The functions of intruments are shown, not the instruments. Only the primary connection to a vessel or line, or devices installed in a line (such as orifice plates and control valves) are indicated.

There is some uniformity, among the larger companies at least, in the way in which instrumentation is shown. There is a willingness to adopt the recommendations of the Instrument Society of America, but adherence is not always complete. The ISA standard is S5.1, titled 'Instrumentation symbols and identification'.

Compliance with the ISA scheme is to some extent international. This is beneficial when drawings go from one country to another, as there is then no difficulty in understanding the instrumentation.

INSTRUMENT FUNCTIONS   5.5.2

Although instruments are used for many purposes, their basic functions are few in number:

(1) *To sense* a 'condition' of the process material, most commonly its pressure, temperature, flow rate or level. These 'conditions' are termed process variables. The piece of equipment that does the sensing is termed a 'primary element', 'sensor', or 'detector'.

(2) *To transmit* a measure of the process variable from a primary element.

(3) *To indicate* a measure of a process variable to the plant operator, by showing the measured value by a dial and pointer, pen and paper roll or digital display. Another form of indicator is an alarm which gives audible or visual warning when a process variable such as temperature approaches an unsafe or undesired value.

(4) *To record* the measure of a process variable. Most recorders are electrically-operated pen-and-paper-roll types which record either the instantaneous value or the average over a time period.

(5) *To control* the process variable. An instrument initiating this function is termed a 'controller'. A controller sustains or changes the value of the process variable by actuating a 'final control element' (this element is usually a valve, in process piping).

Many instruments combine two or more of these five functions, and may also have mechanical parts integrated — the commonest example of this is the self-contained control valve (see 3.1.10, under 'Pressure regulator', and chart 3.1).

HOW INSTRUMENTATION IS IDENTIFIED   5.5.3

The most-used instruments are pressure and temperature gages ('indicators') and are shown as in figure 5.18 (a) and (b). An example 'instrument identification number' (or 'tag number') is shown in figure 5.18 (c). The balloon around the number is usually drawn 7/16-inch diameter.

INSTRUMENT IDENTIFICATION NUMBERS   **FIGURE 5.18**

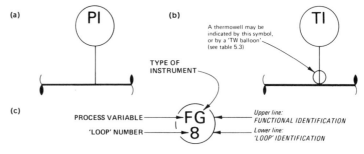

In figure 5.18, 'P', 'T', and 'F' denote process variables pressure, temperature, and flow respectively. 'I' and 'G' show the type of instrument; indicator and gage respectively. Table 5.3 gives other letters denoting process variable, type of instrument, etc. The number '8', labeled 'loop number', is an example sequential number (allocated by an instrumentation engineer).

INSTRUMENT MOUNTING, 5.5.4
& MULTIPLE-FUNCTION INSTRUMENTS

A horizontal line in the ISA balloon shows that the instrument performing the function is to be 'board mounted' in a console, etc. Absence of this line shows 'local mounting', in or near the piping, vessel, etc.

BOARD MOUNTING   LOCAL MOUNTING

The ISA scheme shows instrument functions, not instruments. However, a multiple-function instrument can be indicated by drawing the balloons showing the separate functions so that the circles touch.

Sometimes, a multiple-function instrument will be indicated by a single balloon symbol, with a function identification, such as 'TRC' for a temperature recorder-controller. This practice is not preferred—it is better to draw (in this example) separate 'TR' and 'TC' balloons, touching.

INTERCONNECTED INSTRUMENTS ('LOOPS') 5.5.5

The ISA standard uses the term 'loop' to describe an interconnected group of instruments, which is not necessarily a closed-loop arrangement: that is, instrumentation used in a feedback (or feedforward) arrangement.

If several instruments are interconnected, they may be all allocated the same number for 'loop' identification. Figure 5.19 shows a process line served by one group of instruments (loop number 73) to sense, transmit and indicate temperature, and a second group (loop number 74) to sense, transmit, indicate, record and control flow rate.

EXAMPLE INSTRUMENT 'LOOPS'   FIGURE 5.19

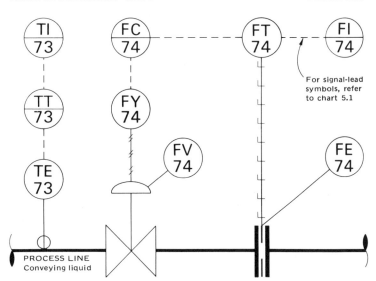

For signal-lead symbols, refer to chart 5.1

PROCESS LINE
Conveying liquid

SIGNAL LEADS 5.5.6

Elements, transmitters, recorders, indicators and controllers communicate with each other by means of signal leads — which are represented by lines on the drawing. The signal can be a voltage, the pressure of a fluid, etc.—these are the most common signals.

Symbols for instrument signal leads are given in chart 5.1.

INSTRUMENTATION CODING : ISA CODING   TABLE 5.3

PROCESS VARIABLE		TYPE OF INSTRUMENT	
ANALYSIS	A	ALARM	A
BURNER (Flame)	B	USER'S CHOICE	B
COMBUSTION	B	CONTROLLER	C
USER'S CHOICE	C	CONTROL VALVE	CV
USER'S CHOICE	D	TRAP	CV
VOLTAGE	E	SENSOR (Primary Element)	E
FLOW RATE	F	RUPTURE DISC	E
USER'S CHOICE	G	SIGHT or GAGE GLASS	G
CURRENT (Electric)	I	TELEVISION MONITOR	G
POWER	J	INDICATOR	I
TIME (Time Control/Clock)	K	CONTROL STATION	K
LEVEL	L	LIGHT (Pilot/Operation)	L
USER'S CHOICE	M	USER'S CHOICE	N
USER'S CHOICE	N	FLOW RESTRICTION ORIFICE	O
USER'S CHOICE	O	TEST POINT (Sample Point)	P
PRESSURE/VACUUM	P	RECORDER	R
RADIATION	R	SWITCH	S
SPEED (or Frequency)	S	TRANSMITTER	T
TEMPERATURE	T	MULTIFUNCTION	U
MULTIVARIABLE	U	VALVE/DAMPER	V
VIBRATION	V	WELL	W
WEIGHT (or Force)	W	UNCLASSIFIED	X
UNCLASSIFIED	X	RELAY	Y
EVENT (Response to)	Y	DRIVER	Z
POSITION, DIMENSION	Z	ACTUATOR	Z

QUALIFYING LETTER AFTER THE 'PROCESS VARIABLE' LETTER		
		THE QUALIFYING LETTER IS USED:—
DIFFERENTIAL	D	When the difference between two values of the process variable is involved
TOTAL	Q	When the process variable is to be summed over a period of time. For example, flow rate can be summed to give total volume
RATIO	F	When the ratio of two values of the process variable is involved
SAFETY ITEM	S	To denote an item such as a relief valve or rupture disc
'HAND'	H	To denote a hand-operated or hand-started item

QUALIFYING LETTER AFTER THE 'TYPE OF INSTRUMENT' LETTER		
HIGH	H	To denote instrument action on 'high' set value of the process variable
INTERMEDIATE	M	To denote instrument action on 'intermediate' set value of the process variable
LOW	L	To denote instrument action on 'low' set value of the process variable

LISTING PIPING MATERIEL ON DRAWINGS   5.6

In the engineering construction industry, it is usual for piping components to be given a code number which appears in the piping specification. In companies not primarily engaged in plant construction, materiel is frequently listed on drawings.

DIFFERENT FORMS OF LIST   5.6.1

This list is usually titled 'list of material', or preferably, 'list of materiel', as items of hardware are referred to. 'Parts list' and 'Bill of materiel' are alternate headings.

Either a separate list can be made for materiel on several drawings, or each drawing sheet can include a list for items on the particular drawing. Lists on drawings are written in the space above the title block. Column headings normally used for the list are:

LIST OF MATERIEL			
ITEM NUMBER	QUANTITY	DESCRIPTION	REMARK, REQUISITION NUMBER, OR COMPANY CODE

SUGGESTED LISTING SCHEME   5.6.2

Vessels, pumps, machinery and instruments are normally listed separately from piping hardware. However, it is not uncommon, on small projects or revamp work, to list all materiel on a drawing.

CLASSIFICATION FOR PIPING COMPONENTS   CHART 5.11

CLASS	INTENDED DUTY OF HARDWARE WITH RESPECT TO FLUID		EXAMPLE HARDWARE
I	CONVEYANCE: *To provide a path for fluid flow*		Pipe, fittings, ordinary flanges, bolt and gasket sets
II	FLOW CONTROL: *To produce a large change in flow rate or pressure of fluid*	(A) Non-powered	In-line valve, orifice plate, venturi
		(B) Powered	Pump, ejector
III	SEPARATION: *To remove material by mechanical means from the fluid*		Steam trap, discharge valve, safety or relief valve, screen, strainer
IV	HEATING OR COOLING: *To change the temperature of the fluid by adding or removing heat*		Jacketed pipe, tracer
V	MEASUREMENT: *To measure a variable of the fluid, such as flow rate, temperature, pressure, density, viscosity, turbidity, color*		Gages (all types), thermometers (all types), flow meter, densitometer, sensor housing (such as a thermowell) and other special fittings for instruments
VI	NONE: *Ancillary hardware*		Insulation, reinforcement, hanger, support

Haphazard listing of items makes reference troublesome. The scheme suggested in chart 5.11 is based on the duty of the hardware and can be extended to listing equipment if desired. Items of higher pressure rating and larger size can be listed first within each class.

LISTING SPECIFIC ITEMS   5.6.3

Under the heading DESCRIPTION, often on drawings the size of the item is stated first. A typical order is: SIZE (NPS), RATING (class, schedule number, etc.), NAME (of item), MATERIAL (ASTM or other material specification), and FEATURE (design feature).

Descriptions are best headed by the NAME of the item, followed by the SIZE, RATING, FEATURE(S), and MATERIAL. As material listings are commonly handled by data-processing equipment, beginning the description of an item by name is of assistance in handling the data. The description for 'pipe' is detailed.

EXAMPLE LISTING FOR PIPE

- NAME: State 'PIPE'

- SIZE: Specify nominal pipe size. See 2.1.3 and tables P-1

- RATING: Specify wall thickness as either a schedule number, a manufacturers' weight, etc. See tables P-1. SCH=schedule, STD= standard, XS= extra-strong, XXS= double-extra-strong, API= American Petroleum Institute.

- FEATURE: Specify design feature(s) unless covered by a pipe specification for the project.

    Pipe is available seamless or with a welded seam—examples of designations are: SMLS = seamless, FBW = furnace-butt-welded, ERW = electric-resistance-welded GALV = galvanized. Specify ends: T&C = threaded and coupled, BE = beveled end, PE = plain end.

- MATERIAL: Carbon-steel pipe is often ordered to ASTM A53 or A106, Grade A or B. Other specifications are given in tables 7.5 and 2.1.

POINTS TO CHECK WHEN MAKING THE LIST   5.6.4

- See that all items in the list have been given a sequential item number

- Label the items appearing on the piping drawings with the item number from the list. Write the item number in a circle with a fine line or arrow pointing to the item on the drawing. Each item in the list of materiel is indicated in this way once on the plan or elevational piping drawings

- Verify that all data on the list agree with:
  (1) Requirements set out in piping drawings
  (2) Available hardware in the manufacturers' catalogs

DESIGN OF PIPING SYSTEMS :
Including Arrangement, Supporting, Insulation, Heating, Venting and Draining of Piping, Vessels and Equipment

6 .1
.1.1

ARRANGING PIPING   **6.1**

GUIDELINES & NOTES   **6.1.1**

Simple arrangements and short lines minimize pressure drops and lower pumping costs.

Designing piping so that the arrangement is 'flexible' reduces stresses due to mechanical or thermal movement—refer to figure 6.1 and 'Stresses on piping', this section.

Inside buildings, piping is usually arranged parallel to building steelwork to simplify supporting and improve appearance.

Outside buildings, piping can be arranged: (1) On pipracks. (2) Near grade on sleepers. (3) In trenches. (4) Vertically against steelwork or large items of equipment.

PIPING ARRANGEMENT

- Use standard available items wherever possible

- Do not use miters unless directed to do so

- Do not run piping under foundations. (Pipes may be run under grade beams)

- Piping may have to go thru concrete floors or walls. Establish these points of penetration as early as possible and inform the group concerned (architectural or civil) to avoid cutting existing reinforcing bars

- Preferably lay piping such as lines to outside storage, loading and receiving facilities, at grade on pipe sleepers (see figure 6.3) if there is no possibility of future roads or site development

- Avoid burying steam lines that pocket, due to the difficulty of collecting condensate. Steam lines may be run below grade in trenches provided with covers or (for short runs) in sleeves

- Lines that are usually buried include drains and lines bringing in water or gas. Where long cold winters freeze the soil, burying lines below the frost line may avoid the freezing of water and solutions, saving the expense of tracing long horizontal parts of the lines

- Include removable flanged spools to aid maintenance, especially at pumps, turbines, and other equipment that will have to be removed for overhaul

- Take gas and vapor branch lines from tops of headers where it is necessary to reduce the chance of drawing off condensate (if present) or sediment which may damage rotating equipment

- Avoid pocketing lines—arrange piping so that lines drain back into equipment or into lines that can be drained

- Vent all high points and drain all low points on lines — see figure 6.47. Indicate vents and drains using symbols in chart 5.7. Carefully-placed drains and valved vents permit lines to be easily drained or purged during shutdown periods: this is especially important in freezing climates and can reduce winterizing costs

ARRANGE FOR SUPPORTING

- Group lines in pipeways, where practicable
- Support piping from overhead, in preference to underneath
- Run piping beneath platforms, rather than over them

REMOVING EQUIPMENT & CLEANING LINES

- Provide union- and flanged joints as necessary, and in addition use crosses instead of elbows, to permit removing material that may solidify

CHART
5.11

CLEARANCES & ACCESS

- Route piping to obtain adequate clearance for maintaining and removing equipment

- Locate within reach, or make accessible, all equipment subject to periodic operation or inspection — with special reference to check valves, pressure relief valves, traps, strainers and instruments

- Take care to not obstruct access ways — doorways, escape panels, truckways, walkways, lifting wells, etc.

- Position equipment with adequate clearance for operation and maintenance. Clearances often adopted are given in table 6.1. In some circumstances, these clearances may be inadequate—for example, with shell-and-tube heat exchangers, space must be provided to permit withdrawal of the tubes from the shell

CLEARANCES & DIMENSIONS — TABLE 6.1

MINIMUM CLEARANCES

HORIZONTAL CLEARANCES:	Operating space around equipment †	2ft	6in.
	Centerline of railroad to nearest obstruction: (1) Straight track	8ft	6in.
	(2) Curved track	9ft	6in.
	Manhole to railing or obstruction	3ft	0in.
VERTICAL CLEARANCES:	Over walkway, platform, or operating area	6ft	6in.
	Over stairway	7ft	0in.
	Over high point of plant roadway: (1) Minor roadway	17ft	0in.
	(2) Major roadway	20ft	0in.
	Over railroad from top of rail	22ft	6in.

MINIMUM HORIZONTAL DIMENSIONS

Width of walkway at floor level	3ft	0in.
Width of elevated walkway or stairway	2ft	6in.
Width of rung of fixed ladder *See chart P-2.*		16in.
Width of way for forklift truck	8ft	0in.

VERTICAL DIMENSIONS

Railing. Top of floor, platform, or stair, to: (1) Lower rail		1ft	9in.
(2) Upper rail		3ft	6in.
Manhole centerline to floor		3ft	0in.
Valves:	*See table 6.2 and chart P-2.*		

†Equipment such as heat exchangers, compressors and turbines will require additional clearance. Check manufacturers' drawings to determine particular space requirements. Refer to figure 6.33 and table 6.5 for spacing heat exchangers.

- Ensure very hot lines are not run adjacent to lines carrying temperature sensitive fluids, or elsewhere, where heat might be undesirable

- Establish sufficient headroom for ductwork, essential electrical runs, and at least two elevations for pipe run north—south and east—west (based on clearance of largest lines, steelwork, ductwork, etc.—see figure 6.49)

- Elevations of lines are usually changed when changing horizontal direction where lines are grouped together or are in a congested area, so as not to block space where future lines may have to be routed

- Stagger flanges, with 12-inch minimum clearance from supporting steel

- Keep field welds and other joints at least 3 inches from supporting steel, building siding or other obstruction. Allow room for the joint to be made

- Allow room for loops and other pipe arrangements to cope with expansion by early consultation with staff concerned with pipe stressing. Notify the structural group of any additional steel required to support such loops

THERMAL MOVEMENT

Maximum and minimum lengths of a pipe run will correspond to the temperature extremes to which it is subjected. The amount of expansion or shrinkage in steel per degree change in temperature ('coefficient of expansion') is approximately the same — that is, the expansion from 40F to 41F is about the same as from 132 F to 133 F, or from 179 F to 180 F, etc. Chart 6.1 gives changes in line length for changes in temperature.

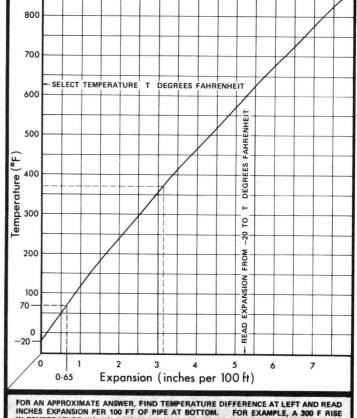

EXPANSION OF CARBON-STEEL PIPE — **CHART 6.1**

FOR AN APPROXIMATE ANSWER, FIND TEMPERATURE DIFFERENCE AT LEFT AND READ INCHES EXPANSION PER 100 FT OF PIPE AT BOTTOM. FOR EXAMPLE, A 300 F RISE IN TEMPERATURE WOULD GIVE EXPANSION PER 100 FT AS 2.5 INCHES. (AN ACCURATE READING FROM 70 F TO 370 F IS 3.15 − 0.65 = 2.40 INCHES.)

STRESSES ON PIPING

THERMAL STRESSES Changes in temperature of piping, due either to change in temperature of the environment or of the conveyed fluid, cause changes in length of the piping. This expansion or contraction in turn causes strains in piping, supports and attached equipment.

SETTLEMENT STRAINS Foundations of large tanks and heavy equipment may settle or tilt slightly in the course of time. Connected piping and equipment not on a common foundation will be stressed by the displacement unless the piping is arranged in a configuration flexible enough to accommodate multiple-plane movement. This problem should not arise in new construction but could occur in a modification to a plant unit or process.

FLEXIBILITY IN PIPING

To reduce strains in piping caused by substantial thermal movement, flexible and expansion joints may be used. However, the use of these joints may be minimized by arranging piping in a flexible manner, as illustrated in figure 6.1. Pipe can flex in a direction perpendicular to its length: thus, the longer an offset, or the deeper a loop, the more flexibility is gained.

COLD SPRING

Cold springing of lines should be avoided if an alternate method can be used. A line may be cold sprung to reduce the amplitude of movement from thermal expansion or contraction in order: (a) To reduce stress on connections. (b) To avoid an interference.

Figure 6.2 schematically illustrates the use of cold springing for both purposes. Cold springing in example (a) consists of making the branch in the indicated cold position, which divides thermal movement between the cold and hot positions. In example (b) the cold spring is made equal to the thermal movement.

COLD SPRINGING   **FIGURE 6.2**

(a) TO REDUCE STRESS

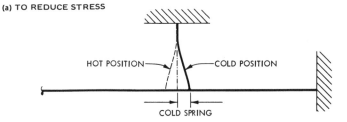

(b) TO AVOID AN INTERFERENCE

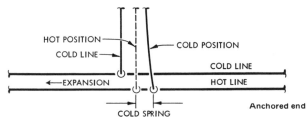

FLEXIBILITY — FIGURE 6.1

RIGID	FLEXIBLE	
support		In current practice, loops are made from straight pipe and elbows in nearly all circumstances. The legs perpendicular to the run give flexibility.
		On piperacks, arrange lines subject to thermal movement to one side with larger lines outermost so that larger loops can be provided for them, and so that all loops can be made over the piperack to save space.
		Offsetting the run gives flexibility which increases with the length of the offset
		Rigid connections between tanks and vessels or other connected equipment are to be avoided if: (1) There is likely to be large changes in temperature due to the process or to climate (2) The tanks or equipment are placed in the open on separate foundations which are liable to settle
		In both arrangements, the pump is used to circulate liquid in the tanks or vessels. The flexible arrangement reduces stresses on nozzles and also permits access between the units.
SOME FLEXIBILITY	MORE FLEXIBILITY	In turning corners, an offset limb gives a limited flexibility to the piping. The longer the offset, the greater the flexibility.
		The extra limb in the more flexible arrangement allows greater thermal movement between branch and run.
		These are two arrangements using a loop at a corner. Greater flexibility is gained by making one of the runs form one limb of the loop—this arrangement also saves an elbow and two welds.

CHART
6.1

FIGURES
6.1—6.2

TABLE
6.1

In the following example, cold springing is employed solely to reduce a stress:

A long pipe connected by a 90-degree elbow and flange to a nozzle may on heating expand so that it imposes a load on the nozzle in excess of that recommended. Assume that piping to the nozzle has been installed at ambient temperature, and that the pipe expands 0.75 inch when hot material flows thru it, putting a lateral (sideways) load of 600 lb on the nozzle.

If the pipe had 0.375 inch of its length removed before connection, the room-temperature lateral load on the nozzle would be about 300 lb (instead of zero), and the hot load would be reduced to about 300 lb.

The fraction of the expansion taken up can be varied. A cold spring of 50% of the expansion between the temperature extremes gives the most benefit in reducing stress. Cold springing is not recommended if an alternate solution can be used. Refer to the Code for Pressure Piping ANSI B31 and to table 7.2.

RESISTANCE OF PIPING TO FLOW

All piping has resistance to flow. The smaller the flow cross section and the more abrupt the change in direction of flow, the greater is the resistance and loss of pressure. For a particular line size the resistance is proportional to the length of pipe, and the resistance of fittings, valves, etc. may be expressed as a length of pipe having the same resistance to flow. Table F-10 gives such equivalent lengths of pipe for fittings, valves, etc.

Table F-11 gives pressure drops for water flowing thru SCH 40 pipe at various rates. Charts to determine the economic size (NPS) of piping are given in the Chemical Engineer's Handbook and other sources.

SLIDERULE FOR FLOW PROBLEMS

Problems of resistance to flow can be quickly solved with the aid of the slide-rule calculator obtainable from Tube Turns Division of Chemetron Corporation, PO Box 32160, Louisville, KY 40232.

PIPERACKS   6.1.2

A 'pipeway' is the space allocated for routing several parallel adjacent lines. A 'piperack' is a structure in the pipeway for carrying pipes and is usually fabricated from steel, or concrete and steel, consisting of connected ⊓-shaped frames termed 'bents' on top of which the pipes rest. The vertical members of the bents are termed 'stanchions'. Figure 6.3 shows two piperacks using this form of construction, one of which is 'double-decked'. Piperacks for only two or three pipes are made from 'T'-shaped members, termed 'tee-head supports'.

Piperacks are expensive, but are necessary for arranging the main process and service lines around the plant site. They are made use of in secondary ways, principally to provide a protected location for ancillary equipment.

Pumps, utility stations, manifolds, fire-fighting and first-aid stations can be located under the piperack. Lighting and other fixtures can be fitted to stanchions. Air-cooled heat exchangers can be supported above the piperack.

The smallest size of pipe run on a piperack without additional support is usually 2 inch. It may be more economic to change proposed small lines to 2-inch pipe, or to suspend them from 4-inch or larger lines, instead of providing additional support.

Table S-1 and charts S-2 give stress and support data for spans of horizontal pipe.

KEY FOR FIGURE 6.3

(1) WHEN USING A DOUBLE DECK, IT IS CONVENTIONAL TO PLACE UTILITY AND SERVICE PIPING ON THE UPPER LEVEL OF THE PIPERACK

(2) DO NOT RUN PIPING OVER STANCHIONS AS THIS WILL PREVENT ADDING ANOTHER DECK

(3) PLACE LARGE LIQUID-FILLED PIPES NEAR STANCHIONS TO REDUCE STRESS ON HORIZONTAL MEMBERS OF BENTS. HEAVY LIQUID-FILLED PIPES (12-in AND LARGER) ARE MORE ECONOMICALLY RUN AT GRADE—SEE NOTE (12)

(4) PROVIDE DISTRIBUTED SPACE FOR FUTURE PIPES—APPROXIMATELY AN ADDITIONAL 25 PERCENT (THAT IS, 20 PERCENT OF FINAL WIDTH—SEE TABLES A-1)

(5) HOT PIPES ARE USUALLY INSULATED AND MOUNTED ON SHOES

(6) WARM PIPES MAY HAVE INSULATION LOCALLY REMOVED AT SUPPORTS

(7) THE HEIGHT OF A RELIEF HEADER IS FIXED BY ITS POINT OF ORIGIN AND THE SLOPE REQUIRED TO DRAIN THE LINE TO A TANK, Etc.

(8) ELECTRICAL AND INSTRUMENT TRAYS (FOR CONDUIT AND CABLES) ARE BEST PLACED ON OUTRIGGERS OR BRACKETS AS SHOWN, TO PRESENT THE LEAST PROBLEM WITH PIPES LEAVING THE PIPEWAY. ALTERNATELY, TRAYS MAY BE ATTACHED TO THE STANCHIONS

(9) WHEN CHANGE IN DIRECTION OF A HORIZONTAL LINE IS MADE, IT IS BEST ALSO TO MAKE A CHANGE OF ELEVATION (EITHER UP OR DOWN). THIS AVOIDS BLOCKING SPACE FOR FUTURE LINES. 90-DEGREE CHANGES IN DIRECTION OF THE WHOLE PIPEWAY OFFER THE OPPORTUNITY TO CHANGE THE ORDER OF LINES. A SINGLE DECK IS SHOWN AT AN INTERMEDIATE ELEVATION

(10) SOMETIMES INTERFACES ARE ESTABLISHED TO DEFINE BREAKPOINTS FOR CONTRACTED WORK (WHERE ONE CONTRACTOR'S PIPING HAS TO JOIN WITH ANOTHERS). AN INTERFACE IS AN IMAGINARY PLANE WHICH MAY BE ESTABLISHED FAR ENOUGH FROM A WALL, SIDING, PROCESS UNIT, OR STORAGE UNIT TO ENABLE CONNECTIONS TO BE MADE

(11) PIPES SHOULD BE RACKED ON A SINGLE DECK IF SPACE PERMITS

(12) PIPING SHOULD BE SUPPORTED ON SLEEPERS AT GRADE IF ROADS, WALKWAYS, Etc., WILL NOT BE REQUIRED OVER THE PIPEWAY AT A LATER DATE. PIPING 'AT GRADE' SHOULD BE 12 INCHES OR MORE ABOVE GRADE

(13) CURRENT PRACTICE IS TO SPACE BENTS 20–25 FEET APART. THIS SPACING IS A COMPROMISE BETWEEN THE ACCEPTABLE DEFLECTIONS OF THE SMALLER PIPES AND THE MOST ECONOMIC BEAM SECTION DESIRED FOR THE PIPERACK. PIPERACKS ARE USUALLY NOT OVER 25 FEET IN WIDTH. IF MORE ROOM IS NEEDED, THE PIPERACK IS DOUBLE- OR TRIPLE-DECKED

(14) MINIMUM CLEARANCE UNDERNEATH THE PIPERACK IS DETERMINED BY AVAILABLE MOBILE LIFTING EQUIPMENT REQUIRING ACCESS UNDER THE PIPERACK. VERTICAL CLEARANCES SHOULD BE AS SET OUT IN TABLE 6.1, BUT CANNOT NECESSARILY BE ADHERED TO AS ELEVATIONS OF PIPES AT INTERFACES ARE SOMETIMES FIXED BY PLANT SUBCONTRACTORS. IF THIS SITUATION ARISES, THE PIPING GROUP SHOULD ESTABLISH MAXIMUM AND MINIMUM ELEVATIONS WHICH THE PIPING SUBCONTRACTORS MUST WORK TO—THIS HELPS TO AVOID PROBLEMS AT A LATER DATE. CHECK THE MINIMUM HEIGHT REQUIRED FOR ACCESS WHERE THE PIPERACK RUNS PAST A UNIT OR PLANT ENTRANCE

(15) WHEN SETTING ELEVATIONS FOR THE PIPERACK, TRY TO AVOID POCKETS IN THE PIPING. LINES SHOULD BE ABLE TO DRAIN INTO EQUIPMENT OR LINES THAT CAN BE DRAINED

(16) GROUP HOT LINES REQUIRING EXPANSION LOOPS AT ONE SIDE OF THE PIPERACK FOR EASE OF SUPPORT—SEE FIGURE 6.1

(17) LOCATE UTILITY STATIONS, CONTROL (VALVE) STATIONS, AND FIREHOSE POINTS ADJACENT TO STANCHIONS FOR SUPPORTING

(18) LEAVE SPACE FOR DOWNCOMERS TO PUMPS, Etc., BETWEEN PIPERACK AND ADJACENT BUILDING OR STRUCTURE

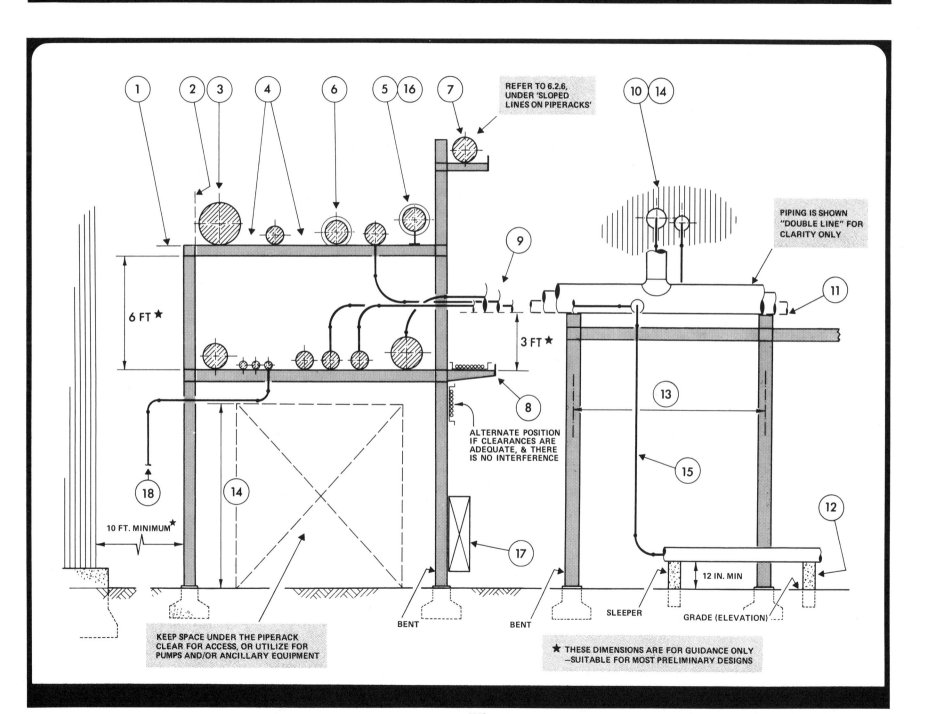

REFER TO 6.2.6, UNDER 'SLOPED LINES ON PIPERACKS'

PIPING IS SHOWN "DOUBLE LINE" FOR CLARITY ONLY

6 FT ★

3 FT ★

ALTERNATE POSITION IF CLEARANCES ARE ADEQUATE, & THERE IS NO INTERFERENCE

10 FT. MINIMUM ★

12 IN. MIN

KEEP SPACE UNDER THE PIPERACK CLEAR FOR ACCESS, OR UTILIZE FOR PUMPS AND/OR ANCILLARY EQUIPMENT

BENT

BENT

SLEEPER

GRADE (ELEVATION)

★ THESE DIMENSIONS ARE FOR GUIDANCE ONLY —SUITABLE FOR MOST PRELIMINARY DESIGNS

FIGURE 6.3

Valves are used for these purposes:

(1)   Process control during operation

(2)   Controlling services and utilities—steam, water, air, gas and oil

(3)   Isolating equipment or instruments, for maintenance

(4)   Discharging gas, vapor or liquid

(5)   Draining piping and equipment on shutdown

(6)   Emergency shutdown in the event of plant mishap or fire

WHICH SIZE VALVE TO USE ?

Nearly all valves will be line size — one exception is control valves, which are usually one or two sizes smaller than line size; never larger.

At control stations and pumps it has been almost traditional to use line-size isolating valves. However, some companies are now using isolating valves at control stations the same size as the control valve, and at pumps are using 'pump size' isolating valves at suction and discharge. The choice is usually an economic one made by a project engineer.

The sizes of bypass valves for control stations are given in 6.1.4, under 'Control (valve) stations'.

WHERE TO PLACE VALVES

See 6.3.1 for valving pumps, under 'Pump emplacement & connections'.

● Preferably, place valves in lines from headers (on piperacks) in horizontal rather than vertical runs, so that lines can drain when the valves are closed. (In cold climates, water held in lines may freeze and rupture the piping: such lines should be traced — see 6.8.2)

● To avoid spooling unnecessary lengths of pipe, mount valves directly onto flanged equipment, if the flange is correctly pressure-rated. See 6.5.1 under 'Nozzle loading'

● A relief valve that discharges into a header should be placed higher than the header in order to drain into it

● Locate heavy valves near suitable support points. Flanges should be not closer than 12 inches to the nearest support, so that installation is not hampered

● For appearance, if practicable, keep centerlines of valves at the same height above floor, and in-line on plan view

OPERATING ACCESS TO VALVES

● Consider frequency of operation when locating manually-operated valves

● Locate frequently-operated valves so they are accessible to an operator from grade or platform. Above this height and up to 20 ft, use chain operators or extension stem. Over 20 ft, consider a platform or remote operation

VALVE OPERATING HEIGHTS *   TABLE 6.2

ORDER OF PREFERENCE FOR VALVE LOCATION	STEM CENTERLINE ELEVATION FOR HORIZONTAL VALVES		HANDWHEEL ELEVATION FOR VERTICAL VALVES	MINIMUM ELEVATION OF HANDWHEEL RIM FOR TILTED VALVES (handwheel overhead)	
	OPERATING	MAINTENANCE	(upright, closed)	ANGLE OF STEM FROM VERTICAL	MINIMUM ELEVATION
1st	3'-6" to 4'-6"	3'-6" to 4'-6"	3'-9" to 4'-3"		
2nd †	2'-0" to 3'-6"	1'-0" to 3'-6"	2'-0" to 3'-9"		
3rd † (HEAD HAZARD)	4'-6" to 6'-6"+ ½ handwheel diameter	4'-6" to 7'-9"		30°	5'-0"
				45°	5'-6"
				60°	6'-0"
ACCEPTABLE FOR 1-INCH AND SMALLER VALVES	0'-6" to 2'-0" and 6'-9" to 7'-6"				

```
* REFER TO CHART P-2 IN PART II
† TO MINIMIZE HAZARD TO PERSONNEL IF VALVES ARE TO BE LOCATED AT HEIGHTS
  WITHIN 2nd AND 3rd CHOICES, AVOID POINTING STEMS INTO WALKWAYS AND WORKING
  AREAS. TRY TO PLACE VALVES CLOSE TO WALLS OR LARGE ITEMS WHICH ARE CLEARLY
  SEEN.
```

● Infrequently-used valves can be reached by a ladder—but consider alternatives

● Do not locate valves on piperacks, unless unavoidable

● Group valves which would be out of reach so that all can be operated by providing a platform, if automatic operators are not used

● If a chain is used on a horizontally-mounted valve, take the bottom of the loop to within 3 ft of floor level for safety, and provide a hook nearby to hold the chain out of the way —see 3.1.2, under 'Chain'

● Do not use chain operators on screwed valves, or on any valve 1½-inches and smaller

● With lines handling dangerous materials it is better to place valves at a suitably low level above grade, floor, platform, etc., so that the operator does not have to reach above head height

ACCESS TO VALVES IN HAZARDOUS AREAS

● Locate main isolating valves where they can be reached in an emergency such as an outbreak of fire or a plant mishap. Make sure that personnel will be able to reach valves easily by walkway or automobile

● Locate manually-operated valves at the plant perimeter, or outside the hazardous area

● Ensure that automatic operators and their control lines will be protected from the effects of fire

● Make use of brick or concrete walls as possible fire shields for valve stations

● Inside a plant, place isolating valves in accessible positions to shut feed lines for equipment and processes having a fire risk

● Consider the use of automatic valves in fire-fighting systems to release water, foam and other fire-fighting agents, responding to heat-fusible links, smoke detectors, etc., triggered by fire or undue rise in temperature —advice may be obtained from the insurer and the local fire department

MAKE MAINTENANCE SIMPLE

● Provide access for mobile lifting equipment to handle heavy valves

● Consider providing lifting davits for heavy valves difficult to move by other means, if access is restricted

● If possible, arrange valves so that supports will not be on removable spools:

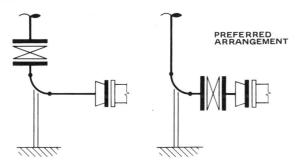

PREFERRED ARRANGEMENT

● A plug valve requiring lubrication must be easily accessible, even though it may not be frequently operated

MAKE MAINTENANCE SAFE

● Use line-blind valves, spectacle plates or the 'double block and bleed' where positive shutoff is required either for maintenance or process needs — see 2.7

ORIENTATION OF VALVE STEMS

● Do not point valve stems into walkways. truckways, ladder space, etc.

● Unless necessary, do not arrange gate and globe valves with their stems pointing downward (at *any* angle below the horizontal), as:—

    (1) Sediment may collect in the gland packing and score the stem.

    (2) A projecting stem may be a hazard to personnel.

● If an inverted position is necessary, consider employing a dripshield:

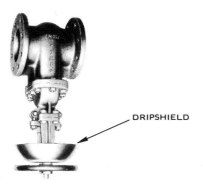

DRIPSHIELD

CLOSING DOWN LINES

Consider valve-closing time in shutting down or throttling large lines. Rapid closure of the valve requires rapid dissipation of the liquid's kinetic energy, with a risk of rupturing the line. Long-distance pipelines present an example of this problem.

A liquid line fitted with a fast-closing valve should be provided with a stand-pipe upstream and close to the valve to absorb the kinetic energy of the liquid. A standpipe is a closed vertical branch on a line: air or other gas is trapped in this branch to form a pneumatic cushion.

IF THERE IS NO P&ID .....

● Provide valves at headers, pumps, equipment, etc., to ensure that the system will be pressure-tight for hydrostatic testing, and to allow equipment to be removed for maintenance without shutting down the system

● Provide isolating valves in all small lines branching from headers—for example, see figure 6.12

● Provide isolating valves at all instrument pressure points for removal of instruments under operating conditions

● Provide valved drains on all tanks, vessels, etc., and other equipment which may contain or collect liquids

● Protect sensitive equipment by using a fast-closing check valve to stop backflow before it can gather momentum

● Consider butt-welding or ring-joint flanged valves for lines containing hazardous or 'searching' fluids. Hydrogen is especially liable to leak

● Consider seal welding screwed valves if used in hydrocarbon service —see chart 2.3 (inset sketch)

● Provide sufficient valves to control flows

● Consider providing a concrete pit (usually about 4 ft x 4 ft) for a valve which is to be located below grade

● Consider use of temporary closures for positive shutoff—see 2.7

● Provide a bypass if necessary for equipment which may be taken out of service

● Provide a bypass valve around control stations if continuous flow is required. See 6.1.4 and figure 6.6. The bypass should be at least as large as the control valve, and is usually globe type, unless 6-inch or larger, when a gate valve is normally used (see 3.1.4, under 'Gate valve')

● Provide an upstream isolating valve with a small valved bypass to equipment which may be subject to fracture if heat is too rapidly applied on opening the isolating valve. Typical use is in steam systems to lessen the risk of fracture of such things as castings, vitreous-lined vessels, etc.

● Consider providing large gate valves with a valved bypass to equalize pressure on either side of the disc to reduce effort needed to open the valve

TABLE 6.2

- Refer to 3.1.9 for valve orientation

- Extend safety-valve discharge risers that discharge to atmosphere at least 10 ft above the roof line or platform for safety. Support the vent pipe so as not to strain the valve or the piping to the valve. Pointing the discharge line upward (see figure 6.4) imposes less stress when the valve discharges than does the horizontal arrangement

- The downstream side of a safety valve should be unobstructed and involve the minimum of piping. The downstream side of a relief or safety-relief valve is piped to a relief header or knockout drum—see 6.11.3, under 'Venting gases', and 6.12, under 'Relieving pressure—liquids'

- Pipe exhausting to atmosphere is cut square, not at a slant as formerly done, as no real advantage is gained for the cost involved

- Normally, do not instal a valve upstream of a pressure-relief valve protecting a vessel or system from excessive pressure. However, if an isolating valve is used to facilitate maintenance of a pressure-relief valve, the isolating valve is 'locked open'—sometimes termed 'car sealed open' (CSO)

- In critical applications, two pressure-relief valves provided with isolating valves can be used

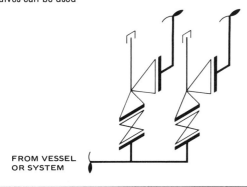

FROM VESSEL OR SYSTEM

The installation of pressure-relieving devices and the use of isolating valves in lines to and from such devices is governed by the Code for Pressure Piping, ANSI B31 and the ASME Boiler and Pressure Vessel Code.

INSTALLING BUTTERFLY VALVES

- Ensure that the disc has room to rotate when the valve is installed, as the disc enters the piping in the open position

- Place butterfly valves with integral gaskets between welding-neck or socket-welding flanges—see 3.1.6, under 'Butterfly valve'. The usual method of welding a slip-on flange (see figure 2.7) will not give an adequate seal, unless the pipe is finished smooth with the face of the flange

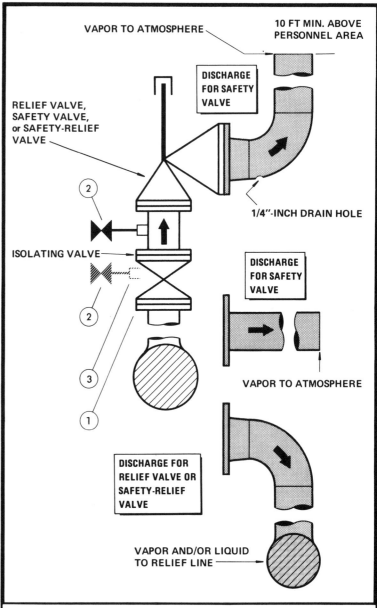

VAPOR TO ATMOSPHERE

10 FT MIN. ABOVE PERSONNEL AREA

DISCHARGE FOR SAFETY VALVE

RELIEF VALVE, SAFETY VALVE, or SAFETY-RELIEF VALVE

1/4"-INCH DRAIN HOLE

ISOLATING VALVE

DISCHARGE FOR SAFETY VALVE

VAPOR TO ATMOSPHERE

DISCHARGE FOR RELIEF VALVE OR SAFETY-RELIEF VALVE

VAPOR AND/OR LIQUID TO RELIEF LINE

KEY

(1) REFER TO 6.1.3 UNDER 'PIPING SAFETY AND RELIEF VALVES' REGARDING USE OF AN ISOLATING VALVE IN THIS POSITION

(2) IF AN ISOLATING VALVE IS PROVIDED, IT IS ALSO NECESSARY TO PROVIDE A BLEED VALVE TO RELIEVE PRESSURE BETWEEN THE ISOLATING VALVE AND THE PRESSURE RELIEF VALVE (FOR MAINTENANCE PURPOSES)

(3) IF A SPOOL BETWEEN THE TWO VALVES IS NOT USED, THE BLEED VALVE MAY BE PLACED AS SHOWN IF THE VALVE'S BODY CAN BE TAPPED

CONTROL (VALVE) STATIONS 6.1.4

A control station is an arrangement of piping in which a control valve is used to reduce and regulate the pressure or rate of flow of steam, gas, or liquid.

◆

Control stations should be designed so that the control valve can be isolated and removed for servicing. To facilitate this, the piping of the stations should be as flexible as circumstances permit. Figure 6.5 shows ways of permitting control valve removal in welded or screwed systems. Figure 6.6 shows the basic arrangement for control station piping.

The two isolating valves permit servicing of the control valve. The emergency bypass valve is used for manual regulation if the control valve is out of action.

The bypass valve is usually a globe valve of the same size and pressure rating as the control valve. For manual regulation in lines 6-inch and larger, a gate valve is often the more economic choice for the bypass line—refer to 3.1.4, under 'Gate valve'.

Figures 6.7–11 show other ways of arranging control stations—many more designs than these are possible. These illustrations are all schematic and can be adapted to both welded and screwed systems.

DESIGN POINTS

- For best control, place the control station close to the equipment it serves, and locate it at grade or operating platform level

- Provide a pressure-gage connection downstream of the station's valves. Depending on the operation of the plant, this connection may either be fitted with a permanent pressure indicating gage, or be used to attach a gage temporarily (for checking purposes)

- Preferably, do not 'sandwich' valves. Place at least one of the isolating valves in a vertical line so that a spool can be taken out allowing the control valve to be removed

- If the equipment and piping downstream of the station is of lower pressure rating than piping upstream, it may be necessary to protect the downstream system with a pressure-relief valve

- Provide a valved drain near to and upstream of the control valve. To save space, the drain is placed on the reducer. The drain valve allows pressure between the isolating valve(s) and control valve to be released. One drain is used if the control valve fails open, and two drains (one each side of the control valve) if the control valve fails closed

- Locate stations in rack piping at grade, next to a bent or column for easy supporting

DRAFTING THE STATION

In plan view, instead of drawing the valves, etc., the station is shown as a rectangle labeled 'SEE DETAIL "X" ' or 'DWG "Y"—DETAIL "X" ', if the elevational detail appears on another sheet. See chart 5.7.

UTILITY STATIONS 6.1.5

A utility station usually comprises three service lines carrying steam, compressed air and water. The steam line is normally ¾-inch minimum, and the other two services are usually carried in 1-inch lines. These services are for cleaning local equipment and hosing floors. (Firewater is taken from points fed from an independent water supply .)

The steam line is fitted with a globe valve and the air and water lines with gate valves. All are terminated with hose connections about 3½ ft above floor or grade. A utility station should be located at some convenient steel column for supporting, and all areas it is to serve should be reachable with a 50-ft hose.

Most companies have a standard design for a utility station. Figure 6.12 shows a design for a standard station which can be copied onto one of the design drawings for reference, or otherwise supplied with the drawings to the erecting contractor who usually runs the necessary lines. A notation used on plan views to indicate the station and services required is:

SERVICES:	STEAM, AIR, WATER	AIR, WATER	STEAM, WATER	STEAM, AIR
STATION SYMBOL:	S A W	A W	S W	S A

UTILITY STATION **FIGURE 6.12**

KEY to figure 6.12:

(1) GATE VALVE NPS 1
(2) GLOBE VALVE NPS 1
(3) GLOBE VALVE NPS 3/4
(4) HOSE COUPLING NPS 3/4
(5) HOSE COUPLING NPS 1
(6) PIPE NPS 1 SCH 80
(7) PIPE NPS 3/4 SCH 80
(8) TRAP (optional)

If subject to freezing conditions, utility station steam lines are usually trapped (otherwise, the trap can be omitted). Water is sometimes run underground in cold climates using an additional underground cock or plug valve with an extended key for operating, and a self-draining valve at the base of the riser. Another method to prevent freezing, is to run the water and steam lines in a common insulation.

SCHEMATIC CONTROL STATION ARRANGEMENTS

PIPING FITTINGS ALLOWING CONTROL VALVE REMOVAL

FLANGED CONTROL VALVES

THREADED CONTROL VALVES

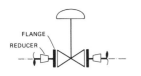

FIGURE 6.5

FLANGE
REDUCER

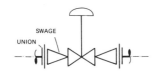

SWAGE
UNION

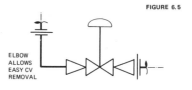

ELBOW
ALLOWS
EASY CV
REMOVAL

BASIC ARRANGEMENT

FIGURE 6.6

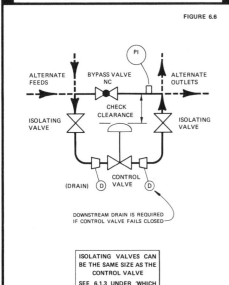

ALTERNATE
FEEDS

BYPASS VALVE
NC

ALTERNATE
OUTLETS

CHECK
CLEARANCE

ISOLATING
VALVE

ISOLATING
VALVE

(DRAIN)

CONTROL
VALVE

DOWNSTREAM DRAIN IS REQUIRED
IF CONTROL VALVE FAILS CLOSED

> ISOLATING VALVES CAN
> BE THE SAME SIZE AS THE
> CONTROL VALVE
>
> SEE 6.1.3 UNDER 'WHICH
> SIZE VALVE TO USE'

ARRANGEMENTS FOR ANGLE CV's

FIGURE 6.7

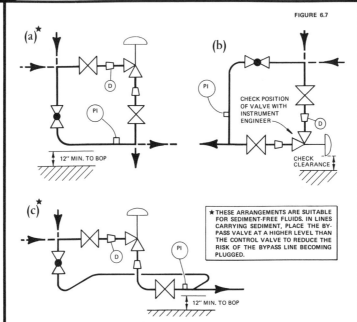

(a)★

(b)

CHECK POSITION
OF VALVE WITH
INSTRUMENT
ENGINEER

12" MIN. TO BOP

CHECK
CLEARANCE

(c)★

★ THESE ARRANGEMENTS ARE SUITABLE
FOR SEDIMENT-FREE FLUIDS. IN LINES
CARRYING SEDIMENT, PLACE THE BY-
PASS VALVE AT A HIGHER LEVEL THAN
THE CONTROL VALVE TO REDUCE THE
RISK OF THE BYPASS LINE BECOMING
PLUGGED.

12" MIN. TO BOP

STATIONS FOR LIQUIDS HARMFUL TO PERSONNEL

FIGURE 6.8

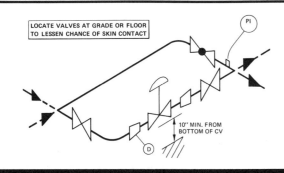

LOCATE VALVES AT GRADE OR FLOOR
TO LESSEN CHANCE OF SKIN CONTACT

10" MIN. FROM
BOTTOM OF CV

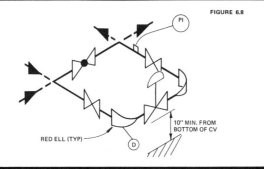

RED ELL (TYP)

10" MIN. FROM
BOTTOM OF CV

STEAM STATIONS

STATION SUITABLE FOR TURBINE & OTHER STEAM USERS   FIGURE 6.9

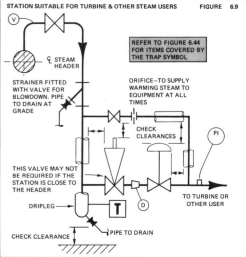

REFER TO FIGURE 6.44
FOR ITEMS COVERED BY
THE TRAP SYMBOL

STEAM
HEADER

STRAINER FITTED
WITH VALVE FOR
BLOWDOWN. PIPE
TO DRAIN AT
GRADE

ORIFICE—TO SUPPLY
WARMING STEAM TO
EQUIPMENT AT ALL
TIMES

CHECK
CLEARANCES

THIS VALVE MAY NOT
BE REQUIRED IF THE
STATION IS CLOSE TO
THE HEADER

DRIPLEG

TO TURBINE OR
OTHER USER

CHECK CLEARANCE

PIPE TO DRAIN

CONTINUOUSLY-OPERATING STATION SUITABLE FOR   FIGURE 6.10
ALL CONDITIONS INCLUDING FREEZING

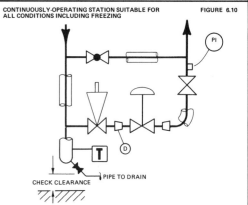

CHECK CLEARANCE

PIPE TO DRAIN

STATION FOR INTERMITTENT USE SUITABLE   FIGURE 6.11
FOR OPEN-AIR USE IN FREEZING CLIMATES

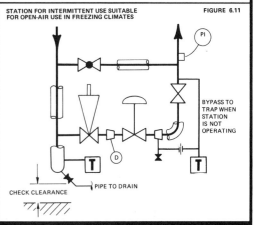

BYPASS TO
TRAP WHEN
STATION
IS NOT
OPERATING

CHECK CLEARANCE

PIPE TO DRAIN

ARRANGING SUPPORTS FOR PIPING 6.2

Pipe is held either from above by hangers or by supports of various types on which it rests. Hangers are also referred to as supports. Refer to 2.12 for typical hardware.

In the open, single pipes are usually routed so that they may be supported by fixtures to buildings or structures. A group of parallel pipes in the open is normally supported on a piperack—see 6.1.2.

Within a building, piping is routed primarily with regard to its process duty and secondarily with regard to existing structural steelwork, or to structural steel which may be conveniently added. Separate pipe-holding structures inside buildings are rare.

FUNCTIONS OF THE SYSTEM OF SUPPORT 6.2.1

The mechanical requirements of the piping support system are:

(1) To carry the weight of the piping filled with water (or other liquid involved) and insulation if used, with an ample safety margin — use a factor of 3 (= ratio of load just causing failure of support or hanger to actual load) or the safety factor specified for the project. External loading factors to be considered are the wind loads, the probable weight of ice buildup in cold climates, and seismic shock in some areas

(2) To ensure that the material from which the pipe is made is not stressed beyond a safe limit. In continuous runs of pipe, maximum tensile stress occurs in the pipe cross sections at the supports. Table S-1 gives spans for water-filled steel and aluminum pipe at the respective stress limits 4000 and 2000 psi. Charts S-2 give the maximum overhangs if a 3-ft riser is included in the span. The system of supports should minimize the introduction of twisting forces in the piping due to offset loads on the supports; the method of cantilevered sections set out in 6.2.4 substantially eliminates torsional forces

(3) To allow for draining. Holdup of liquid can occur due to pipes sagging between supports. Complete draining is ensured by making adjacent supports adequately tilt the pipe—see 6.2.6

(4) To permit thermal expansion and contraction of the piping—see 6.1.1, under 'Stresses on piping'

(5) To withstand and dampen vibrational forces applied to the piping by compressors, pumps, etc.

PIPING SUPPORT GROUP RESPONSIBILITIES 6.2.2

A large company will usually have a specialist piping support group responsible for designing and arranging supports. This group will note all required supports on the piping drawings (terminal job) and will add drawings of any special details.

The piping support group works in cooperation with a stress analysis group— or the two may be combined as one group—which investigates areas of stress due to thermal movement, vibration, etc., and makes recommendations to the piping group. The stress group should be supplied with preliminary layouts for this purpose by the piping group, as early as possible.

Supports for lines smaller than 2-inch and non-critical lines are often left to the 'field' to arrange, by noting 'FIELD SUPPORT' on the piping drawings.

LOADS ON SUPPORTS

Refer to tables P-1, which list the weights per foot of pipe and contained water (see 6.11.2). Weights of fittings, flanges, valves, bolts and insulation are given in tables W-1, compiled from suppliers' data.

ARRANGING POINTS OF SUPPORT 6.2.3

Pipe supports should be arranged bearing in mind all five points in 6.2.1. Inside buildings, it is usually necessary to arrange supports relative to existing structural steelwork, and this restricts choice of support points.

The method of support set out in 6.2.4 is ideal: In practice, some compromise may be necessary. The use of dummy legs and the addition of pieces of structural steel may be needed to obtain optimal support arrangements.

CALCULATING PREFERRED POINTS OF SUPPORT 6.2.4

Ideally, each point of support would be at the center of gravity of an associated length of piping. Carrying this scheme thru the entire piping system would substantially relieve the system from twisting forces, and supports would be only stressed vertically. A method of balancing sections of pipe at single support points is illustrated for a straight run of pipe in figure 6.13.

BALANCING SECTIONS OF PIPE   **FIGURE 6.13**

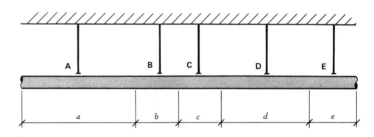

Consider hanger **B** associated with a length of pipe b. This length of pipe is supported by **B**, located at its center of gravity, which is at the midway point for a straight length of uniform pipe. Hangers **A**, **C**, **D** and **E** are likewise placed at the respective centers of gravity of lengths of pipe a, c, d and e. If any length of pipe is removed, the balance of the rest of the line would be unaffected. Each of the hangers must be designed to adequately support the load of the associated piping—see 6.2.1, point (1).

The presence of heavy flanges, valves, etc., in the piping will set the center of gravity away from the midpoint of the associated length. Calculation of support points and loadings is more quickly done using simple algebra. Answers may be found by making trial-and-error calculations, but this is much more tedious.

Correct location of piping supports can be determined by the use of 'moments of force'. Multiplying a force by the distance of its line of action from a point gives the 'moment' of the force about that point. A moment of force can be expressed in lb-ft (pounds weight times feet distance). The forces involved in support calculations either are the reactions at supports and nozzles, or are the downward-acting forces due to the weight of pipe, fittings, valves, etc.

In figure 6.14(a), the moment about the support of the two flanges is $(30 + 20)(16) = 800$ lb-ft, counter-clockwise. The moment of the 100-lb valve about the support is $(100)(8) = 800$ lb-ft, clockwise. As the lengths of pipe each side of the support are about the same, they may be omitted from the moment equation. The problem is simplified to balancing the valve and flanges.

USE OF MOMENTS   **FIGURE 6.14**

(a)

| 16ft | 8ft | 8ft |

30lb BLIND FLANGE | 20lb SO FLANGE   100-lb VALVE

SUPPORT

800 lb-ft   800 lb-ft

(b)

| 16ft | x ft | $(16-x)$ ft |

30 lb BLIND FLANGE | 20 lb SO FLANGE   120-lb VALVE

SUPPORT

800 lb-ft $(120)(x)$ lb-ft

Suppose it was required to balance this length of piping with a 120 lb valve on the right—where should the 120 lb valve be placed?

Referring to figure 6.14(b), if x represents the unknown distance of the 120 lb valve from the support, the piping section would be in balance if:

$$(50)(16) = (120)(x).$$

That is, if $x = (50)(16)/(120) = (800)/(120) = 6$ ft 8 in.

A more involved example follows:—

Figure 6.15 shows a length of 4-inch piping held by the hangers **F**, **G**, and **H**, and support **J**. The lengths of associated piping are shown by dashed separation lines. The weights of pipe and fittings are shown on the drawing. The 4-inch pipe is assumed to weigh 15 lb per foot of length. Welded elbows and tees are assumed to weigh the same as line pipe.

First consider the section associated with hanger **F**. The weight of pipe to the left of **F** is $(15)(20-x)$ lb, and as its center of gravity is at $(20-x)/(2)$ ft, its moment on the hanger is $(15)(20-x)^2/(2)$ lb-ft. The heavy valve and flanges are assumed to have their mass center 5 ft from the end, and their moment is $(x-5)(360)$ lb-ft. Ignoring the pipe 'replaced' by the valve, the weight of pipe to the right of **F** is $(15)(x)$ lb and its moment about **F** is $(15)(x)(x)/(2)$ lb-ft. As the associated length is in balance:

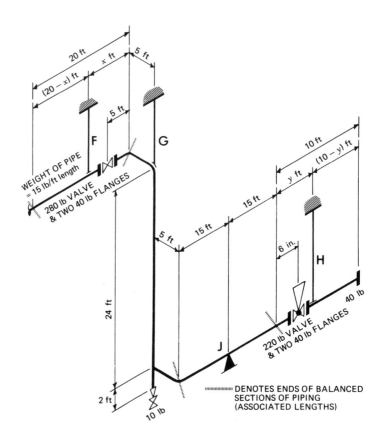

$$(15)(20-x)^2/(2) = (360)(x-5) + (15)(x^2)/(2)$$

$$x = (80)/(11), \text{ or about } 7 \text{ ft } 3 \text{ in.}$$

The x^2 terms canceled—this must be so, as there can physically be only one value for x. The load on hanger **F** is $(20)(15) + 360$ or 660 lb.

The support **J** should be at the center of the associated length of pipe, as already shown in figure 6.15, and the load on the support is $(30)(15)$, or 450 lb.

The hanger **G** is easily seen to be suitably placed, as there is 5 ft of 4-inch pipe overhanging each side. Only the load on the hanger need be calculated, which is $(5 + 5 + 24 + 2)(15) + (10)$, or 550 lb.

The location of hanger **H** has to be found by a calculation like that for hanger **F**, except that the heavy terminal flange has also to be taken into account. The moment equation in lb-ft is:

$$(300)(y-0.5) + (15)(y^2)/(2) = (15)(10-y)^2/(2) + (40)(10-y)$$

which gives y as nearly 2 ft 8 in.

The load on hanger **H** is about $(220)+(3)(40)+(15)(10)$, or 490 lb.

PROBLEM OF THE END

The supported length at one end of a run of piping may be cantilevered in the same way as the other lengths, and this has the advantage that if the piping terminates at a nozzle the load on the nozzle is minimal. However, it may be necessary to use or arrange a support at or near the end of a piping run. If the end of the run is vertical, the end support should be designed to carry the vertical run. The problem is usually more complex when the end of the run is horizontal.

The locations of fittings and support points will usually be already defined, and the problem is to calculate the reaction on the terminal support, and to see that the support is designed to withstand the load on it. In calculating the load on the terminal support, it should be made certain that the load *is* downward—with some arrangements, the piping would tend to raise itself off the terminal support (negative load) and if this type of arrangement is not changed, the terminal support will have to anchor the piping.

The sketch shows a horizontal end arrangement. Taking moments in lb-ft about the support **A**:

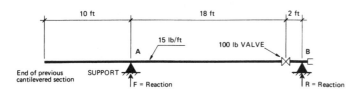

$$(15)(10)(\tfrac{1}{2})(10) = (15)(18+2)(\tfrac{1}{2})(18+2) + (100)(18) - (R)(18+2)$$

which gives $\qquad$ R = 202½ lb.

The reaction, F, on the support **A** can be calculated by taking moments about the support **B** or another axis, or more simply by equating vertical forces:

$$F + 202\tfrac{1}{2} = (15)(10+18+2) + 100 = 550, \text{ which gives } F = 347\tfrac{1}{2}\text{ lb.}$$

PROBLEM OF THE RISER

Supports for lines changing in direction can be calculated by the cantilever method. Sketch (a) below shows that the weight of the vertical part of the piping can be divided between two cantilevered sections in any proportion suited to the available support points. Sketches (b) and (c) show the vertical piping associated wholly with the left- or right-hand cantilevered sections. The piping may be supported by means of a dummy leg, if direct support is not practicable.

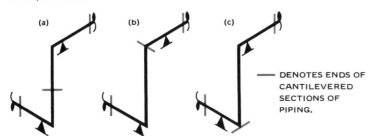

----- DENOTES ENDS OF CANTILEVERED SECTIONS OF PIPING.

GRAPHIC METHOD FOR FINDING LOADS ON SUPPORTS

The following graphical method permits quick calculation of bearing loads for 'corner' piping arrangements.

PROBLEM To find the load to be taken by a support to be placed at point 'E' in the piping arrangement shown:

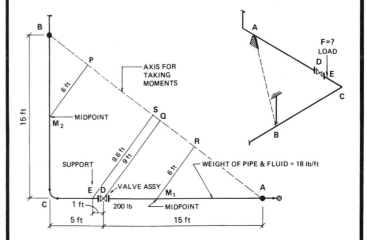

SOLUTION

[1] Draw the plan view to any convenient scale (as above)

[2] Add the axis line AB (this must pass thru points of support)

[3] Divide the run of piping into parts. Piping between the support points A and B is considered in three parts: (1) The valve. (2) The length of pipe BC. (3) The length of pipe AC—the short piece of line omitted for the valve is ignored, and the effect of the elbow neglected.

[4] Drop perpendiculars from midpoints M_1 and M_2, the valve and support point E to the axis line.

[5] Take moments about the axis line, measuring the lengths of perpendiculars M_2P, ES, DQ and M_1R directly from the plan view (these lengths are noted on the sketch):

PIPE LENGTH AC		PIPE LENGTH CB		VALVE ASSY.		LOAD ON SUPPORT
(20)(18)(6)	+	(15)(18)(6)	+	(200)(9)	=	(F)(9.6)

which gives the load on the support at E as:

$$F = 581\text{ lb}$$

EXTENSION OF THE METHOD

The same method can be used if the angle at the corner is different from 90 degrees, or if vertical lines are included in the piping.

NOTES

[1] The axis line must pass thru points of support. If the axis line is not horizontal, the lengths of the perpendiculars are still measured directly from the plan view.

[2] This method does not take into account additional moments due to bending and torsion of pipe. However, it is legitimate to calculate loads on supports as if the pipe is rigid.

This problem often occurs when running pipes from one piperack to another, with a change in elevation, as in figure 6.15. Too much overhang will stress the material of the pipe beyond a safe limit near one of the supports adjacent to the bend, and the designer needs to know the allowable overhang.

The stresses set up in the material of the pipe set practical limits on the overhangs allowed at corners. The problem is like that for spans of straight pipe allowable between supports. Overhangs permitted by stated limits for stress are given in charts S-2.

PIPE SUPPORTS ALLOWING THERMAL MOVEMENT 6.2.5

Piping subject to large temperature changes should be routed so as to flex under the changes in length—see figure 6.1. However, hangers and supports must permit these changes in length. Figures 2.72 A & B show a selection of hangers and supports able to accommodate movement. For single pipes hung from rod or bar hangers, the hanger should be sufficiently long to limit total movement to 10 degrees of arc.

SPRING SUPPORTS

There are two basic types of spring support: (1) Variable load. (2) Constant load—refer to 2.12.2. Apart from cost, the choice between the two types depends on how critical the circumstances are. For example, if a vertical line supported on a rigid support at floor level is subject to thermal movement, a variable-spring hanger or support at the top of the line is suitable—see figure 6.16 (a) and (b).

If a hot line comes down to a nozzle connected to a vessel or machine, and it is necessary to keep the nozzle substantially free from vertical loading, a constant-load hanger can be used—see figure 6.16(c). Cheaper alternate methods of supporting the load are by a cable-held weight working over a pulley, as illustrated in figure 6.16(d), or by a cantilevered weight.

VARIABLE- & CONSTANT-LOAD HANGERS & SUPPORTS   **FIGURE 6.16**

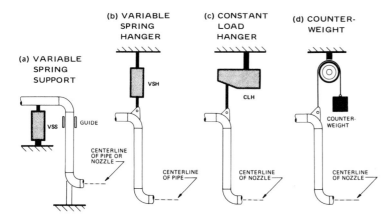

SLOPED LINES AVOID POCKETING AND AID DRAINING 6.2.6

As pipe is not completely rigid, sagging between points of support must occur. In many instances, sagging is acceptable, but in others it must be restricted.

The nature of the conveyed material, the process, and flow requirements determine how much sagging can be accepted. Sagging is reduced by bringing adjacent points of support closer. Pocketing of liquid due to sagging can be eliminated by sloping the line so that the difference in height between adjacent supports is at least equal to triple the deflection (sag) at the mid-point. Lines which require sloping include blowdown headers, pressure-relief lines, and some process, condensate and air lines. (Air lines are discussed in 6.3.2, and draining of compressed-air lines in 6.11.4.)

Complete draining may be required for lines used in batch processing to avoid contamination, or where a product held in a line may degenerate or polymerize, or where solids may settle and become a problem.

In freezing conditions, lines conveying condensate from traps to drains are sloped; condensate headers may be sloped (as an alternative to steam tracing), depending on the rate of flow.

In the past, steam lines were sloped to assist in clearing condensate, but the improved draining is now not considered to be worth the difficulty and expense involved.

SLOPED LINES ON PIPERACKS

Sloped lines can be carried on brackets attached to the piperack stanchions (see figure 6.3). To obtain the required change in elevation at each bent, the brackets may be attached at the required elevations; alternately, a series of brackets can be arranged at the same elevation and the slope obtained by using shoes of different sizes—this method leads to fewer construction problems.

Shoes of graded sizes are also the best method for sloping smaller lines on the piperack. It is not usual or desirable to hang lines from the piperack unless necessary vertical clearances can be maintained.

SLOPED LINES IN BUILDINGS

Inside a building, both large and small sloped lines can rest on steel brackets, or be held with hangers. Rods with turnbuckles are used for hangers on lines required to be sloped. Otherwise, drilled flat bar can be used. (Adjustable brackets are available from the Unistrut and Kindorf ranges of support hardware.)

SUPPORTING PIPE MADE FROM PLASTICS OR GLASS 6.2.7

Pipe made either from flexible or rigid plastics cannot sustain the same span loads as metal pipe, and requires a greater number of support points. One way of providing support is to lay the pipe upon lengths of steel channel sections or half sections of pipe, or by suspending it from other steel pipes. The choice of steel section would depend on the span loads and the size and type of plastic pipe.

For glass process and drain lines, hangers for steel pipe are used, provided that they hold the pipe without causing local strains and are padded so as not to crack the pipe.   Rubber and asbestos paddings are suitable. Uninsulated horizontal lines from 1 to 6 inch in size containing gas or liquid of specific gravity less than 1.3 should be supported at 8 to 10 ft intervals. Couplings and fittings should be about 1 ft from a point of support.

Terms such as 'dummy leg', 'anchor', 'shoe', etc., used in detailing supporting hardware are explained in 2.12.2. Refer to chart 5.7 for symbols.

GENERAL

- Design hangers for 2½-inch and larger pipe to permit adjustment after installation

- If piping is to be connected to equipment, a valve, etc., or piping assembly that will require removal for maintenance, support the piping so that temporary supports are not needed

- Base load calculations for variable-spring and constant-load supports on the operating conditions of the piping (do not include the weight of hydrostatic test fluid)

- If necessary, suspend pipes smaller than 2-inch nominal size from 4-inch and larger pipes

DUMMY LEGS

Table 6.3 suggests sizes for dummy legs. The allowable stress on the wall of the elbow or line pipe to which the dummy leg is attached sets a maximum length for the leg. The advice of the stress group should be sought.

APPROXIMATE SIZES FOR DUMMY LEGS TABLE 6.3

NPS of Piping (inches)	2	3	4	6	8	10	12	14
NPS of Pipe forming Leg (in.)	1½	2	3	4	6	8	8	10
Size of W-Flange (in.)					5	8	8	10

ANCHORS

Anchors are required as stated in the following two points. However, advice from the stress and/or piping support groups should be obtained:

- Provide anchors as necessary to prevent thermal or mechanical movement overloading nozzles on vessels or machinery, branch connections, cast-iron valves, etc.

- Provide anchors to control direction of expansion; for example, at battery limits and on piping leaving units, so that movement is not transmitted to piping on a piperack

SHOES, GUIDES, & SADDLES

- Do not use shoes on uninsulated pipes, unless required for sloping purposes. For reduced friction where lines are long and subject to movement, slide plates are an alternative—see 2.12.2.

- Use of wye-type shoes enables pipes to be placed on the shoe before welding and makes construction easier — see figure 2.72A

- Welding the pipe directly to shoes is not always acceptable; for example with rubber-lined pipe. Bolted or strapped shoes are more suitable

- Check the code pertinent to the project, as it may prohibit 'partial' welds for supports—that is, welds that do not encircle the pipe

- Provide slots in shoes to accept the straps or wires used to hold insulation to pipe

- Provide guides for long straight pipes subject to thermal movement, either by guiding the shoe or by guiding pipe support saddles attached to the pipe, as shown:

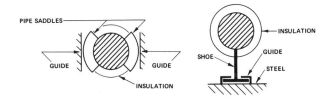

- For better stress distribution in the pipe wall, pipe support saddles are usually used on large lines. They can also be used for lines that may twist over when moving

SUPPORTING VALVES

- Provide support as close as possible to heavy valves, or try to get valves moved close to a suitable point where support can be provided

- Large valves and equipment such as meters located at grade will usually require a concrete foundation for support

WELDING PIPE-SUPPORT & PLATFORM BRACKETS TO VESSELS, Etc.

- Instruct the vendor to add brackets required on pressure vessels prior to stress-relieving and testing—otherwise, retesting and recertification may be obligatory

- It is permissible to specify brackets to be welded to non-pressure vessels provided that the strength of the vessel is not degraded

SUPPORTING PIPE AT NOZZLES

Ensure that nozzles on machinery, compressors, pumps, turbines, etc., are substantially free from loads transmitted by the piping, which may be due to the weight of the piping, or to movement in the piping resulting from contraction, expansion, twisting, vibration or surging. Equipment suppliers will sometimes state maximum loadings permissible at nozzles. *Excessive loads applied to nozzles on machinery can force it from alignment and may cause damage.*

Piping to pumps, turbines, etc., should be supported adequately, but should allow the equipment to be removed. Supports for this piping are best made integral with the concrete foundations, especially if thermal movement occurs and should be on the same level as the base of the equipment, so that on heating or cooling, vertical differential expansion and contraction between supports and equipment will be minimized.

FIGURE 6.16

TABLE 6.3

PUMP EMPLACEMENT & CONNECTIONS   6.3.1

TYPICAL PIPING FOR CENTRIFUGAL PUMPS

Most pumps used in industry are of the centrifugal type. Figures 6.17 and 6.18 show typical piping and fittings required at a centrifugal pump together with the valves necessary to isolate the pump from the system.

The check valve is required to prevent possible flow reversal in the discharge line. A permanent in-line strainer is normally used for screwed suction piping and a temporary strainer for butt-welded/flanged piping. The temporary strainer is installed between flanges—see figure 2.69. A spool is usually required to facilitate removal.

Although centrifugal pumps are provided with suction and discharge ports of cross-sectional area large enough to cope with the full rated capacity of the pump, it is often necessary with thick fluids or with long suction lines to use an inlet pipe of larger size than the inlet port, to avoid cavitation. Cavitation is the pulling by the pump of vapor spaces in the pumped liquid, causing reduction of pumping efficiency, noisy running, and possible impellor and bearing damage. Refer to 6.1.3, under 'Which size valve to use?'.

Most pumps have end suction and top discharge. Limitations on space may require another configuration, such as top suction with top discharge, side suction with side discharge, etc. Determination of nozzle orientation takes place when equipment layout and piping studies are made.

AUXILIARY, TRIM, or HARNESS PIPING

Pumps, compressors and turbines may require water for cooling bearings, for mechanical seals, or for quenching vapors to prevent their escape to atmosphere. Piping for cooling water or seal fluid is usually referred to as auxiliary, trim, or harness piping, and the requirement for this piping is normally shown on the P&ID. This piping is usually shown in isometric view on one of the piping drawings.

In order to cool the gland or seal of a centrifugal pump and ensure proper sealing, it is usually supplied with liquid from the discharge of the pump, by a built-in arrangement, or piped from a connection on the pump's casing. The gland may be provided with a cooling chamber, requiring piped water. If a pump handles hot or volatile liquid, seal liquid may be piped from an external source.

DRAINING

Each pump is usually provided with a drain hub 4 to 6 inches in diameter, positioned about 9 inches in front of the pump foundation on the centerline of the pump. The drain hub is piped to the correct sewer or effluent line—see 6.13. If two small pumps have a common foundation, they can share the same drain hub.

Most centrifugal pumps have baseplates that collect any leakage from the pump. The baseplate will have a threaded connection which is piped to the drain hub. Waste seal water is also piped to the drain hub—see figure 6.19.

- In outside installations in freezing climates, provide a valved drain from the pump's casing

- Provide a short spool for a 3/4-inch drain between the on/off valve and the check valve, to drain the discharge line. If the valve is large enough, the drain can be made by drilling and tapping a boss on the check valve, as shown in figure 6.17, note (3), in which instance no spool is required.

INSTALLATION

- Do not route piping over the pump, as this interferes with maintenance. It is better to bring the piping forward of the pump as shown in figure 6.17

- Leave vertical clearance over pumps to permit removal for servicing —sufficient headroom must be left for a mobile crane for all but the smaller pumps, unless other handling is planned

- If pumps positioned close to supply tanks are on separate foundations, avoid rigid piping arrangements, as the tanks will 'settle' in the course of time

- Locate the pump as closely as practicable to the source of liquid to be pumped from storage tanks, sumps, etc., with due consideration for flexibility of the piping

- Position valves for ease of operation placing them so they are unlikely to be damaged by traffic and will not be a hazard to personnel—see table 6.2 and chart P-2

- The foundation may be of any material that has rigidity sufficient to support the pump baseplate and withstand vibration. A concrete foundation built on solid ground or a concrete slab floor is usual. The pump is positioned, the height fixed (using packing), and the grout is then poured. Grout thickness is not usually less than one inch—see figure 6.17

- A pit in which a pump is installed should have a drain, or have a sump that can be drained or pumped out

- Make the concrete foundation at least as large as the baseplate, and ensure that concrete extends at least 3 inches from each bolt

VALVES

- Valves are 'line size' unless shown otherwise on the P&ID. See 6.1.3 under 'Which size valve to use?'

- Use tilting disc or swing check valves for preference

- Do not use globe valves for isolating pumps. Suction and discharge line isolating valves are usually gate valves, but may be other valves offering low resistance to flow

SUCTION LINE

To avoid cavitation, the pump must be at the correct elevation, related to the level or head of the liquid being pumped. If the location of the pump has not previously been established on an equipment arrangement drawing, refer to the engineer involved.

Concentric reducers are used in lines 2-inch and smaller. Eccentric reducers are used in lines 2½-inch and larger, and are arranged to avoid: (1) Creating a vapor space. (2) Creating a pocket which would need to be drained. These conditions set the configuration of the reducer—that is, whether it is to be installed 'top flat' or 'bottom flat'.

If a centrifugal pump has the suction nozzle at the end (in line with the drive shaft), an elbow may be connected directly to the nozzle at any orientation.

If a pump has the suction nozzle at the side with split flow to the impellor provide a straight run of pipe equal to 3 to 5 pipe diameters of the suction line to connect to the nozzle. Alternately, an elbow may be connected to the suction nozzle, but it must be arranged in a plane at 90 degrees to the driving shaft, to promote equal flow to both sides of the impellor. If an elbow must be in the same plane as the driving shaft of the pump, consider the use of turning (or splitter) vanes to induce more even flow. Uneven flow causes damage to the impellor and bearings.

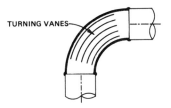

TURNING VANES

- Route suction lines as directly as possible so as not to starve the pump and incur the risk of cavitation
- If the pump draws liquid from a sump at a lower elevation, provide a combined foot valve and strainer. A centrifugal pump working in this situation requires priming initially—provide for this by a valved branch near the inlet port, or by other means
- Provide a strainer in the suction line—see figures 6.17 thru 6.21. Do not place a temporary startup screen immediately downstream of a valve, as debris may back up and prevent the valve from being closed

DISCHARGE LINE

The outlet pipe for centrifugal and other non-positive displacement pumps is in most cases chosen to be of larger bore than the discharge port, in order to reduce velocity and consequent pressure drop in the line. A concentric reducer or reducing elbow is used in the discharge line to increase the diameter. There is no restriction on the placement of elbows in discharge lines as there is in suction lines.

- Provide a pressure connection in the discharge line, close to the pump outlet — see figures 6.17 thru 6.21. It may be necessary to provide a short spool for this purpose if there is no pressure point tapping on the pump discharge nozzle
- For locations of drain connections in the discharge line, see figures 6.17 thru 6.21

PUMPS WITH SCREWED CONNECTIONS

A pump with screwed connections requires unions in the suction and discharge lines to permit removal of the pump.

PIPING FOR POSITIVE-DISPLACEMENT PUMPS

Reciprocating and rotary pumps of this type must be protected against overloading due to restriction in the discharge line. If a positive-displacement pump is not equipped with a relief valve by the manufacturer, provide a relief valve between the pump discharge nozzle and the first valve in the discharge line. The discharge from the relief valve is usually connected to the suction line between the isolating valve and the pump.

As positive displacement pumping does not greatly change the flow velocity, reducers and increasers are not usually required in suction and discharge lines. See figures 6.20 and 6.21. A positive-displacement pump having a pulsating discharge may set the piping into vibration, and to reduce this an air chamber (pneumatic reservoir) such as a standpipe can be provided downstream of the discharge valve.

KEEPING MATERIAL FROM SOLIDIFYING IN THE PUMP

It may be necessary to trace a pump (see 6.8.2) in order to keep the conveyed material in a fluid state, especially after shutdown. This problem arises either with process material having a high melting point, or in freezing conditions. Alternately, jacketed pumps can be employed (such as Foster jacketed pumps available from Parks-Cramer).

FIGURES 6.17 THRU 6.21 ARE ON THE FOLLOWING THREE PAGES, & THE KEY FOR THESE FIGURES IS ON THE THIRD OF THESE PAGES

CENTRIFUGAL PUMP PIPING IN ELEVATION

SCREWED PIPING   FIGURE 6.18

FLANGED BUTT-WELDED PIPING

FIGURE 6.17

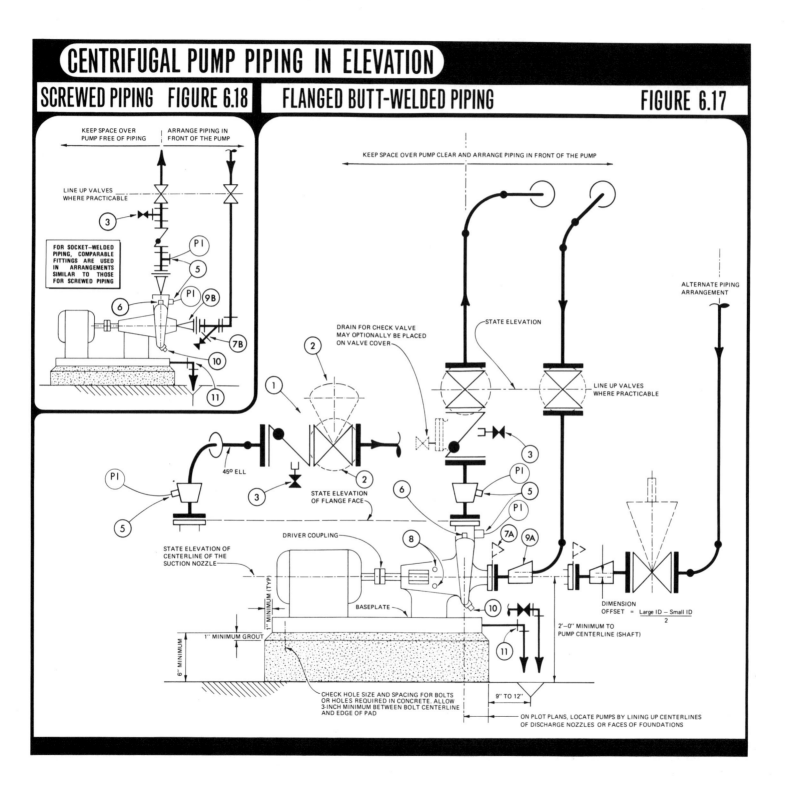

KEEP SPACE OVER PUMP FREE OF PIPING

ARRANGE PIPING IN FRONT OF THE PUMP

LINE UP VALVES WHERE PRACTICABLE

FOR SOCKET–WELDED PIPING, COMPARABLE FITTINGS ARE USED IN ARRANGEMENTS SIMILAR TO THOSE FOR SCREWED PIPING

KEEP SPACE OVER PUMP CLEAR AND ARRANGE PIPING IN FRONT OF THE PUMP

ALTERNATE PIPING ARRANGEMENT

DRAIN FOR CHECK VALVE MAY OPTIONALLY BE PLACED ON VALVE COVER

STATE ELEVATION

LINE UP VALVES WHERE PRACTICABLE

45° ELL

STATE ELEVATION OF FLANGE FACE

STATE ELEVATION OF CENTERLINE OF THE SUCTION NOZZLE

DRIVER COUPLING

BASEPLATE

1" MINIMUM (TYP)

1" MINIMUM GROUT

6" MINIMUM

CHECK HOLE SIZE AND SPACING FOR BOLTS OR HOLES REQUIRED IN CONCRETE. ALLOW 3-INCH MINIMUM BETWEEN BOLT CENTERLINE AND EDGE OF PAD

DIMENSION OFFSET = $\dfrac{\text{Large ID} - \text{Small ID}}{2}$

2'–0" MINIMUM TO PUMP CENTERLINE (SHAFT)

9" TO 12"

ON PLOT PLANS, LOCATE PUMPS BY LINING UP CENTERLINES OF DISCHARGE NOZZLES OR FACES OF FOUNDATIONS

PIPING TO CENTRIFUGAL PUMPS—ALTERNATIVES

● REFER ALSO TO 6.1.3, 'WHICH SIZE VALVE TO USE'

FIGURE 6.19

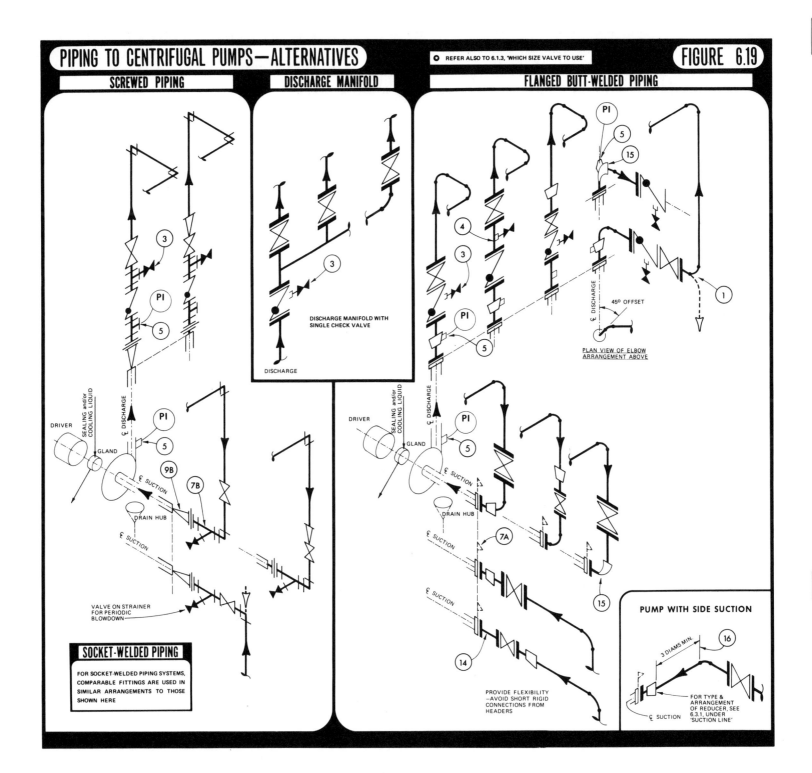

SCREWED PIPING

DISCHARGE MANIFOLD

FLANGED BUTT-WELDED PIPING

DISCHARGE MANIFOLD WITH SINGLE CHECK VALVE

DISCHARGE

DRIVER

SEALING and/or COOLING LIQUID

GLAND

DRAIN HUB

₵ SUCTION

VALVE ON STRAINER FOR PERIODIC BLOWDOWN

SOCKET-WELDED PIPING

FOR SOCKET-WELDED PIPING SYSTEMS, COMPARABLE FITTINGS ARE USED IN SIMILAR ARRANGEMENTS TO THOSE SHOWN HERE

45° OFFSET

PLAN VIEW OF ELBOW ARRANGEMENT ABOVE

DRIVER

SEALING and/or COOLING LIQUID

GLAND

DRAIN HUB

PROVIDE FLEXIBILITY —AVOID SHORT RIGID CONNECTIONS FROM HEADERS

PUMP WITH SIDE SUCTION

3 DIAMS MIN.

FOR TYPE & ARRANGEMENT OF REDUCER, SEE 6.3.1, UNDER 'SUCTION LINE'

₵ SUCTION

FIGURES 6.17–6.19

PIPING FOR POSITIVE-DISPLACEMENT PUMPS

FLANGED BUTT-WELDED PIPING   FIGURE 6.20

PIPING AT NOZZLES IS
SPREAD (AS NECESSARY)
TO ACCOMMODATE VALVES

(OBLIQUE PRESENTATION)

SCREWED PIPING   FIGURE 6.21

FOR SOCKET-WELDED
PIPING, COMPARABLE
FITTINGS ARE USED
IN ARRANGEMENTS
SIMILAR TO THOSE
FOR SCREWED PIPING

(OBLIQUE PRESENTATION)

KEY FOR FIGURES 6.17-6.21

(1)   ALTERNATE HORIZONTAL DISCHARGES, WITH LINE OFFSET AND WITH VALVES LAID OVER AND OFFSET AS NECESSARY—THIS MAY BE NECESSARY IF THE VERTICAL POSITION PLACES HANDWHEEL OUT OF REACH OR IF DISCHARGE NEEDS TO TURN DOWN

(2)   ALTERNATE POSITIONS FOR HANDWHEEL

(3)   PROVIDE 1/2 TO 3/4-INCH DRAIN ON CHECK VALVE ABOVE DISC (A DRAINPOINT OR BOSS IS USUALLY PROVIDED ON 2-INCH AND LARGER VALVES) AND RUN LINE TO DRAIN. OTHERWISE, PLACE DRAIN ON SPOOL BETWEEN CHECK AND ISOLATING VALVES. ON SCREWED AND SOCKET-WELDED PIPING, PROVIDE A TEE FOR THE DRAIN CONNECTION

(4)   SPOOL FOR DRAIN POINT, IF DRAIN CANNOT GO ON CHECK VALVE

(5)   ALTERNATE PRESSURE GAGE POINTS ON DISCHARGE PIPING IF POINT IS NOT PROVIDED ON PUMP BY VENDOR

(6)   CASING VENT. CAN BE USED FOR SEAL LIQUID TAKEOFF

(7 A) TEMPORARY STARTUP STRAINER

(7 B) PERMANENT LINE STRAINER FOR SCREWED OR SOCKET-WELDED PIPING

(8)   CONNECTIONS FOR COOLING OR SEAL LIQUID. USUALLY WATER OR OIL

(9 A) REDUCER   { CONCENTRIC TYPES MAY BE USED ON PUMPS
(9 B) SWAGE (SWAGED NIPPLE)   { WITH INLET PORTS 2-INCH AND SMALLER

(10)   CASING DRAIN PLUG. RUN VALVED LINE IF LIQUID IS LIKELY TO FREEZE

(11)   PIPE BASEPLATE OF PUMP TO DRAIN HUB. PROVIDE HUB AT EACH PUMP. PIPE HUB TO APPROPRIATE DRAIN OR SEWER. IF TWO PUMPS ARE ON A COMMON BASE, THEY CAN SHARE THE SAME HUB

(12)   BYPASS PROTECTS POSITIVE-DISPLACEMENT PUMP AND DRIVER IF AN ATTEMPT IS MADE TO OPERATE PUMP WITH A DISCHARGE VALVE CLOSED

(13)   BYPASSES FOR PUMPS OPERATING IN PARALLEL ALLOW FLOW IN SUCTION AND DISCHARGE LINES TO A HEADER IF A PUMP IS SHUT DOWN

(14)   SPOOL FOR TEMPORARY STRAINER

(15)   REDUCING ELBOW MAY REPLACE REGULAR ELBOW AND REDUCER

(16)   IF A PUMP HAS SIDE SUCTION WITH SPLIT FLOW TO IMPELLOR, PROVIDE 3 OR MORE DIAMETERS OF STRAIGHT PIPE AS SHOWN, OR CONNECT AN ELBOW IN A PLANE AT 90 DEGREES TO THE IMPELLOR SHAFT

Refer to 3.2.2 for a description of compressors and associated equipment. A compressor supplies compressed air or a gas to process or other equipment. A compressor is usually purchased as a 'package unit', which includes coolers, and the designer is left with the problem of installing it and piping auxiliaries to it. These various auxiliaries are shown in figure 6.23.

Compressors may be installed in the open, or within a plant or separate compressor house. An arrangement of compressor, ancillary equipment and distribution lines is shown in figure 6.22 (derived from an illustration by Atlas Copco).

COMPRESSOR HOUSE

● If the compressor is handling a gas heavier than air, eliminate pits or trenches in the compressor house to avoid a suffocation or explosion risk

● Provide air entry louvers if a compressor takes air from within a compressor house or other building

● Provide maintenance facilities, including a lifting rail or access for mobile lifting equipment. Allow adequate floor space for use during maintenance. Additional access may be required for installation

● Prevent transmission of vibration by providing a foundation for the compressor, separate from the compressor-house foundation

● Consider the use of noise-absorbing materials and construction for a compressor house

The vendor's drawings should be examined to determine what auxiliary piping, valves and equipment covered in the following design points are to be supplied with the compressor by the vendor:

● Install the compressor on a concrete pad or elevated structure. Piling is often a necessary part of the foundation

● Large reciprocating compressors are often installed on an elevated structure to allow access to valves and provide space for piping. Provide a platform for operation and maintenance of such an installation

● Keep piping clear of cylinders of reciprocating compressors and provide withdrawal space at cylinder heads

● Use long-radius elbows or bends, not short-radius elbows or miters

● If the compressor and the pressurized gas are cooled with water, route cooling water first to the aftercooler, then to the intercooler (for a two-stage machine), and lastly to the cylinder jackets (or casing jacket, if present, in other types of compressor)

● Arrange an air compressor, associated equipment and piping so that water is able to drain continuously from the system

● Pipe a separate trapped drain for each pressure stage. Ensure that the pressure into which any trap discharges will be lower than that of the system being drained—less the pressure drop over the trap and its associated piping. Do not pipe different pressure stages thru separate check valves to a common trap

● If a toxic or otherwise hazardous gas is to be compressed, vent possible shaft seal leakage to the suction line to avoid a dangerous atmosphere forming around the compressor

● Do not overlook substantial space required for lube oil and seal oil control consoles for compressors

● Discuss piping arrangement with the stress group

AIR-COMPRESSOR PIPING   FIGURE 6.22

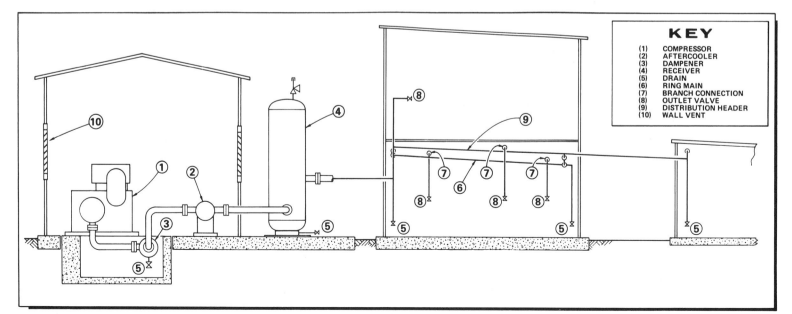

KEY
(1) COMPRESSOR
(2) AFTERCOOLER
(3) DAMPENER
(4) RECEIVER
(5) DRAIN
(6) RING MAIN
(7) BRANCH CONNECTION
(8) OUTLET VALVE
(9) DISTRIBUTION HEADER
(10) WALL VENT

SCHEMATIC ARRANGEMENTS OF COMPRESSED-AIR EQUIPMENT

FIGURE 6.23

(a) SINGLE-STAGE COMPRESSOR

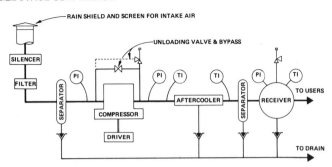

(b) TWO-STAGE COMPRESSOR

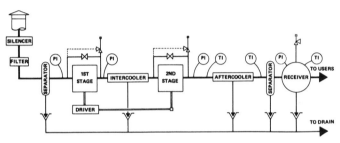

SUCTION PIPING FOR AIR COMPRESSORS

- To reduce damage to a compressor by abrasion or corrosion, the air supply needs to be free from solids and water (water in the air intake does not affect operation of liquid-ring air compressors). Air intakes are best located where the atmosphere is uncontaminated by exhaust gases, industrial operations, or by traffic

- For efficiency the air supply should be taken from the coolest source such as the shaded side of a building, keeping to building clearances shown in figure 6.24

- If the air supply is from outside the building, locate the suction point above the roofline, and away from walls to avoid excessive noise

- Keep suction piping as short as possible. If a line is unavoidably long and condensate likely to form, provide a separator at the compressor intake

- Provide a rain cover and screen as shown in figure 6.24

- Small (and sometimes medium-sized) air compressors usually take air from inside a building. Large air compressors take air from outside a compressor house (figure 6.24): this minimizes effects on the building of pulsations radiated from the air inlet. In both instances, a filter is needed to remove dust, which is always present to some extent

- Filters must have capacity to retain large quantities of impurities with low pressure drop, and must be rugged enough to withstand pulsations from reciprocating compressors

- Provide a pressure gage connection between filter and compressor to allow the pressure drop across the filter to be measured in order to check when cleaning or replacement is needed

- Use a temporary screen at the compressor inlet at startup—see 2.10.4

- Avoid low points in suction lines where moisture and dirt can collect. If low points cannot be avoided, provide a clean-out—see figure 6.24

- If the suction line is taken from a header, take it from the top of the header to reduce the chance of drawing off moisture or sediment

- A line-size isolating valve is required for the suction line if the suction line draws from a header shared with other compressors

- Consider pickling or painting the inside of the suction piping to inhibit rust formation and lessen the risk of drawing rust into the compressor

SUCTION LINES TO AIR COMPRESSORS

FIGURE 6.24

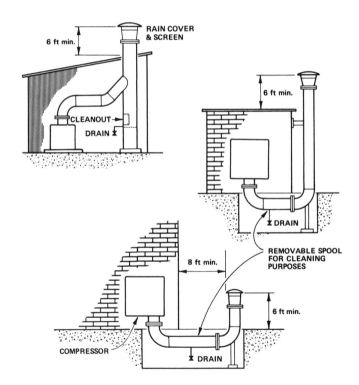

DISCHARGE PIPING (GENERAL)

Discharge piping should be arranged to allow for thermal movement and draining. Anchors and braces should be provided to suppress vibration. The outflow from the aftercooler will usually be wet (from the excess moisture in suction air) and this water must be continually removed.

- An isolating valve in the discharge line is line-size

- Provide discharge piping with connections for temperature and pressure gages

- Provide an unloading valve and bypass circuit connected upstream of the discharge isolating valve, and downstream of the suction isolating valve, so as to ensure circulation thru the compressor during unloading, and to permit equalizing pressure in the compressor—see 3.2.2, under 'Unloading'

- Normally locate a receiver close to the compressor. (Auxiliary receivers may be located near points of heavy use.)

- For draining compressed-air discharge lines, refer to 6.11.4

The use of dampeners and volume bottles in the discharge is discussed in 3.2.2, under 'Equipment for compressors'.

LOADS & VIBRATION

The design of supports for piping to large compressors (especially for reciprocating machines) requires special knowledge. Usually, collaboration is necessary with the piping support group, the stress group, and the compressor manufacturer's representative. A major problem is that the compressor may be forced from alignment with its driver if the piping and supports are not properly arranged.

If a diesel or gasoline engine is used as driver, a flexible joint on the engine's exhaust pipe will reduce transmission of vibration, and protect the exhaust nozzle. Flexible connections are sometimes needed on discharge and suction piping. Pulsation in discharge and—to a lesser extent—suction lines, tends to vibrate piping. This effect is reduced by using bellows, large bends and laterals, instead of elbows and tees.

INSTRUMENTATION & INSTRUMENT CONNECTIONS

Figure 6.23 shows the more useful locations for pressure and temperature gages, but does not show instrumentation for starting, stopping and unloading the compressors. Simple compressor control arrangements using pressure switches have long been used, but result in frequent starting and stopping of the compressor, causing unnecessary wear to equipment.

Automatic control using an unloading valve is superior: table 3.6 gives the working principles—see 3.2.2, under 'Unloading'. Further information can be found in the 'Compressor installation manual' (Atlas-Copco). Unloading valves are allocated instrument numbers.

The air-pressure signals for unloading, starting, loading and stopping a compressor should be free from pulsations. It is best to take these signals from a connection on the receiver or a little downstream of it.

Details of construction of instrument connections are given in 6.7. Instrument branches should be braced to withstand transmission of line vibration.

ISOLATING VALVES FOR COMPRESSOR

Compressors operating in parallel should be provided with isolating valves arranged so that any compressor in the group may be shut down or removed. An isolating valve at the discharge should be placed downstream of the pressure-relief valve and any bypass valve connection. The isolating valve at the suction should be upstream of the bypass valve connection. Isolating valves are not required for a single compressor installation.

PRESSURE-RELIEF VALVES

Pressure-relief valves should be installed on interstage piping and on a discharge line from a compressor to the first downstream isolating valve. A pressure-relief valve may be vented to the suction line—see figure 6.23. Each pressure-relief valve should be able to discharge the full capacity of the compressor.

CHECK VALVE

Unless supplied with (or integral with) a compressor, a check valve must be provided to prevent backflow of stored compressed air or other gas.

DISTRIBUTION OF COMPRESSED AIR

Headers larger than 2-inch are often butt welded. Distribution lines are screwed and usually incorporate malleable-iron fittings, as explained in 2.5.1. Equipment used in distribution piping is described in 3.2.2.

A more efficient layout for compressed air lines is the ring main with auxiliary receivers placed as near as possible to points of heavy intermittent demand. The loop provides two-way air flow to any user.

COMPRESSED AIR USAGE

The compressed air provided for use in plants is designated 'instrument air', 'plant air' or 'process air'. Instrument air is cleaned and dried compressed air, used to prevent corrosion in some instruments. Plant air is compressed air but is usually neither cleaned nor dried, although most of the moisture and oil, etc., can be collected by a separator close to the compressor, especially if adequate cooling can take place. Plant air is used for cleaning, power tools, blowing out vessels, etc: if used for air-powered tools exclusively, some suspended oil is advantageous for lubrication, although filter/lube units are usually installed in the air line to the tool.

Process air is compressed air, cleaned and dried, which may be used in the process stream for oxidizing or agitation. The trend is to supply cleaned and dried air for both general process and instrument purposes. This avoids running separate lines for process and instrument air.

Process and instrument air for some applications requires to have an oil content less than 10 parts per million. As almost all oily contaminants are present as extremely small droplets (less than 1 micron in diameter) mechanical filtration may be ineffective; adsorption equipment can efficiently remove the oil.

FIGURES
6.23 & 6.24

PIPING TO STEAM TURBINES 6.4

A turbine is a machine for deriving mechanical power (rotating shaft) from the expansion of a gas or vapor (usually air or steam, in industrial plants).

Steam turbines are used where there is a readily-available source of steam, and are also used to drive standby process pumps in critical service in the event of an electrical power failure, and emergency standby equipment such as firewater pumps and electric generators.

Figure 6.9 shows a schematic arrangement of piping for automatic operation. There are similarities between steam-turbine and pump and compressor piping. Their common requirements are:—

(1) To limit loads on nozzles from weight of piping or from thermal movement
(2) To provide access and overhead clearance
(3) To prevent harmful material from entering the machine

INLET (STEAM FEED) 6.4.1

In order to guard against damage to a steam turbine, protective piping arrangements such as those mentioned in table 6.4 are needed in the steam feed.

PROTECTIVE PIPING FOR FEEDING STEAM TO TURBINE TABLE 6.4

HAZARD TO TURBINE	PROTECTIVE PIPING
FOREIGN MATTER & WATER IN THE STEAM FEED	DRIPLEG & STRAINER, or SEPARATOR, IN THE FEED LINE (See figure 6.9)
EXCESSIVE PRESSURE IN STEAM FEED CAUSING OVER-FAST RUNNING OR CASING RUPTURE	PRESSURE RELIEF VALVE &/OR CONTROL VALVE IN THE FEED LINE
THERMAL SHOCK, DUE TO TOO RAPID HEATING ON STARTUP	ORIFICE BYPASS TO FEED SMALL AMOUNT OF STEAM TO TURBINE AT ALL TIMES

EXHAUST (STEAM DISCHARGE) 6.4.2

Figure 6.25 shows three methods for dealing with the turbine's exhaust. Steam from an intermittently operated turbine may be run to waste and all that is required is a simple run of pipe to the nearest outside wall or up thru the roof. Exhausts should be well clear of the building and arranged so as not to be hazardous to personnel. The turbine discharge will include drops of water and oil from the turbine, which are best collected and run to drain. A device suitable for this purpose is a Swartwout 'exhaust head' shown in figure 6.26. Alternately, steam discharged from a continuously running turbine may be utilized elsewhere, in a lower-pressure system.

TURBINE EXHAUST ARRANGEMENTS FIGURE 6.25

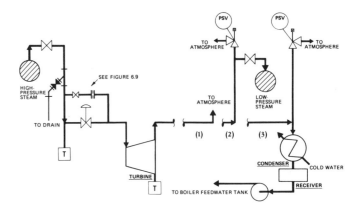

KEY:

(1) Exhaust is discharged directly to atmosphere. Suitable for small turbine in intermittent use.
(2) Exhaust is taken to a low-pressure header for use elsewhere. Suitable for continuously-operating turbine, to avoid wasting steam.
(3) Exhaust is condensed to increase pressure drop across the turbine.

BYPASS STEAM & OTHER PIPING FOR TURBINES 6.4.3

An orifice plate is used as a 'bleed' bypass to ensure that steam constantly passes thru the turbine. An orifice plate is used rather than a straight pipe, as a changeable constriction is needed. Alternately, the small amount of steam needed to keep the turbine warm can be admitted by a cracked-open valve in a bypass—a wasteful and uncertain practice.

A trap is fitted to the casing of the turbine to remove condensate. Piping is provided to supply seal liquid to the turbine's bearings—refer to 6.3.1, under 'Auxiliary, trim, or harness piping'.

SWARTWOUT HEAD FIGURE 6.26

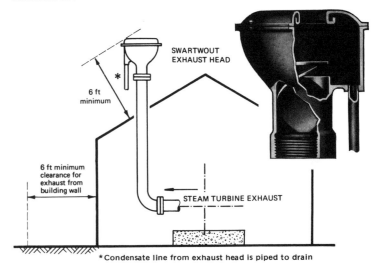

* Condensate line from exhaust head is piped to drain

VESSEL CONNECTIONS   6.5.1

Vessel connections are often made with couplings (for smaller lines), flanged or welding nozzles, and pads fitted with studs, designed to mate with flanged piping. Nozzle outlets are also made by extrusion, to give a shape like that of the branch of a welding tee—this gives a good flow pattern, but is an expensive method usually reserved for such items as manifolds and dished heads. Weldolets, sockolets and thredolets are suitable for vessel connections and are available flat-based for dished heads, tanks, and large vessels.

Almost any type of connection may be made to open vessels or vessels vented to atmosphere, but for pressure vessels, the applicable design code will dictate requirements for connections (and possible reinforcement—see 2.11).

PRESSURE VESSELS

With exceptions and limitations stated in section 8 of the ASME Boiler and Pressure Vessel Code, vessels subject to internal or external operating pressures not exceeding 15 PSI need not be considered to be pressure vessels. A vessel operating under full or partial vacuum and not subject to an external pressure greater than 15 PSI would not require Code certification.

VESSEL DRAWING & REQUIRED NOZZLES

Preliminary piping layouts are made to determine a suitable nozzles arrangement. A sketch of the vessel showing all pertinent information is sent to the vessel fabricator, who then makes a detail drawing. The preliminary studies for pressure vessel piping layouts should indicate where pipe supports and platforms (if required) are to be located. In the event that the vessel has to be stress-relieved, the fabricator can provide clips or brackets—see 6.2.8, under 'Welding pipe-support and platform brackets to vessels, etc.'

Figure 5.14 shows the type of drawing or sketch sent to a vessel fabricator.

NOZZLES NEEDED ON VESSELS

- Nozzles needed on non-pressure vessels include: inlet, outlet, vent (gas or air), manhole, drain, overflow, agitator, temperature element, level instrument, and a 'steamout' connection, sometimes arranged tangentially, for cleaning the vessel

- Nozzles needed on pressure vessels include: inlet, outlet, manhole, drain, pressure relief, agitator, level gage, pressure gage, temperature element, vent, and for 'steamout', as above

- Check whether nozzles are required for an electric heater, coils for heating or cooling, or vessel jacket. A jacket requires a drain and vent

- Check special nozzle needs, such as for flush-bottom tank valves (see 3.1.9)

PIPE FLEXIBLY TO NOZZLES

- Provide additional flexibility in lines to a vessel from pumps and other equipment mounted on a separate foundation (if liable to settle)

- Be cautious in making rigid straight connections between nozzles. Such connections may be acceptable if both items of equipment are on the same foundation, and are not subject to more than normal atmospheric temperature changes (see figure 6.1)

NOZZLE LOADING

- Ensure that a nozzle can take the load imposed on it by connected piping—see 6.2.8, under 'Supporting pipe at nozzles'. Manufacturers often can provide nozzle-loading data for their standard equipment

- Check all connections to ensure that stresses due to thermal movement, and shock pressures ('kicks') from opening pressure relief valves, etc., are safely handled

FRACTIONATION COLUMN PIPING   6.5.2
(OR TOWER PIPING)

As columns and their associated equipment take different forms, according to process needs, the following text gives a simplified explanation of column operation, and outlines basic design considerations.

THE COLUMN'S JOB

A fractionation column is a type of still. A simple still starts with mixed liquids, such as alcohol and water produced by fermenting a grain, etc., and by boiling produces a distillate in which the concentration of alcohol is many times higher than in the feed. In the petroleum industry in particular, mixtures not of two but a great many components are dealt with. Crude oil is a typical feed for a fractionation column, and from it the column can form simultaneously several distillates such as wax distillate, gas oil, heating oil, naphtha and fuel gases. These fractions are termed 'cuts'.

COLUMN OPERATION

The feed is heated (in a 'furnace' or exchanger) before it enters the column. As the feed enters the column, quantities of vapor are given off by 'flashing', due to the release of pressure on the feed.

As the vapors rise up the column, they come into intimate contact with downflowing liquid—see figure 6.29. During this contact, some of the heavier components of the vapor are condensed, and some of the lighter components of the downflowing liquid are vaporized. This process is termed 'refluxing'.

If the composition of the feed remains the same and the column is kept in steady operation, a temperature distribution establishes in the column. The temperature at any tray is the boiling point of the liquid on the tray. 'Cuts' are not taken from every tray. The P&ID shows cuts that are to be made, including alternatives—nozzles on selected trays are piped, and nozzles for alternate operation are provided with line blinds or valves.

FIGURES
6.25 & 6.26

TABLE
6.4

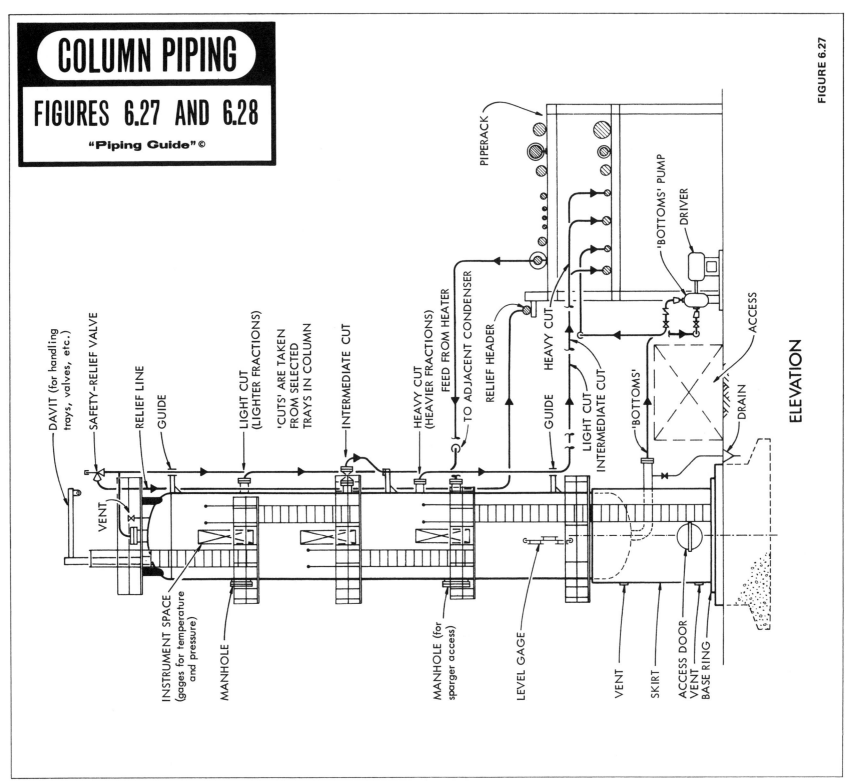

COLUMN PIPING

FIGURES 6.27 AND 6.28

"Piping Guide"©

DAVIT (for handling trays, valves, etc.)

SAFETY-RELIEF VALVE

RELIEF LINE

GUIDE

LIGHT CUT (LIGHTER FRACTIONS)

'CUTS' ARE TAKEN FROM SELECTED TRAYS IN COLUMN

INTERMEDIATE CUT

HEAVY CUT (HEAVIER FRACTIONS)

FEED FROM HEATER

TO ADJACENT CONDENSER

RELIEF HEADER

GUIDE

HEAVY CUT

LIGHT CUT INTERMEDIATE CUT

'BOTTOMS'

PIPERACK

'BOTTOMS' PUMP

DRIVER

ACCESS

VENT

INSTRUMENT SPACE (gages for temperature and pressure)

MANHOLE

MANHOLE (for sparger access)

LEVEL GAGE

VENT

SKIRT

ACCESS DOOR

VENT

BASE RING

DRAIN

ELEVATION

FIGURE 6.27

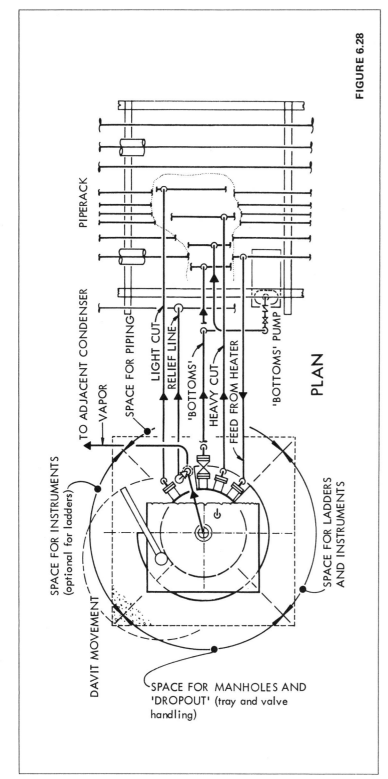

FIGURE 6.28

TO ADJACENT CONDENSER

VAPOR

SPACE FOR PIPING

LIGHT CUT

RELIEF LINE

'BOTTOMS'

HEAVY CUT

FEED FROM HEATER

'BOTTOMS' PUMP

PIPERACK

PLAN

SPACE FOR INSTRUMENTS
(optional for ladders)

DAVIT MOVEMENT

SPACE FOR MANHOLES AND
'DROPOUT' (tray and valve
handling)

SPACE FOR LADDERS
AND INSTRUMENTS

Trays are of various designs. Their purpose is to collect a certain amount of liquid but allow vapors to pass up thru them so that vapor and liquid come into contact. (Refer to figure 6.29, which shows simple bubblecap trays —many tray designs are available.)

TRAYS & BUBBLECAPS   **FIGURE 6.29**

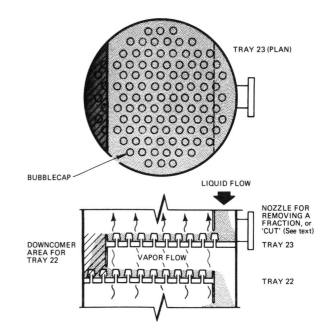

TRAY 23 (PLAN)

BUBBLECAP

LIQUID FLOW

NOZZLE FOR
REMOVING A
FRACTION, or
'CUT' (See text)

TRAY 23

DOWNCOMER
AREA FOR
TRAY 22

VAPOR FLOW

TRAY 22

To produce the required 'cuts', a column operates under steady temperature, feed, and product removal conditions. Starting from cold, products are collected after steady conditions are reached, and the column is then operated continuously.

All materials enter and leave the column thru pipes; therefor columns are located close to piperacks. Figures 6.27 and 6.28 show an arrangement. Products from the column are piped to collecting tanks (termed 'drums', 'accumulators', etc.) and held for further processing, or storage.

If the vapor from the top of the column is condensible, it is piped to a condenser to form a volatile liquid. The condenser may be mounted at grade, or sometimes on the side of the column.

Product from the top of the column may be gaseous at atmospheric pressure after cooling; if the product liquefies under moderate pressure, it may be stored pressurized in containers.

In addition to the condenser for the top product, a steam-heated heat exchanger, termed a 'reboiler', may be used to heat material drawn from a selected level in a column; the heated material is returned to the column. Reboilers are required for tall columns, and for columns operated at high temperatures, which are subject to appreciable loss of heat. Mounting the reboiler on the side of the column minimizes piping.

Material from the bottom of a column is termed 'bottoms', and must be pumped away (see figure 6.27)—this material consists of 'heavier' (higher molecular weight) liquids which either did not vaporize, or had condensed, plus any highly viscous material and solids in the feed.

COLUMN ORIENTATION & REQUIREMENTS

Simultaneously with orientating nozzles and arranging piping to the column, the piping designer decides the positions of manholes, platforms, ladders, davit, and instruments.

COLUMN ORIENTATION   **FIGURE 6.30**

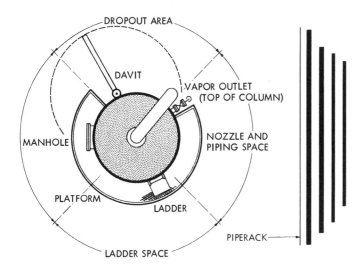

Manholes are necessary to allow installation and removal of tray parts.

Platforms and ladders are required for personnel access to valves on nozzles, to manholes, and to column instruments.

A davit is needed to raise and lower column parts, and a dropout area has to be reserved.

MANHOLES & NOZZLES

For a particular project or column, manholes are preferably of the same type. They should be located away from piping, and within range of the davit.

If required, manholes can be placed off the column centerlines (plan view).

The manhole serving the sparger unit (figure 6.31) should permit easy removal of the unit, which may be angled to place the feed connection in a desired position.

The portions of the column wall available for nozzles are determined by the orientation and type of tray—see figure 6.29. Elevations of nozzles are taken from the column data sheet (normally in the form of a vessel drawing).

SPARGER UNIT   **FIGURE 6.31**

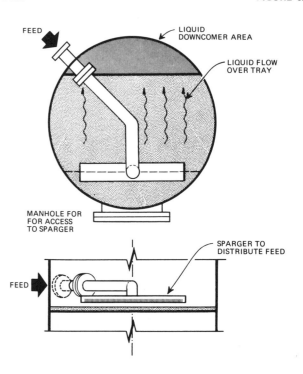

If the cuts are to be taken either from even-numbered trays, or from odd-numbered trays, all nozzles can be located on one side of the column, facing the piperack. If cuts are to come from both even- and odd-numbered trays, it will almost certainly be impossible to arrange all nozzles toward the piperack. (See 'Arranging column piping', this section.)

PLATFORMS & LADDERS

Platforms are required under manholes, valves at nozzles, level gages, controllers if any, and pressure relief valves. Columns may be grouped and sometimes interconnecting platforms between columns are used. Individual platforms for a column are usually shaped as circular segments, as shown in figure 6.30. A platform is required at the top of the column, for operating a davit, a vent on shutdown, and for access to the safety-relief valve. This top platform is often rectangular.

Usual practice is to provide a separate ladder to go from grade past the lowest platform. Ladders are arranged so that the operator steps sideways onto the platforms.

Ladder length is usually restricted to 30 ft between landings. Some States allow 40 ft (check local codes). If operating platforms are further apart than the maximum permissible ladder height, a small intermediate platform is provided.

Ladders and cages should conform to the company standard and satisfy the requirements of the US Department of Labor (OSHA), part 1910.(D).

DAVIT

Referring to figure 6.30, the davit should be located at the top of the column so that it can lower and raise tray parts, piping, valves, etc., between the platforms and the dropout area at grade.

ARRANGING COLUMN PIPING

To achieve simplicity and good arrangement, some trial-and-error working is necessary. Columns are major pieces of equipment, and their piping needs take precedence over other piping.

As lines from nozzles on the column are run down the length of the column, it is logical to start arranging downcomers from the top and proceed down the column. A lower nozzle may need priority, but usually piping can be arranged more efficiently if the space requirements of piping coming from above are already established.

Sometimes tray spacing is increased slightly to permit installation of manholes. It may be possible to rotate trays within limits, to overcome a difficulty in arranging column piping. Such changes in tray spacing and arrangement must be sanctioned by the process engineer and vessel designer.

- Allocate space for vertical lines from lower nozzles, avoiding running these lines thru platforms if possible

- Lines from the tops of columns tend to be larger than others. Allocate space for them first, keeping the lines about 12 inches from the platforms and the wall of the column—this makes supporting easier, and permits access to valves, instruments, etc.

- Allocate space for access (manholes, ladders) clear of piping—especially clear of vertical lines

- Provide a clear space for lowering equipment from the top of a column (for maintenance, etc.)

- Provide access for mobile lifting equipment to condenser and reboiler

- Provide clearance to grade (approximately 8ft) under the suction line, from the column to the bottoms pump

- Arrange vent(s) in the skirt of the column

- Ensure that no low point occurs in the line conveying 'bottoms' to the suction port of the bottoms pump, in order to avoid blocking of this line due to cooling, etc.

INFORMATION NEEDED TO ARRANGE THE COLUMN PIPING

- Plot plan showing space available for column location, and details of equipment which is to connect to the column

- P&ID for nozzle connections, NPSH of bottoms pump, instrumentation, line blinds, relief valves, etc.

- Column data sheets and sketch of column showing elevations of nozzles

- Line designation sheets, to obtain operating temperatures of lines for calculating thermal movement

- Details of trays and other internal parts of the column

- Restrictions on the heights of ladders

- Operational requirements for the plant

BOTTOMS PUMP & ELEVATION OF COLUMN

The elevation of a column is set primarily by the NPSH required by the bottoms pump, the access required under the suction line to the bottoms pump, and by requirements for a thermosyphon reboiler, if used.

VALVES

Valves and blinds which serve the tower should be positioned directly on nozzles, for economy. It is desirable to arrange other valves so that lines are self-draining.

Platforms should be located to give access to large valves. Small valves may be located at the ends of platforms. Control valves should be accessible from operating platforms or from grade.

The pressure-relief valve for the relief line should be placed at the highest point in the line, and should be accessible from the top platform.

Valves should not be located within the skirt of the column.

INSTRUMENTS & CONNECTIONS

Temperature connections should be located to communicate with liquids in the trays, and pressure connections with the vapor spaces below the trays. Access to isolated gages can be provided by ladder.

Gages, and gage and level glasses, must be visible when operating valves, and be accessible for maintenance.

Gages and other instruments should be located clear of manholes and accessways to ladders and platforms. If necessary, temperature and pressure gages may be located for reading from ladders. Locating instruments at one end of a circular platform may allow a narrower platform.

THERMAL INSULATION

Thermal insulation of the exterior of a column may be required in order to reduce heat loss to the atmosphere. Insulation may be inadequate to maintain the required temperature distribution; in these circumstances, a reboiler is used. Thermal insulation is discussed in 6.8.1.

FOUNDATION FOR COLUMN

The base ring of a column's skirt is attached to a reinforced-concrete construction. The lower part of this construction, termed the 'foundation', is below grade, and is square in plan view: the upper part, termed the 'base', to which the base ring is attached, is usually octagonal and projects above grade approximately 6 inches.

Heat exchangers are discussed in 3.3.5.

DATA NEEDED TO PLAN EXCHANGER PIPING   6.6.1

Preliminary exchanger information should be given early to the piping group, so that piping studies can be made with special reference to orientation of nozzles. Before arranging heat-exchanger piping, the following information is needed:

PROCESS FLOW DIAGRAM This will show the fluids that are to be handled by the exchangers, and will state their flow rates, temperatures and pressures.

EXCHANGER DATA SHEETS One of these sheets is compiled for each exchanger design by the project group. The piping group provides nozzle orientation sketches (resulting from the piping studies). The data sheet informs the manufacturer or vendor of the exchanger concerning performance and code stamp requirements, materials, and possible dimensional limitations.

TEMA CODING FOR EXCHANGER TYPE

The Tubular Exchangers Manufacturers Association (TEMA) has devised a method for designating exchanger types, using a letter coding. The exchanger shown in figure 6.32 would have the basic designation AEW. See chart H-1.

Engineering Notes:

- Provide the shell with a pressure-relieving device to protect against excessive shell-side pressure in the event of internal failure

- Put fouling and/or corrosive fluids inside the tubes as these are (except U-type) easily cleaned, and cheaper to replace than the shell

- Put the hotter fluid in the tubes to reduce heat loss to the surroundings

- However, if steam is used to heat a fluid in an exchanger, passing the steam thru the shell has advantages: for example, condensate is far easier to handle shellside. Insulation of the shell is normally required to protect personnel, and to reduce the rates of condensate formation and heat loss

- Pass refrigerant or cooling liquid thru the tubes, if the exchanger is not insulated, for economic operation

- If heat transfer is between two liquids, a countercurrent flow pattern will usually give greater overall heat transfer than a paralleled flow pattern, other factors being the same

- Orientate single-tube spiral, helical and U-tube exchangers (with steam fed thru the tube) to permit outflow of condensate

SHELL-AND-TUBE HEAT EXCHANGER WITH REMOVABLE TUBE BUNDLE   **FIGURE 6.32**

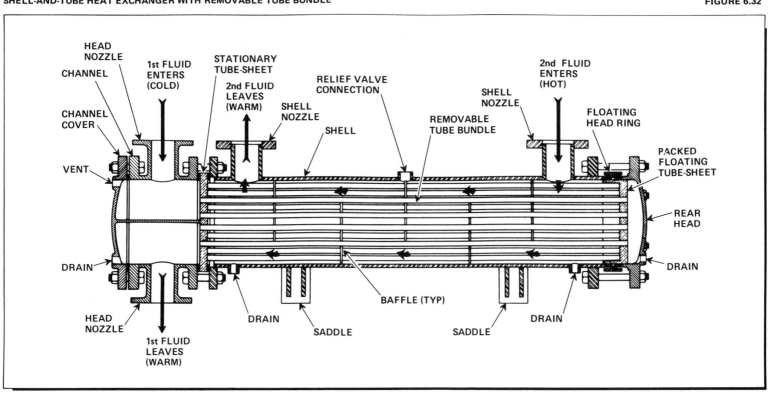

Nozzle Positions:

- Arrange nozzles to suit the best piping and plant layout. Nozzles may be positioned tangentially or on elbows, as well as on vertical or horizontal centerlines (as usually offered at first by vendors). Although a tangential or elbowed nozzle is more expensive, it may permit economies in piping multiple heat exchangers

- Make condensing vapor the descending stream

- Make vaporizing fluid the ascending stream

Locating Exchangers:

- Position exchangers so that piping is as direct and simple as possible. To achieve this, consider alternatives, such as reversing flows, arranging exchangers side-by-side or stacking them, to minimize piping

- Elevate an exchanger to allow piping to the exchanger's nozzles to be arranged above grade or floor level, unless piping is to be brought up thru a floor or from a trench

- Exchangers are sometimes of necessity mounted on structures, process columns and other equipment. Special arrangements for maintenance and tube handling will be required

PIPING TO NOZZLES OF HEAT EXCHANGERS **FIGURE 6.33**

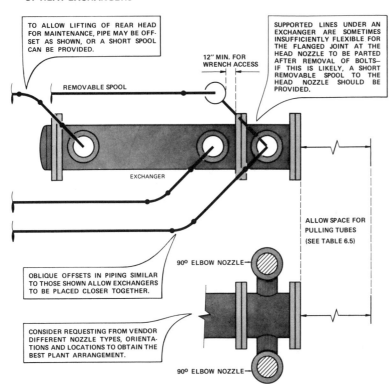

TO ALLOW LIFTING OF REAR HEAD FOR MAINTENANCE, PIPE MAY BE OFFSET AS SHOWN, OR A SHORT SPOOL CAN BE PROVIDED.

SUPPORTED LINES UNDER AN EXCHANGER ARE SOMETIMES INSUFFICIENTLY FLEXIBLE FOR THE FLANGED JOINT AT THE HEAD NOZZLE TO BE PARTED AFTER REMOVAL OF BOLTS— IF THIS IS LIKELY, A SHORT REMOVABLE SPOOL TO THE HEAD NOZZLE SHOULD BE PROVIDED.

12" MIN. FOR WRENCH ACCESS

REMOVABLE SPOOL

EXCHANGER

OBLIQUE OFFSETS IN PIPING SIMILAR TO THOSE SHOWN ALLOW EXCHANGERS TO BE PLACED CLOSER TOGETHER.

CONSIDER REQUESTING FROM VENDOR DIFFERENT NOZZLE TYPES, ORIENTATIONS AND LOCATIONS TO OBTAIN THE BEST PLANT ARRANGEMENT.

ALLOW SPACE FOR PULLING TUBES (SEE TABLE 6.5)

90° ELBOW NOZZLE

90° ELBOW NOZZLE

MINIMUM SPACING & CLEARANCES FOR MULTIPLE HEAT EXCHANGERS **TABLE 6.5**

(a) Exchangers arranged with 2 ft 6 in. operating space between piping

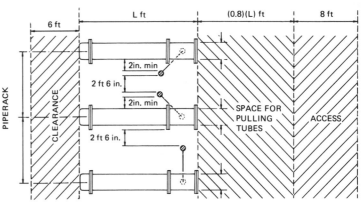

(b) Exchangers arranged with 2 ft 0 in. maintenance space between paired units and 2 ft 6 in. operating space between piping

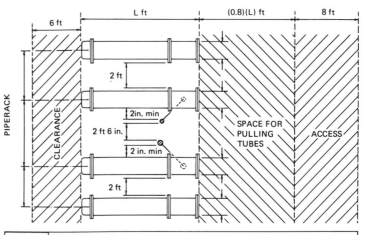

NOTES	(1)	Show outlines of exchanger supports or foundations before arranging piping
	(2)	Add to clearances shown, thicknesses of insulation for exchanger shells and connected piping
	(3)	Provide additional clearance to the 2'–6'' operating space if valve handwheels and valve stems, etc., protrude, depending on piping arrangement

Operating and Maintenance Requirements:

- Access to operating valves and instruments (on one side only suffices)

- Operating space for any davit, monorail or crane, etc., both for movement and to set loads down

- Access to exchanger — space is needed for tube-bundle removal, for cleaning, and around the exchanger's bolted ends (channelcover and rear head) and the bolted channel-to-shell closure

- Access for tube bundle removal is often given on manufacturers' drawings, and is usually about 1½ times the bundle length. 15 to 20 ft clearance should be allocated from the outer side of the last exchanger in a row for mobile lifting equipment access and tube handling

INSTRUMENT CONNECTIONS

INCH SIZES FOR VALVES, FITTINGS & PIPE ARE NOMINAL AND SHOWN ON DRAWINGS AS NOMINAL PIPE SIZES. FOR EXAMPLE, PIPE 4" IS SHOWN ON DRAWINGS AS PIPE NPS 4

CHART 6.2

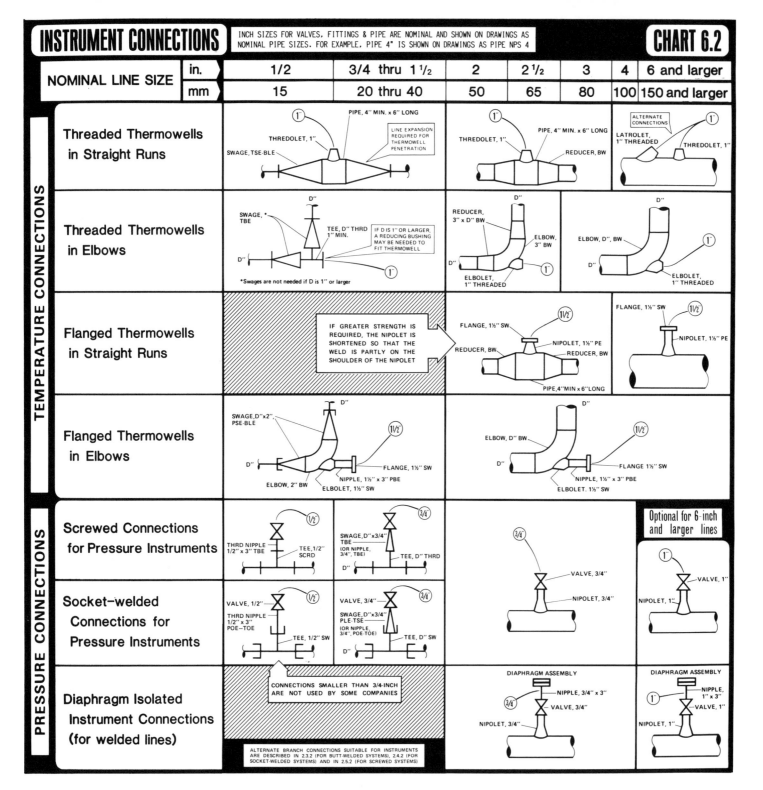

NOMINAL LINE SIZE	in.	1/2	3/4 thru 1 1/2	2	2 1/2	3	4	6 and larger
	mm	15	20 thru 40	50	65	80	100	150 and larger

TEMPERATURE CONNECTIONS

- Threaded Thermowells in Straight Runs
- Threaded Thermowells in Elbows
- Flanged Thermowells in Straight Runs
- Flanged Thermowells in Elbows

PRESSURE CONNECTIONS

- Screwed Connections for Pressure Instruments
- Socket-welded Connections for Pressure Instruments
- Diaphragm Isolated Instrument Connections (for welded lines)

PRIMARY CONNECTIONS TO LINES & EQUIPMENT   6.7.1

Connections will usually be specified by company standards or by the specifications for the project. If no specification exists, full- and half-couplings, swaged nipples, thredolets, nipolets and elbolets, etc., may be used. Chart 6.2 illustrates instrument connections used for lines of various sizes. The fittings shown in chart 6.2 are described in chapter 2. Orifice flange connections are discussed in 6.7.5.

CHOOSING THE CONNECTION   6.7.2

The choice of instrument connection will depend on the conveyed fluid and sometimes on the required penetration of the element into the vessel or pipe. Instrument connections should be designed so that servicing or replacement of instruments can be carried out without interrupting the process. Valves are needed to isolate gages for maintenance during plant operation and during hydrostatic testing of the piping system. These valves are shown in chart 6.2 and are referred to as 'root' or 'primary' valves.

TEMPERATURE & PRESSURE CONNECTIONS   6.7.3

Chart 6.2 illustrates various methods for making temperature and pressure connections. At the bottom of chart 6.2 a method of connecting a diaphragm flange assembly (diaphragm isolator) is shown. Corrosive, abrasive or viscous fluid in the process line presses on one side of the flexible diaphragm, and the neutral fluid (glycol, etc.) on the other side transmits the pressure.

If the conveyed fluid is hazardous or under high pressure a branch fitted with a bleed valve is inserted between the gage and its isolating valve, to relieve pressure and/or drain the liquid before servicing the gage. The bleed valve can also be used to sample, or for adding a comparison gage.

- Position connections for instruments so that the instruments can be seen when operating associated valves, etc.

- Pressure connections for vessels containing liquids are usually best located above liquid level

- A temperature-measuring element is inserted into a metal housing termed a 'thermowell'. Place thermowells so that they are in contact with the fluid—an elbow is a good location, due to the increased turbulence

THERMOWELL CONSTRUCTION (EXAMPLE)

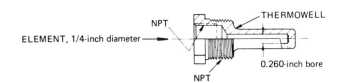

ELEMENT, 1/4-inch diameter →   NPT   THERMOWELL
0.260-inch bore   NPT

- Locate a liquid level controller (float type, for example) clear of any turbulence from nozzles

- More than one level gage, level switch, etc., may be required on a vessel: consider installing a 'strongback' to a horizontal vessel on which instrument connections have to be made—see figure 6.34(c)

LEVEL-GAGE CONNECTIONS   **FIGURE 6.34**

(a) LEVEL GAGE ASSEMBLY   (b) CONNECTIONS FOR A GAGE GLASS

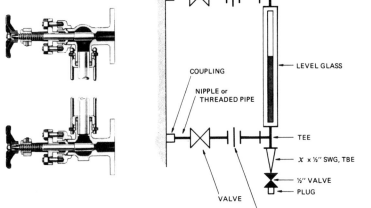

(c) CONNECTIONS ON STRONGBACK

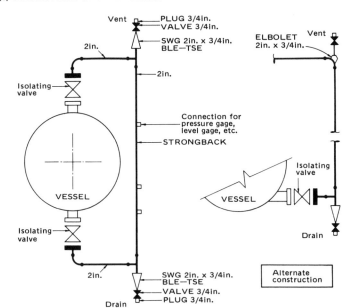

6 .7
.7.4

CHART
6.2

FIGURE
6.34

ROTAMETER CONNECTIONS

A rotameter consists of a transparent tube with tapered and calibrated bore, arranged vertically, wide end up, supported in a casing or framework with end connections. The instrument should be connected so that flow enters at the lower end and leaves at the top. A ball or spinner rides on the rising gas or liquid inside the tapered tube — the greater the flow rate, the higher the ball or spinner rides. Isolating valves and a bypass should be provided, as in figure 6.35

ROTAMETER   **FIGURE 6.35**

(a) PIPING TO ROTAMETER   **(b) INDUSTRIAL ROTAMETER**

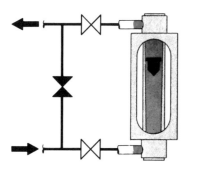 

ORIFICE PLATE ASSEMBLY

An 'orifice plate' is a flat disc with a precisely-made hole at its center. It offers a well-defined obstruction to flow when inserted in a line—see figure 6.36. The resistance of the orifice sets up a pressure difference in the fluid either side of the plate, which can be used to measure the rate of flow.

ORIFICE PLATE ASSEMBLY & GAGE (MANOMETER)   **FIGURE 6.36**

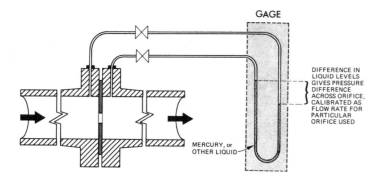

GAGE

DIFFERENCE IN LIQUID LEVELS GIVES PRESSURE DIFFERENCE ACROSS ORIFICE, CALIBRATED AS FLOW RATE FOR PARTICULAR ORIFICE USED

MERCURY, or OTHER LIQUID

The orifice plate is held between special flanges having 'orifice taps'—these are tapped holes made in the flange rims, to which tubing and a pressure gage can be connected, as in figure 6.36. A pressure gage may be termed a 'manometer'.

Manometers for use with orifice plate assemblies are calibrated in terms of differential pressure by the manufacturer. The meter run (that is, the piping in which the orifice plate is to be installed) must correspond with the piping used to calibrate the orifice plate—the readings will be in error if there is very much variation in these two piping arrangements.

Sometimes the orifice assembly includes adjacent piping, ready for welding in place. Otherwise, lengths of straight pipe, free from welds, branches or obstruction, should be provided upstream and downstream of the orifice assembly.

Table 6.6 shows lengths of straight pipe required upstream and downstream of orifice flanges (for different piping arrangements) to sufficiently reduce turbulence in liquids for reliable measurement.

PIPING TO FLANGE TAPS

Figure 6.37 shows a suitable tapping and valving arrangement at orifice flange taps. In horizontal runs, the taps are located at the tops of the flanges in gas, steam and vapor lines. An approximately horizontal position avoids vapor locks in liquid lines. Taps should not be pointed downward, as sediment may collect in pipes and tubes.

CONNECTIONS TO ORIFICE FLANGES   **FIGURE 6.37**
& INSTRUMENT

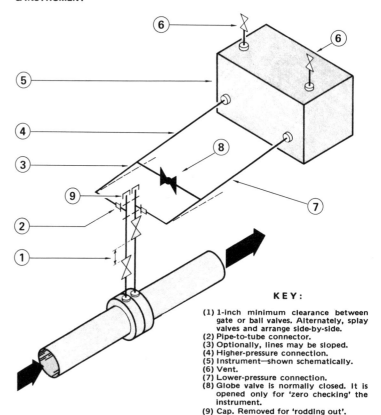

K E Y :

(1) 1-inch minimum clearance between gate or ball valves. Alternately, splay valves and arrange side-by-side.
(2) Pipe-to-tube connector.
(3) Optionally, lines may be sloped.
(4) Higher-pressure connection.
(5) Instrument—shown schematically.
(6) Vent.
(7) Lower-pressure connection.
(8) Globe valve is normally closed. It is opened only for 'zero checking' the instrument.
(9) Cap. Removed for 'rodding out'.

STRAIGHT PIPE RUN TO THE ORIFICE

The arrangement of orifice plate assemblies should be made in consulation with the instrument engineer. Usually, it is preferred to locate orifice plate assemblies in horizontal lines.

Flow conditions consistent with those used to calibrate the instrument are ensured by providing adequately long straight sections of pipe upstream and downstream of the orifice. Table 6.6 gives lengths that have been found satisfactory for liquids.

STRAIGHT PIPE UPSTREAM & DOWNSTREAM OF ORIFICE ASSEMBLY — TABLE 6.6

KEY NUMBER OF PIPING ARRANGEMENT	U=UPSTREAM D=DOWNSTREAM	RATIO OF INTERNAL DIAMETERS OF ORIFICE PLATE AND PIPE					
		1 : 8	1 : 4	3 : 8	1 : 2	5 : 8	3 : 4 *
		MINIMUM RUNS OF STRAIGHT PIPE REQUIRED UPSTREAM AND DOWNSTREAM OF ORIFICE, IN PIPE DIAMETERS (NPS)					
1	U	6	6	6	6¾	10	17
1	D	2½	3	3¼	3¾	4	4½
2	U	13	13	13	15	20	31
2	D	2½	3	3¼	3¾	4	4½
3	U	6	6	6	7½	10¼	13½
3	D	2½	3	3¼	3¾	4	4½
4	U	5	5	5½	6½	8¼	11
4	D	2½	3	3¼	3¾	4	4½
5	U	16½	18½	21½	25	32	44
5	D	2½	3	3¼	3¾	4	4½

* USE THIS COLUMN FOR PRELIMINARY PLANNING

KEY: PIPING ARRANGEMENTS FOR ABOVE RUN LENGTHS

1	Ell or Tee
2	Two 90° Ells
3	Reducer or Increaser
4	Gate Valve
5	Globe Valve

[121]

CLEARANCES

Clear space should be left around an orifice assembly. Figure 6.38 shows minimum clearances required for mounting instruments, seal pots, etc., and for maintenance.

CLEARANCES TO ORIFICE ASSEMBLIES — **FIGURE 6.38**

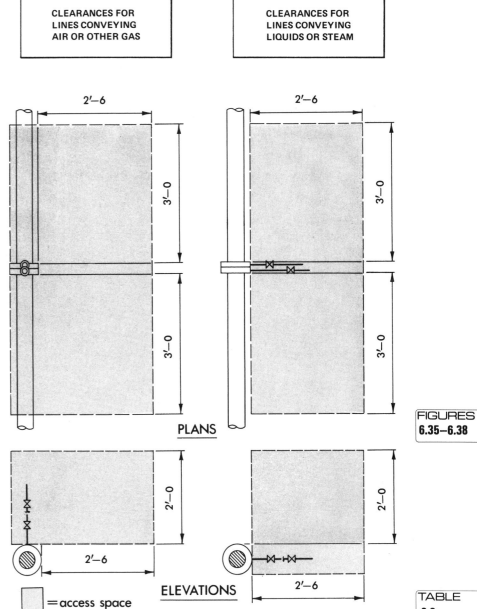

CLEARANCES FOR LINES CONVEYING AIR OR OTHER GAS

CLEARANCES FOR LINES CONVEYING LIQUIDS OR STEAM

PLANS

ELEVATIONS

= access space

To ensure continuity of plant operations it is necessary to maintain some process, service and utility lines within a desired temperature range in order to keep materials in a fluid state, to prevent degradation, and to prevent damage caused by liquids freezing in cold conditions. Piping can be kept warm by insulation, or by applying heat to the insulated piping—this is 'jacketing' or 'tracing', as discussed in 6.8.2 and 6.8.3.

THERMAL INSULATION   6.8.1

INSULATION

'Insulation' is covering material having poor thermal conductivity applied externally to pipe and vessels, and is used: (1) To retain heat in a pipe or vessel so as to maintain process temperature or prevent freezing. (2) To minimize transfer of heat from the surroundings into the line or vessel. (3) To safeguard personnel from hot lines. The choice of insulation is normally included with the piping specification. The method of showing insulation on piping drawings is included in chart 5.7.

Installed insulation normally consists of three parts: (1) The thermal insulating material. (2) The protective covering for it. (3) The metal banding to fasten the covering. Most insulating materials are supplied in formed pieces to fit elbows, etc. Formed coverings are also available. Additionally, it is customary to paint the installed insulation, and to weatherproof it before painting, if for external use.

The principal thermal insulating materials and their accepted approximate maximum line temperatures, where temperature cycling (repetitive heating and cooling periods) occurs are: asbestos (1200 F), calcium silicate (1200 F), cellular glass [foamglas] (800 F), cellular silica (1600 F), diatomaceous silica plus asbestos (1600 F), mineral fiber (250–1200 F, depending on type), mineral wool (1200 F), magnesia (600 F), and polyurethane foam (250 F). Certain foamed plastics have a very low conductivity, and are suitable for insulating lines as cold as −400 F. Rock cork [bonded mineral fiber] is satisfactory down to −250 F, and mineral wool down to −150 F.

HOW THICK SHOULD INSULATION BE ?

Most insulation in a plant will not exceed 2 inches in thickness. A rough guide to insulation thicknesses of the more common materials required on pipe to 8-inch size is:

GUIDE TO INSULATION THICKNESS   TABLE 6.7

APPLICATION	TYPICAL INSULATING MATERIAL	USUAL THICKNESS OF INSULATION
Hot Lines (to 500 F)	Asbestos, Silicate, Magnesia	1 to 2 inches
Cold Lines (to −150 F)	Mineral Wool	1 to 3 inches
Personnel Protection	Asbestos, Silicate, Magnesia	1 inch

For personnel protection insulation should be provided up to a height of about 8 ft above operating floor level. Alternately, wire mesh guards can be provided. The following more detailed table gives insulation thickness for heat conservation, based on 85% magnesia to 600 F, and calcium silicate above 600 F.

INSULATION REQUIRED FOR PIPE
AT VARIOUS TEMPERATURES   TABLE 6.8

NOMINAL PIPE SIZE (in.)	INCHES THICKNESS OF INSULATION FOR STATED TEMPERATURE RANGE					
	below 400	400-549	550-699	700-899	900-1049	1050-1200
to 1	1	1	1.5	2	2	2.5
1.5	1	1.5	1.5	2	2	2.5
2	1	1.5	1.5	2	2.5	3
3	1	1.5	1.5	2.5	2.5	3
4	1	1.5	1.5	2.5	2.5	3.5
6	1	1.5	1.5	2.5	3	3.5
8	1.5	1.5	1.5	2.5	3	3.5
10	1.5	1.5	2	2.5	3	4
12	1.5	1.5	2	2.5	3	4
14	1.5	2	2	3	3	4
16	2	2	2	3	3.5	4
18	2	2	2	3	3.5	4
20	2	2	2	3	3.5	4
24	2	2	2	3	3.5	4

JACKETING & TRACING   6.8.2

The common methods by which temperatures are maintained, other than by simple insulation, are jacketing and tracing (with insulation).

JACKETING

Usually, 'jacketing' refers to double-walled construction of pipe, valves, vessels, hose, etc., designed so that a hot or cold fluid can circulate in the cavity between the walls. Heating media include water, oils, steam, or proprietary high-boiling-point fluids which can be circulated at low pressure, such as Dowtherm or Therminol. Cooling media include water, water mixtures and various alcohols.

Jacketed pipe can be made by the piping fabricator, but an engineered system bought from a specialist manufacturer would be a more reliable choice. The jumpover lines connecting adjacent jackets, thru which the heating or cooling medium flows are factory-made by the specialist manufacturer with less joints than those made on-site, where as many as nine screwed joints may be necessary to make one jumpover. Details of the range of fittings, valves and equipment available and methods of construction for steel jacketed piping systems can be found in Parks-Cramer's and other catalogs.

Another type of jacketing is 'Platecoil' (Tranter Manufacturing Inc.) which is a name given to heat transfer units fabricated from embossed metal sheets, joined together to form internal channeling thru which the heating (or cooling) fluid is passed. The term 'jacketing' is also applied to electric heating pads or mantles which are formed to fit equipment. It also sometimes refers to the spiral winding of electric tracing and fluid tracing lines around pipes, vessels, etc.

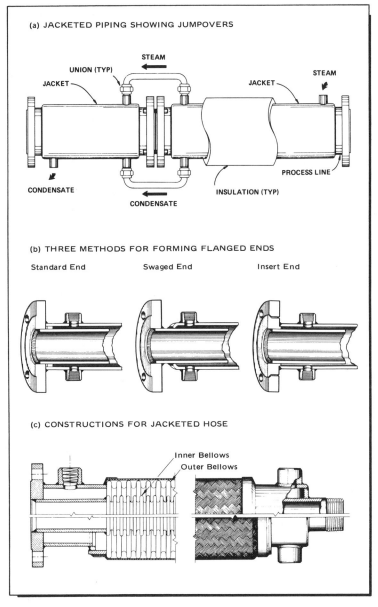

(a) JACKETED PIPING SHOWING JUMPOVERS

STEAM
UNION (TYP)
JACKET
JACKET
STEAM
PROCESS LINE
CONDENSATE
INSULATION (TYP)
CONDENSATE

(b) THREE METHODS FOR FORMING FLANGED ENDS

Standard End   Swaged End   Insert End

(c) CONSTRUCTIONS FOR JACKETED HOSE

Inner Bellows
Outer Bellows

TRACING

External 'tracing' consists in running tubing filled with a hot fluid (usually steam), or electric heating cables, in contact with the outer surface of the pipe to be kept warm. The tubing or cables may be run parallel to the pipe or wound spirally around it. The pipe and tracer(s) are encased in thermal insulation.

An alternative, now little used due to sealing and cleaning problems, is internal tracing by means of tubing fitted inside the line to be heated. An internal tracer is termed a 'gutline'.

'Unitrace' (Aluminum Company of America) is an integral product and tracer pipe extruded in aluminum, which gives excellent heat transfer. The system uses flanges and jumpover fittings similar to those used for jacketed systems to connect adjacent traced sections of the lines.

Electric tracing allows close control of temperature, and can provide a wider range of temperatures than steam heating.

GETTING HEAT TO THE PROCESS LINE (USING STEAM)

If the process line temperature has to approach that of the available steam, jacketing gives the best results. Barton and Williams have stated [4] that the cheaper method of welding steam tracers directly to the process lines has proven adequate. In this unusual method, the welding is 'tack' or continuous depending on how much heat is required to be transferred thru the weld.

A greater rate of heat transfer may be achieved by using two (seldom more) parallel tracers. Sometimes a single tracer is spirally wound about the pipe, but spiral winding should be restricted to vertical lines where condensate can drain by gravity. If the temperature of the conveyed fluid has to be closely maintained, winding the tracer is too inaccurate—but it is a suitable method for getting increased heating in non-critical applications.

To improve heat transfer between the tracer and pipe, they may either be pressed into contact by banding or wiring them together at frequent (1 to 4 ft) intervals, or a heat-conducting cement such as 'Thermon' can be applied. Unless used to anchor the tracer, banding is normally applied sufficiently loosely to permit the tracer to expand.

Hot spots occur at the bands. If this is undesirable for a product line, a thin piece of asbestos may be inserted between tracer and line.

CHOOSING THE SYSTEM

There are advantages and disadvantages with the various systems. Piping which is to be externally traced can be planned with little concern for the tracing.

Fluid-jacketed systems are flanged, and last-minute changes could result in delays. Jacketing offers superior heat transfer and should be seriously considered for product lines, especially for those conveying viscous liquids and material which may solidify, whereas service lines usually just need to be kept from freezing and tracing is quite adequate for them. If process material has to be kept cold in the line, refrigerant-jacketed systems are the only practicable choice.

For process lines, all systems should be evaluated on the criteria of heat distribution, initial cost and long-term operating and maintenance costs before a decision can be made.

WHERE TRACING & JACKETING ARE SHOWN

Using the symbols given in chart 5.7, tracing is shown on the plan and elevation drawings of the plant piping and it will similarly be indicated on the isometric drawings. It will also be indicated on any model used. Tracing is one of the last aspects of plant design, and steam subheaders can either be shown directly on the piping drawings or on sepias or film prints.

FIGURE
6.39

STEAM TRACING

This is a widely-used way of keeping lines warm—surplus steam is usually available for this purpose. Figure 6.40 shows typical tracing arrangements. A steam-tracing system consists of tracer lines separately fed from a steam supply header (or subheader), each tracer terminating with a separate trap. Horizontal pipes are commonly traced along the bottom by a single tracer. Multiply-traced pipe, with more than two tracers, is unusual.

STEAM PRESSURE FOR TRACING

Steam pressures in the range 10 to 200 PSIG are used. Sometimes steam will be available at a suitable pressure for the tracing system, but if the available steam is at too high a pressure, it may be reduced by means of a control (valve) station—see 6.1.4. Low steam pressures may be adequate if tracers are fitted with traps discharging to atmospheric pressure. If a pressurized condensate system is used, steam at 100 to 125 PSIG is preferred.

SIZING HEADERS

The best way to size a steam subheader or condensate header serving several tracers is to calculate the total internal cross-sectional area of all the tracers, and to select the header size offering about the same flow area. Table 6.9 allows quick selection if the tracers are all of the same size:

NUMBER OF TRACERS PER HEADER

TABLE 6.9

HEADER SIZE (IN.)	SIZE OF TRACER (IN.)				
	1/4	3/8	1/2	3/4	1
	NUMBER OF TRACERS				
¾	9	4	2	1	—
1	16	7	4	2	1
1½	36	16	9	4	2
2	64	28	16	7	4

MAXIMUM LENGTHS & RISES

The rate at which condensate forms and fills the line determines the length of the tracer in contact with the pipe. Too many variables are involved to give useful maximum tracer lengths. Most companies have their own design figure (or figures based on experience) for this: usually, length of tracer in contact with pipe does not exceed 250 ft.

1 PSI steam will lift condensate about 2.3 ft, and therefor vertical rises will present no problem unless low-pressure steam is being used. Companies prefer to limit the vertical rise in a tracer at any one place to 6 ft (for 25-49 PSIG steam) or 10 ft (for 50-100 PSIG steam). As a rough guide, the total height, in feet, of all the rises in one tracer may be limited to one quarter of the initial steam pressure, in PSIG. For example, if the initial steam pressure is 100 PSIG, the total height of all risers in the tracer should be limited to 25 ft. The rise for a sloped tracer is the difference in elevations between the ends of the sloping part of the tracer.

EXPANSION OF THE TRACER, & ANCHORING

Expansion can be accommodated by looping the tracer at elbows and/or providing horizontal expansion loops in the tracer. Vertical downward expansion loops obstruct draining and will cause trouble in freezing climates, unless the design includes a drain at the bottom of the loop, or a union to break the loop. It is necessary to anchor tracers to control the amount of expansion that can be tolerated in any one direction. Straight tracers 100 ft or longer are usually anchored at their midpoints.

Expansion at elbows must be limited where no loop is used and excessive movement of the tracer could lift the insulation. In such cases the tracer is anchored not more than 10 to 25 ft away from an elbow which limits start-up expansion to 1/2 to 3/4 inch in most cases. The distance of the anchor from the elbow is best calculated from the ambient and steam temperatures.

EXAMPLE: System traced with copper tubing: coefficient of linear expansion of copper = 0.000009 per deg F. Steam pressure to be used = 50 PSIG (equivalent steam temperature 298F). Lowest ambient temperature = 50 F. If the anchor is located 20 ft from the elbow, the maximum expansion in inches is $(298-50)(0.000009)(20)(12) = 0.53$ in. This expansion will usually be tolerable even for a small line with the tracer construction for elbows shown in figure 6.40.

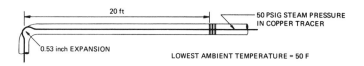

PIPE, TUBE & FITTINGS FOR TRACING

SCH 80 carbon steel pipe, or copper or stainless steel tubing is used for tracers. Selection is based on steam pressure and required tracer size. In practice, tracers are either 1/2 or 3/8-inch size, as smaller sizes involve too much pressure drop, and larger material does not bend well enough for customary field installation.

1/2-inch OD copper tube is the most economic material for tracing straight piping. 3/8-inch OD copper tubing is more useful where small bends are required around valve bodies, etc. Copper tubing can be used for pressures up to 150 PSIG (or to 370 F). Table T-1 gives data for copper tube.

Supply lines from the header are usually socket welded or screwed and seal-welded depending on the pressures involved and the company's practice. A pipe-to-tube connector is used to make the connection between the steel pipe and tracer tube — see figure 2.41.

TRACING VALVES & EQUIPMENT

Different methods are used. Some companies require valves to be wrapped with tracer tubing. Others merely run the tubing in a vertical loop alongside and against the valve body. In either method, room should be left for removing flange bolts, and unions should be placed in the tracer so that the valve or equipment can be removed.

STEAM TRACING DETAILS

FIGURE 6.40

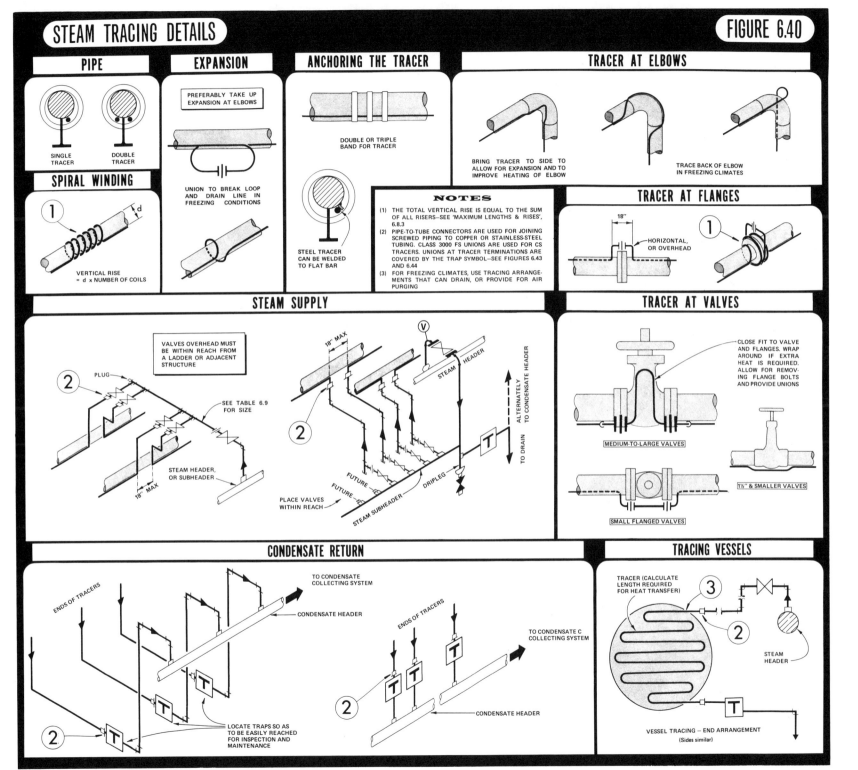

PIPE

SINGLE TRACER

DOUBLE TRACER

SPIRAL WINDING

①

d

VERTICAL RISE = d x NUMBER OF COILS

EXPANSION

PREFERABLY TAKE UP EXPANSION AT ELBOWS

UNION TO BREAK LOOP AND DRAIN LINE IN FREEZING CONDITIONS

ANCHORING THE TRACER

DOUBLE OR TRIPLE BAND FOR TRACER

STEEL TRACER CAN BE WELDED TO FLAT BAR

NOTES

(1) THE TOTAL VERTICAL RISE IS EQUAL TO THE SUM OF ALL RISERS—SEE 'MAXIMUM LENGTHS & RISES', 6.8.3

(2) PIPE-TO-TUBE CONNECTORS ARE USED FOR JOINING SCREWED PIPING TO COPPER OR STAINLESS-STEEL TUBING. CLASS 3000 FS UNIONS ARE USED FOR CS TRACERS. UNIONS AT TRACER TERMINATIONS ARE COVERED BY THE TRAP SYMBOL—SEE FIGURES 6.43 AND 6.44

(3) FOR FREEZING CLIMATES, USE TRACING ARRANGEMENTS THAT CAN DRAIN, OR PROVIDE FOR AIR PURGING

TRACER AT ELBOWS

BRING TRACER TO SIDE TO ALLOW FOR EXPANSION AND TO IMPROVE HEATING OF ELBOW

TRACE BACK OF ELBOW IN FREEZING CLIMATES

TRACER AT FLANGES

18"

HORIZONTAL, OR OVERHEAD

①

STEAM SUPPLY

VALVES OVERHEAD MUST BE WITHIN REACH FROM A LADDER OR ADJACENT STRUCTURE

② PLUG

SEE TABLE 6.9 FOR SIZE

STEAM HEADER, OR SUBHEADER

18" MAX

18" MAX

②

STEAM HEADER

ALTERNATELY TO CONDENSATE HEADER

TO DRAIN

FUTURE

FUTURE

STEAM SUBHEADER

DRIPLEG

PLACE VALVES WITHIN REACH

Ⓥ

T

TRACER AT VALVES

CLOSE FIT TO VALVE AND FLANGES. WRAP AROUND IF EXTRA HEAT IS REQUIRED. ALLOW FOR REMOVING FLANGE BOLTS AND PROVIDE UNIONS

MEDIUM-TO-LARGE VALVES

SMALL FLANGED VALVES

1½" & SMALLER VALVES

CONDENSATE RETURN

ENDS OF TRACERS

TO CONDENSATE COLLECTING SYSTEM

CONDENSATE HEADER

② LOCATE TRAPS SO AS TO BE EASILY REACHED FOR INSPECTION AND MAINTENANCE

ENDS OF TRACERS

②

T T T

TO CONDENSATE C COLLECTING SYSTEM

CONDENSATE HEADER

TRACING VESSELS

TRACER (CALCULATE LENGTH REQUIRED FOR HEAT TRANSFER)

③

②

STEAM HEADER

T

VESSEL TRACING — END ARRANGEMENT (Sides similar)

FIGURE 6.40

TABLE 6.9

DESIGN POINTS FOR STEAM TRACING & INSULATION

- Run tracers parallel to and against the underside of the pipe to be heated

- Ensure that the temperature limit for process material is not exceeded by the temperature of the steam supplying the tracer. Hot spots occur at bands—see 6.8.2, under 'Getting heat to the process line'

- Run a steam subheader from the most convenient source if there is no suitable existing steam supply that can be used either directly or by reducing the pressure of the available steam

- Take tracer lines separately from the top of the subheader, and provide an isolating valve in the horizontal run

- Feed steam first to the highest point of the system of lines to be traced, so that gravity will assist the flow of condensate to trap(s) and condensate header

- Do not split (branch) a tracer and then rejoin—the shorter limb would take most of the steam

- Preferably, absorb expansion of the tracer at elbows. If loops are used in the line, arrange them to drain on shutdown

- Keep loops around flanges horizontal or overhead, and provide unions so that tracers can be disconnected at flanges

- If possible, group supply points and traps, locating traps at grade or platform level

- Do not place a trap at every low point of a tracer (as is the practice with steam lines) but provide a trap at the end of the tracer

- Do not run more than one tracer to a trap

- Increased heating may be obtained:
  - (1) By using more than one tracer
  - (2) By winding the tracer in a spiral around the line
  - (3) By applying heat-transfer cement to the tracer and line
  - (4) By welding the tracer to the line—refer to 6.8.2, under 'Getting heat to the process line'

- Reserve spiral winding of tracers for vertical lines where condensate can drain by gravity flow

- In freezing conditions, provide drains at low points—and at other points where condensate could collect during shutdown

- Provide slots in insulation to accommodate expansion of the tracer where it joins and leaves the line to be traced

- Indicate thickness of insulation to envelop line and tracer, and show whether insulation is also required at flanges

- Indicate limits for insulation for personnel protection—see 6.8.1, under 'How thick should insulation be?', and chart 5.7

- Provide crosses instead of elbows and flanged joints at intervals in heated lines conveying materials which may solidify, to permit cleaning if the heating fails

EXPLANATIONS OF STEAM TERMS 6.9.1

HOW STEAM IS FORMED

Steam is a convenient and easily handled medium for heating, for driving machinery, for cleaning, and for creating vacuum.

After water has reached the boiling point, further addition of heat will convert water into the vapor state: that is, steam. During boiling there is no further rise in temperature of the water, but the vaporization of the water uses up heat. This added heat energy, which is not shown by a rise in temperature, is termed 'latent heat of vaporization', and varies with pressure.

In boiling one pound of water at atmospheric pressure (14.7 PSIA) 970.3 BTU is absorbed. If the steam condenses back into water (still at the boiling temperature and 14.7 PSIA) it will release exactly the amount of heat it absorbed on vaporizing.

The term 'saturated steam' refers to both *dry steam* and *wet steam*, described below. Steam tables give pressure and temperature data applicable to dry and to wet steam. Small amounts of air, carbon dioxide, etc., are present in steam from industrial boilers.

STEAM/WATER/ICE DIAGRAM CHART 6.3

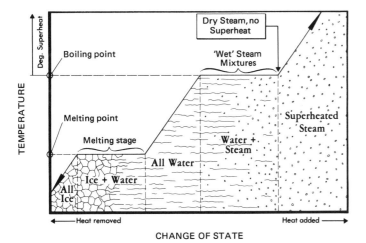

CHANGE OF STATE

DRY STEAM

Dry steam is a gas, consisting of water vapor only. Placed in contact with water at the same temperature, dry steam will not condense, nor will more steam form—liquid and vapor are in equilibrium.

WET STEAM

Wet steam consists of water vapor and suspended water particles at the same temperature as the vapor. Heating ability ('quality') varies with the percentage of dry steam in the mixture (the water particles contain no latent heat of vaporization). Like dry steam, wet steam is in equilibrium with water at the same temperature.

SUPERHEATED STEAM

If heat is added to a quantity of dry steam, the temperature of the steam will rise, and the number of degrees rise in temperature is the 'degrees of superheat'. Thus, superheat is 'sensible' heat — that is, it can be measured by a thermometer.

EFFECT OF PRESSURE CHANGE

Under normal atmospheric pressure (14.7 PSIA) pure water boils at 212 F. Reduction of the pressure over the water will lower the boiling point. Increase in pressure raises the boiling point. Steam tables give boiling points corresponding to particular pressures.

FLASH STEAM

Suppose a quantity of water is being boiled at 300 PSIA (corresponding to 417 F). If the source of heat is removed, boiling ceases. If the pressure over the water is then reduced, say from 300 to 250 PSIA, the water starts boiling on its own, without any outside heat applied, until the temperature drops to 401 F (this temperature corresponds to 250 PSIA). Such spontaneous boiling due to reduction in pressure is termed 'flashing', and the steam produced, 'flash steam'.

The data provided in steam tables enable calculation of the quantity and temperature of steam produced in 'flashing'.

CONDENSATE — WHAT IT IS & HOW IT FORMS

Steam in a line will give up heat to the piping and surroundings, and will gradually become 'wetter', its temperature remaining the same. The change of state of part of the vapor to liquid gives heat to the piping without lowering the temperature in the line. The water that forms is termed 'condensate'. If the line initially contains superheated steam, heat lost to the piping and surroundings will first cause the steam to lose sensible heat until the steam temperature drops to that of dry steam at the line pressure.

AIR IN STEAM

With both dry and wet steam, a certain pressure will correspond to a certain temperature. The temperature of the steam at various pressures can be found in steam tables. If air is mixed with steam, this relationship between pressure and temperature no longer holds. The more air that is admixed, the more the temperature is reduced below that of steam at the same pressure. There is no practicable way to separate air from steam (without condensation) once it is mixed.

LOW-PRESSURE HEATING MEDIA 6.9.2

Special liquid media such as Dowtherms (Dow Chemical Co.) and Therminols (Monsanto Co.) can be boiled like water, but the same vapor temperatures as steam are obtained at lower pressures. Heating systems using these liquids are more complicated than steam systems, and experience with them is necessary in order to design an efficient installation. However, the basic principles of steam-heating systems apply.

STEAM PIPING 6.10

REMOVING AIR FROM STEAM LINES 6.10.1

Air in steam lines lowers the temperature for a given pressure, and calculated rates of heating may not be met. See 6.9.1 under 'Air in steam'.

The most economic means for removing air from steam lines is automatically thru temperature-sensitive traps or traps fitted with temperature-sensitive air-venting devices placed at points remote from the steam supply. When full line temperature is attained the vent valves will close completely. See 6.10.7 under 'Temperature-sensitive (or thermostatic) traps'.

WHY PLACE VENTS AT REMOTE POINTS ?

On start-up, cold lines will be filled with air. Steam issuing from the source will mix with some of this air, but will also act as a piston pushing air to the remote end of each line.

WHY REMOVE CONDENSATE ? 6.10.2

In heating systems using steam with little or no superheat, steam condenses to form water, termed 'condensate', which is essentially distilled water. Too valuable to waste, condensate is returned for use as boiler feedwater unless it is contaminated with oil (usually from a steam engine) or unless it is uneconomic to do so, when it can either be used locally as a source of hot water, or run to a drain. If condensate is not removed:—

- Steam with entrained water droplets will form a dense water film on heat transfer surfaces and interfere with heating
- Condensate can be swept along by the rapidly-moving steam (at 120 ft/sec or more) and the high-velocity impact of slugs of water with fittings, etc. (waterhammer) may cause erosion or damage

UTILIZING CONDENSATE FIGURE 6.41

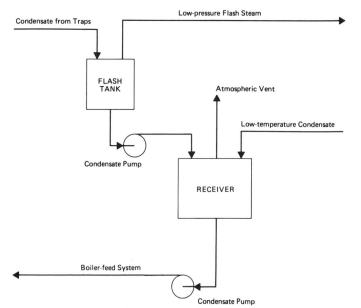

In early steam systems, there was considerable waste of steam and condensate after passing thru heating coils, etc., as steam was merely vented to the open air. Later, the wastefulness of this resulted in closed steam lines from which only the condensed steam was removed and then re-fed to the boiler. The removal of condensate to atmospheric pressure was effected with traps—special automatic discharge valves—see 6.10.7.

This was a much more efficient system, but it still wasted flash steam. On passing thru the traps, the depressurized condensate boiled, generating lower-pressure steam. In modern systems, this flash steam is used and the residual condensate returned to the boiler.

STEAM SEPARATOR OR DRYER 6.10.3

This is an in-line device which provides better drying of steam being immediately fed to equipment. A separator is shown in figure 2.67. It separates droplets entrained in the steam which have been picked up from condensate in the pipe and from the pipe walls, by means of one or more baffles (which cause a large pressure drop). The collected liquid is piped to a trap.

SLOPING & DRAINING STEAM & CONDENSATE LINES 6.10.4

Sloping of steam and condensate lines is discussed in 6.2.6, under 'Sloped lines avoid pocketing and aid draining'.

Condensate is collected from a steam line either by a steam separator (sometimes termed a 'dryer')—see 6.10.3 above—or more cheaply by a dripleg (drip pocket or well — see below) from where it passes to a trap for periodic discharge to a condensate return line or header which will be at a lower pressure than the steam line. The header is either taken to a boiler feedwater tank feeding make-up water to the boiler or to a hotwell for pumping to the boiler feedwater tank.

DRIPLEGS COLLECT CONDENSATE 6.10.5

It is futile to provide a small dripleg or drain pocket on large lines, as the condensate will not be collected efficiently.

Driplegs are made from pipe and fittings. Figure 6.42 shows three methods of construction, and table 6.10 suggests dripleg and valve sizes.

DRIPLEG CONSTRUCTIONS FIGURE 6.42

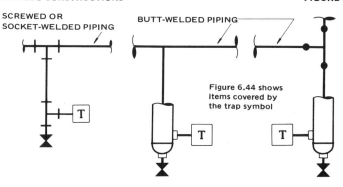

		DIMENSIONS & SIZES (NOMINAL) IN INCHES										
LINE SIZE	*	3	4	6	8	10	12	14	16	18	20	24
DIMENSION 'A'		3	4	6	6	8	8	10	12	12	12	12
DIMENSION 'B'		12	12	14	14	16	16	18	20	21	22	24
SIZE OF V₁		¾	¾	¾	¾	¾	¾	1	1	1	1	1
SIZE OF V₂		¾	¾	¾	¾	¾	1	1	1	1	1	1

TO 2"

*For lines 2-inch and smaller, use ¾-inch pipe, valves and fittings, reducing line size at the trap as necessary

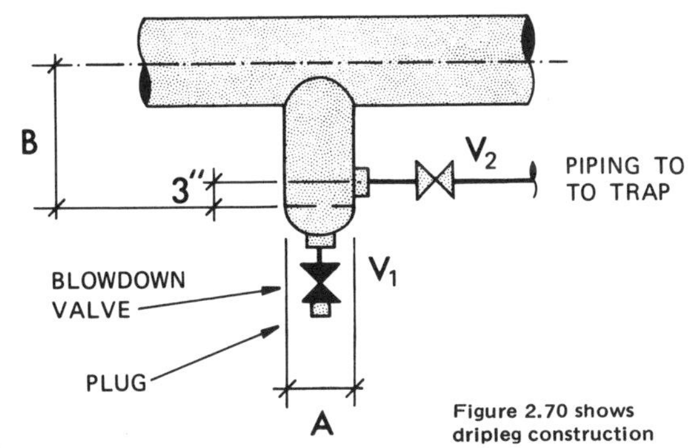

B

3"

V₂   PIPING TO TO TRAP

BLOWDOWN VALVE   V₁

PLUG

A

Figure 2.70 shows dripleg construction

STEAM LINE PRESSURE FORCES CONDENSATE INTO RECOVERY SYSTEM   6.10.6

In almost every steam-heating system where condensate is recovered the trapped condensate has to be lifted to a condensate header and run to a boiler feedwater tank, either directly or via a receiver. Each PSI of steam pressure behind a trap can lift the condensate about two feet vertically. The pressure available for lifting the condensate is the pressure difference between the steam and condensate lines less any pressure drop over pipe, valves, fittings, trap, etc.

STEAM TRAPS   6.10.7

The purpose of fitting traps to steam lines is to obtain fast heating of systems and equipment by freeing the steam lines of condensate and air. A steam trap is a valve device able to discharge condensate from a steam line without also discharging steam. A secondary duty is to discharge air—at start-up, lines are full of air which has to be flushed out by the steam, and in continuous operation a small amount of air and non-condensible gases introduced in the boiler feedwater have also to be vented.

Some traps have built-in strainers to give protection from dirt and scale which may cause the trap to jam in an open position. Traps are also available with checking features to safeguard against backflow of condensate. Refer to the manufacturers' catalogs for details.

Choosing a trap from the many designs should be based on the trap's ability to operate with minimal maintenance, and on its cost. To reduce inventory and aid maintenance, the minimum number of types of trap should be used in a plant. The assistance of manufacturers' representatives should be sought before trap types and sizes are selected.

Steam traps are designed to react to changes in temperature, pressure or density:

TEMPERATURE-SENSITIVE (or 'THERMOSTATIC') TRAPS are of two types: The first type operates by the movement of a liquid-filled bellows, and the second uses a bimetal element. Both types are open when cold and readily discharge air and condensate at start-up. Steam is in direct contact with the closing valve and there is a time delay with both types in operating. A large dripleg allowing time for condensate to cool improves operation. As these traps are actuated by temperature differential, they are economic at steam pressures greater than 6 PSIG. The temperature rating of the bellows and the possibility of damage by waterhammer should be considered—refer to 6.10.8.

IMPULSE TRAPS are also referred to as 'thermodynamic' and 'controlled disc'. These traps are most suited to applications where the pressure downstream of the trap is less than about half the upstream pressure. Waterhammer does not affect operation. They are suitable for steam pressures over 8 PSIG.

DENSITY-SENSITIVE TRAPS are made in 'float' and 'bucket' designs. The *float trap* is able to discharge condensate continuously, but this trap will not discharge air unless fitted with a temperature-sensitive vent (the temperature limitation of the vent should be checked). Float traps sometimes may fail from severe waterhammer. The *inverted bucket trap* (see 3.1.9) is probably the most-used type. The trap is open when cold, but will not discharge large quantities of air at startup unless the bucket is fitted with a temperature-sensitive vent. The action in discharging condensate is rapid. Steam will be discharged if the trap loses its priming water due to an upstream valve being opened; refer to note (9) in the key to figure 6.43. Inverted bucket traps will operate at pressures down to 1/4 PSIG.

FLASHING   6.10.8

Refer to 6.9.1. When hot condensate under pressure is released to a lower pressure return line, the condensate immediately boils. This is referred to as 'flashing' and the steam produced as 'flash steam'.

The hotter the steam line and the colder the condensate discharge line, the more flashing will take place; it can be severe if the condensate comes from high pressure steam. Only part of the condensate forms steam. However, if the header is inadequately sized to cope with the quantity of flash steam produced and backpressure builds up, waterhammer can result.

Often, where a trap is run to a drain, a lot of steam seems to be passing thru the trap, but this is usually only from condensate flashing.

DRAINING SUPERHEATED STEAM LINES   6.10.9

Steam lines with more than a few degrees of superheat will not usually form condensate in operation. During the warming-up period after starting a cold circuit, the large bulk of metal in the piping will nearly always use up the degrees of superheat to produce a quantity of condensate.

FOR COLLECTED CONDENSATE FIGURE 6.43

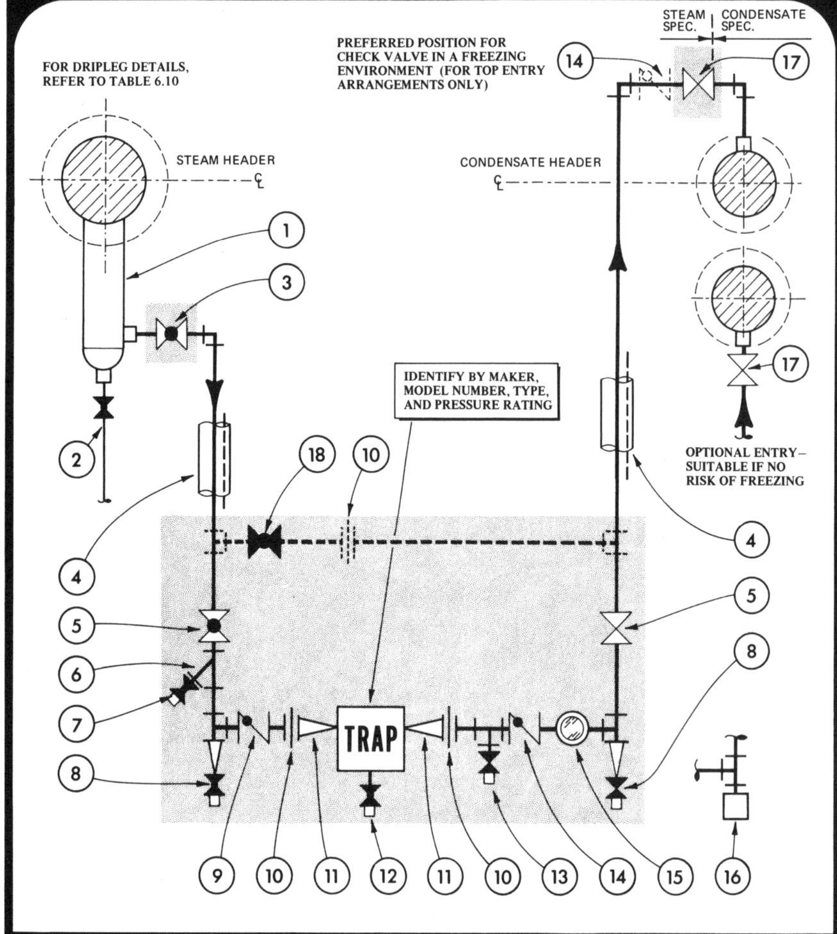

FOR DRIPLEG DETAILS, REFER TO TABLE 6.10

STEAM HEADER

PREFERRED POSITION FOR CHECK VALVE IN A FREEZING ENVIRONMENT (FOR TOP ENTRY ARRANGEMENTS ONLY)

CONDENSATE HEADER

STEAM SPEC.   CONDENSATE SPEC.

IDENTIFY BY MAKER, MODEL NUMBER, TYPE, AND PRESSURE RATING

OPTIONAL ENTRY— SUITABLE IF NO RISK OF FREEZING

TRAP

FOR DRAINED CONDENSATE FIGURE 6.44

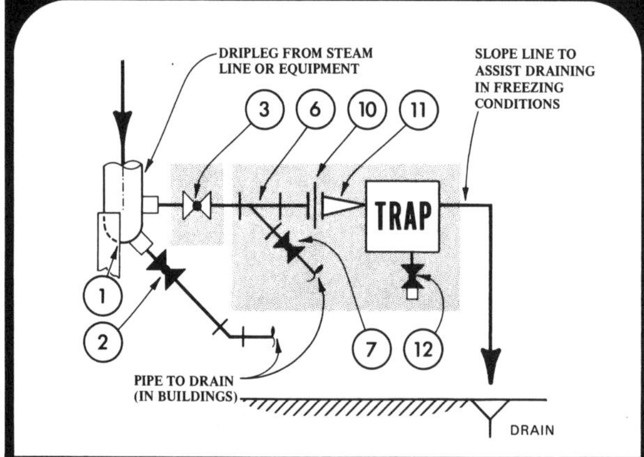

DRIPLEG FROM STEAM LINE OR EQUIPMENT

SLOPE LINE TO ASSIST DRAINING IN FREEZING CONDITIONS

TRAP

PIPE TO DRAIN (IN BUILDINGS)

DRAIN

SYMBOL

T

Pipe, fittings and valves within shaded areas in figures 6.43 and 6.44 are shown on drawings by the above symbol

KEY

FIGURES 6.43 & 6.44 SHOW EQUIPMENT WHICH CAN BE USED IN TRAP PIPING ARRANGEMENTS. ONLY ITEMS OF EQUIPMENT NECESSARY FOR ECONOMIC & SAFE DESIGN NEED BE USED. THE FOLLOWING NOTES WILL AID SELECTION

(1) DRIPLEG FROM STEAM HEADER, OR LINE TO EQUIPMENT, OR LINE FROM STEAM-FED EQUIPMENT

(2) DRIPLEG VALVE FOR PERIODICALLY BLOWING DOWN SEDIMENT. FOR SAFETY, VALVE SHOULD BE PIPED TO A DRAIN OR TO GRADE

(3) ISOLATING VALVE TO BE LOCATED CLOSE TO DRIPLEG

(4) ★ INSULATION. NEEDED IN A COLD ENVIRONMENT IF THERE IS A RISK OF CONDENSATE FREEZING AS A RESULT OF SHUTDOWN OR INTERMITTENT OPERATION. IN EXTREME COLD, TRACING MAY ALSO BE REQUIRED—IF STEAM IS NOT CONSTANTLY AVAILABLE FOR THIS PURPOSE, ELECTRIC TRACING WOULD BE NECESSARY

(5) ★ ISOLATING VALVE. REQUIRED ONLY IF VALVES (3) AND (17) ARE OUT OF REACH, OR IF A BYPASS IS USED—SEE NOTE (18)

(6) STRAINER. NORMALLY FITTED IN LINES TO TRAPS OF LESS THAN 2-INCH SIZE. A STRAINER MAY BE AN INTEGRAL FEATURE OF THE TRAP

(7) ★ VALVE FOR BLOWING STRAINER SEDIMENT TO ATMOSPHERE. PLUG FOR SAFETY

(8) ★ MANUALLY-OPERATED DRAIN VALVE FOR USE IN FREEZING CONDITIONS WHEN THE TRAP IS POSITIONED HORIZONTALLY — SEE NOTE (16)

(9) ★ CHECK VALVE. PRIMARILY REQUIRED IN LINES USING BUCKET TRAPS TO PREVENT LOSS OF SEAL WATER IF DIFFERENTIAL PRESSURE ACROSS TRAP REVERSES DUE TO BLOWING DOWN THE LINE OR STRAINER UPSTREAM OF THE TRAP

(10) UNIONS FOR REMOVING TRAP, ETC

(11) ★ SWAGES FOR ADAPTING TRAP TO SIZE OF LINE

(12) ★ BLOWDOWN VALVE FOR A TRAP WITH A BUILT-IN STRAINER (ALTERNATIVE TO (6))

(13) ★ TEST VALVE SHOWS IF A FAULTY TRAP IS PASSING STEAM. SOMETIMES, BODY OF TRAP HAS A TAPPED PORT FOR FITTING THIS VALVE

(14) ★ CHECK VALVE PREVENTS BACKFLOW THRU TRAP IF CONDENSATE IS BEING RETURNED TO A HEADER FROM MORE THAN ONE TRAP. IN THE LOWER POSITION, THE VALVE HAS THE ASSISTANCE OF A COLUMN OF WATER TO HELP IT CLOSE AND TO GIVE IT A WATER SEAL. REQUIRED IF SEVERAL TRAPS DISCHARGE INTO A SINGLE HEADER WHICH IS OR MAY BE UNDER PRESSURE

(15) ★ SIGHT GLASS ALLOWS VISUAL CHECK THAT TRAP IS DISCHARGING CORRECTLY INTO A PRESSURIZED CONDENSATE RETURN LINE, BUT IS SELDOM USED BECAUSE THE GLASS MAY ERODE, PRESENTING A RISK OF EXPLOSION

(16) ★ TEMPERATURE-SENSITIVE (AUTOMATIC) DRAIN ALLOWS LINE TO EMPTY, PREVENTING DAMAGE TO PIPING IN A COLD ENVIRONMENT (SEE NOTE (4)). IF VALVE (14) IS OVER-HEAD, THE AUTOMATIC DRAIN MAY BE FITTED TO THE TRAP — SOME TRAP BODIES PROVIDE FOR THIS

(17) ISOLATING VALVE AT HEADER

(18) ★ BY-PASS. NOT RECOMMENDED AS IT CAN BE LEFT OPEN. IT IS BETTER TO PROVIDE A STANDBY TRAP

★ ASTERISK INDICATES THAT THE EQUIPMENT IS OPTIONAL AND IS NOT ESSENTIAL TO THE BASIC TRAP PIPING DESIGN

Start-ups are infrequent and with more than a few degrees of superheat it is unnecessary to trap a system which is continuously operated. These superheated steam lines can operate with driplegs only, and are usually fitted with a blowdown line having two valves so that condensate can be manually released from the dripleg after startup.

A superheated steam supply to an intermittently operated piece of equipment will require trapping directly before the controlling valve for the equipment, as the temperature will drop at times allowing condensate to form.

PREVENT TRAPS FROM FREEZING   6.10.10

Insulation and steam or electric tracing of the trap and its piping may also be required in freezing environments. Temperature-sensitive and impulse traps are not subject to freezing trouble if mounted correctly, so that the trap can drain. Bucket traps are always mounted with the bucket vertical and a type with top inlet and bottom outlet should be chosen, unless the trap can be drained by fitting an automatic drain.

GUIDELINES TO STEAM TRAP PIPING   6.10.11

- Figures 6.43 thru 6.45 are a guide to piping traps from driplegs, lines, vessels, etc.

- Try to group traps to achieve an orderly arrangement

- Unless instructed otherwise, pipe, valves and fittings will be the same size as the trap connections, but not smaller than 3/4 in.

- Traps are normally fitted at a level lower than the equipment or dripleg that they serve

- Trap each item of equipment using steam separately, even if the steam pressure is common

- Provide driplegs (and traps on all steam lines with little or no superheat) at low points before or at the bottom of risers, at pockets and other places where condensate collects on starting up a cold system. Table 6.10 gives dripleg sizes

- Locate driplegs at the midpoints of exchanger shells, short headers, etc. If dual driplegs are provided it is better to locate them near each end

- For installations in freezing conditions, where condensate is wasted, preferably choose traps that will not pocket water and which can be installed vertically, to allow draining by gravity. Otherwise, select a trap that can be fitted with an automatic draining device by the manufacturer

- Avoid long horizontal discharge lines in freezing conditions, as ice can form in the line from the trap. Keep discharge lines short and pitch them downward, unless they are returning condensate to a header

- For efficient operation of equipment such as heat exchangers using large amounts of steam, consider installing a separator in the steam feed

- 'Syphon' removal of condensate: In certain instances it is not possible to provide a gravity drain path — for example, where condensate is formed inside a rotating drum. The pressure of the steam is used to force ('syphon') the condensate up a tube and into a trap. Figure 6.45 shows such an arrangement

TRAPPING ARRANGEMENT FOR ROTATING DRUM   **FIGURE 6.45**

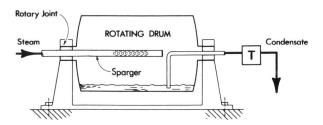

- If condensate is continuously discharging to an open drain in an inside installation a personnel hazard or objectionable atmosphere may be created. To correct this, discharge piping can be connected to an exhaust stack venting to atmosphere and a connection to the main drain provided, as in figure 6.46

CONDENSATE VENT STACK   **FIGURE 6.46**

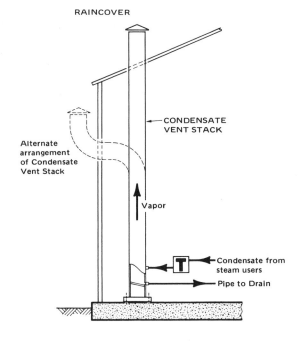

WHY VENTS ARE NEEDED   6.11.1

Vents are needed to let gas (usually air) in and out of systems. When a line or vessel cools, the pressure drops and creates a partial vacuum which can cause syphoning or prevent draining. When pressure rises in storage tanks due to an increase in temperature, it is necessary to release excess pressure. Air must also be released from tanks to allow filling, and admitted to permit draining or pumping out liquids.

Unless air is removed from fuel lines to burners, flame fading can result. In steam lines, air reduces heating efficiency.

After piping has been erected, it is often necessary to subject the system to a hydrostatic test to see if there is any leakage. In compliance with the applicable code, this consists of filling the lines with water or other liquid, closing the line, applying test pressure, and observing how well pressure is maintained for a specified time, while searching for leaks.

As the test pressure is greater than the operating pressure of the system, it is necessary to protect equipment and instruments by closing all relevant valves. Vessels and equipment usually are supplied with a certificate of code compliance. After testing, the valved drains are opened and the vent plugs temporarily removed to allow air into the piping for complete draining.

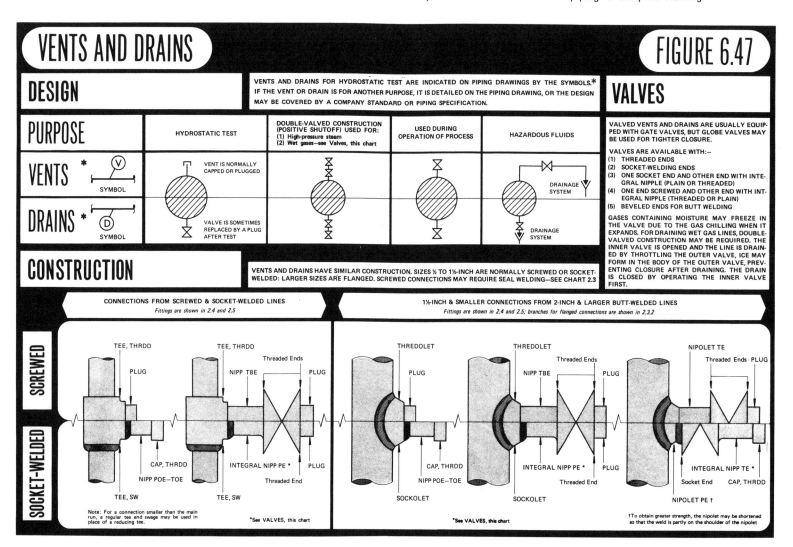

VENTS AND DRAINS   **FIGURE 6.47**

Positions of the required vent and drain points are established on the piping drawings. (P&ID's will show only process vents, such as vacuum breakers, and process drains.) Refer to figure 6.47 for construction details.

VENTING GASES 6.11.3

Quick-opening vents of ample size are needed for gases. Safety and safety-relief valves are the usual venting means. See 3.1.9 for pressure-relieving devices, and 6.1.3, under 'Piping safety and relief valves'.

Gases which offer no serious hazard after some dilution with air may be vented to atmosphere by means ensuring that no direct inhalation can occur. If a (combustible) gas is toxic or has a bad odor, it may be piped to an incinerator or flarestack, and destroyed by burning.

DRAINING COMPRESSED-AIR LINES 6.11.4

Air has a moisture content which is partially carried thru the compressing and cooling stages. It is this moisture that tends to separate, together with any oil, which may have been picked up by the air in passing thru the compressor.

If air for distribution has not been dried, distribution lines should be sloped toward points of use and drains: lines carrying dried air need not be sloped. Sloping is discussed in 6.2.6.

If the compressed-air supply is not dried, provide:—

(1) Traps at all drains from equipment forming or collecting liquid—such as intercooler, aftercooler, separator, receiver.

(2) Driplegs with traps on distribution headers (at low points before rises) and traps or manual drains at the ends of distribution headers.

LIQUID REMOVAL FROM AIR LINES   **FIGURE 6.48**

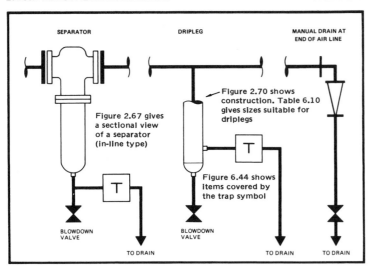

RELIEVING PRESSURE—LIQUIDS 6.12

The buildup of pressure in a liquid is halted by discharging a small amount of liquid. Relieving devices having large ports are not required. Relief valves—see 3.1.9—are used, and need to be piped at the discharge side, but the piping should be kept short. See 6.1.3 under 'Piping safety & relief valves'.

Rarely will the relieved liquid be sufficiently non-hazardous to be piped directly to a sewer. Often the liquid is simply to be reclaimed. Relieved liquid is frequently piped to a 'knockout drum', or to a sump or other receiver for recovery. The P&ID should show what is to be done with the relieved liquid.

RELIEF HEADERS 6.12.1

Headers should be sized to handle adequately the large amounts of vapor and liquid that may be discharged during major mishap. Relief headers taken to knockout drums, receivers or incinerators, are normally sloped, Refer to 6.2.6 and figure 6.3, showing the preferred location of a relief header on a piperack.

WASTES & EFFLUENTS 6.13

Manufacturing processes may generate materials that cannot be recycled, and for which there is no commercial use. These materials are termed 'waste products', or 'wastes'. An 'effluent' is any material flowing from a plant site to the environment. Effluents need not be polluting: for example, properly-treated waste water may be discharged without harming the environment or sewage-treatment plants.

Restrictions on the quantities and nature of effluents discharged into rivers, sewers or the atmosphere, necessitate treatment of wastes prior to discharge. Waste treatment is increasingly a factor in plant design, whether wastes are processed at the plant, or are transported for treatment elsewhere. For in-plant treatment, waste-treatment facilities are described on separate P&ID's (see 5.2.4) and should be designed in consultation with the responsible local authority.

Liquid wastes have to be collected within a plant, usually by a special drainage system. Corrosive and hazardous properties of liquid wastes will affect the choice and design of pipe, fittings, open channels, sumps, holding tanks, settling tanks, etc. Because many watery wastes are acidic and corrosive to carbon steel, collection and drainage piping is often lined or made of alloy or plastic. Sulfates frequently appear in wastes, and special concretes may be necessary for sewers, channels, sumps, etc., because sulfates deteriorate regular concretes.

Flammable wastes may be recovered and/or burned in smokeless incinerators or flarestacks. Vapors from flammable liquids present serious explosion hazards in collection and drainage systems, especially if the liquid is insoluble and floats.

Wastes may be held permanently at the manufacturing site. Solid wastes may be piled in dumps, or buried. Watery wastes containing solids may be pumped into artificial 'ponds' or 'lagoons', where the solids settle.

REFERENCES

'Fire hazard properties of flammable liquids, gases, volatile solids'. 1984. NFPA 325M

'Flammable and combustible liquid code'. 1987. NFPA 30

'Flammable and combustible liquid code handbook'. Third edition. 1987. NFPA

'Fire protection in refineries'. Sixth edition. 1984 American Petroleum Institute. API RP 2001

'Protection against ignitions arising out of static, lightning and stray currents'. Fourth edition. 1982. API RP 2003

'Inspection for fire protection'. First edition. 1984. API RP 2004

'Welding or hot-tapping on equipment containing flammables'. 1985. API RP 2201

'Guide for fighting fire in and around petroleum storage tanks'. 1980 API publication 2021

NFPA address: Batterymarch Park, Quincy MA 02269

TANK SPACINGS (NFPA)   **TABLE 6.11**

CONDITIONS	MINIMUM INTER-TANK CLEARANCE
FLAMMABLE or COMBUSTIBLE LIQUID STORAGE TANKS (Not exceeding 150 ft. dia.)	Whichever is greater:— 3ft (Sum of diameters of adjacent tanks)/6
CRUDE PETROLEUM 126,000 gal max tank size Non-congested locale	3 ft
UNSTABLE FLAMMABLE and UNSTABLE COMBUSTIBLE LIQUID STORAGE TANKS	(Sum of diameters of adjacent tanks)/2
LIQUEFIED PETROLEUM GAS CONTAINER from Flammable or Combustible Liquid Storage Tank	20 ft
LIQUEFIED PETROLEUM GAS CONTAINER outside diked area containing Flammable or Combustible Liquid Storage Tank(s)	10 ft from centerline of dike wall NOTE: If LPG container is smaller than 125 gal (US) and each liquid storage tank is smaller than 660 gal, exemption applies
TANKS surrounded by other Tanks	Authority Limit

For minimum clearances from property lines, public ways and buildings, consult the National Fire Code Vol 1, NFPA 30. 1987. Chap. 2

LPG tanks: Title 29 of the Code of Federal Regulations. 1989. Chapter XVII, part 1910-110, the US Department of Labor's 'Occupational Safety and Health Administration's' tables H-23, H-33, gives clearances. Part 1919-111 advises on the storage and handling of anhydrous ammonia.

SOME GUIDELINES

- Apply the recommendations relating to the project of the NFPA, API or other advisory body

- Check insurer's requirements

- Isolate flammable liquid facilities so that they do not endanger important buildings or equipment. In main buildings, isolate from other areas by firewalls or fire-resistive partitions, with fire doors or openings and with means of drainage

- Confine flammable liquid in closed containers, equipment, and piping systems. Safe design of these should have three primary objectives: (1) To prevent uncontrolled escape of vapor from the liquid. (2) To provide rapid shut-off if liquid accidentally escapes. (3) To confine the spread of escaping liquid to the smallest practicable area

- If tanks containing flammable material are sited in the open, it is good practice to space them according to the minimum separations set out in the NFPA Code (No. 395. 'Farm storage of flammable liquids') and to provide dikes (liquid-retaining walls) around groups of tanks. Additional methods for dealing with tank fires are: (1) To transfer the tank's contents to another tank. (2) To stir the contents to prevent a layer of heated fuel forming

- Locate valves for emergency use in plant mishap or fire—see 6.1.3

- Valves for emergency use should be of fast-acting type

- Provide pressure-relief valves to tanks containing flammable liquid (or liquefied gas) if exposed to strong sunlight and/or high ambient temperature, so that vapor under pressure can escape

- Consider providing water sprays for cooling tanks containing flammable liquid which are exposed to sunlight

- Provide ample ventilation in buildings for all processing operations so that vapor concentration is always below the lower flammability limit. Process ventilation should be interlocked so that the process cannot operate without it

- Install explosion panels in buildings to relieve explosion pressure and reduce structural damage

- Install crash panels for personnel in hazardous areas

- Ensure that the basic protection, automatic sprinklers, is to be installed

- Some hazards require special fixed extinguishing systems—foam, carbon dioxide, dry chemical or water spray—in addition to sprinklers. Seek advice from the fire department responsible for the area, and from the insurers

SPACE BETWEEN FLOORS 6.15.1

To avoid interferences and to simplify design, adequate height is necessary between floors in buildings and plants for piping, electrical trays, and air ducts if required. Figure 6.49 suggests vertical spacings:

VERTICAL SPACING **FIGURE 6.49**
BETWEEN FLOOR & CEILING

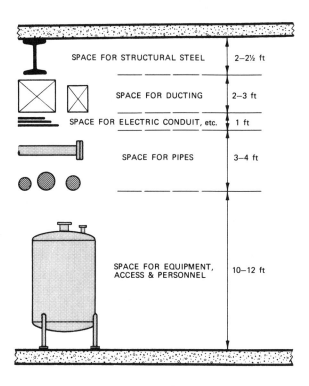

Provision of a services shaft or 'chase' in multi-storied buildings greatly simplifies arrangement of vertical piping, ducting and electric cables communicating between floors. Conceptual arrangements of services and elevator shafts, with fan room for air-conditioning and/or process needs, are shown in figure 6.50. Services shafts can be located in any position suitable to the process, and need not extend the whole height of the building.

SUGGESTED BUILDING LAYOUTS **FIGURE 6.50**

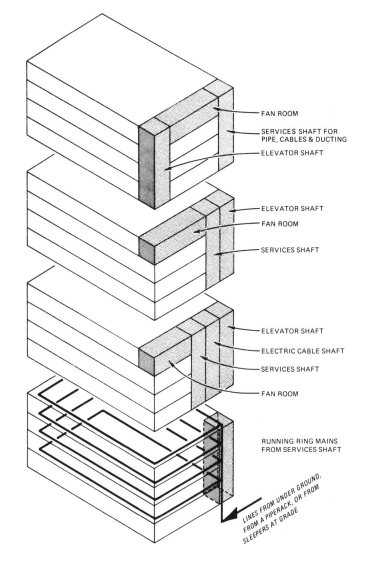

INSTALLATION OF LARGE SPOOLS & EQUIPMENT 6.15.2

Large openings in walls, floors or the roof of a building may be needed for installing equipment. Wall and roof openings are covered when not in use, but sometimes floor openings are permanent and guarded with railings, etc.

BUILDING LAYOUT 6.15.3

RELATION TO PROCESS

Different processes require different types of buildings. Some processes are best housed in single-story buildings with the process beginning at one end and finishing at the other end. Other processes are better assisted by gravity, starting at the top of a building or structure and finishing at or near grade.

STANDARDS AND CODES
for Piping Systems, Pipe, Pipe Supports, Flanges, Gaskets, Fittings, Valves, Traps, Pumps, Vessels, Heat Exchangers, Symbols and Screwthreads

WHAT ARE STANDARDS & CODES ? 7.1

Standards are documents which establish methods for manufacturing and testing. Codes are documents which establish good design practices, including the factors of safety and efficiency. The documents are prepared and periodically updated by committees whose members may include representatives from industry, government, universities, institutes, professional societies, trade associations, and labor unions.

Proven engineering practices form the basis of standards and codes, so that they embody minimum requirements for selection of material, dimensions, design, erection, testing, and inspection, to ensure the safety of piping systems. Periodic revisions are made to reflect developments in the industry.

The terms 'standard' and 'code' have become almost interchangeable, but documents are termed codes when they cover a broad area, have governmental acceptance, and can form a basis for legal obligations. 'Recommendations' document advisable practice. 'Shall' in the wording of standards and codes denotes a requirement or obligation, and 'should' implies recommendation.

FOUR REASONS FOR THEIR USE 7.2

(1) Items of hardware made according to a standard are interchangeable and of known dimensions and characteristics

(2) Compliance with a relevant code or standard guarantees performance, reliability, quality, and provides a basis for contract negotiations, for obtaining insurance, etc.

(3) A lawsuit which may follow a plant mishap, possibly due to failure of some part of a system, is less likely to lead to a punitive judgment if the system has been engineered and built to a code or standard

(4) Codes often supply the substance for Federal, State, and Municipal safety regulations. However, the US Federal Government may, as needed, devise its own regulations, which are sometimes in the form of a code.

WHO ISSUES STANDARDS ? 7.3

The American Standards Association was founded in 1918 to authorize national standards originating from five major engineering societies. Previously a chaotic situation had arisen as many societies and trade associations had been issuing individual standards which sometimes overlapped. In 1967, the name of the ASA was changed to the USA Standards Institute, and in 1969 a second change was made, to American National Standards Institute. Standards previously issued under the prefixes 'ASA' and 'USASI' are now prefixed 'ANSI'.

Not all USA standards and codes are issued directly by the Institute. The American Society of Mechanical Engineers, the Instrument Society of America, and several other organizations issue standards and codes that apply to piping. Table 7.1 lists the principal sources.

ANSI makes available many such standards from other standards-issuing organizations ("sponsors"). Each of these standards is identified by the sponsor's designation (where one exists) preceded by ANSI's and the sponsor's acronym ——— for example, the ASME Code for chemical plant and refinery piping is designated ANSI/ASME B31.3. If the sponsor does not provide a designation, ANSI assigns one. If an American Standards committee developed the standard, the committee designation is used.

The ANSI catalog is available from the American National Standards Institute, 1430 Broadway, New York, NY 10018

Other countries also issue standards. The British Standards Institution (BSI) in the UK, the Deutscher Normenausschuss (DIN) in Germany, and the Swedish national organization (SIS) issue many standards. Copies of foreign standards can be obtained directly, or from the American National Standards Institute.

IDENTIFYING THE SOURCES OF STANDARDS   7.4

The tables in 7.5.6 give the initial letters of the standards-issuing organizations preceeding the number of the standard, thus: 'ASTM N28'. Table 7.1 includes the initials used in tables 7.3 thru 7.14, and gives the full titles of the organizations. (Table 7.1 is not a comprehensive listing.)

PRINCIPAL ORGANIZATIONS ISSUING STANDARDS   TABLE 7.1

INITIALS	FULL TITLE OF ORGANIZATION
AIA	American Insurance Association *
ANSI	American National Standards Institute †
API	American Petroleum Institute
ASME	American Society of Mechanical Engineers
ASTM	American Society for Testing and Materials
AWS	American Welding Society
AWWA	American Waterworks Association
FCI	Fluid Controls Institute
GSA	General Service Administration
ISA	Instrument Society of America
MSS	Manufacturers' Standardization Society of the Valve and Fittings Industry
NFPA	National Fire Protection Association
PFI	Pipe Fabrication Institute
USDC	United States Department of Commerce

*Standards formerly issued by Underwriters' Laboratories Inc.
†Formerly, United States of America Standards Institute, and American Standards Association.

PRINCIPAL DESIGN-ORIENTATED CODES   7.5

ANSI CODE B31   7.5.1

The most important code for land-based pressure-piping systems is ANSI B31. Parts of this code which apply to various types of plant piping are listed in table 7.2.

TITLE	SECTION	APPLICATION
Corrosion Control	B31 Guide	Guidelines for protecting B31 piping systems from corrosion
Power Piping	B31.1-	Piping for industrial plants and marine applications
Chemical Plant and Petroleum Refinery Piping	B31.3-	Design of chemical and petrochemical plants and refineries processing chemicals and hydrocarbons, water and steam
Liquid Petroleum Transportation	B31.4-	Liquid transportation systems for hydrocarbons, LPG, anhydrous ammonia and alcohols
Refrigeration piping	B31.5-	Principally describes the piping of packaged units
Gas Transmission and Distribution Piping Systems	B31.8-	Principally describes overland conveyance of fuel gases and feedstock gases
Building Services Piping Code	B31.9-	High-pressure commercial/sanitary piping
Slurry Transportation Piping	B31.11-	Design, construction, inspection, security requirements of slurry piping systems

AMERICAN PETROLEUM INSTITUTE'S STANDARD 2510   7.5.2

This Standard covers design and construction of liquefied petroleum gas installations at marine and pipeline terminals, natural gas processing plants, refineries, petroleum plants and tank farms

The two following codes are not directly related to piping, but frequently are involved in the piping designer's work:

API 510, PRESSURE VESSEL INSPECTION CODE   7.5.3

This code applies to repairs and alterations made to vessels in petro-chemical service constructed to the former API-ASME Code for Unfired Pressure Vessels for Petroleum Liquids and Gases, Section 8 of the ASME Boiler and Pressure Vessel Code, and to other vessels.

ASME BOILER & PRESSURE VESSEL CODE   7.5.4

The ASME Boiler and Pressure Vessel Code is mandatory in many states with regard to design, material specification, fabrication, erection, and testing procedures. Compliance is required in the USA and Canada to qualify for insurance. The Code consists of the following eleven sections:

ASME BOILER & PRESSURE VESSEL CODE

	section
Power boilers	1
Material specifications	2
Nuclear power plant components	3
Heating boilers	4
Nondestructive examination	5
Recommended rules for care and operation of heating boilers	6
Recommended rules for care of power boilers	7
Pressure vessels	8
Welding qualifications	9
Fiberglass-reinforced plastic pressure vessels	10
Rules for inservice inspection of nuclear reactor coolant systems	11

CODES FOR MARINE PIPING 7.5.5

Requirements for merchant and naval vessels are contained in the following standards:

(1) **American Bureau of Shipping:** 'Rules for building and classing vessels'
(2) **Lloyds' Register of Shipping:** 'Rules'
(3) **US Coast Guard:** 'Marine engineering regulations and material specifications'
(4) **US Navy, Bureau of Ships:** 'General specifications for building naval vessels', 'General machinery specifications'

SELECTED STANDARDS 7.5.6

The following tables are not comprehensive: a selection has been made from standards relating to piping design and technology. Sources of these standards may be found from table 7.1. Addresses of the issuing organizations may be found from the current edition of 'Encyclopedia of associations: Vol 1, National organizations of the United States' (Gale Research Company).

STANDARDS FOR SYMBOLS AND DRAFTING   TABLE 7.3

Piping	Graphic symbols for pipe fittings, valves and piping	ANSI/ASME Y32.2.3
	Graphic symbols for plumbing fixtures	ANSI Y32.4
	Graphic symbols for fluid power diagrams	ANSI Y32.10
	Fluid power diagrams	ANSI Y14.7
Process Engineering	Graphic symbols for process flow diagrams in petroleum and chemical industries	ANSI Y32.11
	Letter symbols for chemical engineering	ANSI Y10.12
	Letter symbols for hydraulics	ANSI Y10.2
Instrumentation	Instrumentation symbols and identification	ISA S5.1
Welding	Symbols for welding and nondestructive testing	AWS A2.4-79
Heating and Ventilating	Graphic symbols for heating, ventilating and air conditioning	ANSI Y32.2.4
Electrical	Electrical and electronics diagrams	ANSI Y14.15
	Graphic symbols for electrical wiring and layout diagrams used for architecture and building construction	ANSI Y32.9
Drafting	Drawing sheet size and format	ANSI Y14.1
	Line conventions, sectioning and lettering	ANSI Y14.2
	Multi and sectional view drawings	ANSI Y14.3
	Pictorial drawing	ANSI Y14.4
	Dimensioning and tolerancing for engineering drawings	ANSI Y14.5
	Screw thread representation	ANSI Y14.6
Safety	Symbols for fire fighting operations	NFPA 178

STANDARDS FOR PIPING (DESIGN AND FABRICATION)   TABLE 7.4

Design	Power piping code (refer to Table 7.2)	ASME B31
Drafting	Method for dimensioning piping assemblies	PFI ES-2
	Minimum length and spacing for welded nozzles	PFI ES-7
Fabrication	Buttwelding ends for pipe, valves, flanges and fittings	ASME B16.25
	Internal machining and solid machined backing rings for circumferential back-welds	PFI ES-1
	Fabricating tolerances	PFI ES-3
Testing	Hydrostatic testing of fabricated piping	PFI ES-4
Cleaning	Cleaning of fabricated piping	PFI ES-5
Color Coding	Scheme for the identification of piping systems	ANSI A13.1
	Recommended practice for color coding of piping materials	PFI ES-22

PRINCIPAL STANDARDS FOR PIPE   TABLE 7.5

Steel or Iron	Specification for welded and seamless steel pipe	ASTM A53
	Specification for seamless carbon-steel pipe for high-temperature service	ASTM A106
	Specification for electric-fusion(arc)-welded steel pipe, NPS 16 and over	ASTM A134
	Specification for electric-resistance-welded steel pipe	ASTM A135
	Specification for seamless and welded austenitic stainless steel pipe	ASTM A312
	Specification for seamless ferritic alloy-steel pipe for high-temperature service	ASTM A335
	Specification for seamless carbon-steel pipe for atmospheric and lower temperatures	ASTM A524
	Specification for line pipe (5L and 5LX)	API 5L
	Welded and seamless wrought-steel pipe	ASME B36.10M
	Stainless steel pipe	ANSI B36.19
	Ductile iron pipe, centrifugally cast, in metal molds or sand-lined molds for water and other liquids	ANSI/AWWA51 C151/A21.51
	Ductile iron pipe, centrifugally cast, in metal molds or sand-lined molds for gas	ANSI A21.52
Nonferrous Alloy	Specification for aluminum and aluminum-alloy seamless pipe and extruded seamless tube	ASTM B241
	Specification for seamless copper pipe, standard sizes	ASTM B42
	Specification for seamless red brass pipe, standard sizes	ASTM B43
	Specification for seamless copper alloy pipe and tube	ASTM B315
	Specification for seamless nickel pipe and tube	ASTM B161
Plastics	Specification for cellulose acetate butyrate (CAB) plastic pipe, SCH 40	ASTM D1503
	Specification for acrylonitrile-butadiene-styrene (ABS) plastic pipe, SCH 40 and 80	ASTM D1527
	Specification for polyvinyl chloride (PVC) plastic pipe, SCH 40, 80 and 120	ASTM D1785
	Specification for polyethylene (PE) plastic pipe, SCH 40	ASTM D2104
	Specification for acrylonitrile-butadiene-styrene (ABS) plastic pipe (SDR-PR)	ASTM D2282
	Specification for polyvinyl chloride (PVC) plastic pipe (SDR series)	ASTM D2241
	Specification for polyethylene (PE) plastic pipe (SIDR-PR) based on controlled inside diameter	ASTM D2239
	Polyvinyl chloride (PVC) pressure pipe for water NPS 4 thru NPS 12	AWWA C900
	Polyethylene (PE) pressure pipe, tubing and fittings for water NPS 1/2 thru NPS 3	AWWA C901
	Polybutylene (PB) pressure pipe, tubing and fittings for water NPS 1/2 thru NPS 3	AWWA C902
	Glass fiber reinforced pipe	AWWA C950

STANDARDS FOR HANGERS AND SUPPORTS   TABLE 7.6

| Application | Pipe hangers and supports - selection and application | MSS SP-69 |
| Production | Pipe hangers and supports - materials, design and manufacture | MSS SP-58 |

STANDARDS FOR GASKETS   TABLE 7.7

Metallic	Ring-joint gaskets and grooves for steel pipe flanges	ASME B16.20
	Metallic gaskets for raised-face pipe flanges and flanged connections (double-jacket corrugated and spiral-wound)	API 601
Nonmetallic	Nonmetallic flat gaskets for pipe flanges	ASME B16.21
	Rubber gasket joints for ductile-iron and gray-iron pressure pipe and fittings	AWWA C111
	Gasketed joints for ductile iron and gray iron pressure pipe and fittings for fire protection service	UL 194
	Standard specification for dense elastomer silicone rubber gaskets and accessories	ASTM C1115

STANDARDS FOR FITTINGS (Also, see Table 7.10) TABLE 7.8

Steel Fittings	Factory-made wrought-steel buttwelding fittings	ASME B16.9
	Wrought-steel buttwelding short-radius elbows and returns	ASME B16.28
	Forged-steel fittings, socketwelding and threaded	ASME B16.11
	Carbon steel pipe unions, socketwelding & threaded	MSS-SP-83
	Factory-made buttwelding fittings for class 1 nuclear piping applications	MSS SP-87
Stainless Steel	Wrought stainless steel buttwelding fittings including reference to other corrosion resistant materials	MSS SP-43
Malleable Iron	Malleable iron threaded fittings	ASME B16.3
Cast Iron	Cast-iron threaded fittings, class 125 and 250	ASME B16.4
	Cast-iron threaded drainage fittings	ANSI B16.12
Ductile Iron	Ductile-iron fittings, NPS 3 thru NPS 24 for gas	ANSI A21.14
	Ductile-iron pipe flanges and flanged fittings	ASME B16.42
Ferrous	Ferrous pipe plugs, bushings and locknuts with pipe threads	ANSI B16.14
Copper Alloy	Cast bronze threaded fittings, class 125 and 250	ASME B16.15
	Cast copper alloy solder joint pressure fittings	ANSI B16.18
	Bronze pipe flanges and flanged fittings, class 150 and 300	ANSI B16.24
	Cast copper alloy solder joint fittngs for Sovent drainage systems	ASME B16.32
	Wrought copper and wrought copper alloy solder-joint drainage fittings for Sovent drainage fittings	ANSI B16.43
Plastics	Specification for socket type acrylonitrile-butadiene-styrene (ABS) plastic pipe fittings SCH 40	ASTM D2468
	Specification for socket type polyvinyl chloride (PVC) plastic pipe fittings SCH 80	ASTM D2467

STANDARDS FOR VALVES TABLE 7.9

General	Face-to-face and end-to-end dimensions of ferrous valves, classes 125 thru 2500 (gate, globe, plug ball, and check valves)	ANSI B16.10
	Manually operated metallic gas valves for use in gas piping systems up to 125 PSIG (sizes NPS 1/2 thru NPS 2)	ANSI B16.33
	Valves, flanged and buttwelding end -- steel, nickel alloy, and other special alloys	ASME B16.34
	Specification for pipeline valves (steel gate, plug, ball and check valves)	API 6D
	Earthquake activated automatic gas shutoff system	AGA Z21.70
Gate Valves	Steel venturi gate valves, flanged and butt-welding ends	API 597
	Steel gate valves, flanged and butt-welding ends	API 600
	Compact steel gate valves	API 602
	Class 150 cast, corrosion-resistant flanged end gate valves	API 603
	Ductile-iron gate valves, flanged ends	API 604
	Gate valves, NPS 3 thru NPS 48, for water and sewage systems	AWWA C500
	Resilient seated gate valves, NPS 3 thru NPS 12, for water and sewage systems	AWWA C509
Butterfly	Butterfly valves	MSS SP-67
	Rubber seated butterfly valves	AWWA C504
	Butterfly valves, lug-type and wafer-type	API 609
Check Valves	Swing check valves for waterworks service, NPS 2 thru NPS 24	AWWA C508
	Wafer check valves	API 594
	Cast-iron swing check valves, flanged and threaded ends	MSS SP-71
Ball Valves	Ball valves-flanged and butt-welding ends	API 608
	Ball valves with flanged or buttwelding ends for general service	MSS SP-72
	Ball valves, NPS 6 thru NPS 48	AWWA C507
Relief	Safety and relief valves	ASME PTC25.3
	Flanged steel safety-relief valves	API 526
Control	Control valve manifold designs -- recommended practice	ISA RP75.06
	Face-to-face dimensions for flanged globe-style control valve bodies (ANSI classes 125, 150, 250, 300 and 600)	ISA S75.03
	Face-to-face dimensions for flangeless control valves (ANSI classes 150, 300 and 600)	ISA S75.04
	Face-to-face dimensions for buttweld-end globe-style control valves (ANSI class 4500)	ISA S75.14

STANDARDS FOR UNFIRED VESSELS AND TANKS TABLE 7.10

Pressure Vessels	Boiler and Pressure Vessel Code, section VIII, "Pressure vessels"	ASME Code
Low Pressure Vessels	Requirements for tank containers for liquids and gases	ASME MH5.1.3
	Specification for bolted tanks for storage of production liquids	API 12B
	Specification for field-welded tanks for storage of production liquids	API 12D
	Specification for shop-welded tanks for storage of production liquids	API 12F
	Recommended rules for design and construction of large welded low-pressure storage tanks	API 620
	Welded steel tanks for oil storage	API 650
	Specification for welded aluminum alloy storage tanks	ANSI B96.1
	Steel aboveground tanks for flammable and combustible liquids	UL 142
	Safety standard for steel inside tanks for oil-burner fuel	UL 80
	Steel underground tanks for flammable and combustible liquids	UL 58
	Factory-coated bolted steel tanks for water storage	AWWA D103-80
	Welded steel tanks for water storage	AWWA D100-79
Lined Vessels	Design, fabrication and surface finish of metal tanks and vessels to be lined for chemical service	NACE RP-01
Calibration	Method for liquid calibration of tanks	ASTM D1406
	Method for measurement and calibration of horizontal tanks	ASTM D1410
	Method for measurement and calibration of spheres and spheroids	ASTM D1408
	Method for measurement and calibration of upright cylindrical tanks	ASTM D1220
Venting and Flame Arresters	Venting atmospheric and low-pressure storage tanks (refrigerated and nonrefrigerated)	API 2000
	Flame arresters for vents of tanks storing petroleum products	API 2210
	Flame arresters for use on vents of storage tanks for petroleum oil and gasoline	UL 525

STANDARDS FOR FLANGES TABLE 7.11

Steel Flanges	Pipe flanges and flanged fittings	ANSI B16.5
	Steel orifice flanges	ANSI B16.36
	Large diameter carbon-steel flanges (NPS 26-60, class 75, 150, 300, 400, 600 and 900	API 605
	Steel pipeline flanges	MSS SP-44
	High-pressure chemical industry flanges and threaded stubs for use with lens gaskets	MSS SP-65
	Steel flanges for waterworks service, NPS 4 thru NPS 144	AWWA C207-78
Cast-iron Flanges	Cast-iron pipe flanges and flanged fittings	ASME B16.1
	Class 150LW corrosion-resistant cast flanges and flanged fittings	MSS SP-51
Ductile Iron	Ductile iron flanges and flanged fittings, class 150 and 300	ASME B16.42
Finishing	Finishes for contact faces of pipe flanges and connecting-end flanges of valves and fittings	MSS SP-6

STANDARDS FOR SCREW THREADS FOR PIPING, NUTS AND BOLTS TABLE 7.12

General	Unified inch screw threads (UN & UNR thread form)	ANSI B1.1
	Pipe threads, general purpose (inch)	ANSI/ ASME B1.20.1
	Nomenclature, definitions and letter symbols for screw threads	ASME B1.7M
Dryseal Pipe Threads	Dryseal pipe threads (inch)	ANSI B1.20.3
	Dryseal pipe threads (metric translation of ANSI B1.20.3)	ANSI B1.20.4
Hose Threads	Hose coupling screw threads for all connections having nominal hose (inside) diameters of 1/2, 5/8, 3/4. 1, 1 1/4, 1 1/2, 2, 2 1/2, 3, 3 1/2 and 4 inches (except fire hose)	ASME B1.20.7
	Screw threads and gaskets for fire hose connections	NFPA 1963-85

STANDARDS FOR HEAT EXCHANGERS AND HEATERS TABLE 7.13

Shell-and-Tube Exchangers	Tubular heat exchangers in chemical process service	ANSI B78.1
	Shell-and-tube exchangers for general refinery services	API 660
	Specification for seamless cold-drawn low-carbon steel heat exchanger and condenser tubes	ASTM A179/M
	Specification for seamless cold-drawn intermediate alloy steel heat exchanger and condenser tubes	ASTM A199/M
	Specification for seamless ferritic and austenitic alloy steel boiler, superheater and heat-exchanger tubes	ASTM A213/M
	Specification for seamless nickel and nickel alloy condenser and heat exchanger tubes	ASTM B163
Air Exchangers	Air cooled heat exchangers for general refinery service	API 661
	Winterizing of air-cooled heat exchangers	API 632
Heaters	Closed feedwater heaters	ASME PTC12.1
	Performance test code -- air heaters	ASME PTC4.3
	Desuperheater/water heaters	ARI 470-80

STANDARDS FOR PRIME MOVERS TABLE 7.14

General	Specification for pumping units	API 11E
	Positive displacement pumps -- reciprocating	API 674
	Positive displacement pumps -- controlled volume	API 675
	Pumps for oil burning appliances	UL 343
Centrifugal Pumps	Centrifugal pumps	ASME PTC8.2
	Specifications for horizontal end suction centrifugal pumps for chemical process	ASME B73.1M
	Specifications for vertical in-line centrifugal pumps for chemical process	ASME B73.2M
	Centrifugal pumps for general refinery service	API 610
Positive Displacement	Displacement pumps (performance test code)	ASME PTC7.1
	Reciprocating steam-driven displacement pumps	ASME PTC7
	Displacement compressors, vacuum pumps and blowers	ASME PTC9
Compressors, exhausters and ejectors	Safety standard for compressors for process industries	ASME B19.3
	Installation of blowers and exhaust systems	NFPA 91
	Centrifugal compressors for general refinery services	API 617
	Compressors and exhausters - performance test code	ASME PTC10
	Ejectors - performance test code	ASME PTC24

ABBREVIATIONS for Piping Drawings and Industrial Chemicals

ABBREVIATIONS USED ON PIPING DRAWINGS, DOCUMENTS, Etc. 8.1

A

A	(1) Air
	(2) Absolute
ABS	Absolute
AGA	American Gas Association
AISI	American Iron and Steel Institute
ANSI	American National Standards Institute
API	American Petroleum Institute
ASTM	American Society for Testing and Materials
AWS	American Welding Society
AWWA	American Waterworks Association

B

BBL	Barrel
BC	Bolt circle
BLE	Beveled large end
BLK	Black
BLVD	Beveled
BOP	Bottom [of outside] of pipe. Used for pipe support location
BS	British Standard
BTU	British thermal unit
BW	(1) Butt weld
	(2) Butt welded

C

C	(1) Centigrade, or Celsius
	(2) Condensate
CENT	Centigrade
CFM	Cubic feet per minute
CHU	Centigrade heat unit
CI	Cast iron
CM	Centimeter
Cr	Chromium
CS	(1) Carbon steel
	(2) Cold spring
CSC	Car-sealed closed. Denotes a valve to be locked in the closed position under all circumstances other than repair to adjacent piping
CSO	Car-sealed open. See CSC
CTR	Center
CU	Cubic

D

DEG	Degree
DIA	Diameter
DIN	Deutsche Industrie Norm [German standard]
DO	Drawing office
DRG	Drawing. [Not preferred]
DWG	Drawing

E

E	East
ECN	Engineering change number
EFW	Electric-fusion-welded
ELL	Elbow
ERW	Electric-resistance-welded

F

F	Fahrenheit
F&D	Faced and drilled
FAHR	Fahrenheit
FBW	Furnace-butt-welded
FCN	Field change number
FD&SF	Faced, drilled and spot-faced
FE	Flanged end
FF	(1) Flat face(d)
	(2) Full face [of gasket]
	(3) Flange face [dimensioning]
FLG	Flange
FLGD	Flanged
FOB	(1) Flat on bottom. [Indicates orientation of eccentric reducer]
	(2) Freight on board. [Indicates location of supply of vendor's freight at the stated price]
	(3) Free on board. [Indicates location of supply of vendor's freight]
FOT	Flat on top. [Indicates orientation of eccentric reducer]
FRP	[Glass-] fiber reinforced pipe
FS	Forged steel
FW	Field weld

G

G	(1) Gas
	(2) Grade
	(3) Gram
GAL	Gallon
GALV	Galvanized
GPH	Gallons per hour
GPM	Gallon per minute

H

H	(1) Horizontal
	(2) Hour
HEX	Hexagon(al)
Hg	Mercury
HPT	Hose-pipe thread
HR	Hour

I

IE	Invert elevation
ID	(1) Inside diameter
	(2) Internal diameter
IMP	Imperial. [British unit]
IPS	Iron pipe size
IS	Inside screw. [Of valve stem]
ISO	Isometric drawing
IS&Y	Inside screw and yoke

K

K	Kilo, times one thousand, x1000
kg	Kilogram

L

L	Liquid
LB,Lb	Pound weight
LT	Light-wall [of Pipe]
LR	Long radius. [Of Elbow]

M

M	(1) Meter
	(2) Mega, times one million, 1 000 000. [On old drawings, x1000]
MACH	Machined
MATL	Material
MAWP	Maximum allowable working pressure
MAX	Maximum
MCC	Motor control center
M/C	Machine
MFR	Manufacturer
MI	Malleable iron
MIN	(1) Minimum
	(2) Minute. [Of time]
mm	Millimeter
Mo	Molybdenum
MSS	Manufacturers' Standardization Society of the Valve and Fittings Industry

N

N	North
NC	Normally closed
NEMA	National Electrical Manufacturers' Assn.
Ni	Nickel
NIC	Not in contract
NO	Normally open
NPSC	2.5.5
NPSF	2.5.5
NPSH	(1) Net positive suction head. [3,2,1]
	(2) 2.5.5
NPSI	2.5.5
NPSL	2.5.5
NPSM	2.5.5

NPT	National pipe thread
NPTF	2.5.5
NRS	Non-rising stem. [Of valve]

O

O	Oil
OD	Outside diameter
OS	Outside screw. [Valve stem]
OS&Y	Outside screw and yoke. [Valve stem]

P

P&ID	Piping and instrumentation diagram
PBE	Plain both ends. [Swage, etc.]
PE	Plain end. [Pipe, etc.]
PFI	Pipe Fabrication Institute
POE	Plain one end. [Nipple, etc.]
PS	(1) Pipe support. [Anchor, guide or shoe, or items combined to form the support]
	(2) Pre-spring
PSI	Pound [weight] per square inch. [Pressure]
PSIA	Pound per square inch absolute
PSIG	Pound per square inch gage

R

RED	Reducing
RF	Raised face
RJ	Ring joint
RPM	Revolutions per minute
RS	Rising stem. [Of valve]

S

S	(1) South
	(2) Steam

SAE	Society of Automotive Engineers
SCH	Schedule. [Of pipe]
SCRD	Screwed
SF	Spot-faced
SKT	Socket
SMLS	Seamless
Si	Silicon
SO	Slip-on
SP	(1) Sample point
	(2) Standard practice. [MSS term]
SR	Short radius. [Of elbow]
SST	Stainless steel
ST	Steam trap
STM	Steam
STD	Standard
STR	Straight
SW	Socket welding
SWG	Swage
SWG / NIPP \	Swaged nipple
SWP	Steam working pressure

T

T	(1) Temperature
	(2) Trap
T&C	Threaded and coupled. [Pipe]
TEMA	Tubular Exchanger Manufacturers' Assn.
TGT	Tangent
TOE	Threaded one end. [Nipple or Swage]
TOS	Top of support
TPI	Threads per inch
TSE	Threaded small end
TYP	Typical. [Used to avoid redrawing similar arrangements]

U

UNC	2.6.3
UNF	2.6.3
UNS	2.6.3

V

V	(1) Vertical
	(2) Vanadium

W

W	(1) West
	(2) Water
WGT	Weight
WLD	Weld(ed)
WN	Welding neck
WOG	Water, oil and gas
WP	(1) Workpoint or reference point
	(2) Markings with this prefix designate certain steels and are used on pipe, fittings and plate. Example: 'WPB' marked on forged fittings denotes A181 grade 2. Refer to ASME SA-234, tables 1 and 2.
WT	Weight

X

XH	Extra-heavy. [See Index]
XS	Extra-strong
XXS	Double-extra-strong

OTHER

℄	Centerline
⌀	Diameter

ABBREVIATIONS FOR COMMERCIAL CHEMICALS

8.2

ABBREVIATION	CHEMICAL NAME	AREA OF USE
A		
ADA	Acetone dicarboxylic acid	Drugs
AEA	Air-entraining agent	Concrete
ANW	83% ammonium nitrate in water	
B		
BAP	Benzyl para-amino phenol	Fuel
BHA	Butylated hydroxyanisole	Food
BHC	Benzene hexachloride	General
BHT	Butylated hydroxytoluene	Food
BOV	77-78% sulfuric acid ('blown oil of vitreol')	General
BzH	Benzaldehyde	General
BzOH	Benzoic acid	General
C		
CO	Carbon monoxide	
COV	95-96% sulfuric acid ('concentrated oil of vitreol')	General
CO2	Carbon dioxide	General

D		
DAP	Diammonium phosphate	Agriculture
DCO	Dehydrated castor oil	Paint
DMC	Dimethylammonium dimethyl carbamate	Refining
DMF	Dimethyl formamide	
DMU	Dimethylurea	
DNA	Dinonyladipate	Plastics
DNM	Dinonyl maleate	Plastics
DNP	Dinonyl phthalate	Plastics
DNT	Dinitrotoluene	Explosives
DOP	Dioctyl phthalate	Plastics
DOV	96% sulfuric acid ('distilled oil of vitreol')	General
DSP	Disodium phosphate	General
DTBP	Ditertiary-butyl peroxide	Plastics
DVB	Divinyl benzene	Plastics
DPG	Diphenyl guanidine	Rubber
DOPA	3,4-dihydroxyphenylaniline	Rubber
E		
EA	Ethylidene aniline	Rubber
EDTA	Ethylene diamine tetra-acetic acid	Food

ABBREVIATION	MEANING	AREA OF USE

F

FA	Furfuryl alcohol	General
FGAN	Ammonium nitrate	Agriculture
FPA	Fluorophosphoric acid	
FREON	One of a large number of chloro- or fluoro- substituted hydrocarbons	Refrigeration, General

H

HCN	Hydrocyanic acid, hydrogen cyanide	Plating
HET	Hexa-ethyl tetraphosphate	Agriculture
HMDT	Hexamethylene triperoxide	
HMT	Hexamethylene tetramine	
HNM	Mannitol hexanitrate	Explosives
HTP	100% hydrogen peroxide ('high test peroxide'), Branched aliphatic alcohols of high b.pt.	Rocketry, General
H2O	Water	

I

IMS	Commercial ethyl alcohol (Brit.)	General
IPA	Isophthalic acid	
IPC	Isopropyl n-phenyl carbonate	
IPS	Isopropyl alcohol (Shell Oil Co.)	General

L

LOX	Liquid oxygen	Rocketry
LPC	Lauryl pyridinium chloride	Soaps
LPG	Liquefied petroleum gases, mainly butane and propane	Fuel

M

MBMC	Monotertiary butyl-methyl-cresol	General
MEK	Methyl-ethyl-ketone	Paint, General
MEP	2-methyl, 5-ethyl pyridine	
MIBC	Methyl isobutyl carbinol	
MIBK	Methyl-isobutyl-ketone	
MNA	Methyl-nonyl acetaldehyde	
MNPT	m-nitro p-toluidine	
MNT	Mononitro toluene	Explosives
MSG	Monosodium glutamate	Food

N

NBA	n-bromacetamide	
NBS	n-bromosuccinamide	
NCA	n-chloracetamide	
NCS	n-chlorosuccinamide	
NH powder	Explosive powder	
N2	Nitrogen	

O

OMPA	Octamethyl pyrophosphoramide	Agriculture
ONB	o-nitrobiphenyl	Plastics
OPE	Octylphenoxyethanol	Refining
O2	Oxygen	General
O3	Ozone	

P

PAS	p-aminosalicylic acid	Drugs
PB	Polybutene	Plastics
PBNA	Phenyl beta-naphthylamine	Rubber
PDB	p-dichlorobenzene	Agriculture
PE	Penta-erythritol	
PETN	Penta-erythritol tetranitrate	Explosives
PTFE	Polytetrafluorethylene	Plastics
PVA or PVAL	Polyvinyl alcohol	
PVAc	Polyvinyl acetate	
PVB	Polyvinyl butyrol	
PVC	Polyvinyl chloride	
PVM	Polyvinyl methyl-ether	

R

| RNV | Sulfuric acid ('refined oil of vitreol') | General |

S

S	Sulfur	General
SAP	Sodium acid pyrophosphate	
SDA	Specially denatured alcohol	General
SO2	Sulfur dioxide	General

T

TCA	Sodium tetrachloracetate	Agriculture
TCE	1,1,1-trichlorethane	Dry cleaning
TCP	Tricresyl phosphate	Fuel, Plastics
TEG	Triethylene glycol	Refining
TEL	Tetraethyl lead	Fuel
TEP	Tetraethyl pyrophosphate	Agriculture
TFA	Tetrahydrofurfuryl alcohol	
TNA	Trinitroaniline	Explosives
TNB	Trinitrobenzene	Explosives
TNG	Trinitroglycerine	Explosives
TNM	Trinitromethane	
TNT	Trinitrotoluene	
TNX	Trinitroxylene	Explosives
TOF	Trioctyl phosphate	Explosives
TPG	Triphenyl guanidine	Plastics
TSP	Trisodium o-phosphate Tetrasodium phosphate	Rubber

V

| VA | Vinyl acetate | |

Z

| ZMA | Zinc methylarsenate | Timber |

INDEX/GLOSSARY

THE 'PIPING GUIDE'

FOR THE DESIGN AND DRAFTING OF INDUSTRIAL PIPING SYSTEMS

PART II

Published by: Construction Trades Press, LLC, 2265 Southeast Blvd
Clinton, North Carolina 28328

Home of Pipefitter.com
http://www.pipefitter.com/

Part II Tables produced by Carmelita E. Bautista

CONTENTS OF PART II

TABLES GIVE THE MINIMUM SPACING. INCREASE DIMENSIONS:
1) FOR INSULATION
2) IF THERMAL MOVEMENT WOULD REDUCE CLEARANCE

DIMENSION – 'C' (LINES WITHOUT FLANGES)

| | NOMINAL PIPE SIZE (NPS) | | | | | | | | | | | |
NPS	2	3	4	6	8	10	12	14	16	18	20	24
2	6	6	7	8	9	10	11	11	12	13	14	16
3	6	7	8	9	10	11	12	12	13	14	15	17
4	7	8	9	10	11	12	13	13	14	15	16	17
6	8	9	10	11	12	13	14	14	15	16	17	18
8	9	10	11	12	13	14	15	15	16	17	18	19
10	10	11	12	13	14	15	16	16	17	18	19	20
12	11	12	13	14	15	16	16	17	18	19	20	21
14	11	12	13	14	15	16	17	18	18	19	20	22
16	12	13	14	15	16	17	18	18	19	20	21	23
18	13	14	15	16	17	18	19	19	20	21	22	24
20	14	15	16	17	18	19	20	20	21	22	23	25
24	16	17	17	18	19	20	21	22	23	24	25	27

SURFACE-TO-CENTER OF PIPE DIMENSION

| | WITHOUT FLANGES 'S' | FLANGE CLASS 'SF' | | |
NOMINAL PIPE SIZE		150	300	600
2	4	4	5	6
3	4	5	6	7
4	5	6	6	8
6	6	7	8	9
8	7	8	9	10
10	8	9	10	11
12	9	11	11	12
14	10	11	12	13
16	11	12	13	14
18	12	13	14	15
20	14	15	15	16
24	16	17	17	17

LINES WITH FLANGES – DIMENSION 'CF'

CLASS 150 & CLASS 150 FLANGES
NOMINAL PIPE SIZE (NPS) OF FLANGED PIPE

150	2	3	4	6	8	10	12	14	16	18	20	24
2	6	6	7	8	8	9	10	10	11	12	13	15
3	6	7	7	8	9	10	11	11	12	13	14	16
4	7	7	8	9	10	11	12	12	13	14	15	16
6	8	8	9	10	11	12	13	13	14	15	16	17
8	9	9	10	11	12	13	14	14	15	16	17	18
10	10	10	11	12	13	14	15	15	16	17	18	19
12	11	11	12	13	14	15	16	16	17	18	19	20
14	11	12	12	13	14	15	16	17	17	18	19	21
16	12	13	13	14	15	16	17	17	18	19	20	21
18	13	14	14	15	16	17	18	18	19	20	21	22
20	14	15	15	16	17	18	19	19	20	21	22	23
24	16	16	16	17	18	19	20	20	21	22	23	25

CLASS 300 & CLASS 300 FLANGES
NOMINAL PIPE SIZE (NPS) OF FLANGED PIPE

300	2	3	4	6	8	10	12	14	16	18	20	24
2	6	7	7	8	9	10	11	11	12	13	14	18
3	7	7	8	9	10	11	12	12	13	14	15	17
4	7	8	8	9	10	11	12	13	14	15	16	18
6	8	9	9	10	11	12	13	14	15	16	17	19
8	9	10	10	11	12	13	14	15	16	17	18	20
10	10	11	11	12	13	14	15	16	17	18	19	21
12	11	12	12	13	14	15	16	17	18	19	20	22
14	11	12	13	14	15	16	17	18	19	20	21	23
16	12	13	14	15	16	17	18	19	20	21	22	24
18	13	14	15	16	17	18	19	20	21	22	23	25
20	14	15	16	17	18	19	20	21	22	23	24	26
24	18	17	18	19	20	21	22	23	24	25	27	29

INSULATION
DIMENSIONS IN THESE TABLES ARE SPACINGS FOR BARE PIPE. FOR INSULATED LINES, ADD THE THICKNESS OF INSULATION AND COVERING TO THESE FIGURES

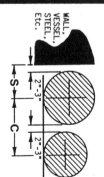

PIPE WITHOUT FLANGES
WALL, VESSEL, STEEL, Etc. — S — C — 2"-3"

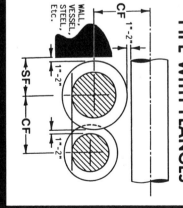

PIPE WITH FLANGES
WALL, VESSEL, STEEL, Etc. — CF — SF — 1"-2"

CLASS 150 & CLASS 300 FLANGES

NPS	2	3	4	6	8	10	12	14	16	18	20	24
2	6	7	7	8	9	10	11	12	12	13	14	16
3	7	8	8	9	10	11	12	13	13	14	15	17
4	8	8	9	10	11	12	13	14	14	15	16	18
6	9	9	10	11	12	13	14	15	16	16	17	19
8	10	11	11	12	13	14	15	16	17	18	18	20
10	11	12	12	13	14	15	16	17	18	19	20	21
12	12	13	13	14	15	16	17	18	19	20	21	22
14	13	14	14	15	16	17	18	19	20	21	22	23
16	14	15	15	16	17	18	19	20	21	22	23	24
18	15	16	16	17	18	19	20	21	22	23	24	25
20	17	17	18	18	19	20	21	22	23	24	25	26
24	19	19	20	21	22	23	24	25	26	27	28	29

(FLANGED PIPE / NOMINAL PIPE SIZE (NPS) OF FLANGED PIPE; classes 150 and 300)

CLASS 300 & CLASS 600 FLANGES

NPS	2	3	4	6	8	10	12	14	16	18	20	24
2	6	7	8	9	10	11	12	13	14	15	16	17
3	7	8	9	10	11	12	13	14	14	15	17	19
4	8	9	10	11	12	13	14	15	15	16	18	20
6	9	10	11	12	13	14	15	16	16	17	19	21
8	10	11	12	13	14	15	16	17	18	19	20	22
10	12	12	13	14	15	16	17	18	19	20	21	23
12	13	13	14	15	16	17	18	19	20	21	22	24
14	14	14	15	16	17	18	19	20	21	22	23	25
16	16	16	16	17	18	19	20	21	22	23	24	26
18	17	17	17	18	19	20	21	22	23	24	25	27
20	18	18	18	19	20	21	22	23	24	25	26	28
24	21	21	22	22	23	24	25	26	27	28	30	32

(FLANGED PIPE / NOMINAL PIPE SIZE (NPS) OF FLANGED PIPE; classes 300 and 600)

CLASS 150 & CLASS 600 FLANGES

NPS	2	3	4	6	8	10	12	14	16	18	20	24
2	6	7	8	9	10	11	11	12	13	14	15	17
3	7	8	9	10	11	12	12	13	14	15	16	18
4	8	9	10	11	12	13	13	14	15	16	17	19
6	9	10	11	12	13	14	14	15	16	17	18	20
8	10	11	12	13	14	15	16	16	17	18	19	21
10	11	12	13	14	15	16	17	18	19	20	21	22
12	13	13	14	15	16	17	18	19	20	21	22	23
14	14	14	15	16	17	18	19	20	21	22	23	24
16	15	15	16	17	18	19	20	21	22	23	24	25
18	16	17	18	18	19	20	21	22	23	24	25	26
20	17	18	19	20	20	21	22	23	24	25	26	27
24	21	22	22	23	24	24	25	26	27	28	30	32

(FLANGED PIPE / NOMINAL PIPE SIZE (NPS) OF FLANGED PIPE; classes 150 and 600)

CLASS 600 & CLASS 600 FLANGES

NPS	2	3	4	6	8	10	12	14	16	18	20	24
2	6	7	8	9	10	11	12	13	14	15	16	18
3	7	8	9	10	11	12	13	14	15	16	17	19
4	8	9	10	11	12	13	14	15	16	17	18	20
6	9	10	11	12	13	14	15	16	17	18	19	21
8	10	11	12	13	14	15	16	17	18	19	20	22
10	11	12	13	14	15	16	17	18	19	20	21	23
12	13	13	14	15	16	17	18	19	20	21	22	24
14	14	14	15	16	17	18	19	20	21	22	23	25
16	15	15	16	17	18	19	20	21	22	23	24	26
18	16	17	18	18	19	20	21	22	23	24	25	27
20	17	18	19	20	20	21	22	23	24	25	26	28
24	22	22	23	23	24	24	25	26	27	28	30	32

(FLANGED PIPE / NOMINAL PIPE SIZE (NPS) OF FLANGED PIPE; classes 600 and 600)

When the order of lines, line sizes, flange classes (for lines with flanges), and insulation thicknesses for insulated lines have been decided, determine pipeway width from Tables A-1, A-2 and A-3, adding 25% so that the final design includes 20% (distributed) space for future piping. Additional space will usually be required for electrical and instrument trays/raceways.

For a **tentative** estimate of the pipeway width required for a selection of lines without flanges, of nominal sizes in the range NPS 2 thru NPS 8, either of the following factors may be used - the first is preferable:

(1) If all pipe sizes are known, add their nominal sizes in inches together and multiply by 0.34 to estimate the width in feet

(2) If only the number of lines is known, multiply number of lines by 1.43 to estimate the width in feet

Either factor gives a pipeway width which includes insulation for 25% of lines, allows 20% of the width for the addition and re-sizing of lines, and allocates a further 20% of the width for future piping.

— PIPEWAY WIDTH —

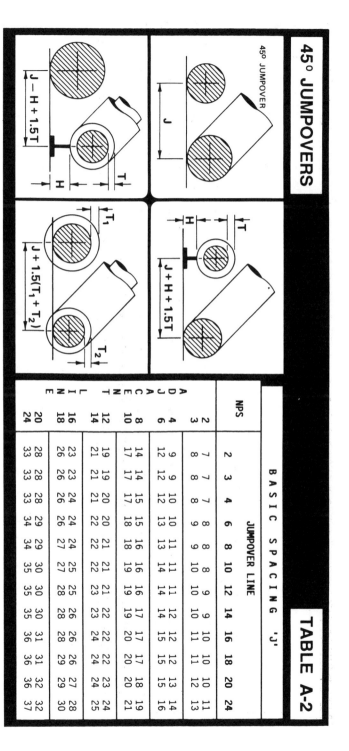

45° JUMPOVERS

(45° JUMPOVER diagrams: dimensions J, J − H + 1.5T, J + 1.5(T₁ + T₂), J + H + 1.5T with H, T, T₁, T₂)

TABLE A-2

BASIC SPACING 'J' — JUMPOVER LINE

NPS	2	3	4	6	8	10	12	14	16	18	20	24
2	7	8	8	9	9	10	10	11	11	12	13	14
3	8	8	9	9	10	10	11	11	12	12	13	13
4	9	9	10	10	11	11	12	12	13	14	14	15
6	12	12	12	13	13	14	14	15	15	15	16	16
8	14	14	15	15	16	16	17	18	18	19	19	20
10	17	17	18	18	19	19	20	20	21	21	22	22
12	19	19	20	20	21	21	22	22	23	23	24	25
14	21	21	22	22	23	23	24	24	25	26	26	28
16	23	23	24	24	25	26	26	27	28	28	29	30
18	26	26	26	26	27	27	28	28	29	29	30	31
20	28	28	28	29	29	30	30	31	31	32	32	32
24	33	33	33	34	34	35	35	36	36	36	36	37

(Row label: A D J A C E N T L I N E)

45° RUNUNDERS

(45° RUNUNDER diagrams: dimensions R, R + H + 1.5T, R − H + 1.5T, R + 1.5(T₁ + T₂) with H, T, T₁, T₂)

TABLE A-3

BASIC SPACING 'R' — RUNUNDER LINE

NPS	2	3	4	6	8	10	12	14	16	18	20	24
2	7	7	8	9	10	11	12	14	16	18	20	24
3	7	8	9	10	12	13	14	15	17	19	22	24
4	8	9	10	12	13	15	16	18	20	22	24	25
6	9	10	11	13	14	16	17	19	21	23	24	27
8	10	11	13	14	16	16	18	19	21	23	25	28
10	8	8	10	11	14	16	19	21	23	26	28	33
12	9	10	11	12	14	16	17	19	22	24	26	29
14	9	10	11	12	14	16	17	19	22	24	26	29
16	10	11	12	13	15	17	18	20	22	24	27	31
18	10	11	12	13	15	17	18	20	23	24	29	31
20	10	12	13	14	16	18	20	22	24	28	30	32
24	11	13	14	16	19	21	24	25	28	30	32	37

(Row label: A D J A C E N T L I N E)

NOTES FOR TABLES A-2 & A-3

(1) SPACING SHOWN IN THE DIAGRAMS ALLOWS A MINIMUM CLEARANCE OF 2-INCHES. COMPARE BASIC SPACING 'J' OR 'R' WITH APPROPRIATE 'C' OR 'CF' SPACING IN TABLE A-1 AND USE THE LARGER DIMENSION

(2) 'H' IS THE EFFECTIVE SHOE HEIGHT AND 'T' IS THE THICKNESS OF INSULATION (WITH COVERING)

(3) FOR SIMPLICITY, THE VALUE 1.5 HAS BEEN SUBSTITUTED FOR THE COEFFICIENT $1/\sin 45$ (1.414.....)

CLASS 600 — BUTT-WELDED PIPING DIMENSIONS — TABLE D-1

DIMENSIONS IN THIS TABLE INCLUDE 0.25-inch RAISED FACE ON FLANGES

FITTINGS
DIMENSIONS FROM ANSI B16.5, B16.9, B16.28 AND MANUFACTURERS DATA

VALVES
DIMENSIONS FROM ANSI B16.10 AND MANUFACTURERS DATA

STRAIGHT TEE — TABLE D-6 FOR REDUCING TEES — BRANCH DIAMETER 2, 3, 4

WELDOLET — STANDARD AND EXTRA-STRONG

REDUCERS — CONCENTRIC & ECCENTRIC — REGULAR & REDUCING

90° LR ELLS — REGULAR & REDUCING

90° SR ELL

45° ELL (LR) — A

OFFSET (TWO 45° ELLS) — B

ROLLED-ELL (45° ELL + 90° LR ELL) — C, D

90° LR ELL + WELDING-NECK RAISED-FACE FLANGE — E

PLUG — VENTURI PATTERN: NPS 2–24, REGULAR PATTERN: NPS 2–16 — F, G

GATE — DIMENSIONS ALSO APPLY TO GATE VALVES WITH BUTT-WELDING ENDS — H, I

BALL — LONG PATTERN — J

GLOBE — DIMENSIONS ALSO APPLY TO GLOBE VALVES WITH BUTT-WELDING ENDS — K, L

CHECK — SWING CHECK: NPS 2–24, TILTING DISC: NPS 2–12, LIFT: NPS 2–12 — FLANGED & BUTT-WELDING — M, N

Dimension Table

Fitting / Valve	Code	2	3	4	6	8	10	12	14	16	18	20	24
STRAIGHT TEE (branch 2)		2.5	3.38	4.12	5.62	7	8.5	10	11	12	13.5	15	17
STRAIGHT TEE (branch 3)		—	—	—	—	—	—	—	—	—	—	—	—
STRAIGHT TEE (branch 4)		—	—	—	—	—	—	—	—	—	—	—	—
WELDOLET		2.69	3.25	3.75	4.81	5.81	6.88	7.88	8.12	8.75	9.5	10.5	13.5
REDUCERS		—	3.5	4	5.06	6.06	7.12	8.12	8.75	9.5	10.5	11.75	13.5
90° LR ELLS		3	4.5	6	9	12	15	18	21	24	27	30	36
90° SR ELL		2	3	4	6	8	10	12	14	16	18	20	24
45° ELL (LR)	A	1.38	2	2.5	3.75	5	6.25	7.5	8.75	10	11.25	12.5	15
OFFSET (TWO 45° ELLS)	B	1.94	2.81	3.56	5.31	7.06	8.81	10.62	12.38	14.12	15.94	17.69	21.19
ROLLED-ELL	C	4.69	6.81	8.56	12.81	17.06	21.31	25.62	29.88	34.12	38.44	42.69	51.19
ROLLED-ELL	D	4.50	6.62	8.5	12.75	17	21.25	25.5	29.81	34.06	38.31	42.56	51.06
90° LR ELL + WN RF FLANGE	E	3.12	3.5	4.25	6	9	12	15	18	21	24	27	32
PLUG (venturi) R	F	6.5	8.25	8.5	14	16.5	20	22	23.75	27	29.25	32	37
PLUG (regular) V	G	6.12	8	10.75	13.88	17.5	21.25	24.38	27.75	31.25	34.5	37.75	44.25
GATE	H	9	12	16	22	24	28	30	36	38	38	42	42
GATE	I	21	26	33	47	53	66	73	81	93	99	107	126
BALL	J	11.5	14	17	22	26	31	33	35	39	43	47	55
GLOBE	K	12	14	18	24	36							
GLOBE	L	21	27	33	44	47							
CHECK (L/S/T)	M	11.5	14	17	22	26	31	33	35	39	43	47	55
CHECK (L/S/T)	N	11.5	14	17	22	26	31	33	35	39	43	47	55

REDUCERS Swage — Table D-4. PLUG rows marked R (regular) / V (venturi). CHECK rows marked L / S / T.

Valve / Installation Notes

- DIMENSIONS FOR COMBINATIONS OF FITTINGS AND INSTALLED WELDOLETS DO NOT INCLUDE THE 'WELD GAP' — REFER TO TEXT: SECTION 5.3.5
- DIMENSIONS IN THIS TABLE ARE NOMINAL AND FOR COMBINATIONS OF FITTINGS ARE ROUNDED TO 1/100-inch
- 'H', 'I', 'K' AND 'L' ARE THE LARGEST DIMENSIONS FOR MANUALLY-OPERATED CAST-STEEL VALVES FROM A SELECTION OF MANUFACTURERS
- GUIDELINES FOR THE USE OF GEAR AND POWERED OPERATORS WITH VALVES ARE GIVEN IN SECTION 3.1.2 OF THE TEXT

Fractional Equivalents

FRACTIONAL	EQUIVALENT
1/16	0.06
1/8	0.12
3/16	0.19
1/4	0.25
5/16	0.31
3/8	0.38
7/16	0.44
1/2	0.50
9/16	0.56
5/8	0.62
11/16	0.69
3/4	0.75
13/16	0.81
7/8	0.88
15/16	0.94

Diagram labels: GATE — H, I (OPEN); GLOBE — K, L, M (OPEN); BALL — J; CHECK — N, SWING CHECK: NPS 2–24, TILTING DISC: NPS 2–12, LIFT: NPS 2–12.

CLASS 300 — BUTT-WELDED PIPING DIMENSIONS — TABLE D-2

DIMENSIONS IN THIS TABLE INCLUDE 0.06-inch RAISED FACE ON FLANGES

FITTINGS
DIMENSIONS FROM ANSI B16.5, B16.9, B16.28 AND MANUFACTURERS DATA

WELDOLET — STANDARD AND EXTRA-STRONG — TABLE D-6 FOR REDUCING TEES — BRANCH DIAMETER 2 3 4

- STRAIGHT TEE
- REDUCERS — CONCENTRIC & ECCENTRIC, REGULAR & REDUCING
- 90° LR ELLS
- 90° SR ELL
- 45° ELL (LR)
- OFFSET (TWO 45° ELLS)
- ROLLED-ELL (45° ELL + 90° LR ELL)
- 90° LR ELL + WELDING-NECK RAISED-FACE FLANGE

VALVES
DIMENSIONS FROM ANSI B16.10 AND MANUFACTURERS DATA

- PLUG — VENTURI PATTERN: NPS 2-24; SHORT PATTERN: NPS 2-12; REGULAR PATTERN: NPS 14-24
- GATE — DIMENSIONS ALSO APPLY TO GATE VALVES WITH BUTT-WELDING ENDS
- BALL — LONG PATTERN: NPS 2-24; SHORT PATTERN: NPS 2-6
- GLOBE — DIMENSIONS ALSO APPLY TO GLOBE VALVES WITH BUTT-WELDING ENDS
- CHECK — SWING: NPS 2-24; TILTING DISC: NPS 2-12; LIFT: NPS 2-6, 10-12; FLANGED & BUTT-WELDING

NOMINAL PIPE SIZE (NPS)

Item	Dim	2	3	4	6	8	10	12	14	16	18	20	24
STRAIGHT TEE		2.5	3.38	4.12	5.62	7	8.5	10	11	12	13.5	15	17
REDUCERS		Swage – Table D-4	3.5	4	5.5	6	7	8	13	14	15	20	24
90° LR ELLS		3	4.5	6	9	12	15	18	21	24	27	30	36
90° SR ELL		2	3	4	6	8	10	12	14	16	18	20	24
45° ELL (LR)		1.38	2	2.5	3.75	5	6.25	7.5	8.75	10	11.25	12.5	15
OFFSET (TWO 45° ELLS)	A	1.94	2.81	3.56	5.31	7.06	8.81	10.62	12.38	14.12	15.94	17.69	21.19
OFFSET	B	4.69	6.81	8.56	12.81	17.06	21.31	25.62	29.88	34.12	38.44	42.69	51.19
ROLLED-ELL	C	4.50	6.62	8.5	12.75	17	21.25	25.5	29.81	34.06	38.31	42.56	51.06
ROLLED-ELL	D	3.12	4.62	6	9	12	15	18	21	24	27	30	36
RAISED-FACE FLANGE	E	5.75	7.62	9.38	12.88	16.38	19.62	23.12	26.62	29.75	33.25	36.38	42.62
RAISED-FACE FLANGE	F	6.5	8.25	10	12.5	15	17.5	20.5	23	25.5	28	30.5	36
RAISED-FACE FLANGE	G	2.75	3.12	3.38	3.88	4.38	4.62	5.12	5.62	5.75	6.25	6.38	6.62
GATE (handwheel)	H	8	10	12	16	20	24	28	28	32	36	36	36
GATE (height)	I	21	25	29	39	49	59	67	76	81	92	102	123
GATE (face-to-face)	J	8.5 S/T/L	11.12	12	15.88	16.5	18	19.75	22.38	25.38	30	33	36
GLOBE (handwheel)	K	10	12	14	22	24	26	28	30	33	34	36	38.5
GLOBE (height)	L	20	24	27	32	41	49	52				40	53
GLOBE (face-to-face)	M	10.5	12.5	14	17.5	22	24	24.5	26	28	30	36	45
CHECK (face-to-face)	N	10.5	12.5	14	17.5	22	24	24.5	28	30	33	34	38.5

Valve cells carry pattern markers (S = short, L = long/butt-welding, T, V, R) for the several valve patterns listed above.

Notes
- DIMENSIONS FOR COMBINATIONS OF FITTINGS AND INSTALLED WELDOLETS DO NOT INCLUDE THE 'WELD GAP' – REFER TO TEXT: SECTION 5.3.5
- DIMENSIONS IN THIS TABLE ARE NOMINAL AND FOR COMBINATIONS OF FITTINGS ARE ROUNDED TO 1/100-inch
- 'H', 'I', 'K', AND 'L' ARE THE LARGEST DIMENSIONS FOR MANUALLY-OPERATED CAST-STEEL VALVES FROM A SELECTION OF MANUFACTURERS
- GUIDELINES FOR THE USE OF GEAR AND POWERED OPERATORS WITH VALVES ARE GIVEN IN SECTION 3.1.2 OF THE TEXT

Fractional / Decimal Equivalents

FRACTIONAL	EQUIVALENT
1/16	0.06
1/8	0.12
3/16	0.19
1/4	0.25
5/16	0.31
3/8	0.38
7/16	0.44
1/2	0.50
9/16	0.56
5/8	0.62
11/16	0.69
3/4	0.75
13/16	0.81
7/8	0.88
15/16	0.94

CLASS 150 — BUTT-WELDED PIPING DIMENSIONS — TABLE D-3

NOMINAL PIPE SIZE (NPS)

DIMENSIONS IN THIS TABLE INCLUDE 0.06-inch RAISED FACE ON FLANGES

FITTINGS
DIMENSIONS FROM ANSI B16.5, B16.9, B16.28 AND MANUFACTURERS DATA

- **WELDOLET** — STANDARD AND EXTRA-STRONG
- **STRAIGHT TEE** — TABLE D-6 FOR REDUCING TEES — BRANCH DIAMETER
- **REDUCERS** — CONCENTRIC & ECCENTRIC, REGULAR & REDUCING
- **90° LR ELLS**
- **90° SR ELL**
- **45° ELL (LR)**
- **OFFSET (TWO 45° ELLS)**
- **ROLLED-ELL** (45° ELL + 90° LR ELL)
- **PLUG** — SHORT PATTERN: NPS 2-12, VENTURI PATTERN: NPS 2-4 & 14-24, REGULAR PATTERN: NPS 14-24
- **RAISED-FACE FLANGE**
- **90° LR ELL + WELDING-NECK**

VALVES
DIMENSIONS FROM ANSI B16.10 AND MANUFACTURERS DATA

- **GATE** — REFER TO TABLE V-1 FOR END-TO-END DIMENSIONS OF GATE VALVES WITH BUTT-WELDING ENDS. USE 'J' ABOVE FOR GATE VALVE.
- **BALL** — LONG PATTERN: NPS 2-24, SHORT PATTERN: NPS 2-16
- **GLOBE** — DIMENSIONS ALSO APPLY TO GLOBE VALVES WITH BUTT-WELDING ENDS
- **CHECK** — SWING: NPS 2-24, TILTING DISC: NPS 2-14, LIFT: NPS 2-4 & 8-14 — FLANGED & BUTT-WELDING

Data Table

Ref	Fitting / Valve	2	3	4	6	8	10	12	14	16	18	20	24
	STRAIGHT TEE (branch 2)	2.5	3.38	4.12	5.62	7	8.5	10	11	12	13.5	15	17
	STRAIGHT TEE (branch 3)	—	3.5	3.75	4.81	5.81	6.88	7.88	8.5	9.5	10.5	11.5	13.5
	STRAIGHT TEE (branch 4)	—	—	4.25	5.31	6.06	7.12	8.12	8.75	9.75	10.75	11.75	13.75
	REDUCERS	Swage- Table D-4											
	90° LR ELLS	3	4.5	6	9	12	15	18	21	24	27	30	36
	90° SR ELL	2	3	4	6	8	10	12	14	16	18	20	24
A	45° ELL (LR)	1.38	2	2.5	3.75	5	6.25	7.5	8.75	10	11.25	12.5	15
B	OFFSET	1.94	2.81	3.56	5.31	7.06	8.81	10.62	12.38	14.12	15.94	17.69	21.19
C	ROLLED-ELL	4.69	6.81	8.56	12.81	17.06	21.31	25.62	29.88	34.12	38.44	42.69	51.19
D	ROLLED-ELL	4.50	6.62	8.5	12.75	17	21.25	25.5	29.81	34.06	38.31	42.56	51.06
E	RAISED-FACE FLANGE	3.12	4.62	6	9	12	15	18	21	24	27	30	36
F	RAISED-FACE FLANGE	6	7.5	9	11	13.5	16	19	21	23.5	25	27.5	32
G	PLUG	2.5 S	2.75 S	3	3.5	4	4.5	5	5.5 R	5.5 R	5.5 R	5.69 R	6 R
H	GATE	8 S	9.5 S	11.5	14	16	18	20	24	26	30	30	36
I	GATE	19	23	28	37	47	53	61	71	80	89	98	113
J	BALL	7	8	9	11.5	11.5	13	14	15	16	17	18	20
K	GLOBE	8	10	12	16	18	21	24	27	30	30	36	42
L	GLOBE	15	19	21	26	33	32	42	49	—	—	—	—
M	GLOBE	8 L/T	9.5 L/T	11.5 L/T	14	16	19.5	24.5	27.5	31	34	38.5	51
N	CHECK	8 S	9.5 S	11.5 S	14	16	19.5	24.5	27.5	31	34	38.5	51

(Pattern codes in valve cells: S = short, V = venturi, R = regular, L = long, T = tilting)

Fractional / Decimal Equivalents

FRACTIONAL	EQUIVALENT
1/16	0.06
1/8	0.12
3/16	0.19
1/4	0.25
5/16	0.31
3/8	0.38
7/16	0.44
1/2	0.50
9/16	0.56
5/8	0.62
11/16	0.69
3/4	0.75
13/16	0.81
7/8	0.88
15/16	0.94

Notes

- DIMENSIONS FOR COMBINATIONS OF FITTINGS AND INSTALLED WELDOLETS DO NOT INCLUDE THE "WELD GAP" — REFER TO TEXT: SECTION 5.3.5
- DIMENSIONS IN THIS TABLE ARE NOMINAL AND FOR COMBINATIONS OF FITTINGS ARE ROUNDED TO 1/100-inch
- 'H', 'I', 'K' AND 'L' ARE THE LARGEST DIMENSIONS FOR MANUALLY-OPERATED VALVES FROM A SELECTION OF MANUFACTURERS
- GUIDELINES FOR THE USE OF GEAR AND POWERED OPERATORS WITH VALVES ARE GIVEN IN SECTION 3.1.2 OF THE TEXT

SWAGES

TABLE D-4

NPS (INCHES)

LARGE END	SMALL END	LENGTHS
2	¼-1½	6.5
2½	¼-2	7.0
3	½-2½	8.0
3½	2-3	8.0

LARGE END	SMALL END	LENGTHS
4	1-3½	8.0
5	2-4	9.0
6	1½-5	11
8	2-6	12
10	4-8	13
12		15

Dimensions in this table are for Mills Iron Works swages, available with ends plain, threaded, beveled, Victaulic grooved, and in any combination of these terminations

---

ELBOLETS: THREADED/SOCKET & BUTT-WELDING

TABLE D-5

DIMENSIONS IN INCHES

NOMINAL PIPE SIZE OF MAIN RUN [NPS]

CLASS 3000 THREADED & SOCKET-WELDING* – STD AND XS BUTT-WELDING

NPS OF BRANCH	2	3	4	6	8	10	12	14	16	18	20	24
1/2	3.53	5.94	7.25	10	12.66	15.38	18.03	20.12	22.75	25.41	28.06	33.34
3/4	4.81	6.22	7.53	10.28	12.94	15.66	18.31	20.41	23.03	25.69	28.34	33.62
1	5.12	6.53	7.84	10.59	13.25	15.97	18.62	20.72	23.34	26	28.66	33.94
1 1/2	5.56	6.97	8.28	11.03	13.69	16.41	19.06	21.16	23.78	26.44	29.09	34.38
2	6.12	7.53	8.84	11.59	14.25	16.97	19.62	21.72	24.34	27	29.97	34.94
3		8.16	9.47	12.22	14.88	17.59	20.25	22.34	24.97	27.62	30.28	35.56
4			10.16	12.91	15.56	18.28	20.94	23.03	25.66	28.31	30.97	36.25
6				14.59	17.25	19.97	22.62	24.72	27.34	30	32.66	37.94
8					18.25	20.97	23.62	25.72	28.34	30.97	33.66	38.94
10						22.78	25.44	27.53	30.16	32.81	35.47	40.75
12							26.44	28.53	31.16	33.81	36.94	41.75

LR ELL

Data provided by BONNEY FORGE. Dimensions for Elbolets are nominal. Size 2-inch Elbolets are designed to fit the different sizes of run pipe; in sizes larger than 2-inches, each size of Elbolet is designed to fit a range of run pipe sizes.

* Threaded and socket-welding Elbolets are not available in sizes 6-inch and larger.

REDUCING BUTT-WELDING TEES

TABLE D-6

NOMINAL PIPE SIZE OF MAIN RUN [NPS]

WEIGHTS: STANDARD and EXTRA-STRONG. SCH 160 thru NPS 12. XXS thru NPS 8.

DIMENSION 'A'

NPS ▶	3	4	6	8	10	12	14	16	18	20	24
	3.38	4.12	5.62	7.00	8.50	10.00	11.00	12.00	13.50	15.00	17.00

DIMENSION 'B'

NPS (branch) ▶ / run	3	4	6	8	10	12	14	16	18	20	24
2	3.00	3.50	4.88	6.12	7.25	8.62	9.38	10.62	11.62	12.62	14.50
3	3.38	3.88	5.12	6.12	7.62	9.00	9.75	10.75	11.75	13.00	15.00
4		4.12	5.12	6.62	8.00	9.38	10.12	11.00	12.00	13.00	15.12
6			5.62	6.62	8.00	9.50	10.12	11.12	12.12	13.12	15.12
8				7.00	8.50	9.75	10.38	11.12	12.12	13.12	15.62
10					8.50	10.00	10.62	11.62	12.62	13.50	16.00
12						10.00	11.00	11.62	12.75	13.62	16.00
14							11.00	12.00	13.00	14.00	16.50
16								12.00	13.12	14.00	16.50
18									13.50	14.50	17.00
20										15.00	17.00

CHECK VALVES - WAFER-TYPE

TABLE D-7

FACE-TO-FACE DIMENSIONS BY CLASS FOR VALVES CONFORMING TO API 594

NPS	FLANGE CLASSES					
	150	300	600	900	1500	2500
2	2.38	2.38	2.38	2.75	2.75	2.75
3	2.88	2.88	2.88	3.25	3.25	3.38
4	2.88	2.88	3.12	4.00	4.00	4.12
6	3.88	3.88	5.38	6.25	6.25	6.25
8	5.00	5.00	6.50	8.12	8.12	8.12
10	5.75	5.75	8.38	9.50	9.75	10.00
12	7.12	7.12	9.00	11.50	12.00	12.00
14	7.25	8.75	10.75	14.00	14.00	
16	7.50	9.12	12.00	15.12	15.12	
18	8.00	10.38	14.25	17.75	18.44	
20	8.62	11.50	14.50	17.75	21.00	
24	8.75	12.50	17.25	19.50	22.00	

SINGLE AND DUAL PLATES

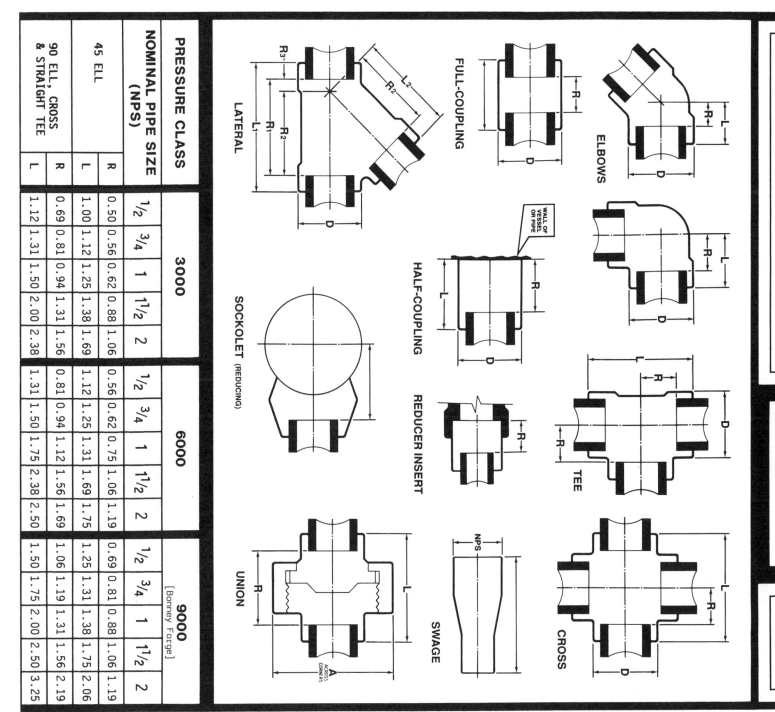

SOCKET WELDING FITTINGS - FORGED STEEL

'R' DIMENSIONS INCLUDE EXPANSION GAP — NOTE 1

TABLE D-8

Figure labels: LATERAL, FULL-COUPLING, ELBOWS, HALF-COUPLING, REDUCER INSERT, TEE, SOCKOLET (REDUCING), UNION, SWAGE, CROSS, WALL OF VESSEL OR PIPE

PRESSURE CLASS			3000					6000					9000 [Bonney Forge]				
NOMINAL PIPE SIZE (NPS)			1/2	3/4	1	1 1/2	2	1/2	3/4	1	1 1/2	2	1/2	3/4	1	1 1/2	2
45 ELL		R	0.50	0.56	0.62	0.88	1.06	0.56	0.62	0.75	1.06	1.19	0.69	0.81	0.88	1.06	1.19
45 ELL		L	1.00	1.12	1.25	1.38	1.69	1.12	1.25	1.31	1.69	1.75	1.25	1.31	1.38	1.75	2.06
90 ELL, CROSS & STRAIGHT TEE		R	0.69	0.81	0.94	1.31	1.56	0.81	0.94	1.12	1.56	1.69	1.06	1.19	1.31	1.56	2.19
90 ELL, CROSS & STRAIGHT TEE		L	1.12	1.31	1.50	2.00	2.38	1.31	1.50	1.75	2.38	2.50	1.50	1.75	2.00	2.50	3.25

Piping Fitting Dimensions

FULL-COUPLING		
R	0.50	0.50
L	1.38	1.50

FULL-COUPLING	R	0.50	0.50	0.50	0.50	0.50	0.50	0.62	0.62	0.62	0.62	0.62	0.62	0.88	0.88	0.88
	L	1.38	1.50	1.75	2.00	2.50	1.38	1.50	1.75	2.00	2.50	1.75	2.00	2.50		

HALF-COUPLING [Bonney Forge] (sub-sizes 1/2, 3/4, 1)

		1/2	3/4	1										
R	0.94	1.00	1.19	1.31	1.69	1.19	1.06	1.12	1.19	1.25	1.31	1.50	1.25	1.44	1.06
	2.12														
L	1.38	1.50	1.75	2.00	2.50	1.38	1.50	1.75	2.00	2.50	1.38	1.50	1.75		2.12

REDUCER INSERT [Bonney Forge] (sub-sizes 1/2, 3/4, 1, 1 1/2)

R	0.94	0.69	0.88	1.06	1.00	0.81	1.00	0.75	0.94	1.12	1.19	1.19	1.25	1.06	1.12	1.19
								1.31					1.94			

LATERAL [Bonney Forge & Ladish]

L2	2.12	2.56	3.00	3.94	4.75	6.88	2.56	3.00	3.50	4.75	6.88	3.00	3.50	3.94	5.00
L1	3.00	3.56	4.12	5.38	6.44	9.00	3.56	4.12	4.81	6.44	9.00	4.12	4.81	5.38	6.62
R3	0.44	0.50	0.56	0.75	0.88	1.31	0.50	0.56	0.69	0.88	1.31	0.62	0.75	0.81	0.81
R2	1.69	2.06	2.44	3.25	3.94	6.06	2.06	2.44	2.88	3.94	6.06	2.50	2.94	3.31	4.19
R1	2.12	2.56	3.00	4.00	4.81	7.38	2.56	3.00	3.56	4.81	6.06	7.38	3.12	3.69	5.00

DIAMETER

D	1.31	1.56	1.88	2.50	3.06	1.56	1.84	2.25	3.06	3.62	1.84	2.22	2.50	3.38	4.00

SOCKOLET (REDUCING) [Bonney Forge] (sub-sizes 1/2, 3/4, 1, 1 1/2)

	1/2	3/4	1	1 1/2									
R	1.22	1.28	1.50	2.00	2.19	1.28	1.50	1.83	2.19	2.50			
1	1.15	1.28	1.58	1.81	1.46	1.60	1.89	2.12	1.57	1.86	2.10		1.43	1.95	2.19
3/4	1.28	1.58	2.44	2.88	3.50	1.60	1.89	2.12	1.95	1.86			1.66	1.95	2.19
B/R/A/N/C/H	1.79	2.03	2.00	1.95	2.19	2.12					2.04	2.28	2.41

UNION [Bonney Forge]

R	1.22	1.28	1.50	2.00	2.19	1.28	1.50	1.83	2.19	2.50			
L	2.00	2.31	2.44	3.00	3.50	2.31	2.44	2.88	3.50	4.12	2.44	2.88	3.50
	2.50	2.78	3.00	3.50	4.12								

SWAGE

A	1.94	2.38	2.78	3.72	4.42	2.38	2.78	3.36	4.42				
D	2.75	3.00	3.50	4.50	6.50	2.75	3.00	3.50	4.50	6.50	2.75	3.00	3.50	4.50	6.50

---

(1) 'R' DIMENSIONS ('REMOVED RUN' OF PIPE) HAVE BEEN ROUNDED TO 1/100-inch AND INCLUDE 0.06-inch EXPANSION GAP(S) FOR WELD-ING. REFER TO 'SOCKET-WELDED PIPING' - CHART 2.2

(2) DIMENSIONS ARE FROM THE FOLLOWING SUPPLIERS' DATA: BONNEY FORGE, ITT GRINNEL, LADISH AND VOGT

(3) UNLESS THE SUPPLIER IS STATED, 'L' & 'D' DIMENSIONS ARE THE LARGEST QUOTED BY BONNEY FORGE, ITT GRINNEL, LADISH AND VOGT

(4) FITTINGS CONFORM TO ANSI B16.11, EXCEPT LATERALS AND REDUCER INSERTS, WHICH ARE MADE TO MANUFACTURERS' STANDARDS

(5) FOR INFORMATION ON THE BORE DIAMETER AND RATING OF FITTINGS, REFER TO 'SOCKET-WELDED PIPING' - CHART 2.2

(6) UNIONS CONFORM TO MSS-SP-83

(7) DIMENSIONS FOR INSTALLED SOCKOLETS EXCLUDE THE 'ROOT GAP' - REFER TO 'DIMENSIONING SPOOLS (WELDED ASSEMBLIES)' - 5.3.5

FRACTIONAL	1/16	1/8	3/16	1/4	5/16	3/8	7/16	1/2	9/16	5/8	11/16	3/4	13/16	7/8	15/16
EQUIVALENT	0.06	0.12	0.19	0.25	0.31	0.38	0.44	0.50	0.56	0.62	0.69	0.75	0.81	0.88	0.94

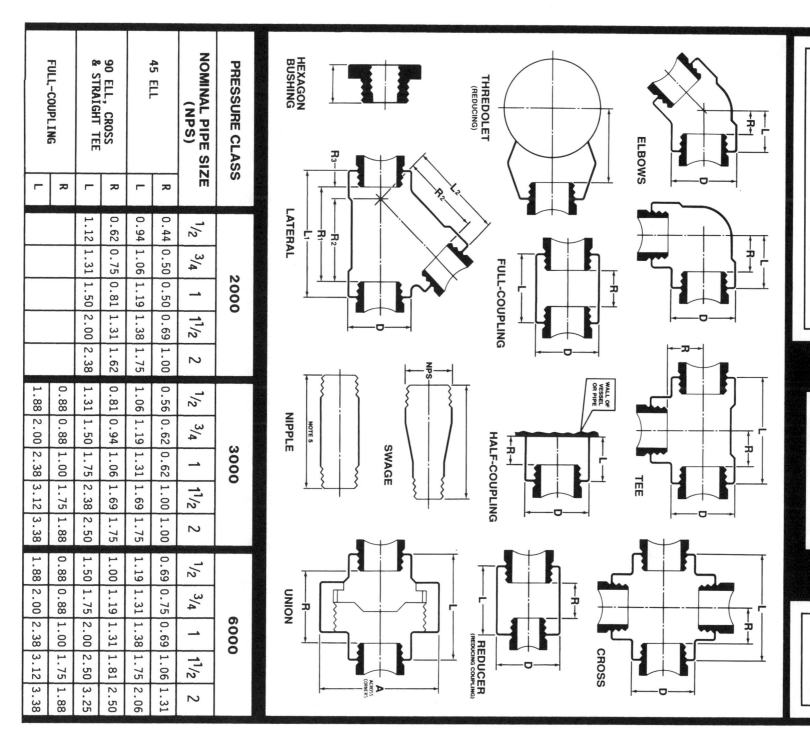

ELBOWS · THREDOLET (REDUCING) · HEXAGON BUSHING · LATERAL · FULL-COUPLING · TEE · NIPPLE (NOTE 5) · SWAGE · HALF-COUPLING (WALL OF VESSEL OR PIPE) · CROSS · REDUCER (REDUCING COUPLING) · UNION (A ACROSS CORNERS)

PRESSURE CLASS	NOMINAL PIPE SIZE (NPS)	45 ELL R	45 ELL L	90 ELL, CROSS & STRAIGHT TEE R	90 ELL, CROSS & STRAIGHT TEE L	FULL-COUPLING R	FULL-COUPLING L
2000	1/2	0.44	0.94	0.62	1.12		
2000	3/4	0.50	1.06	0.75	1.31		
2000	1	0.50	1.19	0.81	1.50		
2000	1 1/2	0.69	1.38	1.31	2.00		
2000	2	1.00	1.75	1.62	2.38		
3000	1/2	0.56	1.06	0.81	1.31	0.88	1.88
3000	3/4	0.62	1.19	0.94	1.50	0.88	2.00
3000	1	0.62	1.31	1.06	1.75	1.00	2.38
3000	1 1/2	1.00	1.69	1.69	2.38	1.75	3.12
3000	2	1.00	1.75	1.75	2.50	1.88	3.38
6000	1/2	0.69	1.19	1.00	1.50	0.88	1.88
6000	3/4	0.75	1.31	1.19	1.75	0.88	2.00
6000	1	0.69	1.38	1.31	2.00	1.00	2.38
6000	1 1/2	1.06	1.75	1.81	2.50	1.75	3.12
6000	2	1.31	2.06	2.50	3.25	1.88	3.38

HALF-COUPLING	R		0.44	0.94				0.44	0.50	0.88	0.94		0.44	0.50	0.88	0.94		
	L		1.00	1.19	1.19	1.56	1.69	1.00	1.19	1.19	1.56	1.69	1.00	1.19	1.19	1.56	1.69	
REDUCER	1/2	R			0.94	1.00				0.94	1.00				0.94	1.00		
	3/4					1.12					1.12					1.12		
	1					1.88	1.94				1.88	1.94				1.88	1.94	
	1 1/2	L				2.06	2.12				2.06	2.12				2.06	2.12	
					1.75	1.94				1.75	1.94				1.75	1.94		
						1.94					1.94					1.94		
LATERAL [Bonney Forge & Ladish]	L		2.00	2.38	3.12	3.38		2.00	2.38	3.12	3.38		2.00	2.38	3.12	3.38		
	R1		2.07	2.49	2.82	4.12	5.06	2.62	3.07	3.56	5.19	7.50	3.19	3.81	4.12	6.19	7.62	
	R2		1.66	2.03	2.34	3.31	4.06	2.09	2.47	2.87	4.12	6.12	2.53	3.00	3.31	6.25		
	R3		0.41	0.47	0.47	0.81	1.00	0.53	0.60	0.69	1.06	1.38	0.66	0.81	0.81	1.44	1.38	
	L1		3.07	3.62	4.19	5.50	6.56	3.62	4.19	4.94	6.56	9.00	4.19	4.94	5.50	9.00	9.12	
	L2		2.16	2.59	3.03	4.00	4.81	2.59	3.03	3.56	4.81	6.88	3.03	3.56	4.00	6.88	7.00	
DIAMETER	D		1.31	1.56	1.84	2.59	3.06	1.56	1.84	2.25	3.06	3.62	1.84	2.25	2.50	3.62	4.31	
THREDOLET (REDUCING) [Bonney Forge]	B R A N C H	1/2						1.03	1.16	1.16	1.58	1.88	1.28	1.53	1.83			
		3/4						1.16	1.45	1.45	1.81		1.41	1.83	1.83	2.06		
		1						1.45	1.69	1.69			1.70	1.94	1.94			
		1 1/2																
		A	1.94	2.41	2.78	3.72	4.42	2.41	2.44	2.78	3.72	4.42	2.41	2.75	3.36	4.42	5.23	
UNION [Bonney Forge]	R		0.94	1.00	1.06	1.31	1.44	1.00	1.06	1.19	1.64	1.95	1.27	1.31	1.50	2.06	2.19	
	L							2.00	2.31	2.44	3.01	3.45	2.27	2.44	2.88	3.44	4.08	
HEX BUSH	R		0.94	1.00	1.31	1.44		1.00	1.06	1.31	1.44		0.94	1.00	1.06	1.31	1.44	
	L							1.84	2.25				1.84	2.25	2.50			
	A												3.03	3.56	4.00	6.88	7.00	
SWAGE			2.75	3.00	3.50	4.50	6.50	2.75	3.00	3.50	4.50	6.50	2.75	3.00	3.50	4.50	6.50	
THREAD ENGAGEMENT			0.50	0.56	0.69	0.69	0.75	0.50	0.56	0.69	0.69	0.75	0.50	0.56	0.69	0.69	0.75	

(1) 'R' DIMENSIONS ('REMOVED RUN' OF PIPE) ARE BASED ON NORMAL THREAD ENGAGEMENT BETWEEN MALE AND FEMALE THREADS TO MAKE TIGHT JOINTS - ROUNDED TO 1/100-inch

(2) DIMENSIONS FOR FITTINGS ARE FROM THE FOLLOWING SUPPLIERS' DATA: BONNEY FORGE, ITT GRINNEL, LADISH AND VOGT

(3) UNLESS THE SUPPLIER IS STATED, 'L' & 'D' DIMENSIONS ARE THE LARGEST QUOTED BY BONNEY FORGE, ITT GRINNELL, LADISH AND VOGT

(4) FITTINGS CONFORM TO ANSI B16.11, EXCEPT LATERALS, WHICH ARE MADE TO MANUFACTURERS' STANDARDS. UNIONS CONFORM TO MSS-SP-83

(5) FOR SIZES AND AVAILABILITIES OF PIPE NIPPLES, REFER TO 'MALLEABLE-IRON PIPE FITTINGS' - TABLE D-11

(6) DIMENSIONS FOR INSTALLED THREDOLETS EXCLUDE THE 'ROOT GAP' - REFER TO 'DIMENSIONING SPOOLS (WELDED ASSEMBLIES)' - 5.3.5

FRACTIONAL	0.06	0.12	0.19	0.25	0.31	0.38	0.44	0.50	0.56	0.62	0.69	0.75	0.81	0.88	0.94
EQUIVALENT	1/16	1/8	3/16	1/4	5/16	3/8	7/16	1/2	9/16	5/8	11/16	3/4	13/16	7/8	15/16

DATA: SMITH VALVE CORPORATION
GATE VALVES: FULL PORT
GLOBE VALVES: CONVENTIONAL PORT

'R' is the 'REMOVED RUN' of
pipe occupied by the valve

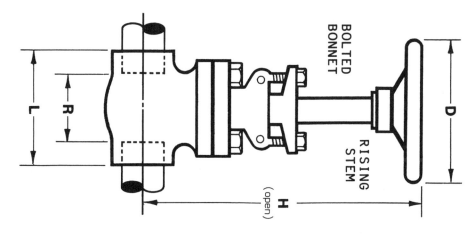

BOLTED BONNET

RISING STEM

H (open)   L   R   D

VALVES WITH THREADED ENDS

		NPS	1/2	3/4	1	1 1/2	2
GATE	D		4.00	4.00	5.50	6.62	6.62
	H		6.38	7.25	8.56	11.00	12.50
	L		3.50	3.88	4.25	5.50	5.69
	R		2.50	2.75	2.88	4.12	4.19
GLOBE	D		4.00	4.00	4.00	4.62	6.62
	H		6.38	6.56	6.81	8.12	10.12
	L*		3.25	3.50	4.50	6.25	7.25
	R*		2.25	2.38	3.12	4.88	5.75

* These dimensions also apply to horizontal lift-check valves

'R' dimensions are based on normal
thread engagement for tight joints

VALVES WITH SOCKET ENDS

		NPS	1/2	3/4	1	1 1/2	2
GATE	D		4.00	4.00	5.50	6.62	6.62
	H		6.38	7.25	8.56	11.00	12.50
	L		3.50	3.88	4.25	5.50	5.69
	R		2.00	2.25	2.75	3.12	3.75
GLOBE	D		4.00	4.00	4.00	4.62	6.62
	H		6.38	6.56	6.81	8.12	10.12
	L*		3.25	3.50	4.50	6.25	7.25
	R*		2.38	2.50	3.38	4.88	5.62

'R' dimensions include 0.06-inch expansion
gaps for welding. Refer to text: Chart 2.2

PRESSURE CLASS — NOMINAL PIPE SIZE [IN]

Fitting	Dim	150 1/2	150 3/4	150 1	150 1 1/2	150 2	150 3	300 1/2	300 3/4	300 1	300 1 1/2	300 2	300 3
45° ELL	A	0.88	1.0	1.12	1.44	1.69	2.19	1.0	1.12	1.31	1.69	2.0	2.5
90° ELL	A	1.12	1.31	1.5	1.94	2.25	3.06	1.12	1.31	1.5	2.0	2.5	3.38
90° STREET ELL	B	1.62	1.88	2.12	2.69	3.25	4.5	2.0	2.19	2.56	3.12	3.69	5.12
90° STREET ELL	A	1.12	1.31	1.5	1.94	2.25	3.06	1.25	1.44	1.62	2.12	2.5	3.38
RETURN BEND CLOSE		1.0	1.25	1.5	2.19	2.62							
RETURN BEND MEDIUM		1.25	1.5	1.88	2.5	3.0							
RETURN BEND OPEN		1.5	2.0	2.5	3.5	4.0							
STRAIGHT TEE	A	1.12	1.31	1.5	1.94	2.25	3.06	1.94	2.25	2.56	3.0	3.88	4.94
LATERAL	A	2.31	2.81	3.31	4.38	5.19	7.25						
LATERAL	C	1.69	2.06	2.44	3.25	3.94	5.56						
UNION	A	1.81	2.0	2.19	2.62	3.06	3.88	1.94	2.25	2.44	3.0	3.88	4.94
COUPLING	A	1.31	1.5	1.69	1.75	2.0	2.62	1.88	2.12	2.38	2.88	3.62	4.12
NIPPLE (CLOSE NIPPLE)		1.12	1.38	1.5	1.75	2.0	2.62	1.12	1.38	1.5	1.75	2.0	2.62
SWAGE		2.75	3.0	3.5	4.5	6.5	8.0	2.75	3.0	3.5	4.5	6.5	8.0
REDUCER		1.25	1.44	1.69	2.31	2.81	3.69	1.69	1.75	2.0	2.69	3.19	4.06
THREAD ENGAGEMENT	TAPER/TAPER	0.5	0.56	0.69	0.69	0.75	1.0	0.5	0.56	0.69	0.69	0.75	1.0

UNION (OCTAGONAL) — SEATS:
S — GRINNELL: COPPER ALLOY-TO-IRON
S — STOCKHAM: BRASS-TO-IRON or ALL-IRON

RETURN BEND: (These data also apply to the center-to-end dimension for straight cross.)

STRAIGHT TEE note applies (center-to-end dimension).

NIPPLE — CARBON-STEEL (TANK NIPPLES ARE 6-IN. LONG)
AVAILABILITIES OF SHORT AND LONG NIPPLES:
AVAILABLE IN 2, 2 1/2, 3, 3 1/2, 4, 4 1/2, 5 1/2, 6, 7, 8, 9, 10, 11 & 12 INCH LENGTHS (1/2 and 3/4 NPS nipples are also available 1 1/2 inches long)

COUPLING — CLOSE NIPPLE

SWAGE — CARBON-STEEL — NPS

REDUCER — CARBON-STEEL — NPS

THREAD ENGAGEMENT — Engagement

FORGED-STEEL FLANGES & LAP-JOINT STUB-ENDS

FLANGE CLASSES 150-2500

TABLES F

DIMENSIONS IN INCHES

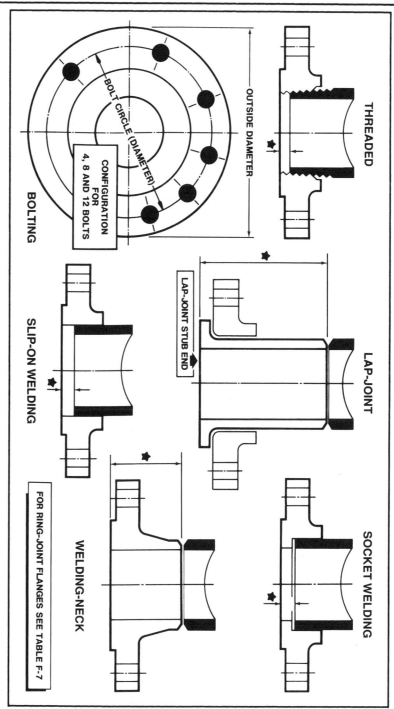

THREADED

BOLTING

OUTSIDE DIAMETER

BOLT CIRCLE (DIAMETER)

CONFIGURATION FOR 4, 8 AND 12 BOLTS

SLIP-ON WELDING

LAP-JOINT

LAP-JOINT STUB END

WELDING-NECK

FOR RING-JOINT FLANGES SEE TABLE F-7

SOCKET WELDING

NOTES

[1] FLANGE DIMENSIONS: ANSI STANDARD B16.5 AND MANUFACTURERS' DATA

[2] BLIND FLANGES: DATA FOR FLANGE DIAMETERS AND BOLTING IN THESE TABLES ALSO APPLIES TO BLIND FLANGES

[3] REDUCING FLANGES: AVAILABLE IN SLIP-ON, THREADED AND WELDING-NECK TYPES

[4] LAP-JOINT STUB-ENDS: ANSI B16.9 (Long Pattern) & MSS SP-43 (Short Pattern)

[5] STUDBOLT THREAD LENGTHS FOR LAP-JOINTS

FLANGE COMBINATION	FLANGE CLASS	INCREASE IN STUDBOLT LENGTH OVER LENGTHS IN TABLES F-1 thru F-6
Lapped to non-lapped	150 or 300	Thickness of lap
	Over 300	Thickness of lap
Lapped to lapped	150 - 2500	Thickness of lap minus 1/4"
		Thickness of two laps

Thickness of lap = Thickness of pipe wall + 0" + 0.06"

CLASS 150 FLANGE DATA — TABLE F-1

• DIMENSIONS INCLUDE 0.06" RAISED FACE ON FLANGES (except lap-joint)
•• DIMENSIONS INCLUDE 0.06" GAP FOR WELDING – REFER TO CHART 2.2

NOMINAL PIPE SIZE: NPS	1/2	3/4	1	1 1/2	2	3	4	6	8	10	12	14	16	18	20	24
OUTSIDE DIAMETER	3.5	3.88	4.25	5	6	7.5	9	11	13.5	16	19	21	23.5	25	27.5	32
FLANGE — END OF PIPE TO FACE OF FLANGE or LAP JOINT: WELD-NECK	1.88	2.06	2.19	2.44	2.5	2.75	3	3.5	4	4	4	4.5	5	5	5.5	6
SLIP-ON	0.62	0.62	0.69	0.88												
SOCKET ••	Wall thickness of pipe + 0.06-inch															
THREADED																
L-J STUB END • — ANSI																
— MSS																
BORE: WELD-NECK & SOCKET	0.62	0.82	1.05	1.61	2.07	3.07	4.03	6.07	7.98	10.02	12	[Order to match pipe ID]				
BOLTS PER FLANGE	4	4	4	4	4	4	8	8	8	12	12	12	16	16	20	20
BOLT CIRCLE DIAMETER	2.38	2.75	3.12	3.88	4.75	6	7.5	9.5	11.75	14.25	17	18.75	21.25	22.75	25	29.5
DIAMETER OF BOLT	1/2	1/2	1/2	1/2	5/8	5/8	5/8	3/4	3/4	7/8	7/8	1	1	1 1/8	1 1/8	1 1/4
STUDBOLT THREAD length - except lap-joint: Note 5 — RF	2.25	2.5	2.5	2.75	3.25	3.5	4	4.25	4.75	5.25	5.75	5.75	6.25	6.25	6.75	7.25
— RJ	–	–	3	3.25	3.75	4	4.5	4.75	5	5.25	5.75	6	6.25	6.75	6.75	7.25

CLASS 300 FLANGE DATA — TABLE F-2

• DIMENSIONS INCLUDE 0.06" RAISED FACE ON FLANGES (except lap-joint)
•• DIMENSIONS INCLUDE 0.06" GAP FOR WELDING – REFER TO CHART 2.2

NOMINAL PIPE SIZE: NPS	1/2	3/4	1	1 1/2	2	3	4	6	8	10	12	14	16	18	20	24
OUTSIDE DIAMETER	3.75	4.62	4.88	6.12	6.5	8.25	10	12.5	15	17.5	20.5	23	25.5	28	30.5	36
FLANGE — END OF PIPE TO FACE OF FLANGE or LAP JOINT: WELD-NECK	2.06	2.25	2.44	2.69	2.75	3.12	3.38	3.88	4.38	4.62	5.12	5.62	5.75	6.25	6.38	6.62
SLIP-ON	0.56	0.62	0.62	0.69	0.94											
SOCKET ••	Wall thickness of pipe + 0.06-inch															
THREADED																
L-J STUB END • — ANSI																
— MSS																
BORE: WELD-NECK & SOCKET	0.62	0.82	1.05	1.61	2.07	3.07	4.03	6.07	7.98	10.02	12	[Order to match pipe ID]				
BOLTS PER FLANGE	4	4	4	4	8	8	8	12	12	16	16	20	20	24	24	24
BOLT CIRCLE DIAMETER	2.62	3.25	3.5	4.5	5	6.62	7.88	10.62	13	15.25	17.75	20.25	22.5	24.75	27	32
DIAMETER OF BOLT	1/2	5/8	5/8	3/4	5/8	3/4	3/4	3/4	7/8	1	1 1/8	1 1/8	1 1/4	1 1/4	1 1/4	1 1/2
STUDBOLT THREAD length - except lap-joint: Note 5 — RF	2.5	3	3	3.5	3.5	4.25	4.5	5	5.5	6.25	6.75	7	7.5	7.75	8	9
— RJ	3	3.5	3.5	4	4	4.75	5	5.5	6	6.75	7.25	7.5	8	8.25	8.75	10

CLASS 600 FLANGE DATA — TABLE F-3

- • DIMENSIONS INCLUDE 0.25" RAISED FACE ON FLANGES (except lap-joint)
- •• DIMENSIONS INCLUDE 0.06" GAP FOR WELDING – REFER TO CHART 2.2

NOMINAL PIPE SIZE: NPS	1/2	3/4	1	1 1/2	2	3	4	6	8	10	12	14	16	18	20	24
OUTSIDE DIAMETER	3.75	4.62	4.88	6.12	6.5	8.25	10.75	14	16.5	20	22	23.75	27	29.25	32	37
FLANGE (★) END OF PIPE TO FACE OF FLANGE — WELD-NECK ••	2.31	2.5	2.69	3	3.12	3.5	4.25	4.88	5.5	6.25	6.38	6.75	7.25	7.5	7.75	8.25
SLIP-ON / SOCKET ••	0.81	0.88	0.88	0.94	1.06	1.31										
THREADED	0.38	0.31	0.31	0.44	0.69	0.50	0.62	0.88	0.94	1						
FLANGE or LAP JOINT STUB END • — L-J STUB END RF	3	3.5	3.5	4.25	4.25	5	5.75	6.75	7.5	8.5	8.75	9.25	9.25	10	10.75	13.25
L-J STUB END ANSI	3	3	3	4	4	4	4	5	5	5	5	6	6	6	6	6
L-J STUB END MSS	2	2	2	2	2.5	2.5	3	3.5	4	5	6	6	6	6	6	6
BORE: WELD-NECK & SOCKET	Order to match Internal Diameter of pipe — Wall thickness of pipe + 0.06-inch															
BOLTS PER FLANGE	4	4	4	4	8	8	8	8	12	12	12	12	20	20	20	24
BOLT CIRCLE DIAMETER	2.62	3.25	3.5	4.5	5	6.62	8.5	11.5	13.75	17	19.25	20.75	23.75	25.75	28.5	33
DIAMETER OF BOLT	1/2	5/8	5/8	3/4	5/8	3/4	7/8	1	1 1/8	1 1/4	1 1/4	1 3/8	1 3/8	1 1/2	1 5/8	1 7/8
STUDBOLT THREAD length – except lap-joint: Note 5 — RF	3	3.5	3.5	4.25	4.25	5	5.75	6.75	7.5	8.5	8.75	9.25	9.25	10	10.75	11.25
RJ	3.5	3.5	3.5	4.25	4.25	5	5.75	6.75	7.5	8.5	8.75	9.25	9.5	10.75	11.25	13.25

CLASS 900 FLANGE DATA — TABLE F-4

- • DIMENSIONS INCLUDE 0.25" RAISED FACE ON FLANGES (except lap-joint)

NOMINAL PIPE SIZE: NPS	1/2	3/4	1	1 1/2	2	3	4	6	8	10	12	14	16	18	20	24
OUTSIDE DIAMETER	4.75	5.12	5.88	7	8.5	9.5	11.5	15	18.5	21.5	24	25.25	27.75	31	33.75	41
FLANGE (★) END OF PIPE TO FACE OF FLANGE — WELD-NECK	2.62	3	3.5	3.5	4.25	4.75	5.75	6.62	7.5	8.12	8.62	8.75	9.25	9.25	10	11.75
SLIP-ON	Not available in this class															
SOCKET	3.12	3.5	4.25	4.5	4.25	5.75	6.62	7.5	8.5							
THREADED	0.62	0.69	0.69	0.81	1.06	0.50	0.62	0.88	0.94	1						
FLANGE or LAP JOINT STUB END • — L-J STUB END THREADED																
BORE: WELD-NECK	Order to match Internal Diameter of pipe — Wall thickness of pipe + 0.06-inch															
BOLTS PER FLANGE	4	4	4	4	8	8	8	12	12	16	20	20	20	20	20	20
BOLT CIRCLE DIAMETER	3.25	3.5	4	4.88	6.5	7.5	9.25	12.5	15.5	18.5	21	22	24.25	27	29.5	35.5
DIAMETER OF BOLT	3/4	3/4	7/8	1	7/8	7/8	1 1/8	1 1/8	1 3/8	1 3/8	1 3/8	1 1/2	1 5/8	1 5/8	1 7/8	2 1/2
STUDBOLT THREAD length – except lap-joint: Note 5 — RF	4.25	4.5	5	5.5	5.75	7.5	9.25	12.5	15.5	18.5	22	27	10.75	11.25	11.5	13.75
RJ	4.25	4.5	5	5.5	5.75	7.75	9.25	12.5	18.5	18.5	21	22	24.25	27	29.5	35.5
—	4.5	4.88	5.5	5.75	7.5	7.5	9.25	12.75	10.75	10.75	11.25	11.25	12.75	13.75	14.25	18

CLASS 1500 FLANGE DATA — TABLE F-5

• DIMENSIONS INCLUDE 0.25" RAISED FACE ON FLANGES (except lap-joint)
•• DIMENSIONS INCLUDE 0.06" GAP FOR WELDING – REFER TO CHART 2.2

NOMINAL PIPE SIZE: NPS	1/2	3/4	1	1 1/2	2	3	4	6	8	10	12	14	16	18	20	24
OUTSIDE DIAMETER	4.75	5.12	5.88	7	8.5	10.5	12.25	15.5	19	23	26.5	29.5	32.5	36	38.75	46
FLANGE – END OF PIPE TO FACE OF FLANGE or LAP JOINT TYPE END •: WELD-NECK	2.62	3	3.12	3.5	4.25	4.88	5.12	7	8.62	10.25	11.38	12	12.5	13.12	14.25	16.25
SLIP-ON																
SOCKET ••	1.19	1.25	1.44	1.44	1.88											
THREADED	0.38	0.31	0.31	0.38	0.62	0.50	0.62	0.88	0.94	0.94	1					
L-J JOINT STUB END: ANSI	2	3	3	4	4	5	6	6	5	6	6	6	6	6	6	6
L-J JOINT STUB END: MSS	2	2	2	2	2.5	3	4	5	6	6						
BORE: WELD-NECK & SOCKET	Order to match Internal Diameter of pipe															
DIAMETER OF BOLT	3/4	3/4	7/8	7/8	1	1 1/8	1 1/4	1 3/8	1 5/8	1 7/8	2	2 1/4	2 1/2	2 3/4	3	3 1/2
BOLT CIRCLE DIAMETER	3.25	3.5	4	4.88	6.5	9.5	12.5	15.5	19	22.5	25	27.75	30.5	32.75		39
BOLTS PER FLANGE	4	4	4	4	4	8	8	8	12	12	12	16	16	16	16	16
STUDBOLT THREAD length – except lap-joint: Note 5 — RF	4.25	4.5	5	5	5.75	7	7.75	10.25	11.5	13.25	14.75	16	17.5	19.5	21.25	24.25
STUDBOLT THREAD length – except lap-joint: Note 5 — RJ	4.25	4.5	5	5.5	5.75	7	7.75	10.5	12.75	13.5	15.25	16.75	18.5	20.75	22.25	25.5

Wall thickness of pipe + 0.06-inch

CLASS 2500 FLANGE DATA — TABLE F-6

• DIMENSIONS INCLUDE 0.25" RAISED FACE ON FLANGES (except lap-joint)

NOMINAL PIPE SIZE: NPS	1/2	3/4	1	1 1/2	2	3	4	6	8	10	12
OUTSIDE DIAMETER	5.25	5.5	6.25	8	9.25	12	14	19	21.75	26.5	30
FLANGE – END OF PIPE TO FACE OF FLANGE or LAP JOINT TYPE END •: WELD-NECK	3.12	3.38	3.75	4.62	5.25	6.88	7.75	11	12.75	16.75	18.5
SLIP-ON											
SOCKET	1.44	1.44									
THREADED	0.31	0.44	0.31	0.69	0.88	0.5	0.62	0.88	0.94	1.06	1
L-J JOINT STUB END: ANSI	2	3	3	4	6	6	8	8	10	10	10
L-J JOINT STUB END: MSS	2	2	2.5	3	3.5	4	5	6			
BORE: WELD-NECK	Order to match Internal Diameter of pipe										
DIAMETER OF BOLT	3/4	3/4	7/8	1	1 1/8	1 1/4	1 1/2	2	2	2 1/2	2 3/4
BOLT CIRCLE DIAMETER	3.5	3.75	4.25	5.75	6.75	9	10.75	14.5	17.25	21.25	24.38
BOLTS PER FLANGE	4	4	4	4	8	8	8	8	12	12	12
STUDBOLT THREAD length – except lap-joint: Note 5 — RF	4.75	5	5.5	6.75	7	8.75	10	13.5	15	19.25	21.25
STUDBOLT THREAD length – except lap-joint: Note 5 — RJ	4.75	5	5.5	6.75	7	9	10.25	14	15.5	20	22

Not available in this class

Wall thickness of pipe + 0.06-inch

RING-JOINT GASKET DATA

TABLE F-7

DIMENSIONS IN INCHES

DATA FOR WELDING-NECK FLANGES

L = LENGTH THRU HUB OF WELDING-NECK FLANGE WITH RING JOINT

G = GAP BETWEEN FLANGE FACES WITH RING IN COMPRESSION

◆ FOR OUTSIDE DIAMETERS OF FLANGES AND BOLTING REFER TO TABLES F-1 THRU F-6

FLANGE CLASSES

NPS	150			300			600			900			1500			2500		
	L	G	RING No	L	G	RING No	L	G	RING No	L	G	RING No	L	G	RING No	L	G	RING No
1/2	-	-	-	2.31	0.12	R 11	2.31	0.12	R 11	2.62	0.16	R 12	2.62	0.16	R 12	3.12	0.16	R 13
3/4	-	-	-	2.50	0.16	R 13	2.50	0.16	R 13	3	0.16	R 14	3.38	0.16	R 14	3.38	0.16	R 16
1	2.44	0.16	R 15	2.69	0.16	R 16	2.69	0.16	R 16	3.12	0.16	R 16	3.12	0.16	R 16	3.75	0.16	R 18
1 1/2	2.69	0.16	R 19	2.94	0.16	R 20	3	0.16	R 20	3.50	0.16	R 20	3.50	0.16	R 20	4.69	0.16	R 23
2	2.75	0.16	R 22	3.06	0.22	R 23	3.19	0.16	R 23	4.31	0.12	R 24	4.31	0.16	R 24	5.31	0.12	R 26
3	3	0.16	R 29	3.44	0.22	R 31	3.56	0.16	R 31	4.13	0.16	R 31	4.94	0.12	R 35	7	0.12	R 32
4	3.25	0.16	R 36	3.69	0.22	R 37	4.31	0.19	R 37	4.81	0.16	R 37	5.19	0.12	R 39	7.94	0.16	R 38
6	3.75	0.16	R 43	4.19	0.22	R 45	4.94	0.19	R 45	5.81	0.16	R 45	7.12	0.12	R 46	11.25	0.16	R 47
8	4.25	0.16	R 48	4.69	0.22	R 49	5.56	0.19	R 49	6.69	0.16	R 49	8.81	0.16	R 50	13.06	0.19	R 51
10	4.25	0.16	R 52	4.94	0.22	R 53	6.31	0.19	R 53	7.56	0.16	R 53	10.44	0.16	R 54	17.19	0.25	R 55
12	4.75	0.16	R 56	5.44	0.22	R 57	6.44	0.19	R 57	8.19	0.16	R 57	11.69	0.19	R 58	18.94	0.31	R 60
14	5.25	0.12	R 59	5.94	0.22	R 61	6.81	0.19	R 61	8.81	0.16	R 62	12.38	0.19	R 63			
16	5.25	0.12	R 64	6.06	0.22	R 65	7.31	0.19	R 65	8.94	0.16	R 66	12.94	0.22	R 67			
18	5.75	0.12	R 68	6.56	0.22	R 69	7.56	0.19	R 69	9.50	0.19	R 70	13.56	0.31	R 71			
20	5.94	0.12	R 72	6.75	0.22	R 73	7.88	0.19	R 73	10.25	0.19	R 74	14.69	0.38	R 75			
24	6.25	0.12	R 76	7.06	0.25	R 77	8.44	0.22	R 77	12.12	0.22	R 78	16.81	0.44	R 79			

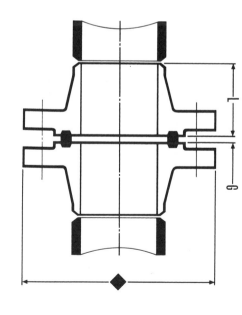

SLIP-ON FLANGES ON BUTT-WELDING ELBOWS   TABLE F-8

FOR USE ON BUTT-WELDING ELBOWS AS PERMITTED BY THE PIPING SPECIFICATION FOR THE PROJECT

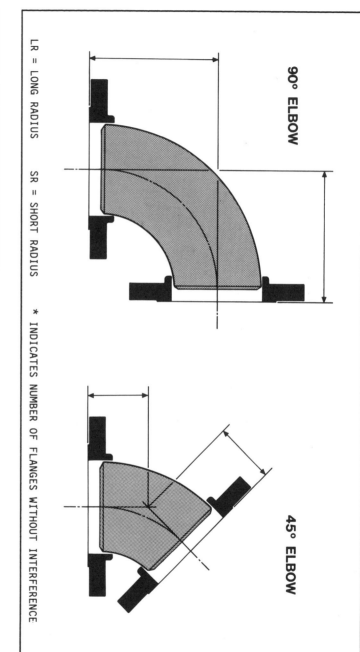

90° ELBOW

45° ELBOW

LR = LONG RADIUS   SR = SHORT RADIUS   * INDICATES NUMBER OF FLANGES WITHOUT INTERFERENCE

NPS	CLASS 150 FLANGES						CLASS 300 FLANGES					
	90 LR	*	90 SR	*	45 LR	*	90 LR	*	90 SR	*	45 LR	*
2	3.50	1	2.69	1	1.88	1	3.81	1	3.00	1	2.19	1
3	5.12	2	3.81	1	2.62	1	5.62	1	4.31	1	3.12	1
4	6.62	2	4.88	1	3.12	1	7.19	2	5.44	1	3.69	1
6	9.56	2	6.88	1	4.31	2	10.06	2	7.38	2	4.81	2
8	12.56	2	8.94	2	5.56	2	13.25	2	9.62	2	6.25	2
10	15.62	2	10.88	2	6.88	2	16.06	2	11.56	2	7.31	2
12	18.62	2	13.06	2	8.12	2	19.19	2	13.75	2	8.69	2
14	21.62	2	14.81	2	9.38	2	22.00	2	15.56	2	9.75	2
16	24.62	2	17.00	2	10.62	2	24.88	2	17.75	2	10.88	2
18	27.62	2	19.06	2	11.88	2	28.00	2	19.88	2	12.25	2
20	30.62	2	21.00	2	13.12	2	31.25	2	21.88	2	13.75	2
24	36.62	2	25.38	2	15.62	2	37.44	2	26.31	2	16.44	2

DIMENSIONS IN INCHES

PRESSURE/TEMPERATURE RATINGS FOR CARBON STEEL FLANGES

TABLE F-9

Maximum Ratings for flanges conforming to ANSI Standard B16.5 dimensions and material specification ASTM A-105

TEMPERATURE FAHRENHEIT	GAGE WORKING PRESSURE IN psi FOR FLANGE CLASSES 150 - 2500 FLANGE CLASSES						
	150	300	400	600	900	1500	2500
-20 to 100	285	740	990	1480	2220	3705	6170
200	260	675	900	1350	2025	3375	5625
300	230	655	875	1315	1970	3280	5470
400	200	635	845	1270	1900	3170	5280
500	170	600	800	1200	1795	2995	4990
600	140	550	730	1095	1640	2735	4560
650	125	535	715	1075	1610	2685	4475
700	110	535	710	1065	1600	2665	4440
750	95	505	670	1010	1510	2520	4200
800	80	410	550	825	1235	2060	3430
850	65	270	355	535	805	1340	2230
900	50	170	230	345	515	860	1430
950	35	105	140	205	310	515	860
1000	20	50	70	105	155	260	430

Standard ANSI B16.5 does not recommend using flanges manufactured from carbon steels made to ASTM specification A-105 at temperatures in excess of 1000F (538C) at any time, or their prolonged usage at temperatures over 800F (427C). [ASTM A-105 carbon steel is included in material group 1.1. of ANSI B16.5.]

THERMAL GRADIENTS, THERMAL CYCLING and EXTERNAL LOADS The suitability of slip-on, socket-welding and threaded flange attachments at 540F (282C) and -50F (-46C) is discussed in ANSI B16.5, which also makes recommendations to prevent leakage from Class 150 flanged joints at 400F (204C), and other classes at higher temperatures, if the above operating conditions are anticipated, and expected to be severe.

Ratings are for non-shock conditions. Values in this table do not prevail over limitations imposed by codes, standards, regulations or other obligations which may pertain to projects.

FLOW RESISTANCE OF FITTINGS & VALVES — TABLE F-10

FLOW RESISTANCE IS STATED AS EQUIVALENT LENGTH OF PIPE IN FEET [REFER TO NOTES (1) & (2)]

FITTINGS & CONNECTIONS

NOMINAL PIPE SIZE (IN.)	1/2	3/4	1	1½	2	3	4	6	8	10	12	14	16	18	20	24
90° LONG-RADIUS ELBOW				2.3	2.8	3.5	4.5	6.8	9.0	11	13	15	17	19	21	25
90° SHORT-RADIUS ELBOW				2.8	3.2	4.9	6.7	9.8	13	19	25	37	44	49	55	66
45° ELBOW (LONG RADIUS)	0.4	0.6	0.8	1.3	1.1	1.8	2.5	4.0	5.5	7.4	9.5	11	12	14	16	20
RETURN, LONG-RADIUS	2.3	2.8	3.5	4.9	5.1	7.5	9.8	15	19	24	29	32	36	41	45	55
RETURN, SHORT-RADIUS	2.3	2.8	3.5	4.9	5.1	7.5	9.8	15	20	25	39	50	63	78	89	100
90° MITERS — 2-PIECE					13	20	25	39	50	63	79	130	240	330	390	270
90° MITERS — 3-PIECE	1.1			2.8	11	14	22	33	48	54	79	130	240	330	390	170
90° MITERS — 4-PIECE					9.0	9.5	13	14	18	21	23	27	30	34	38	47
90° MITERS — 5-PIECE					6.3	8.5	13	13	17	13	16	19	20	23	26	35
VENTURI SWAGE — One listed NPS increase	1.1	2.2	2.9	3.3	2.7	3.0	4.7	3.6	4.2	5.0	5.9	6.3	6.9	7.8	9.1	2.7
REDUCER and SWAGE — One listed NPS reduction	1.1	2.2	2.9	3.1	2.2	3.0	4.7	6.2	8.3	12	17	21	25	49	100	75
STRAIGHT TEE — Thru run	1.0	1.5	2.1	3.7	1.6	2.2	3.0	3.6	4.2	5.0	5.9	6.3	6.9	7.8	8.4	9.1
STRAIGHT TEE — Thru branch and run	2.7	3.3	4.1	6.1	6.6	9.7	13	18	23	29	35	36	41	45	49	59
UNION and COUPLING (Screwed, pipe-to-pipe)	0.2	0.2	0.2	0.2												
REDUCING FLANGE — One listed NPS reduction	0.2	0.2	0.2	0.3	0.6	0.7	1.4	1.6	3.1	4.4	5.8	6.2	7.1	8.4	9.4	12
REDUCING FLANGE — One listed NPS increase	0.1	0.1	0.1	0.2	2.4	2.4	5.8	5.3	5.6	4.9	4.3	3.9	4.6	4.1	6.0	7.5
BELLMOUTH OUTLET (Vessel-to-line)	0.1	0.1	0.1	0.2	0.4	0.7	0.9	1.4	2.1	2.8	3.6	3.9	4.6	5.4	6.0	7.5
INLET, Flush with Wall (Line-to-vessel)	1.1	1.6	2.3	4.1	8.2	14	19	30	42	56	71	78	93	108	120	150
OUTLET, Flush with Wall (Vessel-to-line)	0.6	0.8	1.6	2.0	4.1	6.8	9.3	15	21	28	36	39	46	54	60	75

OPEN VALVES

	1/2	3/4	1	1½	2	3	4	6	8	10	12	14	16	18	20	24
GATE VALVE — Regular Disc	0.4	0.4	0.5	0.8	2.2	2.8	3.0	3.3	3.1	3.3	3.1	3.1	3.1	3.3	3.0	2.7
GLOBE VALVE — Composition Disc	15	16	20	29	70	94	120	170	230	300	380	390				
GLOBE VALVE — Plug-type Disc					82	97	120	190	300	420	560	710	930	1100	1200	1500
CHECK VALVE — Swing	5.2	5.7	6.7	10	16	27	37	59	83	110	140	160	190			
CHECK VALVE — Ball	170	190	220	320	530	880	1200	1900	2700	3600	4600	5100	6000	7000	7800	9800
CHECK VALVE — Fitting-disc					100	140	250	330	440	560						
ROTARY BALL VALVE — Regular Pattern (Walworth Aloyco)	4.0	3.9	4.1	3.3	3.7	4.0	8.0	15	21	28	40	38	45	49	47	57
ROTARY BALL VALVE — Eccentric Pattern (DeZurik)	1.6	1.7	2.2	3.3	3.7	4.0	5.7	8.3	8.4	11	16					
BUTTERFLY VALVE (Walworth 'Pinnacle' Valve)					3.7	5.7	5.5	6.8	9.0	9.5	12	17	21	25	49	
PLUG VALVE 1"-12": Regular pattern; 14"-24": Venturi pattern (W·K·M 'ACF')			1.7	4.1	5.5	8.2	14	56	96	100	140	88	130	120	160	170

NOTES

[1] Hydraulic resistances are for turbulent flow and are given as lengths of SCH 40 pipe having the same resistance. For pipe with a thicker wall use the resistance value for SCH 40 pipe having the closest internal diameter.

[2] Numbers in italics are resistances for threaded valves and fittings. Upright numbers relate to flanged valves and butt-welding fittings.

[3] For reducing and increasing fittings, flow resistance is based on the nominal pipe size at the inflow end.

[4] Tabulated flow resistances are approximate and selected from several sources not all giving comparable values. These sources include the Hydraulic Institute's "Pipe Friction Manual", the Crane Company's "Technical Paper 410", the "Reactor Handbook" (Interscience), the "Chemical Engineer's Handbook", (McGraw-Hill), "Cameron Hydraulic Data" (Ingersoll-Rand), and manufacturers' catalogs.

FRACTIONAL EQUIVALENTS

1/16	0.06	1/8	0.12	3/16	0.19	1/4	0.25
5/16	0.31	3/8	0.38	7/16	0.44	1/2	0.50
9/16	0.56	5/8	0.62	11/16	0.69	3/4	0.75
13/16	0.81	7/8	0.88	15/16	0.94		

FLOW OF WATER THRU SCH 40 PIPE

TABLE F-11

FLOW RATE		PRESSURE DROP (PSI) PER 100 ft SCH 40 PIPE													
GPM	Cu.ft/sec	v Ft/Sec	p psi	v Ft/Sec	p psi	v Ft/Sec	p psi	v Ft/Sec	p psi	v Ft/Sec	p psi	v Ft/Sec	p psi	v Ft/Sec	p psi
		⅛"		¼"											
.1	.00022	.56	.677			⅜"									
.2	.00045	1.14	2.48	.62	.548			½"							
.3	.00067	1.70	5.26	.93	1.16	.50	.255			¾"					
.4	.0089	2.26	9.00	1.24	1.98	.67	.436	.42	.136						
.5	.00111	2.82	13.58	1.55	3.00	.84	.656	.53	.205	.30	.050	1"			
.6	.00134	3.38	19.12	1.85	4.22	1.01	.925	.63	.290	.36	.071			1¼"	
.8	.00178	4.52	32.62	2.47	7.17	1.34	1.58	.84	.494	.48	.121	.30	.036		
1	.00223			3.09	10.91	1.68	2.39	1.06	.749	.60	.183	.37	.055	.21	.014
2	.00446			6.18	39.60	3.36	8.68	2.11	2.72	1.20	.665	.74	.199	.43	.051
3	.00668					5.04	18.46	3.17	5.77	1.80	1.41	1.11	.424	.64	.107
4	.00891					6.72	31.55	4.22	9.86	2.40	2.42	1.49	.724	.86	.183
5	.01114	1½"						5.28	14.92	3.01	3.64	1.86	1.09	1.07	.276
6	.01337							6.33	20.95	3.61	5.13	2.23	1.54	1.29	.390
8	.01782	1.26	.308	2"						4.81	8.76	2.97	2.62	1.71	.667
10	.02228	1.58	.466							6.01	13.28	3.713	3.97	2.142	1.01
15	.03342	2.36	.992	1.43	.285							5.57	8.46	3.21	2.14
20	.04456	3.15	1.69	1.91	.486	2½"						7.43	14.42	4.28	3.66
25	.05570	3.94	2.54	2.39	.736									5.36	5.54
30	.06684	4.73	3.60	2.37	1.03	2.01	.424							6.43	7.79
35	.07798	5.51	4.79	3.35	1.37	2.35	.566	3"						7.50	10.38
40	.08912	6.30	6.14	3.82	1.76	2.68	.724							8.57	13.28
50	.1114	7.88	9.31	4.78	2.67	3.35	1.10	2.17	.371	3½"					
60	.1337	9.45	13.08	5.74	3.75	4.02	1.54	2.61	.520						
70	.1560			6.70	4.99	4.70	2.05	3.04	.693	2.27	.335				
80	.1782			7.65	6.40	5.37	2.63	3.47	.890	2.59	.430	4"			
90	.2005			8.60	7.96	6.04	3.28	3.91	1.10	2.92	.535				
100	.2228			9.56	9.69	6.71	3.98	4.34	1.34	3.24	.650	2.52	.346		
125	.2785					8.38	6.03	5.43	2.01	4.05	.984	3.15	.523	5"	
150	.3342					10.1	8.46	6.52	2.86	4.87	1.38	3.78	.734		
175	.3899					11.7	11.3	7.60	3.81	5.68	1.84	4.41	.978	2.81	.316
200	.4456	6"				13.4	14.4	8.69	4.89	6.49	2.36	5.04	1.25	3.21	.405
225	.5013							9.77	6.09	7.30	2.94	5.67	1.56	3.61	.505
250	.5570	2.78	.245					10.9	7.41	8.11	3.58	6.30	1.90	4.01	.616
275	.6127	3.06	.292					11.9	8.84	8.92	4.27	6.93	2.27	4.41	.734
300	.6684	3.33	.344	8"				13.0	10.4	9.73	5.02	7.56	2.67	4.81	.863
350	.7798	3.89	.457					15.2	13.8	11.4	6.87	8.82	3.55	5.62	1.15
400	.8912	4.44	.587	2.57	.149					13.0	8.58	10.1	4.56	6.41	1.47
450	1.003	5.00	.731	2.89	.185					14.6	10.7	11.3	5.66	7.22	1.83
500	1.114	5.55	.887	3.21	.225					16.2	13.0	12.6	6.89	8.02	2.23
550	1.225	6.11	1.07	3.53	.270	10"				17.8	15.5	13.9	8.25	8.82	2.67
600	1.337	6.66	1.25	3.85	.316					19.5	18.2	15.1	9.68	9.62	3.13
650	1.449	7.22	1.45	4.17	.367	2.65	.118					16.4	11.2	10.4	3.62
700	1.560	7.78	1.66	4.49	.420	2.85	.135					17.6	12.9	11.2	4.16
750	1.671	8.33	1.89	4.81	.480	3.05	.154					18.9	14.7	12.0	4.75
800	1.782	8.89	2.13	5.13	.540	3.26	.173	12"				20.2	16.5	12.8	5.35
850	1.894	9.44	2.38	5.45	.605	3.46	.194					21.4	18.5	13.6	5.98
900	2.005	10.0	2.66	5.77	.627	3.66	.216	2.58	.090			22.7	20.6	14.4	6.65
950	2.117	10.6	2.93	6.09	.744	3.87	.238	2.72	.099			23.9	22.8	15.2	7.36
1000	2.228	11.1	3.23	6.41	.817	4.07	.262	2.87	.109	14"				16.0	8.10
1100	2.451	12.2	3.85	7.06	.975	4.48	.313	3.15	.130					17.6	9.66
1200	2.674	13.3	4.53	7.70	1.15	4.88	.368	3.44	.153	2.85	.096			19.2	11.4
1300	2.896	14.4	5.26	8.34	1.33	5.29	.427	3.73	.178	3.08	.111			20.8	13.2
1400	3.119	15.6	6.01	8.98	1.53	5.70	.490	4.01	.204	3.32	.127	16"		22.4	15.1
1500	3.342	16.7	6.84	9.62	1.74	6.10	.556	4.30	.232	3.56	.145			24.1	17.2
1600	3.565	17.8	7.73	10.3	1.96	6.51	.628	4.59	.262	3.79	.163	2.91	.084		
1800	4.010	20.0	9.64	11.5	2.46	7.32	.782	5.16	.329	4.27	.203	3.27	.104		
2000	4.456	22.2	11.6	12.8	2.97	8.14	.953	5.73	.396	4.74	.247	3.63	.127	18"	
2500	5.570	27.8	17.6	16.0	4.49	10.2	1.44	7.17	.601	5.93	.374	4.54	.192		
3000	6.684			19.2	6.30	12.2	2.02	8.60	.842	7.11	.525	5.45	.270	4.30	.149
3500	7.798			22.4	8.41	14.2	2.70	10.0	1.12	8.30	.700	6.36	.358	5.02	.199
4000	8.912			25.7	10.8	16.3	3.46	11.5	1.44	9.48	.896	7.26	.459	5.74	.255
4500	10.03			28.9	13.4	18.3	4.31	12.9	1.76	10.7	1.12	8.17	.671	6.45	.317
5000	11.14					20.4	5.23	14.3	2.18	11.9	1.36	9.08	.695	7.17	.386
6000	13.37					24.4	7.35	17.2	3.06	14.2	1.91	10.9	.977	8.60	.542
7000	15.60					28.5	9.80	20.1	4.08	16.6	2.54	12.7	1.30	10.0	.723
8000	17.82							22.9	5.22	19.0	3.25	14.5	1.67	11.5	.926
9000	20.05							25.8	6.51	21.3	4.06	16.3	2.08	12.9	1.15
10000	22.28							28.7	7.91	23.7	4.92	18.2	2.53	14.3	1.40
12000	26.74									28.5	6.92	21.8	3.55	17.2	1.97
14000	31.19											25.4	4.72	20.1	2.62
16000	35.65											29.1	6.06	22.9	3.36
18000	40.10											32.7	7.55	25.8	4.18
20000	44.56													28.7	5.08

Reproduced by courtesy of the Lunkenheimer Company. Data are based on the Saph and Schoder formula: $p = LQ^{1.86}/1435D^5$

HEAT EXCHANGER NOMENCLATURE

REPRODUCED BY PERMISSION OF THE TUBULAR EXCHANGER MANUFACTURERS ASSOCIATION

CHART H-1

THREE LETTERS, SUCH AS AEW, BGP, etc. DESIGNATE THE BASIC CONSTRUCTION OF THE EXCHANGER. REFER TO 6.6.1, 'DATA NEEDED TO DESIGN EXCHANGER PIPING'.

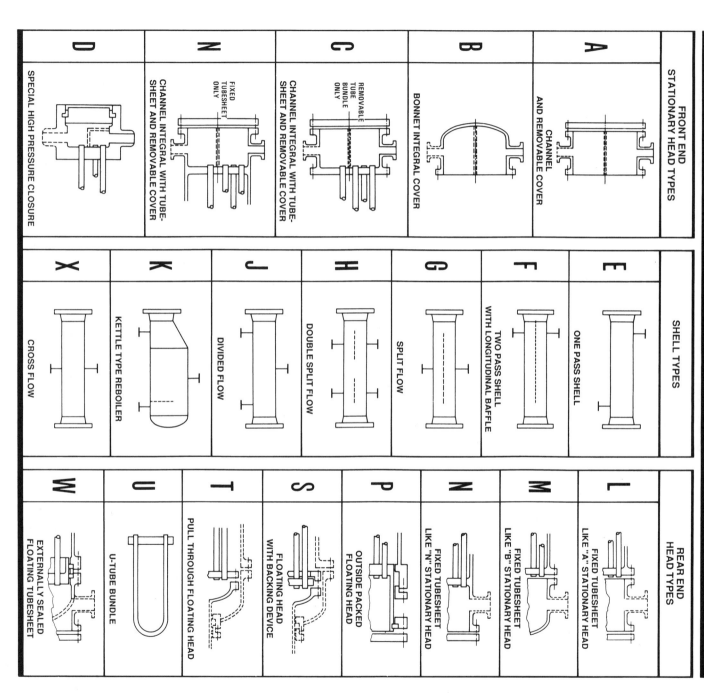

FRONT END STATIONARY HEAD TYPES

- **A** — CHANNEL AND REMOVABLE COVER
- **B** — BONNET INTEGRAL COVER
- **C** — CHANNEL INTEGRAL WITH TUBE-SHEET AND REMOVABLE COVER (REMOVABLE TUBE BUNDLE ONLY)
- **N** — CHANNEL INTEGRAL WITH TUBE-SHEET AND REMOVABLE COVER (FIXED TUBESHEET ONLY)
- **D** — SPECIAL HIGH PRESSURE CLOSURE

SHELL TYPES

- **E** — ONE PASS SHELL
- **F** — TWO PASS SHELL WITH LONGITUDINAL BAFFLE
- **G** — SPLIT FLOW
- **H** — DOUBLE SPLIT FLOW
- **J** — DIVIDED FLOW
- **K** — KETTLE TYPE REBOILER
- **X** — CROSS FLOW

REAR END HEAD TYPES

- **L** — FIXED TUBESHEET LIKE "A" STATIONARY HEAD
- **M** — FIXED TUBESHEET LIKE "B" STATIONARY HEAD
- **N** — FIXED TUBESHEET LIKE "N" STATIONARY HEAD
- **P** — OUTSIDE PACKED FLOATING HEAD
- **S** — FLOATING HEAD WITH BACKING DEVICE
- **T** — PULL THROUGH FLOATING HEAD
- **U** — U-TUBE BUNDLE
- **W** — EXTERNALLY SEALED FLOATING TUBESHEET

COMPOUND ANGLES

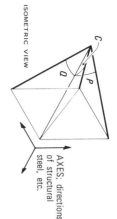

ISOMETRIC VIEW

AXES, directions of structural steel, etc.

Compound angle, C, is given by:

$$(\tan C)^2 = (\tan P)^2 + (\tan Q)^2$$

AREAS & VOLUMES

TRAPEZOID

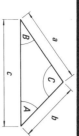

Trapezoid: A four-sided figure with two parallel sides, and the other two sides at any angle. Termed 'trapezium' in U.K.

If $a = b$, this formula applies to any parallelogram or rectangle.

$$\text{AREA} = c(a + b)/2$$

TRIANGLE

$$\text{AREA} = ac/2$$

CIRCLE

Refer to table M-4 for numerical values of circumferences and areas of full circles

FULL CIRCLE

$$\text{CIRCUMFERENCE} = 2\pi r$$
$$= 6.2831853\, r$$
$$\text{AREA} = \pi r^2$$
$$= 3.1415927\, r^2$$

SECTOR (as shown)

$$\text{LENGTH OF ARC} = l =$$
$$= \pi r Q/180$$
$$= 0.0174533\, rQ$$
$$\text{AREA} = \pi r^2 Q/360$$
$$= 0.0087266\, r^2 Q$$

SEGMENT OF CIRCLE

DIAMETER $= a + (b^2/4a)$
RADIUS $= r = (a/2) + (b^2/8a)$
LENGTH OF ARC $^* = l$
$= (\pi r/90).\arccos[1 - (a/r)]$
$= (\pi r/90).\arcsin(b/2r)$
where $\pi/90 = 0.03490659$
AREA $^* = (rl - rb + ab)/2$
NOTE: $\arccos[Q] =$ "angle in degrees whose cosine is Q", and $\arcsin[Q] =$ "angle in degrees whose sine is Q".
*Valid for a positive and less than $2r$.

ELLIPSE

$$\text{AREA} = (\pi/4)(ab)$$
$$= 0.7853982(ab)$$
$$\text{CIRCUMFERENCE} =$$
$$\pi[(a^2 + b^2)/2]^{1/2} \text{ approximately}$$

PRISM

BASE OF ANY SHAPE; UPRIGHT OR SLOPING

$$\text{AREA OF SECTION} = A$$
$$\text{DISTANCE BETWEEN PARALLEL SECTIONS 'A' AND 'A' = } h$$
$$\text{VOLUME} = hA$$

NOTE: THIS FORMULA MAY BE APPLIED TO CYLINDRIC AND RECTANGULAR TANKS.

CONE

BASE OF ANY SHAPE; UPRIGHT OR SLOPING

$$\text{AREA OF BASE} = A$$
$$\text{HEIGHT (measured at right angles to base)} = h$$
$$\text{VOLUME} = hA/3$$

FRUSTUM OF CONE

SECTION OF ANY SHAPE; UPRIGHT OR SLOPING

AREAS OF PARALLEL FLAT SURFACES 'A' AND 'B' = A and B, respectively
DISTANCE BETWEEN SURFACES 'A' AND 'B' = h
$$\text{VOLUME} = (h/3).[A + B + (AB)^{1/2}]$$

TRIANGLES

If Θ is between 90° and 180°,
$\sin \Theta = \sin(180^\circ - \Theta)$, $\cos \Theta = -\cos(180^\circ - \Theta)$
(Thus values may be found in tables.)

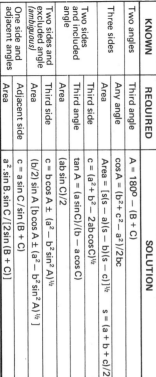

KNOWN	REQUIRED	SOLUTION
Two angles	Third angle	$A = 180^\circ - (B + C)$
Three sides	Any angle	$\cos A = (b^2 + c^2 - a^2)/2bc$
	Area	$\text{Area} = [s(s-a)(s-b)(s-c)]^{1/2}$
		$s = (a + b + c)/2$
Two sides and included angle	Third side	$c = (a^2 + b^2 - 2ab\cos C)^{1/2}$
	Area	$\tan A = (a\sin C)/(b - a\cos C)$
		$(ab \sin C)/2$
Two sides and excluded angle *(ambiguous)*	Third angle	$c = b\cos A \pm (a^2 - b^2\sin^2 A)^{1/2}$
	Third side	$(b/2)\sin A\,[b\cos A \pm (a^2 - b^2\sin^2 A)^{1/2}]$
One side and adjacent angles	Adjacent side	$c = a\sin C/\sin(B + C)$
	Area	$a^2.\sin B.\sin C/[2\sin(B + C)]$

SPHERE

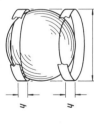

$$\text{RADIUS} = r$$
$$\text{DIAMETER} = d = 2r$$
$$\text{SURFACE AREA} = \pi d^2$$
$$= 3.14159265\, d^2$$
$$\text{VOLUME} = \pi d^3/6$$
$$= 0.5235988\, d^3$$
$$\text{VOLUME OF SEGMENT OF DEPTH } h = (\pi h^2/3)(3r - h)$$
$$= (1.0471976\, h)(3r - h),$$
where h is positive and less than $2r$.

AREA OF SPHERICAL CAP OR SLICE

The area of the curved surface of the cap or the slice equals the area of the cylindric band of the same depth, h; that is, πhd, no matter where the slice is taken, or how thick the slice is positive and less than $2r$, or cap is.

MEASUREMENTS

TABLES M-2

HYPOTENUSE FOR 45° TRIANGLES

VALUES COMPUTED TO NEAREST 1/16-th INCH

SIDE	HYPOTENUSE	SIDE	HYPOTENUSE	SIDE	HYPOTENUSE	SIDE	HYPOTENUSE	SIDE	HYPOTENUSE
0 0-1/16	0 0-1/16	0 3-3/16	0 4-1/2	0 6-5/16	0 8-15/16	0 9-7/16	0 13-3/8	1 0	1 4-31/32
0 0-1/8	0 0-3/16	0 3-1/4	0 4-5/8	0 6-3/8	0 9	0 9-1/2	0 13-7/16	2 0	2 9-15/16
0 0-3/16	0 0-1/4	0 3-5/16	0 4-11/16	0 6-7/16	0 9-1/8	0 9-9/16	0 13-1/2	3 0	4 2-29/32
0 0-1/4	0 0-3/8	0 3-3/8	0 4-3/4	0 6-1/2	0 9-3/16	0 9-5/8	0 13-5/8	4 0	5 7-7/8
0 0-5/16	0 0-7/16	0 3-7/16	0 4-7/8	0 6-9/16	0 9-1/4	0 9-11/16	0 13-11/16	5 0	7 0-27/32
0 0-3/8	0 0-1/2	0 3-1/2	0 4-15/16	0 6-5/8	0 9-3/8	0 9-3/4	0 13-13/16	6 0	8 5-13/16
0 0-7/16	0 0-5/8	0 3-9/16	0 5-1/16	0 6-11/16	0 9-7/16	0 9-13/16	0 13-7/8	7 0	9 10-25/32
0 0-1/2	0 0-11/16	0 3-5/8	0 5-1/8	0 6-3/4	0 9-9/16	0 9-7/8	0 13-15/16	8 0	11 3-3/4
0 0-9/16	0 0-13/16	0 3-11/16	0 5-3/16	0 6-13/16	0 9-5/8	0 9-15/16	0 14-1/16	9 0	12 8-3/4
0 0-5/8	0 0-7/8	0 3-3/4	0 5-5/16	0 6-7/8	0 9-3/4	0 10	0 14-1/8	10 0	14 1-23/32
0 0-11/16	0 1	0 3-13/16	0 5-3/8	0 6-15/16	0 9-7/8	0 10-1/16	0 14-1/4	11 0	15 6-11/16
0 0-3/4	0 1-1/16	0 3-7/8	0 5-1/2	0 7	0 10	0 10-1/8	0 14-5/16	12 0	16 11-21/32
0 0-13/16	0 1-1/8	0 3-15/16	0 5-9/16	0 7-1/16	0 10	0 10-3/16	0 14-7/16	13 0	18 4-5/8
0 0-7/8	0 1-1/4	0 4	0 5-11/16	0 7-1/8	0 10-1/16	0 10-1/4	0 14-1/2	14 0	19 9-19/32
0 0-15/16	0 1-5/16	0 4-1/16	0 5-3/4	0 7-3/16	0 10-3/16	0 10-5/16	0 14-9/16	15 0	21 2-9/16
0 1	0 1-7/16	0 4-1/8	0 5-13/16	0 7-1/4	0 10-1/4	0 10-3/8	0 14-11/16		
0 1-1/16	0 1-1/2	0 4-1/4	0 5-15/16	0 7-3/8	0 10-5/16	0 10-7/16	0 14-3/4	0 8-1/2	1 0
0 1-1/8	0 1-9/16	0 4-5/16	0 6	0 7-7/16	0 10-7/16	0 10-1/2	0 14-7/8	1 4-31/32	2 0
0 1-3/16	0 1-11/16	0 4-3/8	0 6-1/8	0 7-1/2	0 10-1/2	0 10-9/16	0 14-15/16	2 1-15/32	3 0
0 1-1/4	0 1-3/4	0 4-7/16	0 6-3/16	0 7-9/16	0 10-5/8	0 10-5/8	0 15	2 9-15/16	4 0
0 1-5/16	0 1-7/8	0 4-1/2	0 6-3/8	0 7-5/8	0 10-11/16	0 10-11/16	0 15-1/8	3 6-7/16	5 0
0 1-3/8	0 1-15/16	0 4-9/16	0 6-7/16	0 7-11/16	0 10-13/16	0 10-3/4	0 15-3/16	4 2-29/32	6 0
0 1-7/16	0 2-1/16	0 4-5/8	0 6-9/16	0 7-3/4	0 10-7/8	0 10-13/16	0 15-5/16	4 11-13/32	7 0
0 1-1/2	0 2-1/8	0 4-11/16	0 6-5/8	0 7-13/16	0 10-15/16	0 10-7/8	0 15-3/8	5 7-7/8	8 0
0 1-9/16	0 2-3/16	0 4-3/4	0 6-11/16	0 7-7/8	0 11-1/16	0 10-15/16	0 15-7/16	6 4-3/8	9 0
0 1-5/8	0 2-5/16	0 4-13/16	0 6-13/16	0 7-15/16	0 11-1/8	0 11	0 15-9/16	7 0-27/32	10 0
0 1-11/16	0 2-3/8	0 4-7/8	0 6-7/8	0 8	0 11-1/4	0 11-1/16	0 15-5/8	7 9-11/32	11 0
0 1-3/4	0 2-1/2	0 4-15/16	0 7	0 8-1/16	0 11-5/16	0 11-1/8	0 15-3/4	8 5-13/16	12 0
0 1-13/16	0 2-9/16	0 5	0 7-1/16	0 8-1/8	0 11-3/8	0 11-3/16	0 15-13/16	9 2-5/16	13 0
0 1-7/8	0 2-5/8	0 5-1/16	0 7-3/16	0 8-3/16	0 11-1/2	0 11-1/4	0 15-15/16	9 10-25/32	14 0
0 1-15/16	0 2-3/4	0 5-1/8	0 7-1/4	0 8-1/4	0 11-9/16	0 11-5/16	0 16	10 7-9/32	15 0
0 2	0 2-13/16	0 5-3/16	0 7-5/16	0 8-5/16	0 11-11/16	0 11-3/8	0 16-1/16		
0 2-1/16	0 2-15/16	0 5-1/4	0 7-7/16	0 8-3/8	0 11-3/4	0 11-7/16	0 16-3/16		
0 2-1/8	0 3	0 5-5/16	0 7-1/2	0 8-7/16	0 11-7/8	0 11-1/2	0 16-1/4		
0 2-3/16	0 3-1/16	0 5-3/8	0 7-5/8	0 8-1/2	0 12	0 11-9/16	0 16-3/8		
0 2-1/4	0 3-3/16	0 5-7/16	0 7-11/16	0 8-9/16	0 12-1/8	0 11-5/8	0 16-7/16		
0 2-5/16	0 3-1/4	0 5-1/2	0 7-3/4	0 8-5/8	0 12-3/16	0 11-11/16	0 16-1/2		
0 2-3/8	0 3-3/8	0 5-9/16	0 7-7/8	0 8-11/16	0 12-5/16	0 11-3/4	0 16-5/8		
0 2-7/16	0 3-7/16	0 5-5/8	0 7-15/16	0 8-3/4	0 12-3/8	0 11-13/16	0 16-11/16		
0 2-1/2	0 3-9/16	0 5-11/16	0 8-1/16	0 8-13/16	0 12-7/16	0 11-7/8	0 16-13/16		
0 2-9/16	0 3-5/8	0 5-3/4	0 8-1/8	0 8-7/8	0 12-9/16	0 11-15/16	0 16-7/8		
0 2-5/8	0 3-11/16	0 5-13/16	0 8-1/4	0 8 15/16	0 12-5/8	0 12	0 17		
0 2-11/16	0 3-13/16	0 5-7/8	0 8-5/16	0 9	0 12-3/4	0 12-1/16	0 17-1/16		
0 2-3/4	0 3-7/8	0 5-15/16	0 8-3/8	0 9-1/16	0 12-13/16	0 12-3/16	0 17-1/4		
0 2-13/16	0 4	0 6	0 8-1/2	0 9-1/8	0 12-7/8	0 12-1/4	0 17-5/16		
0 2-7/8	0 4-1/16	0 6-1/16	0 8-9/16	0 9-3/16	0 13	0 12-5/16	0 17-3/8		
0 2-15/16	0 4-1/8	0 6-1/8	0 8-11/16	0 9-1/4	0 13-1/16	0 12-3/8	0 17-1/2		
0 3	0 4-1/4	0 6-3/16	0 8-3/4	0 9-5/16	0 13-3/16	0 12-7/16	0 17-9/16		
0 3-1/16	0 4-5/16	0 6-1/4	0 8-13/16	0 9-3/8	0 13-1/4	0 12-1/2	0 17-11/16		
0 3-1/8	0 4-7/16								

FEET

SIDE	HYPOTENUSE
1 0	1 4-31/32
2 0	2 9-15/16
3 0	4 2-29/32
4 0	5 7-7/8
5 0	7 0-27/32
6 0	8 5-13/16
7 0	9 10-25/32
8 0	11 3-3/4
9 0	12 8-3/4
10 0	14 1-23/32
11 0	15 6-11/16
12 0	16 11-21/32
13 0	18 4-5/8
14 0	19 9-19/32
15 0	21 2-9/16

SIDE	HYPOTENUSE
0 8-1/2	1 0
1 4-31/32	2 0
2 1-15/32	3 0
2 9-15/16	4 0
3 6-7/16	5 0
4 2-29/32	6 0
4 11-13/32	7 0
5 7-7/8	8 0
6 4-3/8	9 0
7 0-27/32	10 0
7 9-11/32	11 0
8 5-13/16	12 0
9 2-5/16	13 0
9 10-25/32	14 0
10 7-9/32	15 0

HYPOTENUSE (=1.4142... x SIDE)

SIDE (=0.7071... x HYPOTENUSE)

SIDE (=0.7071... x HYPOTENUSE)

CONSTRUCTION OF MITERS

3-PIECE

22½° · 67½° · 45°

1½×NPS*

4-PIECE

15° · 45° · 60° · 75° · 30°

1½×NPS

5-PIECE

11¼° · 33¾° · 45° · 56¼° · 67½° · 78¾° · 22½°

1½×NPS

* NPS = NOMINAL PIPE SIZE (INCHES)

TANGENT LENGTHS FOR BENDS

t · t · A · r

A = Angle of bend

GENERAL FORMULA

$$t = r.\tan(A/2)$$

(Valid for 'A' less than 180°)

[31]

MILLIMETERS CONVERTED TO FEET AND INCHES

TABLES M-3

mm	ft-in.(fraction)	mm	ft-in.(fraction)	mm	ft-in.(fraction)	mm	ft-in.(fraction)	mm	ft-in.(fraction)	mm	ft-in.(fraction)	mm	ft-in.(fraction)
1	0- 0.04 [3/64]	81	0- 3.19 [3/16]	161	0- 6.34 [11/32]	241	0- 9.49 [31/64]	321	1- 0.64 [41/64]	401	1- 3.79 [25/32]	481	1- 6.94 [15/16]
2	0- 0.08 [5/64]	82	0- 3.23 [15/64]	162	0- 6.38 [3/8]	242	0- 9.53 [17/32]	322	1- 0.68 [43/64]	402	1- 3.83 [53/64]	482	1- 6.98 [31/32]
3	0- 0.12 [1/8]	83	0- 3.27 [17/64]	163	0- 6.42 [27/64]	243	0- 9.57 [9/16]	323	1- 0.72 [23/32]	403	1- 3.87 [55/64]	483	1- 7.02 [1/64]
4	0- 0.16 [5/32]	84	0- 3.31 [5/16]	164	0- 6.46 [29/64]	244	0- 9.61 [39/64]	324	1- 0.76 [3/4]	404	1- 3.91 [29/32]	484	1- 7.06 [1/16]
5	0- 0.20 [13/64]	85	0- 3.35 [11/32]	165	0- 6.50 [1/2]	245	0- 9.65 [41/64]	325	1- 0.80 [51/64]	405	1- 3.94 [15/16]	485	1- 7.09 [3/32]
6	0- 0.24 [15/64]	86	0- 3.39 [25/64]	166	0- 6.54 [17/32]	246	0- 9.69 [11/16]	326	1- 0.83 [53/64]	406	1- 3.98 [63/64]	486	1- 7.13 [9/64]
7	0- 0.28 [9/32]	87	0- 3.43 [27/64]	167	0- 6.57 [37/64]	247	0- 9.72 [23/32]	327	1- 0.87 [7/8]	407	1- 4.02 [1/32]	487	1- 7.17 [11/64]
8	0- 0.31 [5/16]	88	0- 3.46 [15/32]	168	0- 6.61 [39/64]	248	0- 9.76 [49/64]	328	1- 0.91 [29/32]	408	1- 4.06 [1/16]	488	1- 7.21 [7/32]
9	0- 0.35 [23/64]	89	0- 3.50 [1/2]	169	0- 6.65 [21/32]	249	0- 9.80 [51/64]	329	1- 0.95 [61/64]	409	1- 4.10 [7/64]	489	1- 7.25 [1/4]
10	0- 0.39 [25/64]	90	0- 3.54 [35/64]	170	0- 6.69 [11/16]	250	0- 9.84 [27/32]	330	1- 0.99 [63/64]	410	1- 4.14 [9/64]	490	1- 7.29 [19/64]
11	0- 0.43 [7/16]	91	0- 3.58 [37/64]	171	0- 6.73 [47/64]	251	0- 9.88 [7/8]	331	1- 1.03 [1/32]	411	1- 4.18 [3/16]	491	1- 7.33 [21/64]
12	0- 0.47 [15/32]	92	0- 3.62 [5/8]	172	0- 6.77 [49/64]	252	0- 9.92 [59/64]	332	1- 1.07 [5/64]	412	1- 4.22 [7/32]	492	1- 7.37 [3/8]
13	0- 0.51 [33/64]	93	0- 3.66 [21/32]	173	0- 6.81 [13/16]	253	0- 9.96 [61/64]	333	1- 1.11 [7/64]	413	1- 4.26 [17/64]	493	1- 7.41 [13/32]
14	0- 0.55 [35/64]	94	0- 3.70 [45/64]	174	0- 6.85 [27/32]	254	0-10.00	334	1- 1.15 [5/32]	414	1- 4.30 [19/64]	494	1- 7.45 [29/64]
15	0- 0.59 [19/32]	95	0- 3.74 [47/64]	175	0- 6.89 [57/64]	255	0-10.04 [3/64]	335	1- 1.19 [3/16]	415	1- 4.34 [11/32]	495	1- 7.49 [31/64]
16	0- 0.63 [5/8]	96	0- 3.78 [25/32]	176	0- 6.93 [59/64]	256	0-10.08 [5/64]	336	1- 1.23 [15/64]	416	1- 4.38 [3/8]	496	1- 7.53 [17/32]
17	0- 0.67 [43/64]	97	0- 3.82 [13/16]	177	0- 6.97 [31/32]	257	0-10.12 [1/8]	337	1- 1.27 [17/64]	417	1- 4.42 [27/64]	497	1- 7.57 [9/16]
18	0- 0.71 [45/64]	98	0- 3.86 [55/64]	178	0- 7.01 [1/64]	258	0-10.16 [5/32]	338	1- 1.31 [5/16]	418	1- 4.46 [29/64]	498	1- 7.61 [39/64]
19	0- 0.75 [3/4]	99	0- 3.90 [57/64]	179	0- 7.05 [3/64]	259	0-10.20 [13/64]	339	1- 1.35 [11/32]	419	1- 4.50 [1/2]	499	1- 7.65 [41/64]
20	0- 0.79 [25/32]	100	0- 3.94 [15/16]	180	0- 7.09 [3/32]	260	0-10.24 [15/64]	340	1- 1.39 [25/64]	420	1- 4.54 [17/32]	500	1- 7.69 [11/16]
21	0- 0.83 [53/64]	101	0- 3.98 [31/32]	181	0- 7.13 [1/8]	261	0-10.28 [9/32]	341	1- 1.43 [27/64]	421	1- 4.57 [37/64]	501	1- 7.72 [23/32]
22	0- 0.87 [55/64]	102	0- 4.02 [1/64]	182	0- 7.17 [11/64]	262	0-10.31 [5/16]	342	1- 1.46 [15/32]	422	1- 4.61 [39/64]	502	1- 7.76 [49/64]
23	0- 0.91 [29/32]	103	0- 4.06 [1/16]	183	0- 7.20 [13/64]	263	0-10.35 [23/64]	343	1- 1.50 [1/2]	423	1- 4.65 [21/32]	503	1- 7.80 [51/64]
24	0- 0.94 [15/16]	104	0- 4.09 [3/32]	184	0- 7.24 [1/4]	264	0-10.39 [25/64]	344	1- 1.54 [35/64]	424	1- 4.69 [11/16]	504	1- 7.84 [27/32]
25	0- 0.98 [63/64]	105	0- 4.13 [9/64]	185	0- 7.28 [9/32]	265	0-10.43 [7/16]	345	1- 1.58 [37/64]	425	1- 4.73 [47/64]	505	1- 7.88 [7/8]
26	0- 1.02 [1/32]	106	0- 4.17 [11/64]	186	0- 7.32 [21/64]	266	0-10.47 [15/32]	346	1- 1.62 [5/8]	426	1- 4.77 [49/64]	506	1- 7.92 [59/64]
27	0- 1.06 [1/16]	107	0- 4.21 [7/32]	187	0- 7.36 [23/64]	267	0-10.51 [33/64]	347	1- 1.66 [21/32]	427	1- 4.81 [13/16]	507	1- 7.96 [61/64]
28	0- 1.10 [7/64]	108	0- 4.25 [1/4]	188	0- 7.40 [13/32]	268	0-10.55 [35/64]	348	1- 1.70 [45/64]	428	1- 4.85 [27/32]	508	1- 8.00
29	0- 1.14 [9/64]	109	0- 4.29 [19/64]	189	0- 7.44 [7/16]	269	0-10.59 [19/32]	349	1- 1.74 [47/64]	429	1- 4.89 [57/64]	509	1- 8.04 [3/64]
30	0- 1.18 [3/16]	110	0- 4.33 [21/64]	190	0- 7.48 [31/64]	270	0-10.63 [5/8]	350	1- 1.78 [25/32]	430	1- 4.93 [59/64]	510	1- 8.08 [5/64]
31	0- 1.22 [7/32]	111	0- 4.37 [3/8]	191	0- 7.52 [33/64]	271	0-10.67 [43/64]	351	1- 1.82 [13/16]	431	1- 4.97 [31/32]	511	1- 8.12 [1/8]
32	0- 1.26 [17/64]	112	0- 4.41 [13/32]	192	0- 7.56 [9/16]	272	0-10.71 [45/64]	352	1- 1.86 [55/64]	432	1- 5.01 [1/64]	512	1- 8.16 [5/32]
33	0- 1.30 [19/64]	113	0- 4.45 [29/64]	193	0- 7.60 [19/32]	273	0-10.75 [3/4]	353	1- 1.90 [57/64]	433	1- 5.05 [3/64]	513	1- 8.20 [13/64]
34	0- 1.34 [11/32]	114	0- 4.49 [31/64]	194	0- 7.64 [41/64]	274	0-10.79 [25/32]	354	1- 1.94 [15/16]	434	1- 5.09 [3/32]	514	1- 8.24 [15/64]
35	0- 1.38 [3/8]	115	0- 4.53 [17/32]	195	0- 7.68 [43/64]	275	0-10.83 [53/64]	355	1- 1.98 [31/32]	435	1- 5.13 [1/8]	515	1- 8.28 [9/32]
36	0- 1.42 [27/64]	116	0- 4.57 [9/16]	196	0- 7.72 [23/32]	276	0-10.87 [55/64]	356	1- 2.02 [1/64]	436	1- 5.17 [11/64]	516	1- 8.31 [5/16]
37	0- 1.46 [29/64]	117	0- 4.61 [39/64]	197	0- 7.76 [3/4]	277	0-10.91 [29/32]	357	1- 2.06 [1/16]	437	1- 5.20 [13/64]	517	1- 8.35 [23/64]
38	0- 1.50 [1/2]	118	0- 4.65 [41/64]	198	0- 7.80 [51/64]	278	0-10.94 [15/16]	358	1- 2.09 [3/32]	438	1- 5.24 [1/4]	518	1- 8.39 [25/64]
39	0- 1.54 [17/32]	119	0- 4.69 [11/16]	199	0- 7.83 [53/64]	279	0-10.98 [63/64]	359	1- 2.13 [9/64]	439	1- 5.28 [9/32]	519	1- 8.43 [7/16]
40	0- 1.57 [37/64]	120	0- 4.72 [23/32]	200	0- 7.87 [7/8]	280	0-11.02 [1/32]	360	1- 2.17 [11/64]	440	1- 5.32 [21/64]	520	1- 8.47 [15/32]
41	0- 1.61 [39/64]	121	0- 4.76 [49/64]	201	0- 7.91 [29/32]	281	0-11.06 [1/16]	361	1- 2.21 [7/32]	441	1- 5.36 [23/64]	521	1- 8.51 [33/64]
42	0- 1.65 [21/32]	122	0- 4.80 [51/64]	202	0- 7.95 [61/64]	282	0-11.10 [7/64]	362	1- 2.25 [1/4]	442	1- 5.40 [13/32]	522	1- 8.55 [35/64]
43	0- 1.69 [11/16]	123	0- 4.84 [27/32]	203	0- 7.99 [63/64]	283	0-11.14 [9/64]	363	1- 2.29 [19/64]	443	1- 5.44 [7/16]	523	1- 8.59 [19/32]
44	0- 1.73 [47/64]	124	0- 4.88 [7/8]	204	0- 8.03 [1/32]	284	0-11.18 [3/16]	364	1- 2.33 [21/64]	444	1- 5.48 [31/64]	524	1- 8.63 [5/8]
45	0- 1.77 [49/64]	125	0- 4.92 [59/64]	205	0- 8.07 [5/64]	285	0-11.22 [7/32]	365	1- 2.37 [3/8]	445	1- 5.52 [33/64]	525	1- 8.67 [43/64]
46	0- 1.81 [13/16]	126	0- 4.96 [61/64]	206	0- 8.11 [7/64]	286	0-11.26 [17/64]	366	1- 2.41 [13/32]	446	1- 5.56 [9/16]	526	1- 8.71 [45/64]
47	0- 1.85 [27/32]	127	0- 5.00	207	0- 8.15 [5/32]	287	0-11.30 [19/64]	367	1- 2.45 [29/64]	447	1- 5.60 [19/32]	527	1- 8.75 [3/4]
48	0- 1.89 [57/64]	128	0- 5.04 [3/64]	208	0- 8.19 [3/16]	288	0-11.34 [11/32]	368	1- 2.49 [31/64]	448	1- 5.64 [41/64]	528	1- 8.79 [25/32]
49	0- 1.93 [59/64]	129	0- 5.08 [5/64]	209	0- 8.23 [15/64]	289	0-11.38 [3/8]	369	1- 2.53 [17/32]	449	1- 5.68 [43/64]	529	1- 8.83 [53/64]
50	0- 1.97 [31/32]	130	0- 5.12 [1/8]	210	0- 8.27 [17/64]	290	0-11.42 [27/64]	370	1- 2.57 [9/16]	450	1- 5.72 [23/32]	530	1- 8.87 [55/64]
51	0- 2.01 [1/64]	131	0- 5.16 [5/32]	211	0- 8.31 [5/16]	291	0-11.46 [29/64]	371	1- 2.61 [39/64]	451	1- 5.76 [3/4]	531	1- 8.91 [29/32]
52	0- 2.05 [3/64]	132	0- 5.20 [13/64]	212	0- 8.35 [11/32]	292	0-11.50 [1/2]	372	1- 2.65 [41/64]	452	1- 5.80 [51/64]	532	1- 8.94 [15/16]
53	0- 2.09 [3/32]	133	0- 5.24 [15/64]	213	0- 8.39 [25/64]	293	0-11.54 [17/32]	373	1- 2.69 [11/16]	453	1- 5.83 [53/64]	533	1- 8.98 [63/64]
54	0- 2.13 [1/8]	134	0- 5.28 [9/32]	214	0- 8.43 [27/64]	294	0-11.57 [37/64]	374	1- 2.72 [23/32]	454	1- 5.87 [7/8]	534	1- 9.02 [1/32]
55	0- 2.17 [11/64]	135	0- 5.31 [5/16]	215	0- 8.46 [15/32]	295	0-11.61 [39/64]	375	1- 2.76 [49/64]	455	1- 5.91 [29/32]	535	1- 9.06 [1/16]
56	0- 2.20 [13/64]	136	0- 5.35 [23/64]	216	0- 8.50 [1/2]	296	0-11.65 [21/32]	376	1- 2.80 [51/64]	456	1- 5.95 [61/64]	536	1- 9.10 [7/64]
57	0- 2.24 [1/4]	137	0- 5.39 [25/64]	217	0- 8.54 [35/64]	297	0-11.69 [11/16]	377	1- 2.84 [27/32]	457	1- 5.99 [63/64]	537	1- 9.14 [9/64]
58	0- 2.28 [9/32]	138	0- 5.43 [7/16]	218	0- 8.58 [37/64]	298	0-11.73 [47/64]	378	1- 2.88 [7/8]	458	1- 6.03 [1/32]	538	1- 9.18 [3/16]
59	0- 2.32 [21/64]	139	0- 5.47 [15/32]	219	0- 8.62 [5/8]	299	0-11.77 [49/64]	379	1- 2.92 [59/64]	459	1- 6.07 [5/64]	539	1- 9.22 [7/32]
60	0- 2.36 [23/64]	140	0- 5.51 [33/64]	220	0- 8.66 [21/32]	300	0-11.81 [13/16]	380	1- 2.96 [61/64]	460	1- 6.11 [7/64]	540	1- 9.26 [17/64]
61	0- 2.40 [13/32]	141	0- 5.55 [35/64]	221	0- 8.70 [45/64]	301	0-11.85 [27/32]	381	1- 3.00	461	1- 6.15 [5/32]	541	1- 9.30 [19/64]
62	0- 2.44 [7/16]	142	0- 5.59 [19/32]	222	0- 8.74 [47/64]	302	0-11.89 [57/64]	382	1- 3.04 [3/64]	462	1- 6.19 [3/16]	542	1- 9.34 [11/32]
63	0- 2.48 [31/64]	143	0- 5.63 [5/8]	223	0- 8.78 [25/32]	303	0-11.93 [59/64]	383	1- 3.08 [5/64]	463	1- 6.23 [15/64]	543	1- 9.38 [3/8]
64	0- 2.52 [33/64]	144	0- 5.67 [43/64]	224	0- 8.82 [13/16]	304	0-11.97 [31/32]	384	1- 3.12 [1/8]	464	1- 6.27 [17/64]	544	1- 9.42 [27/64]
65	0- 2.56 [9/16]	145	0- 5.71 [45/64]	225	0- 8.86 [55/64]	305	1- 0.01 [1/64]	385	1- 3.16 [5/32]	465	1- 6.31 [5/16]	545	1- 9.46 [29/64]
66	0- 2.60 [19/32]	146	0- 5.75 [3/4]	226	0- 8.90 [57/64]	306	1- 0.05 [3/64]	386	1- 3.20 [13/64]	466	1- 6.35 [11/32]	546	1- 9.50 [1/2]
67	0- 2.64 [41/64]	147	0- 5.79 [25/32]	227	0- 8.94 [15/16]	307	1- 0.09 [3/32]	387	1- 3.24 [15/64]	467	1- 6.39 [25/64]	547	1- 9.54 [17/32]
68	0- 2.68 [43/64]	148	0- 5.83 [53/64]	228	0- 8.98 [31/32]	308	1- 0.13 [1/8]	388	1- 3.28 [9/32]	468	1- 6.43 [27/64]	548	1- 9.57 [37/64]
69	0- 2.72 [23/32]	149	0- 5.87 [55/64]	229	0- 9.02 [1/64]	309	1- 0.17 [11/64]	389	1- 3.31 [5/16]	469	1- 6.46 [15/32]	549	1- 9.61 [39/64]
70	0- 2.76 [3/4]	150	0- 5.91 [29/32]	230	0- 9.06 [1/16]	310	1- 0.20 [13/64]	390	1- 3.35 [23/64]	470	1- 6.50 [1/2]	550	1- 9.65 [21/32]
71	0- 2.80 [51/64]	151	0- 5.94 [15/16]	231	0- 9.09 [3/32]	311	1- 0.24 [1/4]	391	1- 3.39 [25/64]	471	1- 6.54 [35/64]	551	1- 9.69 [11/16]
72	0- 2.83 [53/64]	152	0- 5.98 [63/64]	232	0- 9.13 [9/64]	312	1- 0.28 [9/32]	392	1- 3.43 [7/16]	472	1- 6.58 [37/64]	552	1- 9.73 [47/64]
73	0- 2.87 [7/8]	153	0- 6.02 [1/32]	233	0- 9.17 [11/64]	313	1- 0.32 [21/64]	393	1- 3.47 [15/32]	473	1- 6.62 [5/8]	553	1- 9.77 [49/64]
74	0- 2.91 [29/32]	154	0- 6.06 [1/16]	234	0- 9.21 [7/32]	314	1- 0.36 [23/64]	394	1- 3.51 [33/64]	474	1- 6.66 [21/32]	554	1- 9.81 [13/16]
75	0- 2.95 [61/64]	155	0- 6.10 [7/64]	235	0- 9.25 [1/4]	315	1- 0.40 [13/32]	395	1- 3.55 [35/64]	475	1- 6.70 [45/64]	555	1- 9.85 [27/32]
76	0- 2.99 [63/64]	156	0- 6.14 [9/64]	236	0- 9.29 [19/64]	316	1- 0.44 [7/16]	396	1- 3.59 [19/32]	476	1- 6.74 [47/64]	556	1- 9.89 [57/64]
77	0- 3.03 [1/32]	157	0- 6.18 [3/16]	237	0- 9.33 [21/64]	317	1- 0.48 [31/64]	397	1- 3.63 [5/8]	477	1- 6.78 [25/32]	557	1- 9.93 [59/64]
78	0- 3.07 [5/64]	158	0- 6.22 [7/32]	238	0- 9.37 [3/8]	318	1- 0.52 [33/64]	398	1- 3.67 [43/64]	478	1- 6.82 [13/16]	558	1- 9.97 [31/32]
79	0- 3.11 [7/64]	159	0- 6.26 [17/64]	239	0- 9.41 [13/32]	319	1- 0.56 [9/16]	399	1- 3.71 [45/64]	479	1- 6.86 [55/64]	559	1-10.01 [1/64]
80	0- 3.15 [5/32]	160	0- 6.30 [19/64]	240	0- 9.45 [29/64]	320	1- 0.60 [19/32]	400	1- 3.75 [3/4]	480	1- 6.90 [57/64]	560	1-10.05 [3/64]

MILLIMETERS CONVERTED TO FEET AND INCHES

TABLES M-3

mm	ft-in.(fraction)	mm	ft-in.(fraction)	mm	ft-in.(fraction)	mm	ft-in.(fraction)	mm	ft-in.(fraction)	mm	ft-in.(fraction)	mm	ft-in.(fraction)
561	1-10.09 [3/32]	641	2- 1.24 [15/64]	721	2- 4.39 [25/64]	801	2- 7.54 [17/32]	881	2-10.69 [11/16]	961	3- 1.83 [53/64]	1041	3- 4.98 [63/64]
562	1-10.13 [1/8]	642	2- 1.28 [9/32]	722	2- 4.43 [27/64]	802	2- 7.57 [37/64]	882	2-10.72 [23/32]	962	3- 1.87 [7/8]	1042	3- 5.02 [1/32]
563	1-10.17 [11/64]	643	2- 1.31 [5/16]	723	2- 4.46 [15/32]	803	2- 7.61 [39/64]	883	2-10.76 [49/64]	963	3- 1.91 [29/32]	1043	3- 5.06 [1/16]
564	1-10.20 [13/64]	644	2- 1.35 [23/64]	724	2- 4.50 [1/2]	804	2- 7.65 [21/32]	884	2-10.80 [51/64]	964	3- 1.95 [61/64]	1044	3- 5.10 [7/64]
565	1-10.24 [1/4]	645	2- 1.39 [25/64]	725	2- 4.54 [35/64]	805	2- 7.69 [11/16]	885	2-10.84 [27/32]	965	3- 1.99 [63/64]	1045	3- 5.14 [9/64]
566	1-10.28 [9/32]	646	2- 1.43 [7/16]	726	2- 4.58 [37/64]	806	2- 7.73 [47/64]	886	2-10.88 [7/8]	966	3- 2.03 [1/32]	1046	3- 5.18 [3/16]
567	1-10.32 [21/64]	647	2- 1.47 [15/32]	727	2- 4.62 [5/8]	807	2- 7.77 [49/64]	887	2-10.92 [59/64]	967	3- 2.07 [5/64]	1047	3- 5.22 [7/32]
568	1-10.36 [23/64]	648	2- 1.51 [33/64]	728	2- 4.66 [21/32]	808	2- 7.81 [13/16]	888	2-10.96 [61/64]	968	3- 2.11 [7/64]	1048	3- 5.26 [17/64]
569	1-10.40 [13/32]	649	2- 1.55 [35/64]	729	2- 4.70 [45/64]	809	2- 7.85 [27/32]	889	2-11.00	969	3- 2.15 [5/32]	1049	3- 5.30 [19/64]
570	1-10.44 [7/16]	650	2- 1.59 [19/32]	730	2- 4.74 [47/64]	810	2- 7.89 [57/64]	890	2-11.04 [3/64]	970	3- 2.19 [3/16]	1050	3- 5.34 [11/32]
571	1-10.48 [31/64]	651	2- 1.63 [5/8]	731	2- 4.78 [25/32]	811	2- 7.93 [59/64]	891	2-11.08 [5/64]	971•	3- 2.23 [15/64]	1051	3- 5.38 [3/8]
572	1-10.52 [33/64]	652	2- 1.67 [43/64]	732	2- 4.82 [13/16]	812	2- 7.97 [31/32]	892	2-11.12 [1/8]	972	3- 2.27 [17/64]	1052	3- 5.42 [27/64]
573	1-10.56 [9/16]	653	2- 1.71 [45/64]	733	2- 4.86 [55/64]	813	2- 8.01 [1/64]	893	2-11.16 [5/32]	973	3- 2.31 [5/16]	1053	3- 5.46 [29/64]
574	1-10.60 [19/32]	654	2- 1.75 [3/4]	734	2- 4.90 [57/64]	814	2- 8.05 [3/64]	894	2-11.20 [13/64]	974	3- 2.35 [11/32]	1054	3- 5.50 [1/2]
575	1-10.64 [41/64]	655	2- 1.79 [25/32]	735	2- 4.94 [15/16]	815	2- 8.09 [3/32]	895	2-11.24 [15/64]	975	3- 2.39 [25/64]	1055	3- 5.54 [17/32]
576	1-10.68 [43/64]	656	2- 1.83 [53/64]	736	2- 4.98 [31/32]	816	2- 8.13 [1/8]	896	2-11.28 [9/32]	976	3- 2.43 [27/64]	1056	3- 5.57 [37/64]
577	1-10.72 [23/32]	657	2- 1.87 [55/64]	737	2- 5.02 [1/64]	817	2- 8.17 [11/64]	897	2-11.31 [5/16]	977	3- 2.46 [15/32]	1057	3- 5.61 [39/64]
578	1-10.76 [3/4]	658	2- 1.91 [29/32]	738	2- 5.06 [1/16]	818	2- 8.20 [13/64]	898	2-11.35 [23/64]	978	3- 2.50 [1/2]	1058	3- 5.65 [21/32]
579	1-10.80 [51/64]	659	2- 1.94 [15/16]	739	2- 5.09 [3/32]	819	2- 8.24 [1/4]	899	2-11.39 [25/64]	979	3- 2.54 [35/64]	1059	3- 5.69 [11/16]
580	1-10.83 [53/64]	660	2- 1.98 [63/64]	740	2- 5.13 [9/64]	820	2- 8.28 [9/32]	900	2-11.43 [7/16]	980	3- 2.58 [37/64]	1060	3- 5.73 [47/64]
581	1-10.87 [7/8]	661	2- 2.02 [1/32]	741	2- 5.17 [11/64]	821	2- 8.32 [21/64]	901	2-11.47 [15/32]	981	3- 2.62 [5/8]	1061	3- 5.77 [49/64]
582	1-10.91 [29/32]	662	2- 2.06 [1/16]	742	2- 5.21 [7/32]	822	2- 8.36 [23/64]	902	2-11.51 [33/64]	982	3- 2.66 [21/32]	1062	3- 5.81 [13/16]
583	1-10.95 [61/64]	663	2- 2.10 [7/64]	743	2- 5.25 [1/4]	823	2- 8.40 [13/32]	903	2-11.55 [35/64]	983	3- 2.70 [45/64]	1063	3- 5.85 [27/32]
584	1-10.99 [63/64]	664	2- 2.14 [9/64]	744	2- 5.29 [19/64]	824	2- 8.44 [7/16]	904	2-11.59 [19/32]	984	3- 2.74 [47/64]	1064	3- 5.89 [57/64]
585	1-11.03 [1/32]	665	2- 2.18 [3/16]	745	2- 5.33 [21/64]	825	2- 8.48 [31/64]	905	2-11.63 [5/8]	985	3- 2.78 [25/32]	1065	3- 5.93 [59/64]
586	1-11.07 [5/64]	666	2- 2.22 [7/32]	746	2- 5.37 [3/8]	826	2- 8.52 [33/64]	906	2-11.67 [43/64]	986	3- 2.82 [13/16]	1066	3- 5.97 [31/32]
587	1-11.11 [7/64]	667	2- 2.26 [17/64]	747	2- 5.41 [13/32]	827	2- 8.56 [9/16]	907	2-11.71 [45/64]	987	3- 2.86 [55/64]	1067	3- 6.01 [1/64]
588	1-11.15 [5/32]	668	2- 2.30 [19/64]	748	2- 5.45 [29/64]	828	2- 8.60 [19/32]	908	2-11.75 [3/4]	988	3- 2.90 [57/64]	1068	3- 6.05 [3/64]
589	1-11.19 [3/16]	669	2- 2.34 [11/32]	749	2- 5.49 [31/64]	829	2- 8.64 [41/64]	909	2-11.79 [25/32]	989	3- 2.94 [15/16]	1069	3- 6.09 [3/32]
590	1-11.23 [15/64]	670	2- 2.38 [3/8]	750	2- 5.53 [17/32]	830	2- 8.68 [43/64]	910	2-11.83 [53/64]	990	3- 2.98 [31/32]	1070	3- 6.13 [1/8]
591	1-11.27 [17/64]	671	2- 2.42 [27/64]	751	2- 5.57 [9/16]	831	2- 8.72 [23/32]	911	2-11.87 [55/64]	991	3- 3.02 [1/64]	1071	3- 6.17 [11/64]
592	1-11.31 [5/16]	672	2- 2.46 [29/64]	752	2- 5.61 [39/64]	832	2- 8.76 [3/4]	912	2-11.91 [29/32]	992	3- 3.06 [1/16]	1072	3- 6.20 [13/64]
593	1-11.35 [11/32]	673	2- 2.50 [1/2]	753	2- 5.65 [41/64]	833	2- 8.80 [51/64]	913	2-11.94 [15/16]	993	3- 3.09 [3/32]	1073	3- 6.24 [1/4]
594	1-11.39 [25/64]	674	2- 2.54 [17/32]	754	2- 5.69 [11/16]	834	2- 8.83 [53/64]	914	2-11.98 [63/64]	994	3- 3.13 [9/64]	1074	3- 6.28 [9/32]
595	1-11.43 [27/64]	675	2- 2.57 [37/64]	755	2- 5.72 [23/32]	835	2- 8.87 [7/8]	915	3- 0.02 [1/32]	995	3- 3.17 [11/64]	1075	3- 6.32 [21/64]
596	1-11.46 [15/32]	676	2- 2.61 [39/64]	756	2- 5.76 [49/64]	836	2- 8.91 [29/32]	916	3- 0.06 [1/16]	996	3- 3.21 [7/32]	1076	3- 6.36 [23/64]
597	1-11.50 [1/2]	677	2- 2.65 [21/32]	757	2- 5.80 [51/64]	837	2- 8.95 [61/64]	917	3- 0.10 [7/64]	997	3- 3.25 [1/4]	1077	3- 6.40 [13/32]
598	1-11.54 [35/64]	678	2- 2.69 [11/16]	758	2- 5.84 [27/32]	838	2- 8.99 [63/64]	918	3- 0.14 [9/64]	998	3- 3.29 [19/64]	1078	3- 6.44 [7/16]
599	1-11.58 [37/64]	679	2- 2.73 [47/64]	759	2- 5.88 [7/8]	839	2- 9.03 [1/32]	919	3- 0.18 [3/16]	999	3- 3.33 [21/64]	1079	3- 6.48 [31/64]
600	1-11.62 [5/8]	680	2- 2.77 [49/64]	760	2- 5.92 [59/64]	840	2- 9.07 [5/64]	920	3- 0.22 [7/32]	1000	3- 3.37 [3/8]	1080	3- 6.52 [33/64]
601	1-11.66 [21/32]	681	2- 2.81 [13/16]	761	2- 5.96 [61/64]	841	2- 9.11 [7/64]	921	3- 0.26 [17/64]	1001	3- 3.41 [13/32]	1081	3- 6.56 [9/16]
602	1-11.70 [45/64]	682	2- 2.85 [27/32]	762	2- 6.00	842	2- 9.15 [5/32]	922	3- 0.30 [19/64]	1002	3- 3.45 [29/64]	1082	3- 6.60 [19/32]
603	1-11.74 [47/64]	683	2- 2.89 [57/64]	763	2- 6.04 [3/64]	843	2- 9.19 [3/16]	923	3- 0.34 [11/32]	1003	3- 3.49 [31/64]	1083	3- 6.64 [41/64]
604	1-11.78 [25/32]	684	2- 2.93 [59/64]	764	2- 6.08 [5/64]	844	2- 9.23 [15/64]	924	3- 0.38 [3/8]	1004	3- 3.53 [17/32]	1084	3- 6.68 [43/64]
605	1-11.82 [13/16]	685	2- 2.97 [31/32]	765	2- 6.12 [1/8]	845	2- 9.27 [17/64]	925	3- 0.42 [27/64]	1005	3- 3.57 [9/16]	1085	3- 6.72 [23/32]
606	1-11.86 [55/64]	686	2- 3.01 [1/64]	766	2- 6.16 [5/32]	846	2- 9.31 [5/16]	926	3- 0.46 [29/64]	1006	3- 3.61 [39/64]	1086	3- 6.76 [3/4]
607	1-11.90 [57/64]	687	2- 3.05 [3/64]	767	2- 6.20 [13/64]	847	2- 9.35 [11/32]	927	3- 0.50 [1/2]	1007	3- 3.65 [41/64]	1087	3- 6.80 [51/64]
608	1-11.94 [15/16]	688	2- 3.09 [3/32]	768	2- 6.24 [15/64]	848	2- 9.39 [25/64]	928	3- 0.54 [17/32]	1008	3- 3.69 [11/16]	1088	3- 6.83 [53/64]
609	1-11.98 [31/32]	689	2- 3.13 [1/8]	769	2- 6.28 [9/32]	849	2- 9.43 [27/64]	929	3- 0.57 [37/64]	1009	3- 3.72 [23/32]	1089	3- 6.87 [7/8]
610	2- 0.02 [1/64]	690	2- 3.17 [11/64]	770	2- 6.31 [5/16]	850	2- 9.46 [15/32]	930	3- 0.61 [39/64]	1010	3- 3.76 [49/64]	1090	3- 6.91 [29/32]
611	2- 0.06 [1/16]	691	2- 3.20 [13/64]	771	2- 6.35 [23/64]	851	2- 9.50 [1/2]	931	3- 0.65 [21/32]	1011	3- 3.80 [51/64]	1091	3- 6.95 [61/64]
612	2- 0.09 [3/32]	692	2- 3.24 [1/4]	772	2- 6.39 [25/64]	852	2- 9.54 [35/64]	932	3- 0.69 [11/16]	1012	3- 3.84 [27/32]	1092	3- 6.99 [63/64]
613	2- 0.13 [9/64]	693	2- 3.28 [9/32]	773	2- 6.43 [7/16]	853	2- 9.58 [37/64]	933	3- 0.73 [47/64]	1013	3- 3.88 [7/8]	1093	3- 7.03 [1/32]
614	2- 0.17 [11/64]	694	2- 3.32 [21/64]	774	2- 6.47 [15/32]	854	2- 9.62 [5/8]	934	3- 0.77 [49/64]	1014	3- 3.92 [59/64]	1094	3- 7.07 [5/64]
615	2- 0.21 [7/32]	695	2- 3.36 [23/64]	775	2- 6.51 [33/64]	855	2- 9.66 [21/32]	935	3- 0.81 [13/16]	1015	3- 3.96 [61/64]	1095	3- 7.11 [7/64]
616	2- 0.25 [1/4]	696	2- 3.40 [13/32]	776	2- 6.55 [35/64]	856	2- 9.70 [45/64]	936	3- 0.85 [27/32]	1016	3- 4.00	1096	3- 7.15 [5/32]
617	2- 0.29 [19/64]	697	2- 3.44 [7/16]	777	2- 6.59 [19/32]	857	2- 9.74 [47/64]	937	3- 0.89 [57/64]	1017	3- 4.04 [3/64]	1097	3- 7.19 [3/16]
618	2- 0.33 [21/64]	698	2- 3.48 [31/64]	778	2- 6.63 [5/8]	858	2- 9.78 [25/32]	938	3- 0.93 [59/64]	1018	3- 4.08 [5/64]	1098	3- 7.23 [15/64]
619	2- 0.37 [3/8]	699	2- 3.52 [33/64]	779	2- 6.67 [43/64]	859	2- 9.82 [13/16]	939	3- 0.97 [31/32]	1019	3- 4.12 [1/8]	1099	3- 7.27 [17/64]
620	2- 0.41 [13/32]	700	2- 3.56 [9/16]	780	2- 6.71 [45/64]	860	2- 9.86 [55/64]	940	3- 1.01 [1/64]	1020	3- 4.16 [5/32]	1100	3- 7.31 [5/16]
621	2- 0.45 [29/64]	701	2- 3.60 [19/32]	781	2- 6.75 [3/4]	861	2- 9.90 [57/64]	941	3- 1.05 [3/64]	1021	3- 4.20 [13/64]	1101	3- 7.35 [11/32]
622	2- 0.49 [31/64]	702	2- 3.64 [41/64]	782	2- 6.79 [25/32]	862	2- 9.94 [15/16]	942	3- 1.09 [3/32]	1022	3- 4.24 [15/64]	1102	3- 7.39 [25/64]
623	2- 0.53 [17/32]	703	2- 3.68 [43/64]	783	2- 6.83 [53/64]	863	2- 9.98 [31/32]	943	3- 1.13 [1/8]	1023	3- 4.28 [9/32]	1103	3- 7.43 [27/64]
624	2- 0.57 [9/16]	704	2- 3.72 [23/32]	784	2- 6.87 [55/64]	864	2-10.02 [1/64]	944	3- 1.17 [11/64]	1024	3- 4.31 [5/16]	1104	3- 7.46 [15/32]
625	2- 0.61 [39/64]	705	2- 3.76 [3/4]	785	2- 6.91 [29/32]	865	2-10.06 [1/16]	945	3- 1.20 [13/64]	1025	3- 4.35 [23/64]	1105	3- 7.50 [1/2]
626	2- 0.65 [41/64]	706	2- 3.80 [51/64]	786	2- 6.94 [15/16]	866	2-10.09 [3/32]	946	3- 1.24 [1/4]	1026	3- 4.39 [25/64]	1106	3- 7.54 [35/64]
627	2- 0.69 [11/16]	707	2- 3.83 [53/64]	787	2- 6.98 [63/64]	867	2-10.13 [9/64]	947	3- 1.28 [9/32]	1027	3- 4.43 [7/16]	1107	3- 7.58 [37/64]
628	2- 0.72 [23/32]	708	2- 3.87 [7/8]	788	2- 7.02 [1/32]	868	2-10.17 [11/64]	948	3- 1.32 [21/64]	1028	3- 4.47 [15/32]	1108	3- 7.62 [5/8]
629	2- 0.76 [49/64]	709	2- 3.91 [29/32]	789	2- 7.06 [1/16]	869	2-10.21 [7/32]	949	3- 1.36 [23/64]	1029	3- 4.51 [33/64]	1109	3- 7.66 [21/32]
630	2- 0.80 [51/64]	710	2- 3.95 [61/64]	790	2- 7.10 [7/64]	870	2-10.25 [1/4]	950	3- 1.40 [13/32]	1030	3- 4.55 [35/64]	1110	3- 7.70 [45/64]
631	2- 0.84 [27/32]	711	2- 3.99 [63/64]	791	2- 7.14 [9/64]	871	2-10.29 [19/64]	951	3- 1.44 [7/16]	1031	3- 4.59 [19/32]	1111	3- 7.74 [47/64]
632	2- 0.88 [7/8]	712	2- 4.03 [1/32]	792	2- 7.18 [3/16]	872	2-10.33 [21/64]	952	3- 1.48 [31/64]	1032	3- 4.63 [5/8]	1112	3- 7.78 [25/32]
633	2- 0.92 [59/64]	713	2- 4.07 [5/64]	793	2- 7.22 [7/32]	873	2-10.37 [3/8]	953	3- 1.52 [33/64]	1033	3- 4.67 [43/64]	1113	3- 7.82 [13/16]
634	2- 0.96 [61/64]	714	2- 4.11 [7/64]	794	2- 7.26 [17/64]	874	2-10.41 [13/32]	954	3- 1.56 [9/16]	1034	3- 4.71 [45/64]	1114	3- 7.86 [55/64]
635	2- 1.00	715	2- 4.15 [5/32]	795	2- 7.30 [19/64]	875	2-10.45 [29/64]	955	3- 1.60 [19/32]	1035	3- 4.75 [3/4]	1115	3- 7.90 [57/64]
636	2- 1.04 [3/64]	716	2- 4.19 [3/16]	796	2- 7.34 [11/32]	876	2-10.49 [31/64]	956	3- 1.64 [41/64]	1036	3- 4.79 [25/32]	1116	3- 7.94 [15/16]
637	2- 1.08 [5/64]	717	2- 4.23 [15/64]	797	2- 7.38 [3/8]	877	2-10.53 [17/32]	957	3- 1.68 [43/64]	1037	3- 4.83 [53/64]	1117	3- 7.98 [31/32]
638	2- 1.12 [1/8]	718	2- 4.27 [17/64]	798	2- 7.42 [27/64]	878	2-10.57 [9/16]	958	3- 1.72 [23/32]	1038	3- 4.87 [55/64]	1118	3- 8.02 [1/64]
639	2- 1.16 [5/32]	719	2- 4.31 [5/16]	799	2- 7.46 [29/64]	879	2-10.61 [39/64]	959	3- 1.76 [3/4]	1039	3- 4.91 [29/32]	1119	3- 8.06 [1/16]
640	2- 1.20 [13/64]	720	2- 4.35 [11/32]	800	2- 7.50 [1/2]	880	2-10.65 [41/64]	960	3- 1.80 [51/64]	1040	3- 4.94 [15/16]	1120	3- 8.09 [3/32]

MILLIMETERS CONVERTED TO FEET AND INCHES

TABLES M-3

mm	ft-in.(fraction)	mm	ft-in.(fraction)	mm	ft-in.(fraction)	mm	ft-in.(fraction)	mm	ft-in.(fraction)	mm	ft-in.(fraction)	mm	ft-in.(fraction)
1121	3- 8.13 [9/64]	1201	3-11.28 [9/32]	1281	4- 2.43 [7/16]	1361	4- 5.58 [37/64]	1441	4- 8.73 [47/64]	1521	4-11.88 [7/8]	1601	5- 3.03 [1/32]
1122	3- 8.17 [11/64]	1202	3-11.32 [21/64]	1282	4- 2.47 [15/32]	1362	4- 5.62 [5/8]	1442	4- 8.77 [49/64]	1522	4-11.92 [59/64]	1602	5- 3.07 [5/64]
1123	3- 8.21 [7/32]	1203	3-11.36 [23/64]	1283	4- 2.51 [33/64]	1363	4- 5.66 [21/32]	1443	4- 8.81 [13/16]	1523	4-11.96 [61/64]	1603	5- 3.11 [7/64]
1124	3- 8.25 [1/4]	1204	3-11.40 [13/32]	1284	4- 2.55 [35/64]	1364	4- 5.70 [45/64]	1444	4- 8.85 [27/32]	1524	5- 0.00	1604	5- 3.15 [5/32]
1125	3- 8.29 [19/64]	1205	3-11.44 [7/16]	1285	4- 2.59 [19/32]	1365	4- 5.74 [47/64]	1445	4- 8.89 [57/64]	1525	5- 0.04 [3/64]	1605	5- 3.19 [3/16]
1126	3- 8.33 [21/64]	1206	3-11.48 [31/64]	1286	4- 2.63 [5/8]	1366	4- 5.78 [25/32]	1446	4- 8.93 [59/64]	1526	5- 0.08 [5/64]	1606	5- 3.23 [15/64]
1127	3- 8.37 [3/8]	1207	3-11.52 [33/64]	1287	4- 2.67 [43/64]	1367	4- 5.82 [13/16]	1447	4- 8.97 [31/32]	1527	5- 0.12 [1/8]	1607	5- 3.27 [17/64]
1128	3- 8.41 [13/32]	1208	3-11.56 [9/16]	1288	4- 2.71 [45/64]	1368	4- 5.86 [55/64]	1448	4- 9.01 [1/64]	1528	5- 0.16 [5/32]	1608	5- 3.31 [5/16]
1129	3- 8.45 [29/64]	1209	3-11.60 [19/32]	1289	4- 2.75 [3/4]	1369	4- 5.90 [57/64]	1449	4- 9.05 [3/64]	1529	5- 0.20 [13/64]	1609	5- 3.35 [11/32]
1130	3- 8.49 [31/64]	1210	3-11.64 [41/64]	1290	4- 2.79 [25/32]	1370	4- 5.94 [15/16]	1450	4- 9.09 [3/32]	1530	5- 0.24 [15/64]	1610	5- 3.39 [25/64]
1131	3- 8.53 [17/32]	1211	3-11.68 [43/64]	1291	4- 2.83 [53/64]	1371	4- 5.98 [31/32]	1451	4- 9.13 [1/8]	1531	5- 0.28 [9/32]	1611	5- 3.43 [27/64]
1132	3- 8.57 [9/16]	1212	3-11.72 [23/32]	1292	4- 2.87 [55/64]	1372	4- 6.02 [1/64]	1452	4- 9.17 [11/64]	1532	5- 0.31 [5/16]	1612	5- 3.46 [15/32]
1133	3- 8.61 [39/64]	1213	3-11.76 [3/4]	1293	4- 2.91 [29/32]	1373	4- 6.06 [1/16]	1453	4- 9.20 [13/64]	1533	5- 0.35 [23/64]	1613	5- 3.50 [1/2]
1134	3- 8.65 [41/64]	1214	3-11.80 [51/64]	1294	4- 2.94 [15/16]	1374	4- 6.09 [3/32]	1454	4- 9.24 [1/4]	1534	5- 0.39 [25/64]	1614	5- 3.54 [35/64]
1135	3- 8.69 [11/16]	1215	3-11.83 [53/64]	1295	4- 2.98 [63/64]	1375	4- 6.13 [9/64]	1455	4- 9.28 [9/32]	1535	5- 0.43 [7/16]	1615	5- 3.58 [37/64]
1136	3- 8.72 [23/32]	1216	3-11.87 [7/8]	1296	4- 3.02 [1/32]	1376	4- 6.17 [11/64]	1456	4- 9.32 [21/64]	1536	5- 0.47 [15/32]	1616	5- 3.62 [5/8]
1137	3- 8.76 [49/64]	1217	3-11.91 [29/32]	1297	4- 3.06 [1/16]	1377	4- 6.21 [7/32]	1457	4- 9.36 [23/64]	1537	5- 0.51 [33/64]	1617	5- 3.66 [21/32]
1138	3- 8.80 [51/64]	1218	3-11.95 [61/64]	1298	4- 3.10 [7/64]	1378	4- 6.25 [1/4]	1458	4- 9.40 [13/32]	1538	5- 0.55 [35/64]	1618	5- 3.70 [45/64]
1139	3- 8.84 [27/32]	1219	3-11.99 [63/64]	1299	4- 3.14 [9/64]	1379	4- 6.29 [19/64]	1459	4- 9.44 [7/16]	1539	5- 0.59 [19/32]	1619	5- 3.74 [47/64]
1140	3- 8.88 [7/8]	1220	4- 0.03 [1/32]	1300	4- 3.18 [3/16]	1380	4- 6.33 [21/64]	1460	4- 9.48 [31/64]	1540	5- 0.63 [5/8]	1620	5- 3.78 [25/32]
1141	3- 8.92 [59/64]	1221	4- 0.07 [5/64]	1301	4- 3.22 [7/32]	1381	4- 6.37 [3/8]	1461	4- 9.52 [33/64]	1541	5- 0.67 [43/64]	1621	5- 3.82 [13/16]
1142	3- 8.96 [61/64]	1222	4- 0.11 [7/64]	1302	4- 3.26 [17/64]	1382	4- 6.41 [13/32]	1462	4- 9.56 [9/16]	1542	5- 0.71 [45/64]	1622	5- 3.86 [55/64]
1143	3- 9.00	1223	4- 0.15 [5/32]	1303	4- 3.30 [19/64]	1383	4- 6.45 [29/64]	1463	4- 9.60 [19/32]	1543	5- 0.75 [3/4]	1623	5- 3.90 [57/64]
1144	3- 9.04 [3/64]	1224	4- 0.19 [3/16]	1304	4- 3.34 [11/32]	1384	4- 6.49 [31/64]	1464	4- 9.64 [41/64]	1544	5- 0.79 [25/32]	1624	5- 3.94 [15/16]
1145	3- 9.08 [5/64]	1225	4- 0.23 [15/64]	1305	4- 3.38 [3/8]	1385	4- 6.53 [17/32]	1465	4- 9.68 [43/64]	1545	5- 0.83 [53/64]	1625	5- 3.98 [31/32]
1146	3- 9.12 [1/8]	1226	4- 0.27 [17/64]	1306	4- 3.42 [27/64]	1386	4- 6.57 [9/16]	1466	4- 9.72 [23/32]	1546	5- 0.87 [55/64]	1626	5- 4.02 [1/64]
1147	3- 9.16 [5/32]	1227	4- 0.31 [5/16]	1307	4- 3.46 [29/64]	1387	4- 6.61 [39/64]	1467	4- 9.76 [3/4]	1547	5- 0.91 [29/32]	1627	5- 4.06 [1/16]
1148	3- 9.20 [13/64]	1228	4- 0.35 [11/32]	1308	4- 3.50 [1/2]	1388	4- 6.65 [41/64]	1468	4- 9.80 [51/64]	1548	5- 0.94 [15/16]	1628	5- 4.09 [3/32]
1149	3- 9.24 [15/64]	1229	4- 0.39 [25/64]	1309	4- 3.54 [17/32]	1389	4- 6.69 [11/16]	1469	4- 9.83 [53/64]	1549	5- 0.98 [63/64]	1629	5- 4.13 [9/64]
1150	3- 9.28 [9/32]	1230	4- 0.43 [27/64]	1310	4- 3.57 [37/64]	1390	4- 6.72 [23/32]	1470	4- 9.87 [7/8]	1550	5- 1.02 [1/32]	1630	5- 4.17 [11/64]
1151	3- 9.31 [5/16]	1231	4- 0.46 [15/32]	1311	4- 3.61 [39/64]	1391	4- 6.76 [49/64]	1471	4- 9.91 [29/32]	1551	5- 1.06 [1/16]	1631	5- 4.21 [7/32]
1152	3- 9.35 [23/64]	1232	4- 0.50 [1/2]	1312	4- 3.65 [21/32]	1392	4- 6.80 [51/64]	1472	4- 9.95 [61/64]	1552	5- 1.10 [7/64]	1632	5- 4.25 [1/4]
1153	3- 9.39 [25/64]	1233	4- 0.54 [35/64]	1313	4- 3.69 [11/16]	1393	4- 6.84 [27/32]	1473	4- 9.99 [63/64]	1553	5- 1.14 [9/64]	1633	5- 4.29 [19/64]
1154	3- 9.43 [7/16]	1234	4- 0.58 [37/64]	1314	4- 3.73 [47/64]	1394	4- 6.88 [7/8]	1474	4-10.03 [1/32]	1554	5- 1.18 [3/16]	1634	5- 4.33 [21/64]
1155	3- 9.47 [15/32]	1235	4- 0.62 [5/8]	1315	4- 3.77 [49/64]	1395	4- 6.92 [59/64]	1475	4-10.07 [5/64]	1555	5- 1.22 [7/32]	1635	5- 4.37 [3/8]
1156	3- 9.51 [33/64]	1236	4- 0.66 [21/32]	1316	4- 3.81 [13/16]	1396	4- 6.96 [61/64]	1476	4-10.11 [7/64]	1556	5- 1.26 [17/64]	1636	5- 4.41 [13/32]
1157	3- 9.55 [35/64]	1237	4- 0.70 [45/64]	1317	4- 3.85 [27/32]	1397	4- 7.00	1477	4-10.15 [5/32]	1557	5- 1.30 [19/64]	1637	5- 4.45 [29/64]
1158	3- 9.59 [19/32]	1238	4- 0.74 [47/64]	1318	4- 3.89 [57/64]	1398	4- 7.04 [3/64]	1478	4-10.19 [3/16]	1558	5- 1.34 [11/32]	1638	5- 4.49 [31/64]
1159	3- 9.63 [5/8]	1239	4- 0.78 [25/32]	1319	4- 3.93 [59/64]	1399	4- 7.08 [5/64]	1479	4-10.23 [15/64]	1559	5- 1.38 [3/8]	1639	5- 4.53 [17/32]
1160	3- 9.67 [43/64]	1240	4- 0.82 [13/16]	1320	4- 3.97 [31/32]	1400	4- 7.12 [1/8]	1480	4-10.27 [17/64]	1560	5- 1.42 [27/64]	1640	5- 4.57 [9/16]
1161	3- 9.71 [45/64]	1241	4- 0.86 [55/64]	1321	4- 4.01 [1/64]	1401	4- 7.16 [5/32]	1481	4-10.31 [5/16]	1561	5- 1.46 [29/64]	1641	5- 4.61 [39/64]
1162	3- 9.75 [3/4]	1242	4- 0.90 [57/64]	1322	4- 4.05 [3/64]	1402	4- 7.20 [13/64]	1482	4-10.35 [11/32]	1562	5- 1.50 [1/2]	1642	5- 4.65 [41/64]
1163	3- 9.79 [25/32]	1243	4- 0.94 [15/16]	1323	4- 4.09 [3/32]	1403	4- 7.24 [15/64]	1483	4-10.39 [25/64]	1563	5- 1.54 [17/32]	1643	5- 4.69 [11/16]
1164	3- 9.83 [53/64]	1244	4- 0.98 [31/32]	1324	4- 4.13 [1/8]	1404	4- 7.28 [9/32]	1484	4-10.43 [27/64]	1564	5- 1.57 [37/64]	1644	5- 4.72 [23/32]
1165	3- 9.87 [55/64]	1245	4- 1.02 [1/64]	1325	4- 4.17 [11/64]	1405	4- 7.31 [5/16]	1485	4-10.46 [15/32]	1565	5- 1.61 [39/64]	1645	5- 4.76 [49/64]
1166	3- 9.91 [29/32]	1246	4- 1.06 [1/16]	1326	4- 4.20 [13/64]	1406	4- 7.35 [23/64]	1486	4-10.50 [1/2]	1566	5- 1.65 [21/32]	1646	5- 4.80 [51/64]
1167	3- 9.94 [15/16]	1247	4- 1.09 [3/32]	1327	4- 4.24 [1/4]	1407	4- 7.39 [25/64]	1487	4-10.54 [35/64]	1567	5- 1.69 [11/16]	1647	5- 4.84 [27/32]
1168	3- 9.98 [63/64]	1248	4- 1.13 [9/64]	1328	4- 4.28 [9/32]	1408	4- 7.43 [7/16]	1488	4-10.58 [37/64]	1568	5- 1.73 [47/64]	1648	5- 4.88 [7/8]
1169	3-10.02 [1/32]	1249	4- 1.17 [11/64]	1329	4- 4.32 [21/64]	1409	4- 7.47 [15/32]	1489	4-10.62 [5/8]	1569	5- 1.77 [49/64]	1649	5- 4.92 [59/64]
1170	3-10.06 [1/16]	1250	4- 1.21 [7/32]	1330	4- 4.36 [23/64]	1410	4- 7.51 [33/64]	1490	4-10.66 [21/32]	1570	5- 1.81 [13/16]	1650	5- 4.96 [61/64]
1171	3-10.10 [7/64]	1251	4- 1.25 [1/4]	1331	4- 4.40 [13/32]	1411	4- 7.55 [35/64]	1491	4-10.70 [45/64]	1571	5- 1.85 [27/32]	1651	5- 5.00
1172	3-10.14 [9/64]	1252	4- 1.29 [19/64]	1332	4- 4.44 [7/16]	1412	4- 7.59 [19/32]	1492	4-10.74 [47/64]	1572	5- 1.89 [57/64]	1652	5- 5.04 [3/64]
1173	3-10.18 [3/16]	1253	4- 1.33 [21/64]	1333	4- 4.48 [31/64]	1413	4- 7.63 [5/8]	1493	4-10.78 [25/32]	1573	5- 1.93 [59/64]	1653	5- 5.08 [5/64]
1174	3-10.22 [7/32]	1254	4- 1.37 [3/8]	1334	4- 4.52 [33/64]	1414	4- 7.67 [43/64]	1494	4-10.82 [13/16]	1574	5- 1.97 [31/32]	1654	5- 5.12 [1/8]
1175	3-10.26 [17/64]	1255	4- 1.41 [13/32]	1335	4- 4.56 [9/16]	1415	4- 7.71 [45/64]	1495	4-10.86 [55/64]	1575	5- 2.01 [1/64]	1655	5- 5.16 [5/32]
1176	3-10.30 [19/64]	1256	4- 1.45 [29/64]	1336	4- 4.60 [19/32]	1416	4- 7.75 [3/4]	1496	4-10.90 [57/64]	1576	5- 2.05 [3/64]	1656	5- 5.20 [13/64]
1177	3-10.34 [11/32]	1257	4- 1.49 [31/64]	1337	4- 4.64 [41/64]	1417	4- 7.79 [25/32]	1497	4-10.94 [15/16]	1577	5- 2.09 [3/32]	1657	5- 5.24 [15/64]
1178	3-10.38 [3/8]	1258	4- 1.53 [17/32]	1338	4- 4.68 [43/64]	1418	4- 7.83 [53/64]	1498	4-10.98 [31/32]	1578	5- 2.13 [1/8]	1658	5- 5.28 [9/32]
1179	3-10.42 [27/64]	1259	4- 1.57 [9/16]	1339	4- 4.72 [23/32]	1419	4- 7.87 [55/64]	1499	4-11.02 [1/64]	1579	5- 2.17 [11/64]	1659	5- 5.31 [5/16]
1180	3-10.46 [29/64]	1260	4- 1.61 [39/64]	1340	4- 4.76 [3/4]	1420	4- 7.91 [29/32]	1500	4-11.06 [1/16]	1580	5- 2.20 [13/64]	1660	5- 5.35 [23/64]
1181	3-10.50 [1/2]	1261	4- 1.65 [41/64]	1341	4- 4.80 [51/64]	1421	4- 7.94 [15/16]	1501	4-11.09 [3/32]	1581	5- 2.24 [1/4]	1661	5- 5.39 [25/64]
1182	3-10.54 [17/32]	1262	4- 1.69 [11/16]	1342	4- 4.83 [53/64]	1422	4- 7.98 [63/64]	1502	4-11.13 [9/64]	1582	5- 2.28 [9/32]	1662	5- 5.43 [7/16]
1183	3-10.57 [37/64]	1263	4- 1.72 [23/32]	1343	4- 4.87 [7/8]	1423	4- 8.02 [1/32]	1503	4-11.17 [11/64]	1583	5- 2.32 [21/64]	1663	5- 5.47 [15/32]
1184	3-10.61 [39/64]	1264	4- 1.76 [49/64]	1344	4- 4.91 [29/32]	1424	4- 8.06 [1/16]	1504	4-11.21 [7/32]	1584	5- 2.36 [23/64]	1664	5- 5.51 [33/64]
1185	3-10.65 [21/32]	1265	4- 1.80 [51/64]	1345	4- 4.95 [61/64]	1425	4- 8.10 [7/64]	1505	4-11.25 [1/4]	1585	5- 2.40 [13/32]	1665	5- 5.55 [35/64]
1186	3-10.69 [11/16]	1266	4- 1.84 [27/32]	1346	4- 4.99 [63/64]	1426	4- 8.14 [9/64]	1506	4-11.29 [19/64]	1586	5- 2.44 [7/16]	1666	5- 5.59 [19/32]
1187	3-10.73 [47/64]	1267	4- 1.88 [7/8]	1347	4- 5.03 [1/32]	1427	4- 8.18 [3/16]	1507	4-11.33 [21/64]	1587	5- 2.48 [31/64]	1667	5- 5.63 [5/8]
1188	3-10.77 [49/64]	1268	4- 1.92 [59/64]	1348	4- 5.07 [5/64]	1428	4- 8.22 [7/32]	1508	4-11.37 [3/8]	1588	5- 2.52 [33/64]	1668	5- 5.67 [43/64]
1189	3-10.81 [13/16]	1269	4- 1.96 [61/64]	1349	4- 5.11 [7/64]	1429	4- 8.26 [17/64]	1509	4-11.41 [13/32]	1589	5- 2.56 [9/16]	1669	5- 5.71 [45/64]
1190	3-10.85 [27/32]	1270	4- 2.00	1350	4- 5.15 [5/32]	1430	4- 8.30 [19/64]	1510	4-11.45 [29/64]	1590	5- 2.60 [19/32]	1670	5- 5.75 [3/4]
1191	3-10.89 [57/64]	1271	4- 2.04 [3/64]	1351	4- 5.19 [3/16]	1431	4- 8.34 [11/32]	1511	4-11.49 [31/64]	1591	5- 2.64 [41/64]	1671	5- 5.79 [25/32]
1192	3-10.93 [59/64]	1272	4- 2.08 [5/64]	1352	4- 5.23 [15/64]	1432	4- 8.38 [3/8]	1512	4-11.53 [17/32]	1592	5- 2.68 [43/64]	1672	5- 5.83 [53/64]
1193	3-10.97 [31/32]	1273	4- 2.12 [1/8]	1353	4- 5.27 [17/64]	1433	4- 8.42 [27/64]	1513	4-11.57 [9/16]	1593	5- 2.72 [23/32]	1673	5- 5.87 [55/64]
1194	3-11.01 [1/64]	1274	4- 2.16 [5/32]	1354	4- 5.31 [5/16]	1434	4- 8.46 [29/64]	1514	4-11.61 [39/64]	1594	5- 2.76 [3/4]	1674	5- 5.91 [29/32]
1195	3-11.05 [3/64]	1275	4- 2.20 [13/64]	1355	4- 5.35 [11/32]	1435	4- 8.50 [1/2]	1515	4-11.65 [41/64]	1595	5- 2.80 [51/64]	1675	5- 5.94 [15/16]
1196	3-11.09 [3/32]	1276	4- 2.24 [15/64]	1356	4- 5.39 [25/64]	1436	4- 8.54 [17/32]	1516	4-11.69 [11/16]	1596	5- 2.83 [53/64]	1676	5- 5.98 [63/64]
1197	3-11.13 [1/8]	1277	4- 2.28 [9/32]	1357	4- 5.43 [27/64]	1437	4- 8.57 [37/64]	1517	4-11.72 [23/32]	1597	5- 2.87 [7/8]	1677	5- 6.02 [1/32]
1198	3-11.17 [11/64]	1278	4- 2.31 [5/16]	1358	4- 5.46 [15/32]	1438	4- 8.61 [39/64]	1518	4-11.76 [49/64]	1598	5- 2.91 [29/32]	1678	5- 6.06 [1/16]
1199	3-11.20 [13/64]	1279	4- 2.35 [23/64]	1359	4- 5.50 [1/2]	1439	4- 8.65 [21/32]	1519	4-11.80 [51/64]	1599	5- 2.95 [61/64]	1679	5- 6.10 [7/64]
1200	3-11.24 [1/4]	1280	4- 2.39 [25/64]	1360	4- 5.54 [35/64]	1440	4- 8.69 [11/16]	1520	4-11.84 [27/32]	1600	5- 2.99 [63/64]	1680	5- 6.14 [9/64]

MILLIMETERS CONVERTED TO FEET AND INCHES

TABLES M-3

mm	ft-in.(fraction)	mm	ft-in.(fraction)	mm	ft-in.(fraction)	mm	ft-in.(fraction)	mm	ft-in.(fraction)	mm	ft-in.(fraction)	mm	ft-in.(fraction)
1681	5- 6.18 [3/16]	1761	5- 9.33 [21/64]	1841	6- 0.48 [31/64]	1921	6- 3.63 [5/8]	2001	6- 6.78 [25/32]	2081	6- 9.93 [59/64]	2161	7- 1.08 [5/64]
1682	5- 6.22 [7/32]	1762	5- 9.37 [3/8]	1842	6- 0.52 [33/64]	1922	6- 3.67 [43/64]	2002	6- 6.82 [13/16]	2082	6- 9.97 [31/32]	2162	7- 1.12 [1/8]
1683	5- 6.26 [17/64]	1763	5- 9.41 [13/32]	1843	6- 0.56 [9/16]	1923	6- 3.71 [45/64]	2003	6- 6.86 [55/64]	2083	6-10.01 [1/64]	2163	7- 1.16 [5/32]
1684	5- 6.30 [19/64]	1764	5- 9.45 [29/64]	1844	6- 0.60 [19/32]	1924	6- 3.75 [3/4]	2004	6- 6.90 [57/64]	2084	6-10.05 [3/64]	2164	7- 1.20 [13/64]
1685	5- 6.34 [11/32]	1765	5- 9.49 [31/64]	1845	6- 0.64 [41/64]	1925	6- 3.79 [25/32]	2005	6- 6.94 [15/16]	2085	6-10.09 [3/32]	2165	7- 1.24 [15/64]
1686	5- 6.38 [3/8]	1766	5- 9.53 [17/32]	1846	6- 0.68 [43/64]	1926	6- 3.83 [53/64]	2006	6- 6.98 [31/32]	2086	6-10.13 [1/8]	2166	7- 1.28 [9/32]
1687	5- 6.42 [27/64]	1767	5- 9.57 [9/16]	1847	6- 0.72 [23/32]	1927	6- 3.87 [55/64]	2007	6- 7.02 [1/64]	2087	6-10.17 [11/64]	2167	7- 1.31 [5/16]
1688	5- 6.46 [29/64]	1768	5- 9.61 [39/64]	1848	6- 0.76 [3/4]	1928	6- 3.91 [29/32]	2008	6- 7.06 [1/16]	2088	6-10.20 [13/64]	2168	7- 1.35 [23/64]
1689	5- 6.50 [1/2]	1769	5- 9.65 [41/64]	1849	6- 0.80 [51/64]	1929	6- 3.94 [15/16]	2009	6- 7.09 [3/32]	2089	6-10.24 [1/4]	2169	7- 1.39 [25/64]
1690	5- 6.54 [17/32]	1770	5- 9.69 [11/16]	1850	6- 0.83 [53/64]	1930	6- 3.98 [63/64]	2010	6- 7.13 [9/64]	2090	6-10.28 [9/32]	2170	7- 1.43 [7/16]
1691	5- 6.57 [37/64]	1771	5- 9.72 [23/32]	1851	6- 0.87 [7/8]	1931	6- 4.02 [1/32]	2011	6- 7.17 [11/64]	2091	6-10.32 [21/64]	2171	7- 1.47 [15/32]
1692	5- 6.61 [39/64]	1772	5- 9.76 [49/64]	1852	6- 0.91 [29/32]	1932	6- 4.06 [1/16]	2012	6- 7.21 [7/32]	2092	6-10.36 [23/64]	2172	7- 1.51 [33/64]
1693	5- 6.65 [21/32]	1773	5- 9.80 [51/64]	1853	6- 0.95 [61/64]	1933	6- 4.10 [7/64]	2013	6- 7.25 [1/4]	2093	6-10.40 [13/32]	2173	7- 1.55 [35/64]
1694	5- 6.69 [11/16]	1774	5- 9.84 [27/32]	1854	6- 0.99 [63/64]	1934	6- 4.14 [9/64]	2014	6- 7.29 [19/64]	2094	6-10.44 [7/16]	2174	7- 1.59 [19/32]
1695	5- 6.73 [47/64]	1775	5- 9.88 [7/8]	1855	6- 1.03 [1/32]	1935	6- 4.18 [3/16]	2015	6- 7.33 [21/64]	2095	6-10.48 [31/64]	2175	7- 1.63 [5/8]
1696	5- 6.77 [49/64]	1776	5- 9.92 [59/64]	1856	6- 1.07 [5/64]	1936	6- 4.22 [7/32]	2016	6- 7.37 [3/8]	2096	6-10.52 [33/64]	2176	7- 1.67 [43/64]
1697	5- 6.81 [13/16]	1777	5- 9.96 [61/64]	1857	6- 1.11 [7/64]	1937	6- 4.26 [17/64]	2017	6- 7.41 [13/32]	2097	6-10.56 [9/16]	2177	7- 1.71 [45/64]
1698	5- 6.85 [27/32]	1778	5-10.00	1858	6- 1.15 [5/32]	1938	6- 4.30 [19/64]	2018	6- 7.45 [29/64]	2098	6-10.60 [19/32]	2178	7- 1.75 [3/4]
1699	5- 6.89 [57/64]	1779	5-10.04 [3/64]	1859	6- 1.19 [3/16]	1939	6- 4.34 [11/32]	2019	6- 7.49 [31/64]	2099	6-10.64 [41/64]	2179	7- 1.79 [25/32]
1700	5- 6.93 [59/64]	1780	5-10.08 [5/64]	1860	6- 1.23 [15/64]	1940	6- 4.38 [3/8]	2020	6- 7.53 [17/32]	2100	6-10.68 [43/64]	2180	7- 1.83 [53/64]
1701	5- 6.97 [31/32]	1781	5-10.12 [1/8]	1861	6- 1.27 [17/64]	1941	6- 4.42 [27/64]	2021	6- 7.57 [9/16]	2101	6-10.72 [23/32]	2181	7- 1.87 [55/64]
1702	5- 7.01 [1/64]	1782	5-10.16 [5/32]	1862	6- 1.31 [5/16]	1942	6- 4.46 [29/64]	2022	6- 7.61 [39/64]	2102	6-10.76 [3/4]	2182	7- 1.91 [29/32]
1703	5- 7.05 [3/64]	1783	5-10.20 [13/64]	1863	6- 1.35 [11/32]	1943	6- 4.50 [1/2]	2023	6- 7.65 [41/64]	2103	6-10.80 [51/64]	2183	7- 1.94 [15/16]
1704	5- 7.09 [3/32]	1784	5-10.24 [15/64]	1864	6- 1.39 [25/64]	1944	6- 4.54 [17/32]	2024	6- 7.69 [11/16]	2104	6-10.83 [53/64]	2184	7- 1.98 [63/64]
1705	5- 7.13 [1/8]	1785	5-10.28 [9/32]	1865	6- 1.43 [27/64]	1945	6- 4.57 [37/64]	2025	6- 7.72 [23/32]	2105	6-10.87 [7/8]	2185	7- 2.02 [1/32]
1706	5- 7.17 [11/64]	1786	5-10.31 [5/16]	1866	6- 1.46 [15/32]	1946	6- 4.61 [39/64]	2026	6- 7.76 [49/64]	2106	6-10.91 [29/32]	2186	7- 2.06 [1/16]
1707	5- 7.20 [13/64]	1787	5-10.35 [23/64]	1867	6- 1.50 [1/2]	1947	6- 4.65 [21/32]	2027	6- 7.80 [51/64]	2107	6-10.95 [61/64]	2187	7- 2.10 [7/64]
1708	5- 7.24 [1/4]	1788	5-10.39 [25/64]	1868	6- 1.54 [35/64]	1948	6- 4.69 [11/16]	2028	6- 7.84 [27/32]	2108	6-10.99 [63/64]	2188	7- 2.14 [9/64]
1709	5- 7.28 [9/32]	1789	5-10.43 [7/16]	1869	6- 1.58 [37/64]	1949	6- 4.73 [47/64]	2029	6- 7.88 [7/8]	2109	6-11.03 [1/32]	2189	7- 2.18 [3/16]
1710	5- 7.32 [21/64]	1790	5-10.47 [15/32]	1870	6- 1.62 [5/8]	1950	6- 4.77 [49/64]	2030	6- 7.92 [59/64]	2110	6-11.07 [5/64]	2190	7- 2.22 [7/32]
1711	5- 7.36 [23/64]	1791	5-10.51 [33/64]	1871	6- 1.66 [21/32]	1951	6- 4.81 [13/16]	2031	6- 7.96 [61/64]	2111	6-11.11 [7/64]	2191	7- 2.26 [17/64]
1712	5- 7.40 [13/32]	1792	5-10.55 [35/64]	1872	6- 1.70 [45/64]	1952	6- 4.85 [27/32]	2032	6- 8.00	2112	6-11.15 [5/32]	2192	7- 2.30 [19/64]
1713	5- 7.44 [7/16]	1793	5-10.59 [19/32]	1873	6- 1.74 [47/64]	1953	6- 4.89 [57/64]	2033	6- 8.04 [3/64]	2113	6-11.19 [3/16]	2193	7- 2.34 [11/32]
1714	5- 7.48 [31/64]	1794	5-10.63 [5/8]	1874	6- 1.78 [25/32]	1954	6- 4.93 [59/64]	2034	6- 8.08 [5/64]	2114	6-11.23 [15/64]	2194	7- 2.38 [3/8]
1715	5- 7.52 [33/64]	1795	5-10.67 [43/64]	1875	6- 1.82 [13/16]	1955	6- 4.97 [31/32]	2035	6- 8.12 [1/8]	2115	6-11.27 [17/64]	2195	7- 2.42 [27/64]
1716	5- 7.56 [9/16]	1796	5-10.71 [45/64]	1876	6- 1.86 [55/64]	1956	6- 5.01 [1/64]	2036	6- 8.16 [5/32]	2116	6-11.31 [5/16]	2196	7- 2.46 [29/64]
1717	5- 7.60 [19/32]	1797	5-10.75 [3/4]	1877	6- 1.90 [57/64]	1957	6- 5.05 [3/64]	2037	6- 8.20 [13/64]	2117	6-11.35 [11/32]	2197	7- 2.50 [1/2]
1718	5- 7.64 [41/64]	1798	5-10.79 [25/32]	1878	6- 1.94 [15/16]	1958	6- 5.09 [3/32]	2038	6- 8.24 [15/64]	2118	6-11.39 [25/64]	2198	7- 2.54 [17/32]
1719	5- 7.68 [43/64]	1799	5-10.83 [53/64]	1879	6- 1.98 [31/32]	1959	6- 5.13 [1/8]	2039	6- 8.28 [9/32]	2119	6-11.43 [27/64]	2199	7- 2.57 [37/64]
1720	5- 7.72 [23/32]	1800	5-10.87 [55/64]	1880	6- 2.02 [1/64]	1960	6- 5.17 [11/64]	2040	6- 8.31 [5/16]	2120	6-11.46 [15/32]	2200	7- 2.61 [39/64]
1721	5- 7.76 [3/4]	1801	5-10.91 [29/32]	1881	6- 2.06 [1/16]	1961	6- 5.20 [13/64]	2041	6- 8.35 [23/64]	2121	6-11.50 [1/2]	2201	7- 2.65 [21/32]
1722	5- 7.80 [51/64]	1802	5-10.94 [15/16]	1882	6- 2.09 [3/32]	1962	6- 5.24 [1/4]	2042	6- 8.39 [25/64]	2122	6-11.54 [35/64]	2202	7- 2.69 [11/16]
1723	5- 7.83 [53/64]	1803	5-10.98 [63/64]	1883	6- 2.13 [9/64]	1963	6- 5.28 [9/32]	2043	6- 8.43 [7/16]	2123	6-11.58 [37/64]	2203	7- 2.73 [47/64]
1724	5- 7.87 [7/8]	1804	5-11.02 [1/32]	1884	6- 2.17 [11/64]	1964	6- 5.32 [21/64]	2044	6- 8.47 [15/32]	2124	6-11.62 [5/8]	2204	7- 2.77 [49/64]
1725	5- 7.91 [29/32]	1805	5-11.06 [1/16]	1885	6- 2.21 [7/32]	1965	6- 5.36 [23/64]	2045	6- 8.51 [33/64]	2125	6-11.66 [21/32]	2205	7- 2.81 [13/16]
1726	5- 7.95 [61/64]	1806	5-11.10 [7/64]	1886	6- 2.25 [1/4]	1966	6- 5.40 [13/32]	2046	6- 8.55 [35/64]	2126	6-11.70 [45/64]	2206	7- 2.85 [27/32]
1727	5- 7.99 [63/64]	1807	5-11.14 [9/64]	1887	6- 2.29 [19/64]	1967	6- 5.44 [7/16]	2047	6- 8.59 [19/32]	2127	6-11.74 [47/64]	2207	7- 2.89 [57/64]
1728	5- 8.03 [1/32]	1808	5-11.18 [3/16]	1888	6- 2.33 [21/64]	1968	6- 5.48 [31/64]	2048	6- 8.63 [5/8]	2128	6-11.78 [25/32]	2208	7- 2.93 [59/64]
1729	5- 8.07 [5/64]	1809	5-11.22 [7/32]	1889	6- 2.37 [3/8]	1969	6- 5.52 [33/64]	2049	6- 8.67 [43/64]	2129	6-11.82 [13/16]	2209	7- 2.97 [31/32]
1730	5- 8.11 [7/64]	1810	5-11.26 [17/64]	1890	6- 2.41 [13/32]	1970	6- 5.56 [9/16]	2050	6- 8.71 [45/64]	2130	6-11.86 [55/64]	2210	7- 3.01 [1/64]
1731	5- 8.15 [5/32]	1811	5-11.30 [19/64]	1891	6- 2.45 [29/64]	1971	6- 5.60 [19/32]	2051	6- 8.75 [3/4]	2131	6-11.90 [57/64]	2211	7- 3.05 [3/64]
1732	5- 8.19 [3/16]	1812	5-11.34 [11/32]	1892	6- 2.49 [31/64]	1972	6- 5.64 [41/64]	2052	6- 8.79 [25/32]	2132	6-11.94 [15/16]	2212	7- 3.09 [3/32]
1733	5- 8.23 [15/64]	1813	5-11.38 [3/8]	1893	6- 2.53 [17/32]	1973	6- 5.68 [43/64]	2053	6- 8.83 [53/64]	2133	6-11.98 [31/32]	2213	7- 3.13 [1/8]
1734	5- 8.27 [17/64]	1814	5-11.42 [27/64]	1894	6- 2.57 [9/16]	1974	6- 5.72 [23/32]	2054	6- 8.87 [55/64]	2134	7- 0.02 [1/64]	2214	7- 3.17 [11/64]
1735	5- 8.31 [5/16]	1815	5-11.46 [29/64]	1895	6- 2.61 [39/64]	1975	6- 5.76 [3/4]	2055	6- 8.91 [29/32]	2135	7- 0.06 [1/16]	2215	7- 3.20 [13/64]
1736	5- 8.35 [11/32]	1816	5-11.50 [1/2]	1896	6- 2.65 [41/64]	1976	6- 5.80 [51/64]	2056	6- 8.94 [15/16]	2136	7- 0.09 [3/32]	2216	7- 3.24 [1/4]
1737	5- 8.39 [25/64]	1817	5-11.54 [17/32]	1897	6- 2.69 [11/16]	1977	6- 5.83 [53/64]	2057	6- 8.98 [63/64]	2137	7- 0.13 [9/64]	2217	7- 3.28 [9/32]
1738	5- 8.43 [27/64]	1818	5-11.57 [37/64]	1898	6- 2.72 [23/32]	1978	6- 5.87 [7/8]	2058	6- 9.02 [1/32]	2138	7- 0.17 [11/64]	2218	7- 3.32 [21/64]
1739	5- 8.46 [15/32]	1819	5-11.61 [39/64]	1899	6- 2.76 [49/64]	1979	6- 5.91 [29/32]	2059	6- 9.06 [1/16]	2139	7- 0.21 [7/32]	2219	7- 3.36 [23/64]
1740	5- 8.50 [1/2]	1820	5-11.65 [21/32]	1900	6- 2.80 [51/64]	1980	6- 5.95 [61/64]	2060	6- 9.10 [7/64]	2140	7- 0.25 [1/4]	2220	7- 3.40 [13/32]
1741	5- 8.54 [35/64]	1821	5-11.69 [11/16]	1901	6- 2.84 [27/32]	1981	6- 5.99 [63/64]	2061	6- 9.14 [9/64]	2141	7- 0.29 [19/64]	2221	7- 3.44 [7/16]
1742	5- 8.58 [37/64]	1822	5-11.73 [47/64]	1902	6- 2.88 [7/8]	1982	6- 6.03 [1/32]	2062	6- 9.18 [3/16]	2142	7- 0.33 [21/64]	2222	7- 3.48 [31/64]
1743	5- 8.62 [5/8]	1823	5-11.77 [49/64]	1903	6- 2.92 [59/64]	1983	6- 6.07 [5/64]	2063	6- 9.22 [7/32]	2143	7- 0.37 [3/8]	2223	7- 3.52 [33/64]
1744	5- 8.66 [21/32]	1824	5-11.81 [13/16]	1904	6- 2.96 [61/64]	1984	6- 6.11 [7/64]	2064	6- 9.26 [17/64]	2144	7- 0.41 [13/32]	2224	7- 3.56 [9/16]
1745	5- 8.70 [45/64]	1825	5-11.85 [27/32]	1905	6- 3.00	1985	6- 6.15 [5/32]	2065	6- 9.30 [19/64]	2145	7- 0.45 [29/64]	2225	7- 3.60 [19/32]
1746	5- 8.74 [47/64]	1826	5-11.89 [57/64]	1906	6- 3.04 [3/64]	1986	6- 6.19 [3/16]	2066	6- 9.34 [11/32]	2146	7- 0.49 [31/64]	2226	7- 3.64 [41/64]
1747	5- 8.78 [25/32]	1827	5-11.93 [59/64]	1907	6- 3.08 [5/64]	1987	6- 6.23 [15/64]	2067	6- 9.38 [3/8]	2147	7- 0.53 [17/32]	2227	7- 3.68 [43/64]
1748	5- 8.82 [13/16]	1828	5-11.97 [31/32]	1908	6- 3.12 [1/8]	1988	6- 6.27 [17/64]	2068	6- 9.42 [27/64]	2148	7- 0.57 [9/16]	2228	7- 3.72 [23/32]
1749	5- 8.86 [55/64]	1829	6- 0.01 [1/64]	1909	6- 3.16 [5/32]	1989	6- 6.31 [5/16]	2069	6- 9.46 [29/64]	2149	7- 0.61 [39/64]	2229	7- 3.76 [3/4]
1750	5- 8.90 [57/64]	1830	6- 0.05 [3/64]	1910	6- 3.20 [13/64]	1990	6- 6.35 [11/32]	2070	6- 9.50 [1/2]	2150	7- 0.65 [41/64]	2230	7- 3.80 [51/64]
1751	5- 8.94 [15/16]	1831	6- 0.09 [3/32]	1911	6- 3.24 [15/64]	1991	6- 6.39 [25/64]	2071	6- 9.54 [17/32]	2151	7- 0.69 [11/16]	2231	7- 3.83 [53/64]
1752	5- 8.98 [31/32]	1832	6- 0.13 [1/8]	1912	6- 3.28 [9/32]	1992	6- 6.43 [27/64]	2072	6- 9.57 [37/64]	2152	7- 0.72 [23/32]	2232	7- 3.87 [7/8]
1753	5- 9.02 [1/64]	1833	6- 0.17 [11/64]	1913	6- 3.31 [5/16]	1993	6- 6.46 [15/32]	2073	6- 9.61 [39/64]	2153	7- 0.76 [49/64]	2233	7- 3.91 [29/32]
1754	5- 9.06 [1/16]	1834	6- 0.20 [13/64]	1914	6- 3.35 [23/64]	1994	6- 6.50 [1/2]	2074	6- 9.65 [21/32]	2154	7- 0.80 [51/64]	2234	7- 3.95 [61/64]
1755	5- 9.09 [3/32]	1835	6- 0.24 [1/4]	1915	6- 3.39 [25/64]	1995	6- 6.54 [35/64]	2075	6- 9.69 [11/16]	2155	7- 0.84 [27/32]	2235	7- 3.99 [63/64]
1756	5- 9.13 [9/64]	1836	6- 0.28 [9/32]	1916	6- 3.43 [7/16]	1996	6- 6.58 [37/64]	2076	6- 9.73 [47/64]	2156	7- 0.88 [7/8]	2236	7- 4.03 [1/32]
1757	5- 9.17 [11/64]	1837	6- 0.32 [21/64]	1917	6- 3.47 [15/32]	1997	6- 6.62 [5/8]	2077	6- 9.77 [49/64]	2157	7- 0.92 [59/64]	2237	7- 4.07 [5/64]
1758	5- 9.21 [7/32]	1838	6- 0.36 [23/64]	1918	6- 3.51 [33/64]	1998	6- 6.66 [21/32]	2078	6- 9.81 [13/16]	2158	7- 0.96 [61/64]	2238	7- 4.11 [7/64]
1759	5- 9.25 [1/4]	1839	6- 0.40 [13/32]	1919	6- 3.55 [35/64]	1999	6- 6.70 [45/64]	2079	6- 9.85 [27/32]	2159	7- 1.00	2239	7- 4.15 [5/32]
1760	5- 9.29 [19/64]	1840	6- 0.44 [7/16]	1920	6- 3.59 [19/32]	2000	6- 6.74 [47/64]	2080	6- 9.89 [57/64]	2160	7- 1.04 [3/64]	2240	7- 4.19 [3/16]

MILLIMETERS CONVERTED TO FEET AND INCHES

mm	ft-in.(fraction)	mm	ft-in.(fraction)	mm	ft-in.(fraction)	mm	ft-in.(fraction)	mm	ft-in.(fraction)	mm	ft-in.(fraction)	mm	ft-in.(fraction)
2241	7- 4.23 [15/64]	2321	7- 7.38 [3/8]	2401	7-10.53 [17/32]	2481	8- 1.68 [43/64]	2561	8- 4.83 [53/64]	2641	8- 7.98 [31/32]	2721	8-11.13 [1/8]
2242	7- 4.27 [17/64]	2322	7- 7.42 [27/64]	2402	7-10.57 [9/16]	2482	8- 1.72 [23/32]	2562	8- 4.87 [55/64]	2642	8- 8.02 [1/64]	2722	8-11.17 [11/64]
2243	7- 4.31 [5/16]	2323	7- 7.46 [29/64]	2403	7-10.61 [39/64]	2483	8- 1.76 [3/4]	2563	8- 4.91 [29/32]	2643	8- 8.06 [1/16]	2723	8-11.20 [13/64]
2244	7- 4.35 [11/32]	2324	7- 7.50 [1/2]	2404	7-10.65 [41/64]	2484	8- 1.80 [51/64]	2564	8- 4.94 [15/16]	2644	8- 8.09 [3/32]	2724	8-11.24 [1/4]
2245	7- 4.39 [25/64]	2325	7- 7.54 [17/32]	2405	7-10.69 [11/16]	2485	8- 1.83 [53/64]	2565	8- 4.98 [63/64]	2645	8- 8.13 [9/64]	2725	8-11.28 [9/32]
2246	7- 4.43 [27/64]	2326	7- 7.57 [37/64]	2406	7-10.72 [23/32]	2486	8- 1.87 [7/8]	2566	8- 5.02 [1/32]	2646	8- 8.17 [11/64]	2726	8-11.32 [21/64]
2247	7- 4.46 [15/32]	2327	7- 7.61 [39/64]	2407	7-10.76 [49/64]	2487	8- 1.91 [29/32]	2567	8- 5.06 [1/16]	2647	8- 8.21 [7/32]	2727	8-11.36 [23/64]
2248	7- 4.50 [1/2]	2328	7- 7.65 [21/32]	2408	7-10.80 [51/64]	2488	8- 1.95 [61/64]	2568	8- 5.10 [7/64]	2648	8- 8.25 [1/4]	2728	8-11.40 [13/32]
2249	7- 4.54 [35/64]	2329	7- 7.69 [11/16]	2409	7-10.84 [27/32]	2489	8- 1.99 [63/64]	2569	8- 5.14 [9/64]	2649	8- 8.29 [19/64]	2729	8-11.44 [7/16]
2250	7- 4.58 [37/64]	2330	7- 7.73 [47/64]	2410	7-10.88 [7/8]	2490	8- 2.03 [1/32]	2570	8- 5.18 [3/16]	2650	8- 8.33 [21/64]	2730	8-11.48 [31/64]
2251	7- 4.62 [5/8]	2331	7- 7.77 [49/64]	2411	7-10.92 [59/64]	2491	8- 2.07 [5/64]	2571	8- 5.22 [7/32]	2651	8- 8.37 [3/8]	2731	8-11.52 [33/64]
2252	7- 4.66 [21/32]	2332	7- 7.81 [13/16]	2412	7-10.96 [61/64]	2492	8- 2.11 [7/64]	2572	8- 5.26 [17/64]	2652	8- 8.41 [13/32]	2732	8-11.56 [9/16]
2253	7- 4.70 [45/64]	2333	7- 7.85 [27/32]	2413	7-11.00	2493	8- 2.15 [5/32]	2573	8- 5.30 [19/64]	2653	8- 8.45 [29/64]	2733	8-11.60 [19/32]
2254	7- 4.74 [47/64]	2334	7- 7.89 [57/64]	2414	7-11.04 [3/64]	2494	8- 2.19 [3/16]	2574	8- 5.34 [11/32]	2654	8- 8.49 [31/64]	2734	8-11.64 [41/64]
2255	7- 4.78 [25/32]	2335	7- 7.93 [59/64]	2415	7-11.08 [5/64]	2495	8- 2.23 [15/64]	2575	8- 5.38 [3/8]	2655	8- 8.53 [17/32]	2735	8-11.68 [43/64]
2256	7- 4.82 [13/16]	2336	7- 7.97 [31/32]	2416	7-11.12 [1/8]	2496	8- 2.27 [17/64]	2576	8- 5.42 [27/64]	2656	8- 8.57 [9/16]	2736	8-11.72 [23/32]
2257	7- 4.86 [55/64]	2337	7- 8.01 [1/64]	2417	7-11.16 [5/32]	2497	8- 2.31 [5/16]	2577	8- 5.46 [29/64]	2657	8- 8.61 [39/64]	2737	8-11.76 [3/4]
2258	7- 4.90 [57/64]	2338	7- 8.05 [3/64]	2418	7-11.20 [13/64]	2498	8- 2.35 [11/32]	2578	8- 5.50 [1/2]	2658	8- 8.65 [41/64]	2738	8-11.80 [51/64]
2259	7- 4.94 [15/16]	2339	7- 8.09 [3/32]	2419	7-11.24 [15/64]	2499	8- 2.39 [25/64]	2579	8- 5.54 [17/32]	2659	8- 8.69 [11/16]	2739	8-11.83 [53/64]
2260	7- 4.98 [31/32]	2340	7- 8.13 [1/8]	2420	7-11.28 [9/32]	2500	8- 2.43 [27/64]	2580	8- 5.57 [37/64]	2660	8- 8.72 [23/32]	2740	8-11.87 [7/8]
2261	7- 5.02 [1/64]	2341	7- 8.17 [11/64]	2421	7-11.31 [5/16]	2501	8- 2.46 [15/32]	2581	8- 5.61 [39/64]	2661	8- 8.76 [49/64]	2741	8-11.91 [29/32]
2262	7- 5.06 [1/16]	2342	7- 8.20 [13/64]	2422	7-11.35 [23/64]	2502	8- 2.50 [1/2]	2582	8- 5.65 [21/32]	2662	8- 8.80 [51/64]	2742	8-11.95 [61/64]
2263	7- 5.09 [3/32]	2343	7- 8.24 [1/4]	2423	7-11.39 [25/64]	2503	8- 2.54 [35/64]	2583	8- 5.69 [11/16]	2663	8- 8.84 [27/32]	2743	8-11.99 [63/64]
2264	7- 5.13 [9/64]	2344	7- 8.28 [9/32]	2424	7-11.43 [7/16]	2504	8- 2.58 [37/64]	2584	8- 5.73 [47/64]	2664	8- 8.88 [7/8]	2744	9- 0.03 [1/32]
2265	7- 5.17 [11/64]	2345	7- 8.32 [21/64]	2425	7-11.47 [15/32]	2505	8- 2.62 [5/8]	2585	8- 5.77 [49/64]	2665	8- 8.92 [59/64]	2745	9- 0.07 [5/64]
2266	7- 5.21 [7/32]	2346	7- 8.36 [23/64]	2426	7-11.51 [33/64]	2506	8- 2.66 [21/32]	2586	8- 5.81 [13/16]	2666	8- 8.96 [61/64]	2746	9- 0.11 [7/64]
2267	7- 5.25 [1/4]	2347	7- 8.40 [13/32]	2427	7-11.55 [35/64]	2507	8- 2.70 [45/64]	2587	8- 5.85 [27/32]	2667	8- 9.00	2747	9- 0.15 [5/32]
2268	7- 5.29 [19/64]	2348	7- 8.44 [7/16]	2428	7-11.59 [19/32]	2508	8- 2.74 [47/64]	2588	8- 5.89 [57/64]	2668	8- 9.04 [3/64]	2748	9- 0.19 [3/16]
2269	7- 5.33 [21/64]	2349	7- 8.48 [31/64]	2429	7-11.63 [5/8]	2509	8- 2.78 [25/32]	2589	8- 5.93 [59/64]	2669	8- 9.08 [5/64]	2749	9- 0.23 [15/64]
2270	7- 5.37 [3/8]	2350	7- 8.52 [33/64]	2430	7-11.67 [43/64]	2510	8- 2.82 [13/16]	2590	8- 5.97 [31/32]	2670	8- 9.12 [1/8]	2750	9- 0.27 [17/64]
2271	7- 5.41 [13/32]	2351	7- 8.56 [9/16]	2431	7-11.71 [45/64]	2511	8- 2.86 [55/64]	2591	8- 6.01 [1/64]	2671	8- 9.16 [5/32]	2751	9- 0.31 [5/16]
2272	7- 5.45 [29/64]	2352	7- 8.60 [19/32]	2432	7-11.75 [3/4]	2512	8- 2.90 [57/64]	2592	8- 6.05 [3/64]	2672	8- 9.20 [13/64]	2752	9- 0.35 [11/32]
2273	7- 5.49 [31/64]	2353	7- 8.64 [41/64]	2433	7-11.79 [25/32]	2513	8- 2.94 [15/16]	2593	8- 6.09 [3/32]	2673	8- 9.24 [15/64]	2753	9- 0.39 [25/64]
2274	7- 5.53 [17/32]	2354	7- 8.68 [43/64]	2434	7-11.83 [53/64]	2514	8- 2.98 [31/32]	2594	8- 6.13 [1/8]	2674	8- 9.28 [9/32]	2754	9- 0.43 [27/64]
2275	7- 5.57 [9/16]	2355	7- 8.72 [23/32]	2435	7-11.87 [55/64]	2515	8- 3.02 [1/64]	2595	8- 6.17 [11/64]	2675	8- 9.31 [5/16]	2755	9- 0.46 [15/32]
2276	7- 5.61 [39/64]	2356	7- 8.76 [3/4]	2436	7-11.91 [29/32]	2516	8- 3.06 [1/16]	2596	8- 6.20 [13/64]	2676	8- 9.35 [23/64]	2756	9- 0.50 [1/2]
2277	7- 5.65 [41/64]	2357	7- 8.80 [51/64]	2437	7-11.94 [15/16]	2517	8- 3.09 [3/32]	2597	8- 6.24 [1/4]	2677	8- 9.39 [25/64]	2757	9- 0.54 [35/64]
2278	7- 5.69 [11/16]	2358	7- 8.83 [53/64]	2438	7-11.98 [63/64]	2518	8- 3.13 [9/64]	2598	8- 6.28 [9/32]	2678	8- 9.43 [7/16]	2758	9- 0.58 [37/64]
2279	7- 5.72 [23/32]	2359	7- 8.87 [7/8]	2439	8- 0.02 [1/32]	2519	8- 3.17 [11/64]	2599	8- 6.32 [21/64]	2679	8- 9.47 [15/32]	2759	9- 0.62 [5/8]
2280	7- 5.76 [49/64]	2360	7- 8.91 [29/32]	2440	8- 0.06 [1/16]	2520	8- 3.21 [7/32]	2600	8- 6.36 [23/64]	2680	8- 9.51 [33/64]	2760	9- 0.66 [21/32]
2281	7- 5.80 [51/64]	2361	7- 8.95 [61/64]	2441	8- 0.10 [7/64]	2521	8- 3.25 [1/4]	2601	8- 6.40 [13/32]	2681	8- 9.55 [35/64]	2761	9- 0.70 [45/64]
2282	7- 5.84 [27/32]	2362	7- 8.99 [63/64]	2442	8- 0.14 [9/64]	2522	8- 3.29 [19/64]	2602	8- 6.44 [7/16]	2682	8- 9.59 [19/32]	2762	9- 0.74 [47/64]
2283	7- 5.88 [7/8]	2363	7- 9.03 [1/32]	2443	8- 0.18 [3/16]	2523	8- 3.33 [21/64]	2603	8- 6.48 [31/64]	2683	8- 9.63 [5/8]	2763	9- 0.78 [25/32]
2284	7- 5.92 [59/64]	2364	7- 9.07 [5/64]	2444	8- 0.22 [7/32]	2524	8- 3.37 [3/8]	2604	8- 6.52 [33/64]	2684	8- 9.67 [43/64]	2764	9- 0.82 [13/16]
2285	7- 5.96 [61/64]	2365	7- 9.11 [7/64]	2445	8- 0.26 [17/64]	2525	8- 3.41 [13/32]	2605	8- 6.56 [9/16]	2685	8- 9.71 [45/64]	2765	9- 0.86 [55/64]
2286	7- 6.00	2366	7- 9.15 [5/32]	2446	8- 0.30 [19/64]	2526	8- 3.45 [29/64]	2606	8- 6.60 [19/32]	2686	8- 9.75 [3/4]	2766	9- 0.90 [57/64]
2287	7- 6.04 [3/64]	2367	7- 9.19 [3/16]	2447	8- 0.34 [11/32]	2527	8- 3.49 [31/64]	2607	8- 6.64 [41/64]	2687	8- 9.79 [25/32]	2767	9- 0.94 [15/16]
2288	7- 6.08 [5/64]	2368	7- 9.23 [15/64]	2448	8- 0.38 [3/8]	2528	8- 3.53 [17/32]	2608	8- 6.68 [43/64]	2688	8- 9.83 [53/64]	2768	9- 0.98 [31/32]
2289	7- 6.12 [1/8]	2369	7- 9.27 [17/64]	2449	8- 0.42 [27/64]	2529	8- 3.57 [9/16]	2609	8- 6.72 [23/32]	2689	8- 9.87 [55/64]	2769	9- 1.02 [1/64]
2290	7- 6.16 [5/32]	2370	7- 9.31 [5/16]	2450	8- 0.46 [29/64]	2530	8- 3.61 [39/64]	2610	8- 6.76 [3/4]	2690	8- 9.91 [29/32]	2770	9- 1.06 [1/16]
2291	7- 6.20 [13/64]	2371	7- 9.35 [11/32]	2451	8- 0.50 [1/2]	2531	8- 3.65 [41/64]	2611	8- 6.80 [51/64]	2691	8- 9.94 [15/16]	2771	9- 1.09 [3/32]
2292	7- 6.24 [15/64]	2372	7- 9.39 [25/64]	2452	8- 0.54 [17/32]	2532	8- 3.69 [11/16]	2612	8- 6.83 [53/64]	2692	8- 9.98 [63/64]	2772	9- 1.13 [9/64]
2293	7- 6.28 [9/32]	2373	7- 9.43 [27/64]	2453	8- 0.57 [37/64]	2533	8- 3.72 [23/32]	2613	8- 6.87 [7/8]	2693	8-10.02 [1/32]	2773	9- 1.17 [11/64]
2294	7- 6.31 [5/16]	2374	7- 9.46 [15/32]	2454	8- 0.61 [39/64]	2534	8- 3.76 [49/64]	2614	8- 6.91 [29/32]	2694	8-10.06 [1/16]	2774	9- 1.21 [7/32]
2295	7- 6.35 [23/64]	2375	7- 9.50 [1/2]	2455	8- 0.65 [21/32]	2535	8- 3.80 [51/64]	2615	8- 6.95 [61/64]	2695	8-10.10 [7/64]	2775	9- 1.25 [1/4]
2296	7- 6.39 [25/64]	2376	7- 9.54 [35/64]	2456	8- 0.69 [11/16]	2536	8- 3.84 [27/32]	2616	8- 6.99 [63/64]	2696	8-10.14 [9/64]	2776	9- 1.29 [19/64]
2297	7- 6.43 [7/16]	2377	7- 9.58 [37/64]	2457	8- 0.73 [47/64]	2537	8- 3.88 [7/8]	2617	8- 7.03 [1/32]	2697	8-10.18 [3/16]	2777	9- 1.33 [21/64]
2298	7- 6.47 [15/32]	2378	7- 9.62 [5/8]	2458	8- 0.77 [49/64]	2538	8- 3.92 [59/64]	2618	8- 7.07 [5/64]	2698	8-10.22 [7/32]	2778	9- 1.37 [3/8]
2299	7- 6.51 [33/64]	2379	7- 9.66 [21/32]	2459	8- 0.81 [13/16]	2539	8- 3.96 [61/64]	2619	8- 7.11 [7/64]	2699	8-10.26 [17/64]	2779	9- 1.41 [13/32]
2300	7- 6.55 [35/64]	2380	7- 9.70 [45/64]	2460	8- 0.85 [27/32]	2540	8- 4.00	2620	8- 7.15 [5/32]	2700	8-10.30 [19/64]	2780	9- 1.45 [29/64]
2301	7- 6.59 [19/32]	2381	7- 9.74 [47/64]	2461	8- 0.89 [57/64]	2541	8- 4.04 [3/64]	2621	8- 7.19 [3/16]	2701	8-10.34 [11/32]	2781	9- 1.49 [31/64]
2302	7- 6.63 [5/8]	2382	7- 9.78 [25/32]	2462	8- 0.93 [59/64]	2542	8- 4.08 [5/64]	2622	8- 7.23 [15/64]	2702	8-10.38 [3/8]	2782	9- 1.53 [17/32]
2303	7- 6.67 [43/64]	2383	7- 9.82 [13/16]	2463	8- 0.97 [31/32]	2543	8- 4.12 [1/8]	2623	8- 7.27 [17/64]	2703	8-10.42 [27/64]	2783	9- 1.57 [9/16]
2304	7- 6.71 [45/64]	2384	7- 9.86 [55/64]	2464	8- 1.01 [1/64]	2544	8- 4.16 [5/32]	2624	8- 7.31 [5/16]	2704	8-10.46 [29/64]	2784	9- 1.61 [39/64]
2305	7- 6.75 [3/4]	2385	7- 9.90 [57/64]	2465	8- 1.05 [3/64]	2545	8- 4.20 [13/64]	2625	8- 7.35 [11/32]	2705	8-10.50 [1/2]	2785	9- 1.65 [41/64]
2306	7- 6.79 [25/32]	2386	7- 9.94 [15/16]	2466	8- 1.09 [3/32]	2546	8- 4.24 [15/64]	2626	8- 7.39 [25/64]	2706	8-10.54 [17/32]	2786	9- 1.69 [11/16]
2307	7- 6.83 [53/64]	2387	7- 9.98 [31/32]	2467	8- 1.13 [1/8]	2547	8- 4.28 [9/32]	2627	8- 7.43 [27/64]	2707	8-10.57 [37/64]	2787	9- 1.72 [23/32]
2308	7- 6.87 [55/64]	2388	7-10.02 [1/64]	2468	8- 1.17 [11/64]	2548	8- 4.31 [5/16]	2628	8- 7.46 [15/32]	2708	8-10.61 [39/64]	2788	9- 1.76 [49/64]
2309	7- 6.91 [29/32]	2389	7-10.06 [1/16]	2469	8- 1.20 [13/64]	2549	8- 4.35 [23/64]	2629	8- 7.50 [1/2]	2709	8-10.65 [21/32]	2789	9- 1.80 [51/64]
2310	7- 6.94 [15/16]	2390	7-10.09 [3/32]	2470	8- 1.24 [1/4]	2550	8- 4.39 [25/64]	2630	8- 7.54 [35/64]	2710	8-10.69 [11/16]	2790	9- 1.84 [27/32]
2311	7- 6.98 [63/64]	2391	7-10.13 [9/64]	2471	8- 1.28 [9/32]	2551	8- 4.43 [7/16]	2631	8- 7.58 [37/64]	2711	8-10.73 [47/64]	2791	9- 1.88 [7/8]
2312	7- 7.02 [1/32]	2392	7-10.17 [11/64]	2472	8- 1.32 [21/64]	2552	8- 4.47 [15/32]	2632	8- 7.62 [5/8]	2712	8-10.77 [49/64]	2792	9- 1.92 [59/64]
2313	7- 7.06 [1/16]	2393	7-10.21 [7/32]	2473	8- 1.36 [23/64]	2553	8- 4.51 [33/64]	2633	8- 7.66 [21/32]	2713	8-10.81 [13/16]	2793	9- 1.96 [61/64]
2314	7- 7.10 [7/64]	2394	7-10.25 [1/4]	2474	8- 1.40 [13/32]	2554	8- 4.55 [35/64]	2634	8- 7.70 [45/64]	2714	8-10.85 [27/32]	2794	9- 2.00
2315	7- 7.14 [9/64]	2395	7-10.29 [19/64]	2475	8- 1.44 [7/16]	2555	8- 4.59 [19/32]	2635	8- 7.74 [47/64]	2715	8-10.89 [57/64]	2795	9- 2.04 [3/64]
2316	7- 7.18 [3/16]	2396	7-10.33 [21/64]	2476	8- 1.48 [31/64]	2556	8- 4.63 [5/8]	2636	8- 7.78 [25/32]	2716	8-10.93 [59/64]	2796	9- 2.08 [5/64]
2317	7- 7.22 [7/32]	2397	7-10.37 [3/8]	2477	8- 1.52 [33/64]	2557	8- 4.67 [43/64]	2637	8- 7.82 [13/16]	2717	8-10.97 [31/32]	2797	9- 2.12 [1/8]
2318	7- 7.26 [17/64]	2398	7-10.41 [13/32]	2478	8- 1.56 [9/16]	2558	8- 4.71 [45/64]	2638	8- 7.86 [55/64]	2718	8-11.01 [1/64]	2798	9- 2.16 [5/32]
2319	7- 7.30 [19/64]	2399	7-10.45 [29/64]	2479	8- 1.60 [19/32]	2559	8- 4.75 [3/4]	2639	8- 7.90 [57/64]	2719	8-11.05 [3/64]	2799	9- 2.20 [13/64]
2320	7- 7.34 [11/32]	2400	7-10.49 [31/64]	2480	8- 1.64 [41/64]	2560	8- 4.79 [25/32]	2640	8- 7.94 [15/16]	2720	8-11.09 [3/32]	2800	9- 2.24 [15/64]

MILLIMETERS CONVERTED TO FEET AND INCHES

TABLES M-3

mm	ft-in.(fraction)	mm	ft-in.(fraction)	mm	ft-in.(fraction)	mm	ft-in.(fraction)	mm	ft-in.(fraction)	mm	ft-in.(fraction)	mm	ft-in.(fraction)
2801	9- 2.28 [9/32]	2881	9- 5.43 [27/64]	2961	9- 8.57 [37/64]	3041	9-11.72 [23/32]	3121	10- 2.87 [7/8]	3201	10- 6.02 [1/32]	3281	10- 9.17 [11/64]
2802	9- 2.31 [5/16]	2882	9- 5.46 [15/32]	2962	9- 8.61 [39/64]	3042	9-11.76 [49/64]	3122	10- 2.91 [29/32]	3202	10- 6.06 [1/16]	3282	10- 9.21 [7/32]
2803	9- 2.35 [23/64]	2883	9- 5.50 [1/2]	2963	9- 8.65 [21/32]	3043	9-11.80 [51/64]	3123	10- 2.95 [61/64]	3203	10- 6.10 [7/64]	3283	10- 9.25 [1/4]
2804	9- 2.39 [25/64]	2884	9- 5.54 [35/64]	2964	9- 8.69 [11/16]	3044	9-11.84 [27/32]	3124	10- 2.99 [63/64]	3204	10- 6.14 [9/64]	3284	10- 9.29 [19/64]
2805	9- 2.43 [7/16]	2885	9- 5.58 [37/64]	2965	9- 8.73 [47/64]	3045	9-11.88 [7/8]	3125	10- 3.03 [1/32]	3205	10- 6.18 [3/16]	3285	10- 9.33 [21/64]
2806	9- 2.47 [15/32]	2886	9- 5.62 [5/8]	2966	9- 8.77 [49/64]	3046	9-11.92 [59/64]	3126	10- 3.07 [5/64]	3206	10- 6.22 [7/32]	3286	10- 9.37 [3/8]
2807	9- 2.51 [33/64]	2887	9- 5.66 [21/32]	2967	9- 8.81 [13/16]	3047	9-11.96 [61/64]	3127	10- 3.11 [7/64]	3207	10- 6.26 [17/64]	3287	10- 9.41 [13/32]
2808	9- 2.55 [35/64]	2888	9- 5.70 [45/64]	2968	9- 8.85 [27/32]	3048	10- 0.00	3128	10- 3.15 [5/32]	3208	10- 6.30 [19/64]	3288	10- 9.45 [29/64]
2809	9- 2.59 [19/32]	2889	9- 5.74 [47/64]	2969	9- 8.89 [57/64]	3049	10- 0.04 [3/64]	3129	10- 3.19 [3/16]	3209	10- 6.34 [11/32]	3289	10- 9.49 [31/64]
2810	9- 2.63 [5/8]	2890	9- 5.78 [25/32]	2970	9- 8.93 [59/64]	3050	10- 0.08 [5/64]	3130	10- 3.23 [15/64]	3210	10- 6.38 [3/8]	3290	10- 9.53 [17/32]
2811	9- 2.67 [43/64]	2891	9- 5.82 [13/16]	2971	9- 8.97 [31/32]	3051	10- 0.12 [1/8]	3131	10- 3.27 [17/64]	3211	10- 6.42 [27/64]	3291	10- 9.57 [9/16]
2812	9- 2.71 [45/64]	2892	9- 5.86 [55/64]	2972	9- 9.01 [1/64]	3052	10- 0.16 [5/32]	3132	10- 3.31 [5/16]	3212	10- 6.46 [29/64]	3292	10- 9.61 [39/64]
2813	9- 2.75 [3/4]	2893	9- 5.90 [57/64]	2973	9- 9.05 [3/64]	3053	10- 0.20 [13/64]	3133	10- 3.35 [11/32]	3213	10- 6.50 [1/2]	3293	10- 9.65 [41/64]
2814	9- 2.79 [25/32]	2894	9- 5.94 [15/16]	2974	9- 9.09 [3/32]	3054	10- 0.24 [15/64]	3134	10- 3.39 [25/64]	3214	10- 6.54 [17/32]	3294	10- 9.69 [11/16]
2815	9- 2.83 [53/64]	2895	9- 5.98 [31/32]	2975	9- 9.13 [1/8]	3055	10- 0.28 [9/32]	3135	10- 3.43 [27/64]	3215	10- 6.57 [37/64]	3295	10- 9.72 [23/32]
2816	9- 2.87 [55/64]	2896	9- 6.02 [1/64]	2976	9- 9.17 [11/64]	3056	10- 0.31 [5/16]	3136	10- 3.46 [15/32]	3216	10- 6.61 [39/64]	3296	10- 9.76 [49/64]
2817	9- 2.91 [29/32]	2897	9- 6.06 [1/16]	2977	9- 9.20 [13/64]	3057	10- 0.35 [23/64]	3137	10- 3.50 [1/2]	3217	10- 6.65 [21/32]	3297	10- 9.80 [51/64]
2818	9- 2.94 [15/16]	2898	9- 6.09 [3/32]	2978	9- 9.24 [1/4]	3058	10- 0.39 [25/64]	3138	10- 3.54 [35/64]	3218	10- 6.69 [11/16]	3298	10- 9.84 [27/32]
2819	9- 2.98 [63/64]	2899	9- 6.13 [9/64]	2979	9- 9.28 [9/32]	3059	10- 0.43 [7/16]	3139	10- 3.58 [37/64]	3219	10- 6.73 [47/64]	3299	10- 9.88 [7/8]
2820	9- 3.02 [1/32]	2900	9- 6.17 [11/64]	2980	9- 9.32 [21/64]	3060	10- 0.47 [15/32]	3140	10- 3.62 [5/8]	3220	10- 6.77 [49/64]	3300	10- 9.92 [59/64]
2821	9- 3.06 [1/16]	2901	9- 6.21 [7/32]	2981	9- 9.36 [23/64]	3061	10- 0.51 [33/64]	3141	10- 3.66 [21/32]	3221	10- 6.81 [13/16]	3301	10- 9.96 [61/64]
2822	9- 3.10 [7/64]	2902	9- 6.25 [1/4]	2982	9- 9.40 [13/32]	3062	10- 0.55 [35/64]	3142	10- 3.70 [45/64]	3222	10- 6.85 [27/32]	3302	10-10.00
2823	9- 3.14 [9/64]	2903	9- 6.29 [19/64]	2983	9- 9.44 [7/16]	3063	10- 0.59 [19/32]	3143	10- 3.74 [47/64]	3223	10- 6.89 [57/64]	3303	10-10.04 [3/64]
2824	9- 3.18 [3/16]	2904	9- 6.33 [21/64]	2984	9- 9.48 [31/64]	3064	10- 0.63 [5/8]	3144	10- 3.78 [25/32]	3224	10- 6.93 [59/64]	3304	10-10.08 [5/64]
2825	9- 3.22 [7/32]	2905	9- 6.37 [3/8]	2985	9- 9.52 [33/64]	3065	10- 0.67 [43/64]	3145	10- 3.82 [13/16]	3225	10- 6.97 [31/32]	3305	10-10.12 [1/8]
2826	9- 3.26 [17/64]	2906	9- 6.41 [13/32]	2986	9- 9.56 [9/16]	3066	10- 0.71 [45/64]	3146	10- 3.86 [55/64]	3226	10- 7.01 [1/64]	3306	10-10.16 [5/32]
2827	9- 3.30 [19/64]	2907	9- 6.45 [29/64]	2987	9- 9.60 [19/32]	3067	10- 0.75 [3/4]	3147	10- 3.90 [57/64]	3227	10- 7.05 [3/64]	3307	10-10.20 [13/64]
2828	9- 3.34 [11/32]	2908	9- 6.49 [31/64]	2988	9- 9.64 [41/64]	3068	10- 0.79 [25/32]	3148	10- 3.94 [15/16]	3228	10- 7.09 [3/32]	3308	10-10.24 [15/64]
2829	9- 3.38 [3/8]	2909	9- 6.53 [17/32]	2989	9- 9.68 [43/64]	3069	10- 0.83 [53/64]	3149	10- 3.98 [31/32]	3229	10- 7.13 [1/8]	3309	10-10.28 [9/32]
2830	9- 3.42 [27/64]	2910	9- 6.57 [9/16]	2990	9- 9.72 [23/32]	3070	10- 0.87 [55/64]	3150	10- 4.02 [1/64]	3230	10- 7.17 [11/64]	3310	10-10.31 [5/16]
2831	9- 3.46 [29/64]	2911	9- 6.61 [39/64]	2991	9- 9.76 [3/4]	3071	10- 0.91 [29/32]	3151	10- 4.06 [1/16]	3231	10- 7.20 [13/64]	3311	10-10.35 [23/64]
2832	9- 3.50 [1/2]	2912	9- 6.65 [41/64]	2992	9- 9.80 [51/64]	3072	10- 0.94 [15/16]	3152	10- 4.09 [3/32]	3232	10- 7.24 [1/4]	3312	10-10.39 [25/64]
2833	9- 3.54 [17/32]	2913	9- 6.69 [11/16]	2993	9- 9.83 [53/64]	3073	10- 0.98 [63/64]	3153	10- 4.13 [9/64]	3233	10- 7.28 [9/32]	3313	10-10.43 [7/16]
2834	9- 3.57 [37/64]	2914	9- 6.72 [23/32]	2994	9- 9.87 [7/8]	3074	10- 1.02 [1/32]	3154	10- 4.17 [11/64]	3234	10- 7.32 [21/64]	3314	10-10.47 [15/32]
2835	9- 3.61 [39/64]	2915	9- 6.76 [49/64]	2995	9- 9.91 [29/32]	3075	10- 1.06 [1/16]	3155	10- 4.21 [7/32]	3235	10- 7.36 [23/64]	3315	10-10.51 [33/64]
2836	9- 3.65 [21/32]	2916	9- 6.80 [51/64]	2996	9- 9.95 [61/64]	3076	10- 1.10 [7/64]	3156	10- 4.25 [1/4]	3236	10- 7.40 [13/32]	3316	10-10.55 [35/64]
2837	9- 3.69 [11/16]	2917	9- 6.84 [27/32]	2997	9- 9.99 [63/64]	3077	10- 1.14 [9/64]	3157	10- 4.29 [19/64]	3237	10- 7.44 [7/16]	3317	10-10.59 [19/32]
2838	9- 3.73 [47/64]	2918	9- 6.88 [7/8]	2998	9-10.03 [1/32]	3078	10- 1.18 [3/16]	3158	10- 4.33 [21/64]	3238	10- 7.48 [31/64]	3318	10-10.63 [5/8]
2839	9- 3.77 [49/64]	2919	9- 6.92 [59/64]	2999	9-10.07 [5/64]	3079	10- 1.22 [7/32]	3159	10- 4.37 [3/8]	3239	10- 7.52 [33/64]	3319	10-10.67 [43/64]
2840	9- 3.81 [13/16]	2920	9- 6.96 [61/64]	3000	9-10.11 [7/64]	3080	10- 1.26 [17/64]	3160	10- 4.41 [13/32]	3240	10- 7.56 [9/16]	3320	10-10.71 [45/64]
2841	9- 3.85 [27/32]	2921	9- 7.00	3001	9-10.15 [5/32]	3081	10- 1.30 [19/64]	3161	10- 4.45 [29/64]	3241	10- 7.60 [19/32]	3321	10-10.75 [3/4]
2842	9- 3.89 [57/64]	2922	9- 7.04 [3/64]	3002	9-10.19 [3/16]	3082	10- 1.34 [11/32]	3162	10- 4.49 [31/64]	3242	10- 7.64 [41/64]	3322	10-10.79 [25/32]
2843	9- 3.93 [59/64]	2923	9- 7.08 [5/64]	3003	9-10.23 [15/64]	3083	10- 1.38 [3/8]	3163	10- 4.53 [17/32]	3243	10- 7.68 [43/64]	3323	10-10.83 [53/64]
2844	9- 3.97 [31/32]	2924	9- 7.12 [1/8]	3004	9-10.27 [17/64]	3084	10- 1.42 [27/64]	3164	10- 4.57 [9/16]	3244	10- 7.72 [23/32]	3324	10-10.87 [55/64]
2845	9- 4.01 [1/64]	2925	9- 7.16 [5/32]	3005	9-10.31 [5/16]	3085	10- 1.46 [29/64]	3165	10- 4.61 [39/64]	3245	10- 7.76 [3/4]	3325	10-10.91 [29/32]
2846	9- 4.05 [3/64]	2926	9- 7.20 [13/64]	3006	9-10.35 [11/32]	3086	10- 1.50 [1/2]	3166	10- 4.65 [41/64]	3246	10- 7.80 [51/64]	3326	10-10.94 [15/16]
2847	9- 4.09 [3/32]	2927	9- 7.24 [15/64]	3007	9-10.39 [25/64]	3087	10- 1.54 [17/32]	3167	10- 4.69 [11/16]	3247	10- 7.83 [53/64]	3327	10-10.98 [63/64]
2848	9- 4.13 [1/8]	2928	9- 7.28 [9/32]	3008	9-10.43 [27/64]	3088	10- 1.57 [37/64]	3168	10- 4.72 [23/32]	3248	10- 7.87 [7/8]	3328	10-11.02 [1/32]
2849	9- 4.17 [11/64]	2929	9- 7.31 [5/16]	3009	9-10.46 [15/32]	3089	10- 1.61 [39/64]	3169	10- 4.76 [49/64]	3249	10- 7.91 [29/32]	3329	10-11.06 [1/16]
2850	9- 4.20 [13/64]	2930	9- 7.35 [23/64]	3010	9-10.50 [1/2]	3090	10- 1.65 [21/32]	3170	10- 4.80 [51/64]	3250	10- 7.95 [61/64]	3330	10-11.10 [7/64]
2851	9- 4.24 [1/4]	2931	9- 7.39 [25/64]	3011	9-10.54 [35/64]	3091	10- 1.69 [11/16]	3171	10- 4.84 [27/32]	3251	10- 7.99 [63/64]	3331	10-11.14 [9/64]
2852	9- 4.28 [9/32]	2932	9- 7.43 [7/16]	3012	9-10.58 [37/64]	3092	10- 1.73 [47/64]	3172	10- 4.88 [7/8]	3252	10- 8.03 [1/32]	3332	10-11.18 [3/16]
2853	9- 4.32 [21/64]	2933	9- 7.47 [15/32]	3013	9-10.62 [5/8]	3093	10- 1.77 [49/64]	3173	10- 4.92 [59/64]	3253	10- 8.07 [5/64]	3333	10-11.22 [7/32]
2854	9- 4.36 [23/64]	2934	9- 7.51 [33/64]	3014	9-10.66 [21/32]	3094	10- 1.81 [13/16]	3174	10- 4.96 [61/64]	3254	10- 8.11 [7/64]	3334	10-11.26 [17/64]
2855	9- 4.40 [13/32]	2935	9- 7.55 [35/64]	3015	9-10.70 [45/64]	3095	10- 1.85 [27/32]	3175	10- 5.00	3255	10- 8.15 [5/32]	3335	10-11.30 [19/64]
2856	9- 4.44 [7/16]	2936	9- 7.59 [19/32]	3016	9-10.74 [47/64]	3096	10- 1.89 [57/64]	3176	10- 5.04 [3/64]	3256	10- 8.19 [3/16]	3336	10-11.34 [11/32]
2857	9- 4.48 [31/64]	2937	9- 7.63 [5/8]	3017	9-10.78 [25/32]	3097	10- 1.93 [59/64]	3177	10- 5.08 [5/64]	3257	10- 8.23 [15/64]	3337	10-11.38 [3/8]
2858	9- 4.52 [33/64]	2938	9- 7.67 [43/64]	3018	9-10.82 [13/16]	3098	10- 1.97 [31/32]	3178	10- 5.12 [1/8]	3258	10- 8.27 [17/64]	3338	10-11.42 [27/64]
2859	9- 4.56 [9/16]	2939	9- 7.71 [45/64]	3019	9-10.86 [55/64]	3099	10- 2.01 [1/64]	3179	10- 5.16 [5/32]	3259	10- 8.31 [5/16]	3339	10-11.46 [29/64]
2860	9- 4.60 [19/32]	2940	9- 7.75 [3/4]	3020	9-10.90 [57/64]	3100	10- 2.05 [3/64]	3180	10- 5.20 [13/64]	3260	10- 8.35 [11/32]	3340	10-11.50 [1/2]
2861	9- 4.64 [41/64]	2941	9- 7.79 [25/32]	3021	9-10.94 [15/16]	3101	10- 2.09 [3/32]	3181	10- 5.24 [15/64]	3261	10- 8.39 [25/64]	3341	10-11.54 [17/32]
2862	9- 4.68 [43/64]	2942	9- 7.83 [53/64]	3022	9-10.98 [31/32]	3102	10- 2.13 [1/8]	3182	10- 5.28 [9/32]	3262	10- 8.43 [27/64]	3342	10-11.57 [37/64]
2863	9- 4.72 [23/32]	2943	9- 7.87 [55/64]	3023	9-11.02 [1/64]	3103	10- 2.17 [11/64]	3183	10- 5.31 [5/16]	3263	10- 8.46 [15/32]	3343	10-11.61 [39/64]
2864	9- 4.76 [3/4]	2944	9- 7.91 [29/32]	3024	9-11.06 [1/16]	3104	10- 2.20 [13/64]	3184	10- 5.35 [23/64]	3264	10- 8.50 [1/2]	3344	10-11.65 [21/32]
2865	9- 4.80 [51/64]	2945	9- 7.94 [15/16]	3025	9-11.09 [3/32]	3105	10- 2.24 [1/4]	3185	10- 5.39 [25/64]	3265	10- 8.54 [35/64]	3345	10-11.69 [11/16]
2866	9- 4.83 [53/64]	2946	9- 7.98 [63/64]	3026	9-11.13 [9/64]	3106	10- 2.28 [9/32]	3186	10- 5.43 [7/16]	3266	10- 8.58 [37/64]	3346	10-11.73 [47/64]
2867	9- 4.87 [7/8]	2947	9- 8.02 [1/32]	3027	9-11.17 [11/64]	3107	10- 2.32 [21/64]	3187	10- 5.47 [15/32]	3267	10- 8.62 [5/8]	3347	10-11.77 [49/64]
2868	9- 4.91 [29/32]	2948	9- 8.06 [1/16]	3028	9-11.21 [7/32]	3108	10- 2.36 [23/64]	3188	10- 5.51 [33/64]	3268	10- 8.66 [21/32]	3348	10-11.81 [13/16]
2869	9- 4.95 [61/64]	2949	9- 8.10 [7/64]	3029	9-11.25 [1/4]	3109	10- 2.40 [13/32]	3189	10- 5.55 [35/64]	3269	10- 8.70 [45/64]	3349	10-11.85 [27/32]
2870	9- 4.99 [63/64]	2950	9- 8.14 [9/64]	3030	9-11.29 [19/64]	3110	10- 2.44 [7/16]	3190	10- 5.59 [19/32]	3270	10- 8.74 [47/64]	3350	10-11.89 [57/64]
2871	9- 5.03 [1/32]	2951	9- 8.18 [3/16]	3031	9-11.33 [21/64]	3111	10- 2.48 [31/64]	3191	10- 5.63 [5/8]	3271	10- 8.78 [25/32]	3351	10-11.93 [59/64]
2872	9- 5.07 [5/64]	2952	9- 8.22 [7/32]	3032	9-11.37 [3/8]	3112	10- 2.52 [33/64]	3192	10- 5.67 [43/64]	3272	10- 8.82 [13/16]	3352	10-11.97 [31/32]
2873	9- 5.11 [7/64]	2953	9- 8.26 [17/64]	3033	9-11.41 [13/32]	3113	10- 2.56 [9/16]	3193	10- 5.71 [45/64]	3273	10- 8.86 [55/64]	3353	11- 0.01 [1/64]
2874	9- 5.15 [5/32]	2954	9- 8.30 [19/64]	3034	9-11.45 [29/64]	3114	10- 2.60 [19/32]	3194	10- 5.75 [3/4]	3274	10- 8.90 [57/64]	3354	11- 0.05 [3/64]
2875	9- 5.19 [3/16]	2955	9- 8.34 [11/32]	3035	9-11.49 [31/64]	3115	10- 2.64 [41/64]	3195	10- 5.79 [25/32]	3275	10- 8.94 [15/16]	3355	11- 0.09 [3/32]
2876	9- 5.23 [15/64]	2956	9- 8.38 [3/8]	3036	9-11.53 [17/32]	3116	10- 2.68 [43/64]	3196	10- 5.83 [53/64]	3276	10- 8.98 [31/32]	3356	11- 0.13 [1/8]
2877	9- 5.27 [17/64]	2957	9- 8.42 [27/64]	3037	9-11.57 [9/16]	3117	10- 2.72 [23/32]	3197	10- 5.87 [55/64]	3277	10- 9.02 [1/64]	3357	11- 0.17 [11/64]
2878	9- 5.31 [5/16]	2958	9- 8.46 [29/64]	3038	9-11.61 [39/64]	3118	10- 2.76 [3/4]	3198	10- 5.91 [29/32]	3278	10- 9.06 [1/16]	3358	11- 0.20 [13/64]
2879	9- 5.35 [11/32]	2959	9- 8.50 [1/2]	3039	9-11.65 [41/64]	3119	10- 2.80 [51/64]	3199	10- 5.94 [15/16]	3279	10- 9.09 [3/32]	3359	11- 0.24 [1/4]
2880	9- 5.39 [25/64]	2960	9- 8.54 [17/32]	3040	9-11.69 [11/16]	3120	10- 2.83 [53/64]	3200	10- 5.98 [63/64]	3280	10- 9.13 [9/64]	3360	11- 0.28 [9/32]

MILLIMETERS CONVERTED TO FEET AND INCHES

TABLES M-3

mm	ft-in.(fraction)	mm	ft-in.(fraction)	mm	ft-in.(fraction)	mm	ft-in.(fraction)	mm	ft-in.(fraction)	mm	ft-in.(fraction)	mm	ft-in.(fraction)
3361	11- 0.32 [21/64]	3441	11- 3.47 [15/32]	3521	11- 6.62 [5/8]	3601	11- 9.77 [49/64]	3681	12- 0.92 [59/64]	3761	12- 4.07 [5/64]	3841	12- 7.22 [7/32]
3362	11- 0.36 [23/64]	3442	11- 3.51 [33/64]	3522	11- 6.66 [21/32]	3602	11- 9.81 [13/16]	3682	12- 0.96 [61/64]	3762	12- 4.11 [7/64]	3842	12- 7.26 [17/64]
3363	11- 0.40 [13/32]	3443	11- 3.55 [35/64]	3523	11- 6.70 [45/64]	3603	11- 9.85 [27/32]	3683	12- 1.00	3763	12- 4.15 [5/32]	3843	12- 7.30 [19/64]
3364	11- 0.44 [7/16]	3444	11- 3.59 [19/32]	3524	11- 6.74 [47/64]	3604	11- 9.89 [57/64]	3684	12- 1.04 [3/64]	3764	12- 4.19 [3/16]	3844	12- 7.34 [11/32]
3365	11- 0.48 [31/64]	3445	11- 3.63 [5/8]	3525	11- 6.78 [25/32]	3605	11- 9.93 [59/64]	3685	12- 1.08 [5/64]	3765	12- 4.23 [15/64]	3845	12- 7.38 [3/8]
3366	11- 0.52 [33/64]	3446	11- 3.67 [43/64]	3526	11- 6.82 [13/16]	3606	11- 9.97 [31/32]	3686	12- 1.12 [1/8]	3766	12- 4.27 [17/64]	3846	12- 7.42 [27/64]
3367	11- 0.56 [9/16]	3447	11- 3.71 [45/64]	3527	11- 6.86 [55/64]	3607	11-10.01 [1/64]	3687	12- 1.16 [5/32]	3767	12- 4.31 [5/16]	3847	12- 7.46 [29/64]
3368	11- 0.60 [19/32]	3448	11- 3.75 [3/4]	3528	11- 6.90 [57/64]	3608	11-10.05 [3/64]	3688	12- 1.20 [13/64]	3768	12- 4.35 [11/32]	3848	12- 7.50 [1/2]
3369	11- 0.64 [41/64]	3449	11- 3.79 [25/32]	3529	11- 6.94 [15/16]	3609	11-10.09 [3/32]	3689	12- 1.24 [15/64]	3769	12- 4.39 [25/64]	3849	12- 7.54 [17/32]
3370	11- 0.68 [43/64]	3450	11- 3.83 [53/64]	3530	11- 6.98 [31/32]	3610	11-10.13 [1/8]	3690	12- 1.28 [9/32]	3770	12- 4.43 [27/64]	3850	12- 7.57 [37/64]
3371	11- 0.72 [23/32]	3451	11- 3.87 [55/64]	3531	11- 7.02 [1/64]	3611	11-10.17 [11/64]	3691	12- 1.31 [5/16]	3771	12- 4.46 [15/32]	3851	12- 7.61 [39/64]
3372	11- 0.76 [3/4]	3452	11- 3.91 [29/32]	3532	11- 7.06 [1/16]	3612	11-10.20 [13/64]	3692	12- 1.35 [23/64]	3772	12- 4.50 [1/2]	3852	12- 7.65 [21/32]
3373	11- 0.80 [51/64]	3453	11- 3.94 [15/16]	3533	11- 7.09 [3/32]	3613	11-10.24 [1/4]	3693	12- 1.39 [25/64]	3773	12- 4.54 [35/64]	3853	12- 7.69 [11/16]
3374	11- 0.83 [53/64]	3454	11- 3.98 [63/64]	3534	11- 7.13 [9/64]	3614	11-10.28 [9/32]	3694	12- 1.43 [7/16]	3774	12- 4.58 [37/64]	3854	12- 7.73 [47/64]
3375	11- 0.87 [7/8]	3455	11- 4.02 [1/32]	3535	11- 7.17 [11/64]	3615	11-10.32 [21/64]	3695	12- 1.47 [15/32]	3775	12- 4.62 [5/8]	3855	12- 7.77 [49/64]
3376	11- 0.91 [29/32]	3456	11- 4.06 [1/16]	3536	11- 7.21 [7/32]	3616	11-10.36 [23/64]	3696	12- 1.51 [33/64]	3776	12- 4.66 [21/32]	3856	12- 7.81 [13/16]
3377	11- 0.95 [61/64]	3457	11- 4.10 [7/64]	3537	11- 7.25 [1/4]	3617	11-10.40 [13/32]	3697	12- 1.55 [35/64]	3777	12- 4.70 [45/64]	3857	12- 7.85 [27/32]
3378	11- 0.99 [63/64]	3458	11- 4.14 [9/64]	3538	11- 7.29 [19/64]	3618	11-10.44 [7/16]	3698	12- 1.59 [19/32]	3778	12- 4.74 [47/64]	3858	12- 7.89 [57/64]
3379	11- 1.03 [1/32]	3459	11- 4.18 [3/16]	3539	11- 7.33 [21/64]	3619	11-10.48 [31/64]	3699	12- 1.63 [5/8]	3779	12- 4.78 [25/32]	3859	12- 7.93 [59/64]
3380	11- 1.07 [5/64]	3460	11- 4.22 [7/32]	3540	11- 7.37 [3/8]	3620	11-10.52 [33/64]	3700	12- 1.67 [43/64]	3780	12- 4.82 [13/16]	3860	12- 7.97 [31/32]
3381	11- 1.11 [7/64]	3461	11- 4.26 [17/64]	3541	11- 7.41 [13/32]	3621	11-10.56 [9/16]	3701	12- 1.71 [45/64]	3781	12- 4.86 [55/64]	3861	12- 8.01 [1/64]
3382	11- 1.15 [5/32]	3462	11- 4.30 [19/64]	3542	11- 7.45 [29/64]	3622	11-10.60 [19/32]	3702	12- 1.75 [3/4]	3782	12- 4.90 [57/64]	3862	12- 8.05 [3/64]
3383	11- 1.19 [3/16]	3463	11- 4.34 [11/32]	3543	11- 7.49 [31/64]	3623	11-10.64 [41/64]	3703	12- 1.79 [25/32]	3783	12- 4.94 [15/16]	3863	12- 8.09 [3/32]
3384	11- 1.23 [15/64]	3464	11- 4.38 [3/8]	3544	11- 7.53 [17/32]	3624	11-10.68 [43/64]	3704	12- 1.83 [53/64]	3784	12- 4.98 [31/32]	3864	12- 8.13 [1/8]
3385	11- 1.27 [17/64]	3465	11- 4.42 [27/64]	3545	11- 7.57 [9/16]	3625	11-10.72 [23/32]	3705	12- 1.87 [55/64]	3785	12- 5.02 [1/64]	3865	12- 8.17 [11/64]
3386	11- 1.31 [5/16]	3466	11- 4.46 [29/64]	3546	11- 7.61 [39/64]	3626	11-10.76 [3/4]	3706	12- 1.91 [29/32]	3786	12- 5.06 [1/16]	3866	12- 8.20 [13/64]
3387	11- 1.35 [11/32]	3467	11- 4.50 [1/2]	3547	11- 7.65 [41/64]	3627	11-10.80 [51/64]	3707	12- 1.94 [15/16]	3787	12- 5.09 [3/32]	3867	12- 8.24 [1/4]
3388	11- 1.39 [25/64]	3468	11- 4.54 [17/32]	3548	11- 7.69 [11/16]	3628	11-10.83 [53/64]	3708	12- 1.98 [63/64]	3788	12- 5.13 [9/64]	3868	12- 8.28 [9/32]
3389	11- 1.43 [27/64]	3469	11- 4.57 [37/64]	3549	11- 7.72 [23/32]	3629	11-10.87 [7/8]	3709	12- 2.02 [1/32]	3789	12- 5.17 [11/64]	3869	12- 8.32 [21/64]
3390	11- 1.46 [15/32]	3470	11- 4.61 [39/64]	3550	11- 7.76 [49/64]	3630	11-10.91 [29/32]	3710	12- 2.06 [1/16]	3790	12- 5.21 [7/32]	3870	12- 8.36 [23/64]
3391	11- 1.50 [1/2]	3471	11- 4.65 [21/32]	3551	11- 7.80 [51/64]	3631	11-10.95 [61/64]	3711	12- 2.10 [7/64]	3791	12- 5.25 [1/4]	3871	12- 8.40 [13/32]
3392	11- 1.54 [35/64]	3472	11- 4.69 [11/16]	3552	11- 7.84 [27/32]	3632	11-10.99 [63/64]	3712	12- 2.14 [9/64]	3792	12- 5.29 [19/64]	3872	12- 8.44 [7/16]
3393	11- 1.58 [37/64]	3473	11- 4.73 [47/64]	3553	11- 7.88 [7/8]	3633	11-11.03 [1/32]	3713	12- 2.18 [3/16]	3793	12- 5.33 [21/64]	3873	12- 8.48 [31/64]
3394	11- 1.62 [5/8]	3474	11- 4.77 [49/64]	3554	11- 7.92 [59/64]	3634	11-11.07 [5/64]	3714	12- 2.22 [7/32]	3794	12- 5.37 [3/8]	3874	12- 8.52 [33/64]
3395	11- 1.66 [21/32]	3475	11- 4.81 [13/16]	3555	11- 7.96 [61/64]	3635	11-11.11 [7/64]	3715	12- 2.26 [17/64]	3795	12- 5.41 [13/32]	3875	12- 8.56 [9/16]
3396	11- 1.70 [45/64]	3476	11- 4.85 [27/32]	3556	11- 8.00	3636	11-11.15 [5/32]	3716	12- 2.30 [19/64]	3796	12- 5.45 [29/64]	3876	12- 8.60 [19/32]
3397	11- 1.74 [47/64]	3477	11- 4.89 [57/64]	3557	11- 8.04 [3/64]	3637	11-11.19 [3/16]	3717	12- 2.34 [11/32]	3797	12- 5.49 [31/64]	3877	12- 8.64 [41/64]
3398	11- 1.78 [25/32]	3478	11- 4.93 [59/64]	3558	11- 8.08 [5/64]	3638	11-11.23 [15/64]	3718	12- 2.38 [3/8]	3798	12- 5.53 [17/32]	3878	12- 8.68 [43/64]
3399	11- 1.82 [13/16]	3479	11- 4.97 [31/32]	3559	11- 8.12 [1/8]	3639	11-11.27 [17/64]	3719	12- 2.42 [27/64]	3799	12- 5.57 [9/16]	3879	12- 8.72 [23/32]
3400	11- 1.86 [55/64]	3480	11- 5.01 [1/64]	3560	11- 8.16 [5/32]	3640	11-11.31 [5/16]	3720	12- 2.46 [29/64]	3800	12- 5.61 [39/64]	3880	12- 8.76 [3/4]
3401	11- 1.90 [57/64]	3481	11- 5.05 [3/64]	3561	11- 8.20 [13/64]	3641	11-11.35 [11/32]	3721	12- 2.50 [1/2]	3801	12- 5.65 [41/64]	3881	12- 8.80 [51/64]
3402	11- 1.94 [15/16]	3482	11- 5.09 [3/32]	3562	11- 8.24 [15/64]	3642	11-11.39 [25/64]	3722	12- 2.54 [17/32]	3802	12- 5.69 [11/16]	3882	12- 8.83 [53/64]
3403	11- 1.98 [31/32]	3483	11- 5.13 [1/8]	3563	11- 8.28 [9/32]	3643	11-11.43 [27/64]	3723	12- 2.57 [37/64]	3803	12- 5.72 [23/32]	3883	12- 8.87 [7/8]
3404	11- 2.02 [1/64]	3484	11- 5.17 [11/64]	3564	11- 8.31 [5/16]	3644	11-11.46 [15/32]	3724	12- 2.61 [39/64]	3804	12- 5.76 [49/64]	3884	12- 8.91 [29/32]
3405	11- 2.06 [1/16]	3485	11- 5.20 [13/64]	3565	11- 8.35 [23/64]	3645	11-11.50 [1/2]	3725	12- 2.65 [21/32]	3805	12- 5.80 [51/64]	3885	12- 8.95 [61/64]
3406	11- 2.09 [3/32]	3486	11- 5.24 [1/4]	3566	11- 8.39 [25/64]	3646	11-11.54 [35/64]	3726	12- 2.69 [11/16]	3806	12- 5.84 [27/32]	3886	12- 8.99 [63/64]
3407	11- 2.13 [9/64]	3487	11- 5.28 [9/32]	3567	11- 8.43 [7/16]	3647	11-11.58 [37/64]	3727	12- 2.73 [47/64]	3807	12- 5.88 [7/8]	3887	12- 9.03 [1/32]
3408	11- 2.17 [11/64]	3488	11- 5.32 [21/64]	3568	11- 8.47 [15/32]	3648	11-11.62 [5/8]	3728	12- 2.77 [49/64]	3808	12- 5.92 [59/64]	3888	12- 9.07 [5/64]
3409	11- 2.21 [7/32]	3489	11- 5.36 [23/64]	3569	11- 8.51 [33/64]	3649	11-11.66 [21/32]	3729	12- 2.81 [13/16]	3809	12- 5.96 [61/64]	3889	12- 9.11 [7/64]
3410	11- 2.25 [1/4]	3490	11- 5.40 [13/32]	3570	11- 8.55 [35/64]	3650	11-11.70 [45/64]	3730	12- 2.85 [27/32]	3810	12- 6.00	3890	12- 9.15 [5/32]
3411	11- 2.29 [19/64]	3491	11- 5.44 [7/16]	3571	11- 8.59 [19/32]	3651	11-11.74 [47/64]	3731	12- 2.89 [57/64]	3811	12- 6.04 [3/64]	3891	12- 9.19 [3/16]
3412	11- 2.33 [21/64]	3492	11- 5.48 [31/64]	3572	11- 8.63 [5/8]	3652	11-11.78 [25/32]	3732	12- 2.93 [59/64]	3812	12- 6.08 [5/64]	3892	12- 9.23 [15/64]
3413	11- 2.37 [3/8]	3493	11- 5.52 [33/64]	3573	11- 8.67 [43/64]	3653	11-11.82 [13/16]	3733	12- 2.97 [31/32]	3813	12- 6.12 [1/8]	3893	12- 9.27 [17/64]
3414	11- 2.41 [13/32]	3494	11- 5.56 [9/16]	3574	11- 8.71 [45/64]	3654	11-11.86 [55/64]	3734	12- 3.01 [1/64]	3814	12- 6.16 [5/32]	3894	12- 9.31 [5/16]
3415	11- 2.45 [29/64]	3495	11- 5.60 [19/32]	3575	11- 8.75 [3/4]	3655	11-11.90 [57/64]	3735	12- 3.05 [3/64]	3815	12- 6.20 [13/64]	3895	12- 9.35 [11/32]
3416	11- 2.49 [31/64]	3496	11- 5.64 [41/64]	3576	11- 8.79 [25/32]	3656	11-11.94 [15/16]	3736	12- 3.09 [3/32]	3816	12- 6.24 [15/64]	3896	12- 9.39 [25/64]
3417	11- 2.53 [17/32]	3497	11- 5.68 [43/64]	3577	11- 8.83 [53/64]	3657	11-11.98 [31/32]	3737	12- 3.13 [1/8]	3817	12- 6.28 [9/32]	3897	12- 9.43 [27/64]
3418	11- 2.57 [9/16]	3498	11- 5.72 [23/32]	3578	11- 8.87 [55/64]	3658	12- 0.02 [1/64]	3738	12- 3.17 [11/64]	3818	12- 6.31 [5/16]	3898	12- 9.46 [15/32]
3419	11- 2.61 [39/64]	3499	11- 5.76 [3/4]	3579	11- 8.91 [29/32]	3659	12- 0.06 [1/16]	3739	12- 3.20 [13/64]	3819	12- 6.35 [23/64]	3899	12- 9.50 [1/2]
3420	11- 2.65 [41/64]	3500	11- 5.80 [51/64]	3580	11- 8.94 [15/16]	3660	12- 0.09 [3/32]	3740	12- 3.24 [1/4]	3820	12- 6.39 [25/64]	3900	12- 9.54 [35/64]
3421	11- 2.69 [11/16]	3501	11- 5.83 [53/64]	3581	11- 8.98 [63/64]	3661	12- 0.13 [9/64]	3741	12- 3.28 [9/32]	3821	12- 6.43 [7/16]	3901	12- 9.58 [37/64]
3422	11- 2.72 [23/32]	3502	11- 5.87 [7/8]	3582	11- 9.02 [1/32]	3662	12- 0.17 [11/64]	3742	12- 3.32 [21/64]	3822	12- 6.47 [15/32]	3902	12- 9.62 [5/8]
3423	11- 2.76 [49/64]	3503	11- 5.91 [29/32]	3583	11- 9.06 [1/16]	3663	12- 0.21 [7/32]	3743	12- 3.36 [23/64]	3823	12- 6.51 [33/64]	3903	12- 9.66 [21/32]
3424	11- 2.80 [51/64]	3504	11- 5.95 [61/64]	3584	11- 9.10 [7/64]	3664	12- 0.25 [1/4]	3744	12- 3.40 [13/32]	3824	12- 6.55 [35/64]	3904	12- 9.70 [45/64]
3425	11- 2.84 [27/32]	3505	11- 5.99 [63/64]	3585	11- 9.14 [9/64]	3665	12- 0.29 [19/64]	3745	12- 3.44 [7/16]	3825	12- 6.59 [19/32]	3905	12- 9.74 [47/64]
3426	11- 2.88 [7/8]	3506	11- 6.03 [1/32]	3586	11- 9.18 [3/16]	3666	12- 0.33 [21/64]	3746	12- 3.48 [31/64]	3826	12- 6.63 [5/8]	3906	12- 9.78 [25/32]
3427	11- 2.92 [59/64]	3507	11- 6.07 [5/64]	3587	11- 9.22 [7/32]	3667	12- 0.37 [3/8]	3747	12- 3.52 [33/64]	3827	12- 6.67 [43/64]	3907	12- 9.82 [13/16]
3428	11- 2.96 [61/64]	3508	11- 6.11 [7/64]	3588	11- 9.26 [17/64]	3668	12- 0.41 [13/32]	3748	12- 3.56 [9/16]	3828	12- 6.71 [45/64]	3908	12- 9.86 [55/64]
3429	11- 3.00	3509	11- 6.15 [5/32]	3589	11- 9.30 [19/64]	3669	12- 0.45 [29/64]	3749	12- 3.60 [19/32]	3829	12- 6.75 [3/4]	3909	12- 9.90 [57/64]
3430	11- 3.04 [3/64]	3510	11- 6.19 [3/16]	3590	11- 9.34 [11/32]	3670	12- 0.49 [31/64]	3750	12- 3.64 [41/64]	3830	12- 6.79 [25/32]	3910	12- 9.94 [15/16]
3431	11- 3.08 [5/64]	3511	11- 6.23 [15/64]	3591	11- 9.38 [3/8]	3671	12- 0.53 [17/32]	3751	12- 3.68 [43/64]	3831	12- 6.83 [53/64]	3911	12- 9.98 [31/32]
3432	11- 3.12 [1/8]	3512	11- 6.27 [17/64]	3592	11- 9.42 [27/64]	3672	12- 0.57 [9/16]	3752	12- 3.72 [23/32]	3832	12- 6.87 [55/64]	3912	12-10.02 [1/64]
3433	11- 3.16 [5/32]	3513	11- 6.31 [5/16]	3593	11- 9.46 [29/64]	3673	12- 0.61 [39/64]	3753	12- 3.76 [3/4]	3833	12- 6.91 [29/32]	3913	12-10.06 [1/16]
3434	11- 3.20 [13/64]	3514	11- 6.35 [11/32]	3594	11- 9.50 [1/2]	3674	12- 0.65 [41/64]	3754	12- 3.80 [51/64]	3834	12- 6.94 [15/16]	3914	12-10.09 [3/32]
3435	11- 3.24 [15/64]	3515	11- 6.39 [25/64]	3595	11- 9.54 [17/32]	3675	12- 0.69 [11/16]	3755	12- 3.83 [53/64]	3835	12- 6.98 [63/64]	3915	12-10.13 [9/64]
3436	11- 3.28 [9/32]	3516	11- 6.43 [27/64]	3596	11- 9.57 [37/64]	3676	12- 0.72 [23/32]	3756	12- 3.87 [7/8]	3836	12- 7.02 [1/32]	3916	12-10.17 [11/64]
3437	11- 3.31 [5/16]	3517	11- 6.46 [15/32]	3597	11- 9.61 [39/64]	3677	12- 0.76 [49/64]	3757	12- 3.91 [29/32]	3837	12- 7.06 [1/16]	3917	12-10.21 [7/32]
3438	11- 3.35 [23/64]	3518	11- 6.50 [1/2]	3598	11- 9.65 [21/32]	3678	12- 0.80 [51/64]	3758	12- 3.95 [61/64]	3838	12- 7.10 [7/64]	3918	12-10.25 [1/4]
3439	11- 3.39 [25/64]	3519	11- 6.54 [35/64]	3599	11- 9.69 [11/16]	3679	12- 0.84 [27/32]	3759	12- 3.99 [63/64]	3839	12- 7.14 [9/64]	3919	12-10.29 [19/64]
3440	11- 3.43 [7/16]	3520	11- 6.58 [37/64]	3600	11- 9.73 [47/64]	3680	12- 0.88 [7/8]	3760	12- 4.03 [1/32]	3840	12- 7.18 [3/16]	3920	12-10.33 [21/64]

MILLIMETERS CONVERTED TO FEET AND INCHES

TABLES M-3

mm	ft-in.(fraction)	mm	ft-in.(fraction)	mm	ft-in.(fraction)	mm	ft-in.(fraction)	mm	ft-in.(fraction)	mm	ft-in.(fraction)	mm	ft-in.(fraction)
3921	12-10.37 [3/8]	4001	13- 1.52 [33/64]	4081	13- 4.67 [43/64]	4161	13- 7.82 [13/16]	4241	13-10.97 [31/32]	4321	14- 2.12 [1/8]	4401	14- 5.27 [17/64]
3922	12-10.41 [13/32]	4002	13- 1.56 [9/16]	4082	13- 4.71 [45/64]	4162	13- 7.86 [55/64]	4242	13-11.01 [1/64]	4322	14- 2.16 [5/32]	4402	14- 5.31 [5/16]
3923	12-10.45 [29/64]	4003	13- 1.60 [19/32]	4083	13- 4.75 [3/4]	4163	13- 7.90 [57/64]	4243	13-11.05 [3/64]	4323	14- 2.20 [13/64]	4403	14- 5.35 [11/32]
3924	12-10.49 [31/64]	4004	13- 1.64 [41/64]	4084	13- 4.79 [25/32]	4164	13- 7.94 [15/16]	4244	13-11.09 [3/32]	4324	14- 2.24 [15/64]	4404	14- 5.39 [25/64]
3925	12-10.53 [17/32]	4005	13- 1.68 [43/64]	4085	13- 4.83 [53/64]	4165	13- 7.98 [31/32]	4245	13-11.13 [1/8]	4325	14- 2.28 [9/32]	4405	14- 5.43 [27/64]
3926	12-10.57 [9/16]	4006	13- 1.72 [23/32]	4086	13- 4.87 [55/64]	4166	13- 8.02 [1/64]	4246	13-11.17 [11/64]	4326	14- 2.31 [5/16]	4406	14- 5.46 [15/32]
3927	12-10.61 [39/64]	4007	13- 1.76 [3/4]	4087	13- 4.91 [29/32]	4167	13- 8.06 [1/16]	4247	13-11.20 [13/64]	4327	14- 2.35 [23/64]	4407	14- 5.50 [1/2]
3928	12-10.65 [41/64]	4008	13- 1.80 [51/64]	4088	13- 4.94 [15/16]	4168	13- 8.09 [3/32]	4248	13-11.24 [1/4]	4328	14- 2.39 [25/64]	4408	14- 5.54 [35/64]
3929	12-10.69 [11/16]	4009	13- 1.83 [53/64]	4089	13- 4.98 [63/64]	4169	13- 8.13 [9/64]	4249	13-11.28 [9/32]	4329	14- 2.43 [7/16]	4409	14- 5.58 [37/64]
3930	12-10.72 [23/32]	4010	13- 1.87 [7/8]	4090	13- 5.02 [1/32]	4170	13- 8.17 [11/64]	4250	13-11.32 [21/64]	4330	14- 2.47 [15/32]	4410	14- 5.62 [5/8]
3931	12-10.76 [49/64]	4011	13- 1.91 [29/32]	4091	13- 5.06 [1/16]	4171	13- 8.21 [7/32]	4251	13-11.36 [23/64]	4331	14- 2.51 [33/64]	4411	14- 5.66 [21/32]
3932	12-10.80 [51/64]	4012	13- 1.95 [61/64]	4092	13- 5.10 [7/64]	4172	13- 8.25 [1/4]	4252	13-11.40 [13/32]	4332	14- 2.55 [35/64]	4412	14- 5.70 [45/64]
3933	12-10.84 [27/32]	4013	13- 1.99 [63/64]	4093	13- 5.14 [9/64]	4173	13- 8.29 [19/64]	4253	13-11.44 [7/16]	4333	14- 2.59 [19/32]	4413	14- 5.74 [47/64]
3934	12-10.88 [7/8]	4014	13- 2.03 [1/32]	4094	13- 5.18 [3/16]	4174	13- 8.33 [21/64]	4254	13-11.48 [31/64]	4334	14- 2.63 [5/8]	4414	14- 5.78 [25/32]
3935	12-10.92 [59/64]	4015	13- 2.07 [5/64]	4095	13- 5.22 [7/32]	4175	13- 8.37 [3/8]	4255	13-11.52 [33/64]	4335	14- 2.67 [43/64]	4415	14- 5.82 [13/16]
3936	12-10.96 [61/64]	4016	13- 2.11 [7/64]	4096	13- 5.26 [17/64]	4176	13- 8.41 [13/32]	4256	13-11.56 [9/16]	4336	14- 2.71 [45/64]	4416	14- 5.86 [55/64]
3937	12-11.00	4017	13- 2.15 [5/32]	4097	13- 5.30 [19/64]	4177	13- 8.45 [29/64]	4257	13-11.60 [19/32]	4337	14- 2.75 [3/4]	4417	14- 5.90 [57/64]
3938	12-11.04 [3/64]	4018	13- 2.19 [3/16]	4098	13- 5.34 [11/32]	4178	13- 8.49 [31/64]	4258	13-11.64 [41/64]	4338	14- 2.79 [25/32]	4418	14- 5.94 [15/16]
3939	12-11.08 [5/64]	4019	13- 2.23 [15/64]	4099	13- 5.38 [3/8]	4179	13- 8.53 [17/32]	4259	13-11.68 [43/64]	4339	14- 2.83 [53/64]	4419	14- 5.98 [31/32]
3940	12-11.12 [1/8]	4020	13- 2.27 [17/64]	4100	13- 5.42 [27/64]	4180	13- 8.57 [9/16]	4260	13-11.72 [23/32]	4340	14- 2.87 [55/64]	4420	14- 6.02 [1/64]
3941	12-11.16 [5/32]	4021	13- 2.31 [5/16]	4101	13- 5.46 [29/64]	4181	13- 8.61 [39/64]	4261	13-11.76 [3/4]	4341	14- 2.91 [29/32]	4421	14- 6.06 [1/16]
3942	12-11.20 [13/64]	4022	13- 2.35 [11/32]	4102	13- 5.50 [1/2]	4182	13- 8.65 [41/64]	4262	13-11.80 [51/64]	4342	14- 2.94 [15/16]	4422	14- 6.09 [3/32]
3943	12-11.24 [15/64]	4023	13- 2.39 [25/64]	4103	13- 5.54 [17/32]	4183	13- 8.69 [11/16]	4263	13-11.83 [53/64]	4343	14- 2.98 [63/64]	4423	14- 6.13 [9/64]
3944	12-11.28 [9/32]	4024	13- 2.43 [27/64]	4104	13- 5.57 [37/64]	4184	13- 8.72 [23/32]	4264	13-11.87 [7/8]	4344	14- 3.02 [1/32]	4424	14- 6.17 [11/64]
3945	12-11.31 [5/16]	4025	13- 2.46 [15/32]	4105	13- 5.61 [39/64]	4185	13- 8.76 [49/64]	4265	13-11.91 [29/32]	4345	14- 3.06 [1/16]	4425	14- 6.21 [7/32]
3946	12-11.35 [23/64]	4026	13- 2.50 [1/2]	4106	13- 5.65 [21/32]	4186	13- 8.80 [51/64]	4266	13-11.95 [61/64]	4346	14- 3.10 [7/64]	4426	14- 6.25 [1/4]
3947	12-11.39 [25/64]	4027	13- 2.54 [35/64]	4107	13- 5.69 [11/16]	4187	13- 8.84 [27/32]	4267	13-11.99 [63/64]	4347	14- 3.14 [9/64]	4427	14- 6.29 [19/64]
3948	12-11.43 [7/16]	4028	13- 2.58 [37/64]	4108	13- 5.73 [47/64]	4188	13- 8.88 [7/8]	4268	14- 0.03 [1/32]	4348	14- 3.18 [3/16]	4428	14- 6.33 [21/64]
3949	12-11.47 [15/32]	4029	13- 2.62 [5/8]	4109	13- 5.77 [49/64]	4189	13- 8.92 [59/64]	4269	14- 0.07 [5/64]	4349	14- 3.22 [7/32]	4429	14- 6.37 [3/8]
3950	12-11.51 [33/64]	4030	13- 2.66 [21/32]	4110	13- 5.81 [13/16]	4190	13- 8.96 [61/64]	4270	14- 0.11 [7/64]	4350	14- 3.26 [17/64]	4430	14- 6.41 [13/32]
3951	12-11.55 [35/64]	4031	13- 2.70 [45/64]	4111	13- 5.85 [27/32]	4191	13- 9.00	4271	14- 0.15 [5/32]	4351	14- 3.30 [19/64]	4431	14- 6.45 [29/64]
3952	12-11.59 [19/32]	4032	13- 2.74 [47/64]	4112	13- 5.89 [57/64]	4192	13- 9.04 [3/64]	4272	14- 0.19 [3/16]	4352	14- 3.34 [11/32]	4432	14- 6.49 [31/64]
3953	12-11.63 [5/8]	4033	13- 2.78 [25/32]	4113	13- 5.93 [59/64]	4193	13- 9.08 [5/64]	4273	14- 0.23 [15/64]	4353	14- 3.38 [3/8]	4433	14- 6.53 [17/32]
3954	12-11.67 [43/64]	4034	13- 2.82 [13/16]	4114	13- 5.97 [31/32]	4194	13- 9.12 [1/8]	4274	14- 0.27 [17/64]	4354	14- 3.42 [27/64]	4434	14- 6.57 [9/16]
3955	12-11.71 [45/64]	4035	13- 2.86 [55/64]	4115	13- 6.01 [1/64]	4195	13- 9.16 [5/32]	4275	14- 0.31 [5/16]	4355	14- 3.46 [29/64]	4435	14- 6.61 [39/64]
3956	12-11.75 [3/4]	4036	13- 2.90 [57/64]	4116	13- 6.05 [3/64]	4196	13- 9.20 [13/64]	4276	14- 0.35 [11/32]	4356	14- 3.50 [1/2]	4436	14- 6.65 [41/64]
3957	12-11.79 [25/32]	4037	13- 2.94 [15/16]	4117	13- 6.09 [3/32]	4197	13- 9.24 [15/64]	4277	14- 0.39 [25/64]	4357	14- 3.54 [17/32]	4437	14- 6.69 [11/16]
3958	12-11.83 [53/64]	4038	13- 2.98 [31/32]	4118	13- 6.13 [1/8]	4198	13- 9.28 [9/32]	4278	14- 0.43 [27/64]	4358	14- 3.57 [37/64]	4438	14- 6.72 [23/32]
3959	12-11.87 [55/64]	4039	13- 3.02 [1/64]	4119	13- 6.17 [11/64]	4199	13- 9.31 [5/16]	4279	14- 0.46 [15/32]	4359	14- 3.61 [39/64]	4439	14- 6.76 [49/64]
3960	12-11.91 [29/32]	4040	13- 3.06 [1/16]	4120	13- 6.20 [13/64]	4200	13- 9.35 [23/64]	4280	14- 0.50 [1/2]	4360	14- 3.65 [21/32]	4440	14- 6.80 [51/64]
3961	12-11.94 [15/16]	4041	13- 3.09 [3/32]	4121	13- 6.24 [1/4]	4201	13- 9.39 [25/64]	4281	14- 0.54 [35/64]	4361	14- 3.69 [11/16]	4441	14- 6.84 [27/32]
3962	12-11.98 [63/64]	4042	13- 3.13 [9/64]	4122	13- 6.28 [9/32]	4202	13- 9.43 [7/16]	4282	14- 0.58 [37/64]	4362	14- 3.73 [47/64]	4442	14- 6.88 [7/8]
3963	13- 0.02 [1/32]	4043	13- 3.17 [11/64]	4123	13- 6.32 [21/64]	4203	13- 9.47 [15/32]	4283	14- 0.62 [5/8]	4363	14- 3.77 [49/64]	4443	14- 6.92 [59/64]
3964	13- 0.06 [1/16]	4044	13- 3.21 [7/32]	4124	13- 6.36 [23/64]	4204	13- 9.51 [33/64]	4284	14- 0.66 [21/32]	4364	14- 3.81 [13/16]	4444	14- 6.96 [61/64]
3965	13- 0.10 [7/64]	4045	13- 3.25 [1/4]	4125	13- 6.40 [13/32]	4205	13- 9.55 [35/64]	4285	14- 0.70 [45/64]	4365	14- 3.85 [27/32]	4445	14- 7.00
3966	13- 0.14 [9/64]	4046	13- 3.29 [19/64]	4126	13- 6.44 [7/16]	4206	13- 9.59 [19/32]	4286	14- 0.74 [47/64]	4366	14- 3.89 [57/64]	4446	14- 7.04 [3/64]
3967	13- 0.18 [3/16]	4047	13- 3.33 [21/64]	4127	13- 6.48 [31/64]	4207	13- 9.63 [5/8]	4287	14- 0.78 [25/32]	4367	14- 3.93 [59/64]	4447	14- 7.08 [5/64]
3968	13- 0.22 [7/32]	4048	13- 3.37 [3/8]	4128	13- 6.52 [33/64]	4208	13- 9.67 [43/64]	4288	14- 0.82 [13/16]	4368	14- 3.97 [31/32]	4448	14- 7.12 [1/8]
3969	13- 0.26 [17/64]	4049	13- 3.41 [13/32]	4129	13- 6.56 [9/16]	4209	13- 9.71 [45/64]	4289	14- 0.86 [55/64]	4369	14- 4.01 [1/64]	4449	14- 7.16 [5/32]
3970	13- 0.30 [19/64]	4050	13- 3.45 [29/64]	4130	13- 6.60 [19/32]	4210	13- 9.75 [3/4]	4290	14- 0.90 [57/64]	4370	14- 4.05 [3/64]	4450	14- 7.20 [13/64]
3971	13- 0.34 [11/32]	4051	13- 3.49 [31/64]	4131	13- 6.64 [41/64]	4211	13- 9.79 [25/32]	4291	14- 0.94 [15/16]	4371	14- 4.09 [3/32]	4451	14- 7.24 [15/64]
3972	13- 0.38 [3/8]	4052	13- 3.53 [17/32]	4132	13- 6.68 [43/64]	4212	13- 9.83 [53/64]	4292	14- 0.98 [31/32]	4372	14- 4.13 [1/8]	4452	14- 7.28 [9/32]
3973	13- 0.42 [27/64]	4053	13- 3.57 [9/16]	4133	13- 6.72 [23/32]	4213	13- 9.87 [55/64]	4293	14- 1.02 [1/64]	4373	14- 4.17 [11/64]	4453	14- 7.31 [5/16]
3974	13- 0.46 [29/64]	4054	13- 3.61 [39/64]	4134	13- 6.76 [3/4]	4214	13- 9.91 [29/32]	4294	14- 1.06 [1/16]	4374	14- 4.20 [13/64]	4454	14- 7.35 [23/64]
3975	13- 0.50 [1/2]	4055	13- 3.65 [41/64]	4135	13- 6.80 [51/64]	4215	13- 9.94 [15/16]	4295	14- 1.09 [3/32]	4375	14- 4.24 [1/4]	4455	14- 7.39 [25/64]
3976	13- 0.54 [17/32]	4056	13- 3.69 [11/16]	4136	13- 6.83 [53/64]	4216	13- 9.98 [63/64]	4296	14- 1.13 [9/64]	4376	14- 4.28 [9/32]	4456	14- 7.43 [7/16]
3977	13- 0.57 [37/64]	4057	13- 3.72 [23/32]	4137	13- 6.87 [7/8]	4217	13-10.02 [1/32]	4297	14- 1.17 [11/64]	4377	14- 4.32 [21/64]	4457	14- 7.47 [15/32]
3978	13- 0.61 [39/64]	4058	13- 3.76 [49/64]	4138	13- 6.91 [29/32]	4218	13-10.06 [1/16]	4298	14- 1.21 [7/32]	4378	14- 4.36 [23/64]	4458	14- 7.51 [33/64]
3979	13- 0.65 [21/32]	4059	13- 3.80 [51/64]	4139	13- 6.95 [61/64]	4219	13-10.10 [7/64]	4299	14- 1.25 [1/4]	4379	14- 4.40 [13/32]	4459	14- 7.55 [35/64]
3980	13- 0.69 [11/16]	4060	13- 3.84 [27/32]	4140	13- 6.99 [63/64]	4220	13-10.14 [9/64]	4300	14- 1.29 [19/64]	4380	14- 4.44 [7/16]	4460	14- 7.59 [19/32]
3981	13- 0.73 [47/64]	4061	13- 3.88 [7/8]	4141	13- 7.03 [1/32]	4221	13-10.18 [3/16]	4301	14- 1.33 [21/64]	4381	14- 4.48 [31/64]	4461	14- 7.63 [5/8]
3982	13- 0.77 [49/64]	4062	13- 3.92 [59/64]	4142	13- 7.07 [5/64]	4222	13-10.22 [7/32]	4302	14- 1.37 [3/8]	4382	14- 4.52 [33/64]	4462	14- 7.67 [43/64]
3983	13- 0.81 [13/16]	4063	13- 3.96 [61/64]	4143	13- 7.11 [7/64]	4223	13-10.26 [17/64]	4303	14- 1.41 [13/32]	4383	14- 4.56 [9/16]	4463	14- 7.71 [45/64]
3984	13- 0.85 [27/32]	4064	13- 4.00	4144	13- 7.15 [5/32]	4224	13-10.30 [19/64]	4304	14- 1.45 [29/64]	4384	14- 4.60 [19/32]	4464	14- 7.75 [3/4]
3985	13- 0.89 [57/64]	4065	13- 4.04 [3/64]	4145	13- 7.19 [3/16]	4225	13-10.34 [11/32]	4305	14- 1.49 [31/64]	4385	14- 4.64 [41/64]	4465	14- 7.79 [25/32]
3986	13- 0.93 [59/64]	4066	13- 4.08 [5/64]	4146	13- 7.23 [15/64]	4226	13-10.38 [3/8]	4306	14- 1.53 [17/32]	4386	14- 4.68 [43/64]	4466	14- 7.83 [53/64]
3987	13- 0.97 [31/32]	4067	13- 4.12 [1/8]	4147	13- 7.27 [17/64]	4227	13-10.42 [27/64]	4307	14- 1.57 [9/16]	4387	14- 4.72 [23/32]	4467	14- 7.87 [55/64]
3988	13- 1.01 [1/64]	4068	13- 4.16 [5/32]	4148	13- 7.31 [5/16]	4228	13-10.46 [29/64]	4308	14- 1.61 [39/64]	4388	14- 4.76 [3/4]	4468	14- 7.91 [29/32]
3989	13- 1.05 [3/64]	4069	13- 4.20 [13/64]	4149	13- 7.35 [11/32]	4229	13-10.50 [1/2]	4309	14- 1.65 [41/64]	4389	14- 4.80 [51/64]	4469	14- 7.94 [15/16]
3990	13- 1.09 [3/32]	4070	13- 4.24 [15/64]	4150	13- 7.39 [25/64]	4230	13-10.54 [17/32]	4310	14- 1.69 [11/16]	4390	14- 4.83 [53/64]	4470	14- 7.98 [63/64]
3991	13- 1.13 [1/8]	4071	13- 4.28 [9/32]	4151	13- 7.43 [27/64]	4231	13-10.57 [37/64]	4311	14- 1.72 [23/32]	4391	14- 4.87 [7/8]	4471	14- 8.02 [1/32]
3992	13- 1.17 [11/64]	4072	13- 4.31 [5/16]	4152	13- 7.46 [15/32]	4232	13-10.61 [39/64]	4312	14- 1.76 [49/64]	4392	14- 4.91 [29/32]	4472	14- 8.06 [1/16]
3993	13- 1.20 [13/64]	4073	13- 4.35 [23/64]	4153	13- 7.50 [1/2]	4233	13-10.65 [21/32]	4313	14- 1.80 [51/64]	4393	14- 4.95 [61/64]	4473	14- 8.10 [7/64]
3994	13- 1.24 [1/4]	4074	13- 4.39 [25/64]	4154	13- 7.54 [35/64]	4234	13-10.69 [11/16]	4314	14- 1.84 [27/32]	4394	14- 4.99 [63/64]	4474	14- 8.14 [9/64]
3995	13- 1.28 [9/32]	4075	13- 4.43 [7/16]	4155	13- 7.58 [37/64]	4235	13-10.73 [47/64]	4315	14- 1.88 [7/8]	4395	14- 5.03 [1/32]	4475	14- 8.18 [3/16]
3996	13- 1.32 [21/64]	4076	13- 4.47 [15/32]	4156	13- 7.62 [5/8]	4236	13-10.77 [49/64]	4316	14- 1.92 [59/64]	4396	14- 5.07 [5/64]	4476	14- 8.22 [7/32]
3997	13- 1.36 [23/64]	4077	13- 4.51 [33/64]	4157	13- 7.66 [21/32]	4237	13-10.81 [13/16]	4317	14- 1.96 [61/64]	4397	14- 5.11 [7/64]	4477	14- 8.26 [17/64]
3998	13- 1.40 [13/32]	4078	13- 4.55 [35/64]	4158	13- 7.70 [45/64]	4238	13-10.85 [27/32]	4318	14- 2.00	4398	14- 5.15 [5/32]	4478	14- 8.30 [19/64]
3999	13- 1.44 [7/16]	4079	13- 4.59 [19/32]	4159	13- 7.74 [47/64]	4239	13-10.89 [57/64]	4319	14- 2.04 [3/64]	4399	14- 5.19 [3/16]	4479	14- 8.34 [11/32]
4000	13- 1.48 [31/64]	4080	13- 4.63 [5/8]	4160	13- 7.78 [25/32]	4240	13-10.93 [59/64]	4320	14- 2.08 [5/64]	4400	14- 5.23 [15/64]	4480	14- 8.38 [3/8]

MILLIMETERS CONVERTED TO FEET AND INCHES

TABLES M-3

mm	ft-in.(fraction)	mm	ft-in.(fraction)	mm	ft-in.(fraction)	mm	ft-in.(fraction)	mm	ft-in.(fraction)	mm	ft-in.(fraction)	mm	ft-in.(fraction)
4481	14- 8.42 [27/64]	4561	14-11.57 [9/16]	4641	15- 2.72 [23/32]	4721	15- 5.87 [55/64]	4801	15- 9.02 [1/64]	4881	16- 0.17 [11/64]	4961	16- 3.31 [5/16]
4482	14- 8.46 [29/64]	4562	14-11.61 [39/64]	4642	15- 2.76 [3/4]	4722	15- 5.91 [29/32]	4802	15- 9.06 [1/16]	4882	16- 0.20 [13/64]	4962	16- 3.35 [23/64]
4483	14- 8.50 [1/2]	4563	14-11.65 [41/64]	4643	15- 2.80 [51/64]	4723	15- 5.94 [15/16]	4803	15- 9.09 [3/32]	4883	16- 0.24 [1/4]	4963	16- 3.39 [25/64]
4484	14- 8.54 [17/32]	4564	14-11.69 [11/16]	4644	15- 2.83 [53/64]	4724	15- 5.98 [63/64]	4804	15- 9.13 [9/64]	4884	16- 0.28 [9/32]	4964	16- 3.43 [7/16]
4485	14- 8.57 [37/64]	4565	14-11.72 [23/32]	4645	15- 2.87 [7/8]	4725	15- 6.02 [1/32]	4805	15- 9.17 [11/64]	4885	16- 0.32 [21/64]	4965	16- 3.47 [15/32]
4486	14- 8.61 [39/64]	4566	14-11.76 [49/64]	4646	15- 2.91 [29/32]	4726	15- 6.06 [1/16]	4806	15- 9.21 [7/32]	4886	16- 0.36 [23/64]	4966	16- 3.51 [33/64]
4487	14- 8.65 [21/32]	4567	14-11.80 [51/64]	4647	15- 2.95 [61/64]	4727	15- 6.10 [7/64]	4807	15- 9.25 [1/4]	4887	16- 0.40 [13/32]	4967	16- 3.55 [35/64]
4488	14- 8.69 [11/16]	4568	14-11.84 [27/32]	4648	15- 2.99 [63/64]	4728	15- 6.14 [9/64]	4808	15- 9.29 [19/64]	4888	16- 0.44 [7/16]	4968	16- 3.59 [19/32]
4489	14- 8.73 [47/64]	4569	14-11.88 [7/8]	4649	15- 3.03 [1/32]	4729	15- 6.18 [3/16]	4809	15- 9.33 [21/64]	4889	16- 0.48 [31/64]	4969	16- 3.63 [5/8]
4490	14- 8.77 [49/64]	4570	14-11.92 [59/64]	4650	15- 3.07 [5/64]	4730	15- 6.22 [7/32]	4810	15- 9.37 [3/8]	4890	16- 0.52 [33/64]	4970	16- 3.67 [43/64]
4491	14- 8.81 [13/16]	4571	14-11.96 [61/64]	4651	15- 3.11 [7/64]	4731	15- 6.26 [17/64]	4811	15- 9.41 [13/32]	4891	16- 0.56 [9/16]	4971	16- 3.71 [45/64]
4492	14- 8.85 [27/32]	4572	15- 0.00	4652	15- 3.15 [5/32]	4732	15- 6.30 [19/64]	4812	15- 9.45 [29/64]	4892	16- 0.60 [19/32]	4972	16- 3.75 [3/4]
4493	14- 8.89 [57/64]	4573	15- 0.04 [3/64]	4653	15- 3.19 [3/16]	4733	15- 6.34 [11/32]	4813	15- 9.49 [31/64]	4893	16- 0.64 [41/64]	4973	16- 3.79 [25/32]
4494	14- 8.93 [59/64]	4574	15- 0.08 [5/64]	4654	15- 3.23 [15/64]	4734	15- 6.38 [3/8]	4814	15- 9.53 [17/32]	4894	16- 0.68 [43/64]	4974	16- 3.83 [53/64]
4495	14- 8.97 [31/32]	4575	15- 0.12 [1/8]	4655	15- 3.27 [17/64]	4735	15- 6.42 [27/64]	4815	15- 9.57 [9/16]	4895	16- 0.72 [23/32]	4975	16- 3.87 [55/64]
4496	14- 9.01 [1/64]	4576	15- 0.16 [5/32]	4656	15- 3.31 [5/16]	4736	15- 6.46 [29/64]	4816	15- 9.61 [39/64]	4896	16- 0.76 [3/4]	4976	16- 3.91 [29/32]
4497	14- 9.05 [3/64]	4577	15- 0.20 [13/64]	4657	15- 3.35 [11/32]	4737	15- 6.50 [1/2]	4817	15- 9.65 [41/64]	4897	16- 0.80 [51/64]	4977	16- 3.94 [15/16]
4498	14- 9.09 [3/32]	4578	15- 0.24 [15/64]	4658	15- 3.39 [25/64]	4738	15- 6.54 [17/32]	4818	15- 9.69 [11/16]	4898	16- 0.83 [53/64]	4978	16- 3.98 [63/64]
4499	14- 9.13 [1/8]	4579	15- 0.28 [9/32]	4659	15- 3.43 [27/64]	4739	15- 6.57 [37/64]	4819	15- 9.72 [23/32]	4899	16- 0.87 [7/8]	4979	16- 4.02 [1/32]
4500	14- 9.17 [11/64]	4580	15- 0.31 [5/16]	4660	15- 3.46 [15/32]	4740	15- 6.61 [39/64]	4820	15- 9.76 [49/64]	4900	16- 0.91 [29/32]	4980	16- 4.06 [1/16]
4501	14- 9.20 [13/64]	4581	15- 0.35 [23/64]	4661	15- 3.50 [1/2]	4741	15- 6.65 [21/32]	4821	15- 9.80 [51/64]	4901	16- 0.95 [61/64]	4981	16- 4.10 [7/64]
4502	14- 9.24 [1/4]	4582	15- 0.39 [25/64]	4662	15- 3.54 [35/64]	4742	15- 6.69 [11/16]	4822	15- 9.84 [27/32]	4902	16- 0.99 [63/64]	4982	16- 4.14 [9/64]
4503	14- 9.28 [9/32]	4583	15- 0.43 [7/16]	4663	15- 3.58 [37/64]	4743	15- 6.73 [47/64]	4823	15- 9.88 [7/8]	4903	16- 1.03 [1/32]	4983	16- 4.18 [3/16]
4504	14- 9.32 [21/64]	4584	15- 0.47 [15/32]	4664	15- 3.62 [5/8]	4744	15- 6.77 [49/64]	4824	15- 9.92 [59/64]	4904	16- 1.07 [5/64]	4984	16- 4.22 [7/32]
4505	14- 9.36 [23/64]	4585	15- 0.51 [33/64]	4665	15- 3.66 [21/32]	4745	15- 6.81 [13/16]	4825	15- 9.96 [61/64]	4905	16- 1.11 [7/64]	4985	16- 4.26 [17/64]
4506	14- 9.40 [13/32]	4586	15- 0.55 [35/64]	4666	15- 3.70 [45/64]	4746	15- 6.85 [27/32]	4826	15-10.00	4906	16- 1.15 [5/32]	4986	16- 4.30 [19/64]
4507	14- 9.44 [7/16]	4587	15- 0.59 [19/32]	4667	15- 3.74 [47/64]	4747	15- 6.89 [57/64]	4827	15-10.04 [3/64]	4907	16- 1.19 [3/16]	4987	16- 4.34 [11/32]
4508	14- 9.48 [31/64]	4588	15- 0.63 [5/8]	4668	15- 3.78 [25/32]	4748	15- 6.93 [59/64]	4828	15-10.08 [5/64]	4908	16- 1.23 [15/64]	4988	16- 4.38 [3/8]
4509	14- 9.52 [33/64]	4589	15- 0.67 [43/64]	4669	15- 3.82 [13/16]	4749	15- 6.97 [31/32]	4829	15-10.12 [1/8]	4909	16- 1.27 [17/64]	4989	16- 4.42 [27/64]
4510	14- 9.56 [9/16]	4590	15- 0.71 [45/64]	4670	15- 3.86 [55/64]	4750	15- 7.01 [1/64]	4830	15-10.16 [5/32]	4910	16- 1.31 [5/16]	4990	16- 4.46 [29/64]
4511	14- 9.60 [19/32]	4591	15- 0.75 [3/4]	4671	15- 3.90 [57/64]	4751	15- 7.05 [3/64]	4831	15-10.20 [13/64]	4911	16- 1.35 [11/32]	4991	16- 4.50 [1/2]
4512	14- 9.64 [41/64]	4592	15- 0.79 [25/32]	4672	15- 3.94 [15/16]	4752	15- 7.09 [3/32]	4832	15-10.24 [15/64]	4912	16- 1.39 [25/64]	4992	16- 4.54 [17/32]
4513	14- 9.68 [43/64]	4593	15- 0.83 [53/64]	4673	15- 3.98 [31/32]	4753	15- 7.13 [1/8]	4833	15-10.28 [9/32]	4913	16- 1.43 [27/64]	4993	16- 4.57 [37/64]
4514	14- 9.72 [23/32]	4594	15- 0.87 [55/64]	4674	15- 4.02 [1/64]	4754	15- 7.17 [11/64]	4834	15-10.31 [5/16]	4914	16- 1.46 [15/32]	4994	16- 4.61 [39/64]
4515	14- 9.76 [3/4]	4595	15- 0.91 [29/32]	4675	15- 4.06 [1/16]	4755	15- 7.20 [13/64]	4835	15-10.35 [23/64]	4915	16- 1.50 [1/2]	4995	16- 4.65 [21/32]
4516	14- 9.80 [51/64]	4596	15- 0.94 [15/16]	4676	15- 4.09 [3/32]	4756	15- 7.24 [1/4]	4836	15-10.39 [25/64]	4916	16- 1.54 [35/64]	4996	16- 4.69 [11/16]
4517	14- 9.83 [53/64]	4597	15- 0.98 [63/64]	4677	15- 4.13 [9/64]	4757	15- 7.28 [9/32]	4837	15-10.43 [7/16]	4917	16- 1.58 [37/64]	4997	16- 4.73 [47/64]
4518	14- 9.87 [7/8]	4598	15- 1.02 [1/32]	4678	15- 4.17 [11/64]	4758	15- 7.32 [21/64]	4838	15-10.47 [15/32]	4918	16- 1.62 [5/8]	4998	16- 4.77 [49/64]
4519	14- 9.91 [29/32]	4599	15- 1.06 [1/16]	4679	15- 4.21 [7/32]	4759	15- 7.36 [23/64]	4839	15-10.51 [33/64]	4919	16- 1.66 [21/32]	4999	16- 4.81 [13/16]
4520	14- 9.95 [61/64]	4600	15- 1.10 [7/64]	4680	15- 4.25 [1/4]	4760	15- 7.40 [13/32]	4840	15-10.55 [35/64]	4920	16- 1.70 [45/64]	5000	16- 4.85 [27/32]
4521	14- 9.99 [63/64]	4601	15- 1.14 [9/64]	4681	15- 4.29 [19/64]	4761	15- 7.44 [7/16]	4841	15-10.59 [19/32]	4921	16- 1.74 [47/64]	5001	16- 4.89 [57/64]
4522	14-10.03 [1/32]	4602	15- 1.18 [3/16]	4682	15- 4.33 [21/64]	4762	15- 7.48 [31/64]	4842	15-10.63 [5/8]	4922	16- 1.78 [25/32]	5002	16- 4.93 [59/64]
4523	14-10.07 [5/64]	4603	15- 1.22 [7/32]	4683	15- 4.37 [3/8]	4763	15- 7.52 [33/64]	4843	15-10.67 [43/64]	4923	16- 1.82 [13/16]	5003	16- 4.97 [31/32]
4524	14-10.11 [7/64]	4604	15- 1.26 [17/64]	4684	15- 4.41 [13/32]	4764	15- 7.56 [9/16]	4844	15-10.71 [45/64]	4924	16- 1.86 [55/64]	5004	16- 5.01 [1/64]
4525	14-10.15 [5/32]	4605	15- 1.30 [19/64]	4685	15- 4.45 [29/64]	4765	15- 7.60 [19/32]	4845	15-10.75 [3/4]	4925	16- 1.90 [57/64]	5005	16- 5.05 [3/64]
4526	14-10.19 [3/16]	4606	15- 1.34 [11/32]	4686	15- 4.49 [31/64]	4766	15- 7.64 [41/64]	4846	15-10.79 [25/32]	4926	16- 1.94 [15/16]	5006	16- 5.09 [3/32]
4527	14-10.23 [15/64]	4607	15- 1.38 [3/8]	4687	15- 4.53 [17/32]	4767	15- 7.68 [43/64]	4847	15-10.83 [53/64]	4927	16- 1.98 [31/32]	5007	16- 5.13 [1/8]
4528	14-10.27 [17/64]	4608	15- 1.42 [27/64]	4688	15- 4.57 [9/16]	4768	15- 7.72 [23/32]	4848	15-10.87 [55/64]	4928	16- 2.02 [1/64]	5008	16- 5.17 [11/64]
4529	14-10.31 [5/16]	4609	15- 1.46 [29/64]	4689	15- 4.61 [39/64]	4769	15- 7.76 [3/4]	4849	15-10.91 [29/32]	4929	16- 2.06 [1/16]	5009	16- 5.20 [13/64]
4530	14-10.35 [11/32]	4610	15- 1.50 [1/2]	4690	15- 4.65 [41/64]	4770	15- 7.80 [51/64]	4850	15-10.94 [15/16]	4930	16- 2.09 [3/32]	5010	16- 5.24 [1/4]
4531	14-10.39 [25/64]	4611	15- 1.54 [17/32]	4691	15- 4.69 [11/16]	4771	15- 7.83 [53/64]	4851	15-10.98 [63/64]	4931	16- 2.13 [9/64]	5011	16- 5.28 [9/32]
4532	14-10.43 [27/64]	4612	15- 1.57 [37/64]	4692	15- 4.72 [23/32]	4772	15- 7.87 [7/8]	4852	15-11.02 [1/32]	4932	16- 2.17 [11/64]	5012	16- 5.32 [21/64]
4533	14-10.46 [15/32]	4613	15- 1.61 [39/64]	4693	15- 4.76 [49/64]	4773	15- 7.91 [29/32]	4853	15-11.06 [1/16]	4933	16- 2.21 [7/32]	5013	16- 5.36 [23/64]
4534	14-10.50 [1/2]	4614	15- 1.65 [21/32]	4694	15- 4.80 [51/64]	4774	15- 7.95 [61/64]	4854	15-11.10 [7/64]	4934	16- 2.25 [1/4]	5014	16- 5.40 [13/32]
4535	14-10.54 [35/64]	4615	15- 1.69 [11/16]	4695	15- 4.84 [27/32]	4775	15- 7.99 [63/64]	4855	15-11.14 [9/64]	4935	16- 2.29 [19/64]	5015	16- 5.44 [7/16]
4536	14-10.58 [37/64]	4616	15- 1.73 [47/64]	4696	15- 4.88 [7/8]	4776	15- 8.03 [1/32]	4856	15-11.18 [3/16]	4936	16- 2.33 [21/64]	5016	16- 5.48 [31/64]
4537	14-10.62 [5/8]	4617	15- 1.77 [49/64]	4697	15- 4.92 [59/64]	4777	15- 8.07 [5/64]	4857	15-11.22 [7/32]	4937	16- 2.37 [3/8]	5017	16- 5.52 [33/64]
4538	14-10.66 [21/32]	4618	15- 1.81 [13/16]	4698	15- 4.96 [61/64]	4778	15- 8.11 [7/64]	4858	15-11.26 [17/64]	4938	16- 2.41 [13/32]	5018	16- 5.56 [9/16]
4539	14-10.70 [45/64]	4619	15- 1.85 [27/32]	4699	15- 5.00	4779	15- 8.15 [5/32]	4859	15-11.30 [19/64]	4939	16- 2.45 [29/64]	5019	16- 5.60 [19/32]
4540	14-10.74 [47/64]	4620	15- 1.89 [57/64]	4700	15- 5.04 [3/64]	4780	15- 8.19 [3/16]	4860	15-11.34 [11/32]	4940	16- 2.49 [31/64]	5020	16- 5.64 [41/64]
4541	14-10.78 [25/32]	4621	15- 1.93 [59/64]	4701	15- 5.08 [5/64]	4781	15- 8.23 [15/64]	4861	15-11.38 [3/8]	4941	16- 2.53 [17/32]	5021	16- 5.68 [43/64]
4542	14-10.82 [13/16]	4622	15- 1.97 [31/32]	4702	15- 5.12 [1/8]	4782	15- 8.27 [17/64]	4862	15-11.42 [27/64]	4942	16- 2.57 [9/16]	5022	16- 5.72 [23/32]
4543	14-10.86 [55/64]	4623	15- 2.01 [1/64]	4703	15- 5.16 [5/32]	4783	15- 8.31 [5/16]	4863	15-11.46 [29/64]	4943	16- 2.61 [39/64]	5023	16- 5.76 [3/4]
4544	14-10.90 [57/64]	4624	15- 2.05 [3/64]	4704	15- 5.20 [13/64]	4784	15- 8.35 [11/32]	4864	15-11.50 [1/2]	4944	16- 2.65 [41/64]	5024	16- 5.80 [51/64]
4545	14-10.94 [15/16]	4625	15- 2.09 [3/32]	4705	15- 5.24 [15/64]	4785	15- 8.39 [25/64]	4865	15-11.54 [17/32]	4945	16- 2.69 [11/16]	5025	16- 5.83 [53/64]
4546	14-10.98 [31/32]	4626	15- 2.13 [1/8]	4706	15- 5.28 [9/32]	4786	15- 8.43 [27/64]	4866	15-11.57 [37/64]	4946	16- 2.72 [23/32]	5026	16- 5.87 [7/8]
4547	14-11.02 [1/64]	4627	15- 2.17 [11/64]	4707	15- 5.31 [5/16]	4787	15- 8.46 [15/32]	4867	15-11.61 [39/64]	4947	16- 2.76 [49/64]	5027	16- 5.91 [29/32]
4548	14-11.06 [1/16]	4628	15- 2.20 [13/64]	4708	15- 5.35 [23/64]	4788	15- 8.50 [1/2]	4868	15-11.65 [21/32]	4948	16- 2.80 [51/64]	5028	16- 5.95 [61/64]
4549	14-11.09 [3/32]	4629	15- 2.24 [1/4]	4709	15- 5.39 [25/64]	4789	15- 8.54 [35/64]	4869	15-11.69 [11/16]	4949	16- 2.84 [27/32]	5029	16- 5.99 [63/64]
4550	14-11.13 [9/64]	4630	15- 2.28 [9/32]	4710	15- 5.43 [7/16]	4790	15- 8.58 [37/64]	4870	15-11.73 [47/64]	4950	16- 2.88 [7/8]	5030	16- 6.03 [1/32]
4551	14-11.17 [11/64]	4631	15- 2.32 [21/64]	4711	15- 5.47 [15/32]	4791	15- 8.62 [5/8]	4871	15-11.77 [49/64]	4951	16- 2.92 [59/64]	5031	16- 6.07 [5/64]
4552	14-11.21 [7/32]	4632	15- 2.36 [23/64]	4712	15- 5.51 [33/64]	4792	15- 8.66 [21/32]	4872	15-11.81 [13/16]	4952	16- 2.96 [61/64]	5032	16- 6.11 [7/64]
4553	14-11.25 [1/4]	4633	15- 2.40 [13/32]	4713	15- 5.55 [35/64]	4793	15- 8.70 [45/64]	4873	15-11.85 [27/32]	4953	16- 3.00	5033	16- 6.15 [5/32]
4554	14-11.29 [19/64]	4634	15- 2.44 [7/16]	4714	15- 5.59 [19/32]	4794	15- 8.74 [47/64]	4874	15-11.89 [57/64]	4954	16- 3.04 [3/64]	5034	16- 6.19 [3/16]
4555	14-11.33 [21/64]	4635	15- 2.48 [31/64]	4715	15- 5.63 [5/8]	4795	15- 8.78 [25/32]	4875	15-11.93 [59/64]	4955	16- 3.08 [5/64]	5035	16- 6.23 [15/64]
4556	14-11.37 [3/8]	4636	15- 2.52 [33/64]	4716	15- 5.67 [43/64]	4796	15- 8.82 [13/16]	4876	15-11.97 [31/32]	4956	16- 3.12 [1/8]	5036	16- 6.27 [17/64]
4557	14-11.41 [13/32]	4637	15- 2.56 [9/16]	4717	15- 5.71 [45/64]	4797	15- 8.86 [55/64]	4877	16- 0.01 [1/64]	4957	16- 3.16 [5/32]	5037	16- 6.31 [5/16]
4558	14-11.45 [29/64]	4638	15- 2.60 [19/32]	4718	15- 5.75 [3/4]	4798	15- 8.90 [57/64]	4878	16- 0.05 [3/64]	4958	16- 3.20 [13/64]	5038	16- 6.35 [11/32]
4559	14-11.49 [31/64]	4639	15- 2.64 [41/64]	4719	15- 5.79 [25/32]	4799	15- 8.94 [15/16]	4879	16- 0.09 [3/32]	4959	16- 3.24 [15/64]	5039	16- 6.39 [25/64]
4560	14-11.53 [17/32]	4640	15- 2.68 [43/64]	4720	15- 5.83 [53/64]	4800	15- 8.98 [31/32]	4880	16- 0.13 [1/8]	4960	16- 3.28 [9/32]	5040	16- 6.43 [27/64]

CIRCLES: DIAMETER, CIRCUMFERENCE & AREA

TABLE M-4

DIAM. IN.	CIRCUM. IN.	AREA SQ. IN.	DIAM. IN.	CIRCUM. IN.	AREA SQ. IN.	DIAM. IN.	CIRCUM. IN.	AREA SQ. IN.	DIAM. IN.	CIRCUM. IN.	AREA SQ. IN.	DIAM. IN.	CIRCUM. IN.	AREA SQ. IN.
1/64	.04909	.00019	2-7/8	9.0321	6.4918	7-5/8	23.955	45.664	21	65.973	346.36	37	116.239	1075.2
1/32	.09818	.00077	2-15/16	9.2284	6.7771	7-3/4	24.347	47.173	21-1/4	66.759	354.66	37-1/4	117.024	1089.8
3/64	.14726	.00173	3	9.4248	7.0686	7-7/8	24.740	48.707	21-1/2	67.544	363.05	37-1/2	117.810	1104.5
1/16	.19635	.00307	3-1/16	9.6211	7.3662	8	25.133	50.265	21-3/4	68.330	371.54	37-3/4	118.596	1119.2
3/32	.29452	.00690	3-1/8	9.8175	7.6699	8-1/8	25.525	51.849	22	69.115	380.13	38	119.381	1134.1
1/8	.39270	.01227	3-3/16	10.014	7.9798	8-1/4	25.918	53.456	22-1/4	69.900	388.82	38-1/4	120.166	1149.1
5/32	.49087	.01917	3-1/4	10.210	8.2958	8-3/8	26.311	55.088	22-1/2	70.686	397.61	38-1/2	120.951	1164.2
3/16	.58905	.02761	3-5/16	10.407	8.6179	8-1/2	26.704	56.745	22-3/4	71.471	406.49	38-3/4	121.737	1179.3
7/32	.68722	.03758	3-3/8	10.603	8.9462	8-5/8	27.096	58.426	23	72.257	415.48	39	122.522	1194.6
1/4	.78540	.04909	3-7/16	10.799	9.2806	8-3/4	27.489	60.132	23-1/4	73.042	424.56	39-1/4	123.308	1210.0
9/32	.88357	.06213	3-1/2	10.996	9.6211	8-7/8	27.882	61.862	23-1/2	73.827	433.74	39-1/2	124.093	1225.4
5/16	.98175	.07670	3-9/16	11.192	9.9678	9	28.274	63.617	23-3/4	74.613	443.01	39-3/4	124.878	1241.0
11/32	1.0799	.09281	3-5/8	11.388	10.321	9-1/8	28.667	65.397	24	75.398	452.39	40	125.664	1256.6
3/8	1.1781	.11045	3-11/16	11.585	10.680	9-1/4	29.060	67.201	24-1/4	76.184	461.86	40-1/4	126.449	1272.4
13/32	1.2763	.12962	3-3/4	11.781	11.045	9-3/8	29.452	69.029	24-1/2	76.969	471.44	40-1/2	127.235	1288.2
7/16	1.3744	.15033	3-13/16	11.977	11.416	9-1/2	29.845	70.882	24-3/4	77.754	481.11	40-3/4	128.020	1304.2
15/32	1.4726	.17257	3-7/8	12.174	11.793	9-5/8	30.238	72.760	25	78.540	490.87	41	128.805	1320.3
1/2	1.5708	.19635	3-15/16	12.370	12.177	9-3/4	30.631	74.662	25-1/4	79.325	500.74	41-1/4	129.591	1336.4
17/32	1.6690	.22166	4	12.566	12.566	9-7/8	31.023	76.589	25-1/2	80.111	510.71	41-1/2	130.376	1352.7
9/16	1.7671	.24850	4-1/16	12.763	12.962	10	31.416	78.540	25-3/4	80.896	520.77	41-3/4	131.161	1369.0
19/32	1.8653	.27688	4-1/8	12.959	13.364	10-1/4	32.201	82.516	26	81.681	530.93	42	131.947	1385.4
5/8	1.9635	.30680	4-3/16	13.155	13.772	10-1/2	32.987	86.590	26-1/4	82.467	541.19	42-1/4	132.732	1402.0
21/32	2.0617	.33824	4-1/4	13.352	14.186	10-3/4	33.772	90.763	26-1/2	83.252	551.55	42-1/2	133.518	1418.6
11/16	2.1598	.37122	4-5/16	13.548	14.607	11	34.558	95.033	26-3/4	84.038	562.00	42-3/4	134.303	1435.4
23/32	2.2580	.40574	4-3/8	13.744	15.033	11-1/4	35.343	99.402	27	84.823	572.56	43	135.088	1452.2
3/4	2.3562	.44179	4-7/16	13.941	15.466	11-1/2	36.128	103.87	27-1/4	85.608	583.21	43-1/4	135.874	1469.1
25/32	2.4544	.47937	4-1/2	14.137	15.904	11-3/4	36.914	108.43	27-1/2	86.394	593.96	43-1/2	136.659	1486.2
13/16	2.5525	.51849	4-9/16	14.334	16.349	12	37.699	113.10	27-3/4	87.179	604.81	43-3/4	137.445	1503.3
27/32	2.6507	.55914	4-5/8	14.530	16.800	12-1/4	38.485	117.86	28	87.965	615.75	44	138.230	1520.5
7/8	2.7489	.60132	4-11/16	14.726	17.257	12-1/2	39.270	122.72	28-1/4	88.750	626.80	44-1/4	139.015	1537.9
29/32	2.8471	.64504	4-3/4	14.923	17.721	12-3/4	40.055	127.68	28-1/2	89.535	637.94	44-1/2	139.801	1555.3
15/16	2.9452	.69029	4-13/16	15.119	18.190	13	40.841	132.73	28-3/4	90.321	649.18	44-3/4	140.586	1572.8
31/32	3.0434	.73708	4-7/8	15.315	18.665	13-1/4	41.626	137.89	29	91.106	660.52	45	141.372	1590.4
1	3.1416	.7854	4-15/16	15.512	19.147	13-1/2	42.412	143.14	29-1/4	91.892	671.96	45-1/4	142.157	1608.2
1-1/16	3.3379	.8866	5	15.708	19.635	13-3/4	43.197	148.49	29-1/2	92.677	683.49	45-1/2	142.942	1626.0
1-1/8	3.5343	.9940	5-1/16	15.904	20.129	14	43.982	153.94	29-3/4	93.462	695.13	45-3/4	143.728	1643.9
1-3/16	3.7306	1.1075	5-1/8	16.101	20.629	14-1/4	44.768	159.48	30	94.248	706.86	46	144.513	1661.9
1-1/4	3.9270	1.2272	5-3/16	16.297	21.135	14-1/2	45.553	165.13	30-1/4	95.033	718.69	46-1/4	145.299	1680.0
1-5/16	4.1233	1.3530	5-1/4	16.493	21.648	14-3/4	46.338	170.87	30-1/2	95.819	730.62	46-1/2	146.084	1698.2
1-3/8	4.3197	1.4849	5-5/16	16.690	22.166	15	47.124	176.71	30-3/4	96.604	742.64	46-3/4	146.869	1716.5
1-7/16	4.5160	1.6230	5-3/8	16.886	22.691	15-1/4	47.909	182.65	31	97.389	754.77	47	147.655	1734.9
1-1/2	4.7124	1.7671	5-7/16	17.082	23.221	15-1/2	48.695	188.69	31-1/4	98.175	766.99	47-1/4	148.440	1753.5
1-9/16	4.9087	1.9175	5-1/2	17.279	23.758	15-3/4	49.480	194.83	31-1/2	98.960	779.31	47-1/2	149.226	1772.1
1-5/8	5.1051	2.0739	5-9/16	17.475	24.301	16	50.265	201.06	31-3/4	99.746	791.73	47-3/4	150.011	1790.8
1-11/16	5.3014	2.2365	5-5/8	17.671	24.850	16-1/4	51.051	207.39	32	100.531	804.25	48	150.796	1809.6
1-3/4	5.4978	2.4053	5-11/16	17.868	25.406	16-1/2	51.836	213.82	32-1/4	101.316	816.86	48-1/4	151.582	1828.5
1-13/16	5.6941	2.5802	5-3/4	18.064	25.967	16-3/4	52.622	220.35	32-1/2	102.102	829.58	48-1/2	152.367	1847.5
1-7/8	5.8905	2.7612	5-13/16	18.261	26.535	17	53.407	226.98	32-3/4	102.887	842.39	48-3/4	153.153	1866.5
1-15/16	6.0868	2.9483	5-7/8	18.457	27.109	17-1/4	54.192	233.71	33	103.673	855.30	49	153.938	1885.7
2	6.2832	3.1416	5-15/16	18.653	27.688	17-1/2	54.978	240.53	33-1/4	104.458	868.31	49-1/4	154.723	1905.0
2-1/16	6.4795	3.3410	6	18.850	28.274	17-3/4	55.763	247.45	33-1/2	105.243	881.41	49-1/2	155.509	1924.4
2-1/8	6.6759	3.5466	6-1/8	19.242	29.465	18	56.549	254.47	33-3/4	106.029	894.62	49-3/4	156.294	1943.9
2-3/16	6.8722	3.7583	6-1/4	19.635	30.680	18-1/4	57.334	261.59	34	106.814	907.92	50	157.080	1963.5
2-1/4	7.0686	3.9761	6-3/8	20.028	31.919	18-1/2	58.119	268.80	34-1/4	107.600	921.32	50-1/4	157.865	1983.2
2-5/16	7.2649	4.2000	6-1/2	20.420	33.183	18-3/4	58.905	276.12	34-1/2	108.385	934.82	50-1/2	158.650	2003.0
2-3/8	7.4613	4.4301	6-5/8	20.813	34.472	19	59.690	283.53	34-3/4	109.170	948.42	50-3/4	159.436	2022.8
2-7/16	7.6576	4.6664	6-3/4	21.206	35.785	19-1/4	60.476	291.04	35	109.956	962.11	51	160.221	2042.8
2-1/2	7.8540	4.9087	6-7/8	21.598	37.122	19-1/2	61.261	298.65	35-1/4	110.741	975.91	51-1/4	161.007	2062.9
2-9/16	8.0503	5.1572	7	21.991	38.485	19-3/4	62.046	306.35	35-1/2	111.527	989.80	51-1/2	161.792	2083.1
2-5/8	8.2467	5.4119	7-1/8	22.384	39.871	20	62.832	314.16	35-3/4	112.312	1003.80	51-3/4	162.577	2103.3
2-11/16	8.4430	5.6727	7-1/4	22.776	41.282	20-1/4	63.617	322.06	36	113.097	1017.90	52	163.363	2123.7
2-3/4	8.6394	5.9396	7-3/8	23.169	42.718	20-1/2	64.403	330.06	36-1/4	113.883	1032.10	52-1/4	164.148	2144.2
2-13/16	8.8357	6.2126	7-1/2	23.562	44.179	20-3/4	65.188	338.16	36-1/2	114.668	1046.30	52-1/2	164.934	2164.8
									36-3/4	115.454	1060.70			

DECIMALS OF AN INCH & OF A FOOT — TABLE M-5

FRAC-TIONS OF AN INCH	DECIMAL EQUIVALENTS	FRAC-TIONS OF A FOOT	FRAC-TIONS OF AN INCH	DECIMAL EQUIVALENTS	FRAC-TIONS OF A FOOT	FRAC-TIONS OF AN INCH	DECIMAL EQUIVALENTS	FRAC-TIONS OF A FOOT	FRAC-TIONS OF AN INCH	DECIMAL EQUIVALENTS	FRAC-TIONS OF A FOOT
1/64	.0052	1/16"	17/64	.2552	3 1/16"	33/64	.5052	6 1/16"	49/64	.7552	9 1/16"
	.0104	1/8		.2604	3 1/8		.5104	6 1/8		.7604	9 1/8
1/32	.015625	3/16	9/32	.265625	3 3/16	17/32	.515625	6 3/16	25/32	.765625	9 3/16
	.0208	1/4		.2708	3 1/4		.5208	6 1/4		.7708	9 1/4
	.0260	5/16		.2760	3 5/16		.5260	6 5/16		.7760	9 5/16
3/64	.03125	3/8	19/64	.28125	3 3/8	35/64	.53125	6 3/8	51/64	.78125	9 3/8
	.0365	7/16		.2865	3 7/16		.5365	6 7/16		.7865	9 7/16
	.0417	1/2		.2917	3 1/2		.5417	6 1/2		.7917	9 1/2
1/16	.046875	9/16	5/16	.296875	3 9/16	9/16	.546875	6 9/16	13/16	.796875	9 9/16
	.0521	5/8		.3021	3 5/8		.5521	6 5/8		.8021	9 5/8
	.0573	11/16		.3073	3 11/16		.5573	6 11/16		.8073	9 11/16
5/64	.0625	3/4	21/64	.3125	3 3/4	37/64	.5625	6 3/4	53/64	.8125	9 3/4
	.0677	13/16		.3177	3 13/16		.5677	6 13/16		.8177	9 13/16
	.0729	7/8		.3229	3 7/8		.5729	6 7/8		.8229	9 7/8
3/32	.078125	15/16	11/32	.328125	3 15/16	19/32	.578125	6 15/16	27/32	.828125	9 15/16
	.0833	1		.3333	4		.5833	7		.8333	10
	.0885	1 1/16		.3385	4 1/16		.5885	7 1/16		.8385	10 1/16
7/64	.09375	1 1/8	23/64	.34375	4 1/8	39/64	.59375	7 1/8	55/64	.84375	10 1/8
	.0990	1 3/16		.3490	4 3/16		.5990	7 3/16		.8490	10 3/16
	.1042	1 1/4		.3542	4 1/4		.6042	7 1/4		.8542	10 1/4
1/8	.109375	1 5/16	3/8	.359375	4 5/16	5/8	.609375	7 5/16	7/8	.859375	10 5/16
	.1146	1 3/8		.3646	4 3/8		.6146	7 3/8		.8646	10 3/8
	.1198	1 7/16		.3698	4 7/16		.6198	7 7/16		.8698	10 7/16
9/64	.1250	1 1/2	25/64	.3750	4 1/2	41/64	.6250	7 1/2	57/64	.8750	10 1/2
	.1302	1 9/16		.3802	4 9/16		.6302	7 9/16		.8802	10 9/16
	.1354	1 5/8		.3854	4 5/8		.6354	7 5/8		.8854	10 5/8
5/32	.140625	1 11/16	13/32	.390625	4 11/16	21/32	.640625	7 11/16	29/32	.890625	10 11/16
	.1458	1 3/4		.3958	4 3/4		.6458	7 3/4		.8958	10 3/4
	.1510	1 13/16		.4010	4 13/16		.6510	7 13/16		.9010	10 13/16
11/64	.15625	1 7/8	27/64	.40625	4 7/8	43/64	.65625	7 7/8	59/64	.90625	10 7/8
	.1615	1 15/16		.4115	4 15/16		.6615	7 15/16		.9115	10 15/16
	.1667	2		.4167	5		.6667	8		.9167	11
3/16	.171875	2 1/16	7/16	.421875	5 1/16	11/16	.671875	8 1/16	15/16	.921875	11 1/16
	.1771	2 1/8		.4271	5 1/8		.6771	8 1/8		.9271	11 1/8
	.1823	2 3/16		.4328	5 3/16		.6823	8 3/16		.9323	11 3/16
13/64	.1875	2 1/4	29/64	.4375	5 1/4	45/64	.6875	8 1/4	61/64	.9375	11 1/4
	.1927	2 5/16		.4427	5 5/16		.6927	8 5/16		.9427	11 5/16
	.1979	2 3/8		.4479	5 3/8		.6979	8 3/8		.9479	11 3/8
7/32	.203125	2 7/16	15/32	.453125	5 7/16	23/32	.703125	8 7/16	31/32	.953125	11 7/16
	.2083	2 1/2		.4583	5 1/2		.7083	8 1/2		.9583	11 1/2
	.2135	2 9/16		.4635	5 9/16		.7135	8 9/16		.9635	11 9/16
15/64	.21875	2 5/8	31/64	.46875	5 5/8	47/64	.71875	8 5/8	63/64	.96875	11 5/8
	.2240	2 11/16		.4740	5 11/16		.7240	8 11/16		.9740	11 11/16
	.2292	2 3/4		.4792	5 3/4		.7292	8 3/4		.9792	11 3/4
1/4	.234375	2 13/16	1/2	.484375	5 13/16	3/4	.734375	8 13/16	1	.984375	11 13/16
	.2396	2 7/8		.4896	5 7/8		.7396	8 7/8		.9896	11 7/8
	.2448	2 15/16		.4948	5 15/16		.7448	8 15/16		.9948	11 15/16
	.2500	3		.5000	6		.7500	9		1.000	12

TABLE M-6

°F/°C TEMPERATURE CONVERSION

−459.4 TO 0

°C.	Given Temp.	°F.
−273	−459.4	—
−268	−450	—
−262	−440	—
−257	−430	—
−251	−420	—
−246	−410	—
−240	−400	—
−234	−390	—
−229	−380	—
−223	−370	—
−218	−360	—
−212	−350	—
−207	−340	—
−201	−330	—
−196	−320	—
−190	−310	—
−184	−300	—
−179	−290	—
−173	−280	—
−169	−273	−459.4
−168	−270	−454
−162	−260	−436
−157	−250	−418
−151	−240	−400
−146	−230	−382
−140	−220	−364
−134	−210	−346
−129	−200	−328
−123	−190	−310
−118	−180	−292
−112	−170	−274
−107	−160	−256
−101	−150	−238
−96	−140	−220
−90	−130	−202
−84	−120	−184
−79	−110	−166
−73	−100	−148
−68	−90	−130
−62	−80	−112
−57	−70	−94
−51	−60	−76
−46	−50	−58
−40	−40	−40
−34	−30	−22
−29	−20	−4
−23	−10	+14
−17.8	0	+32

0 TO 100

°C.	Given Temp.	°F.
−17.8	0	32
−17.2	1	33.8
−16.7	2	35.6
−16.1	3	37.4
−15.6	4	39.2
−15.0	5	41.0
−14.4	6	42.8
−13.9	7	44.6
−13.3	8	46.4
−12.8	9	48.2
−12.2	10	50.0
−11.7	11	51.8
−11.1	12	53.6
−10.6	13	55.4
−10.0	14	57.2
−9.4	15	59.0
−8.9	16	60.8
−8.3	17	62.6
−7.8	18	64.4
−7.2	19	66.2
−6.7	20	68.0
−6.1	21	69.8
−5.6	22	71.6
−5.0	23	73.4
−4.4	24	75.2
−3.9	25	77.0
−3.3	26	78.8
−2.8	27	80.6
−2.2	28	82.4
−1.7	29	84.2
−1.1	30	86.0
−0.6	31	87.8
0.0	32	89.6
0.6	33	91.4
1.1	34	93.2
1.7	35	95.0
2.2	36	96.8
2.8	37	98.6
3.3	38	100.4
3.9	39	102.2
4.4	40	104.0
5.0	41	105.8
5.6	42	107.6
6.1	43	109.4
6.7	44	111.2
7.2	45	113.0
7.8	46	114.8
8.3	47	116.6
8.9	48	118.4
9.4	49	120.2
10.0	50	122.0
10.6	51	123.8
11.1	52	125.6
11.7	53	127.4
12.2	54	129.2
12.8	55	131.0
13.3	56	132.8
13.9	57	134.6
14.4	58	136.4
15.0	59	138.2
15.6	60	140.0
16.1	61	141.8
16.7	62	143.6
17.2	63	145.4
17.8	64	147.2
18.3	65	149.0
18.9	66	150.8
19.4	67	152.6
20.0	68	154.4
20.6	69	156.2
21.1	70	158.0
21.7	71	159.8
22.2	72	161.6
22.8	73	163.4
23.3	74	165.2
23.9	75	167.0
24.4	76	168.8
25.0	77	170.6
25.6	78	172.4
26.1	79	174.2
26.7	80	176.0
27.2	81	177.8
27.8	82	179.6
28.3	83	181.4
28.9	84	183.2
29.4	85	185.0
30.0	86	186.8
30.6	87	188.6
31.1	88	190.4
31.7	89	192.2
32.2	90	194.0
32.8	91	195.8
33.3	92	197.6
33.9	93	199.4
34.4	94	201.2
35.0	95	203.0
35.6	96	204.8
36.1	97	206.6
36.7	98	208.4
37.2	99	210.2
37.8	100	212.0

110 TO 1110

°C.	Given Temp.	°F.
43	110	230
49	120	248
54	130	266
60	140	284
66	150	302
71	160	320
77	170	338
82	180	356
88	190	374
93	200	392
99	210	410
100	212	413.6
104	220	428
110	230	446
116	240	464
121	250	482
127	260	500
132	270	518
138	280	536
143	290	554
149	300	572
154	310	590
160	320	608
166	330	626
171	340	644
177	350	662
182	360	680
188	370	698
193	380	716
199	390	734
204	400	752
210	410	770
216	420	788
221	430	806
227	440	824
232	450	842
238	460	860
243	470	878
249	480	896
254	490	914
260	500	932
266	510	950
271	520	968
277	530	986
282	540	1004
288	550	1022
293	560	1040
299	570	1058
304	580	1076
310	590	1094
316	600	1112
321	610	1130
327	620	1148
332	630	1166
338	640	1184
343	650	1202
349	660	1220
354	670	1238
360	680	1256
366	690	1274
371	700	1292
377	710	1310
382	720	1328
388	730	1346
393	740	1364
399	750	1382
404	760	1400
410	770	1418
416	780	1436
421	790	1454
427	800	1472
432	810	1490
438	820	1508
443	830	1526
449	840	1544
454	850	1562
460	860	1580
466	870	1598
471	880	1616
477	890	1634
482	900	1652
488	910	1670
493	920	1688
499	930	1706
504	940	1724
510	950	1742
516	960	1760
521	970	1778
527	980	1796
532	990	1814
538	1000	1832
543	1010	1850
549	1020	1868
554	1030	1886
560	1040	1904
566	1050	1922
571	1060	1940
577	1070	1958
582	1080	1976
588	1090	1994
593	1100	2012
599	1110	2030

1120 TO 3000

°C.	Given Temp.	°F.
604	1120	2048
610	1130	2066
616	1140	2084
621	1150	2102
627	1160	2120
632	1170	2138
638	1180	2156
643	1190	2174
649	1200	2192
654	1210	2210
660	1220	2228
666	1230	2246
671	1240	2264
677	1250	2282
682	1260	2300
688	1270	2318
693	1280	2336
699	1290	2354
704	1300	2372
710	1310	2390
716	1320	2408
721	1330	2426
727	1340	2444
732	1350	2462
738	1360	2480
743	1370	2498
749	1380	2516
754	1390	2534
760	1400	2552
766	1410	2570
771	1420	2588
777	1430	2606
782	1440	2624
788	1450	2642
793	1460	2660
799	1470	2678
804	1480	2696
810	1490	2714
816	1500	2732
821	1510	2750
827	1520	2768
832	1530	2786
838	1540	2804
843	1550	2822
849	1560	2840
854	1570	2858
860	1580	2876
866	1590	2894
871	1600	2912
877	1610	2930
882	1620	2948
888	1630	2966
893	1640	2984
899	1650	3002
904	1660	3020
910	1670	3038
916	1680	3056
921	1690	3074
927	1700	3092
932	1710	3110
938	1720	3128
943	1730	3146
949	1740	3164
954	1750	3182
960	1760	3200
966	1770	3218
971	1780	3236
977	1790	3254
982	1800	3272
988	1810	3290
993	1820	3308
999	1830	3326
1004	1840	3344
1010	1850	3362
1016	1860	3380
1021	1870	3398
1027	1880	3416
1032	1890	3434
1038	1900	3452
1043	1910	3470
1049	1920	3488
1054	1930	3506
1060	1940	3524
1066	1950	3542
1071	1960	3560
1077	1970	3578
1082	1980	3596
1088	1990	3614
1093	2000	3632
1121	2050	3722
1149	2100	3812
1204	2200	3992
1232	2250	4082
1260	2300	4172
1316	2400	4352
1371	2500	4532
1427	2600	4712
1482	2700	4892
1538	2800	5072
1593	2900	5252
1649	3000	5432

Reproduced by courtesy of Jenkins Bros., valve manufacturers. Find the temperature it is required to convert in the center column. If this temperature is in degrees F, the centigrade equivalent is in the left column; if this temperature is in degrees C, the fahrenheit equivalent is in the right column.

CONVERSIONS - MULTIPLIERS FOR CUSTOMARY & METRIC UNITS

TABLE M-7

* indicates value is exact. Units in pounds are avoirdupois. Abbreviations include: Btu = British thermal unit; C = Centigrade &/or Celsius; Chu = Centigrade heat unit; cu = cubic; EEC = European Economic Community; F = Fahrenheit; ft = feet or foot; Hg = Mercury; in = inch(es); k = kelvin; kgf = kilogram-force; lbf = pound-force; liq = liquid; R = Rankine; sq = square; UK = United Kingdom; US = United States.

MULTIPLY	BY	TO OBTAIN
acre	43 560	square feet
	4 840	square yards
	4 046.856 4	square meters
	0.404 685 6	hectare
	0.001 562 5*	square mile
acre foot	43 560	cubic feet
	1 233.481 8	cubic meters
	325 851.43	gallons (US)
are [a]	100	square meters
	119.599 01	square yards
	0.024 710 54	acre
atmosphere	1.013 25*	bar
	101 325*	pascals
	759.999 81	mm of Hg @ 32F
	29.921 252	inches of Hg @ 32F
	33.932 447	ft of water @ 60F
	14.695 949	pounds/square inch
bar	100 000	pascals
	100 000	newtons/sq meter
	0.1*	newton/sq mm
	14.503 774 1	pounds/sq inch
barrel [bbl] (petroleum)	42	gallons (US)
	5.614 583 3	cubic feet
	0.158 987 3	cubic meter
Btu	778.169 4	foot-pounds
	107.585 76	kilogram-meters
	0.000 293 07	kilowatt-hour
	1 055.056	joules
(International Table):	1 054.350	joules
(thermochemical U.S.): 1		
Btu/hour	0.216 158 2	foot-pound/second
	0.293 071 1	watt
bushel [bu] (US)	1.244 456	cubic feet
	0.035 239 07	cubic meter
bushel [bu] (UK)	1.032 06	bushels (US)
cable (US)	120	fathoms
	720	feet
	219.456*	meters
Celsius	1	Centigrade
Centigrade	1	Celsius
centimeter [cm]	0.393 700 8	inch
	10	millimeters
chain (gunter or surveyors)	66	feet
	22	yards
	20.116 8*	meters
Chu (obsolete unit)	1.8	Btu

MULTIPLY	BY	TO OBTAIN
cubic decimeter	1	liter
	1 000	cubic cm
cubic inch	16.387 064*	cubic cm
	0.016 387 064	liter
cubic foot	28 316.846 6	cubic cm
	0.028 316 85	cubic meter
	1 728	cubic inches
	28.316 846 6	liters
	7.480 519 5	gallons (US)
	6.228 835 6	gallons (UK)
cubic meter	35.314 667	cubic feet
	1.307 950 6	cubic yards
	264.172 052	gallons (US)
	1 000	liters
	2 113.376 42	pints (US)
cubic foot of water	62.365 578	pounds @ 15.6C
cubic ft/acre	0.069 972 3	cu m/hectare
cubic yard	0.764 554 9	cubic meter
	764.554 86	liters
	201.974 03	gallons(US)
degree (angle)	0.017 453 29	radian
decimeter [dm]	3.937 007 9	inches
	100	millimeters
	10	centimeters
dekameter [dam]	10	meters
fathom	6	feet
	1.828 8*	meters
feet/minute	0.005 08*	meter/second
	0.304 8*	meter/minute
foot	0.304 8*	meter
	304.8*	millimeters
	12	inches
foot-pound/sec	1.355 817 9	joules
foot (head) of water @ 15.6C	2.986 08	pascals
	0.433 094	pound/sq inch
	62.365 578	pounds/sq foot
furlong	660	feet
	201.168*	meters
	220	yards
	0.125*	mile
gallon (US) -liquid	3.785 411 78	liters
	3.785 411 78	cubic cms
	0.003 785 4	cubic meter
	0.133 680 56	cubic foot

MULTIPLY	BY	TO OBTAIN
gallon (US) -liquid	231	cubic inches
	8	pints
	4	quarts
	0.832 674 18	gallon (UK)
	8.336 7	pounds of water @ 15.6C [60F]
gallon (UK) -liquid	1.200 949 9	gallons (US)
	4.546 09*	liters
	4.546 09*	cubic decimeters
	277.419 43	cubic inches
	8	pints
	4	quarts
	10.012	pounds of water @ 15.6C [60F]
gram	0.001*	kilogram
	0.035 273 96	ounce
	15.432 36	grains
gravity: std free fall	32.174	feet/second/second
	9.806 65*	m/second/second
grain	0.064 798 91	gram
	1/700 0	pound
hectare [ha]	2.471 053 8	acres
	107 639.1	square feet
	0.003 861	square mile
horsepower	42.407 219	Btu/minute
	2 544.433 1	Btu/hour
	33 000	foot-pounds/minute
	550	foot-pounds/second
	745.699 87	watts
horsepower (metric)	735.499	watts
horsepower (boiler)	33 471.439 8	Btu/hour
	9 809.5	watts
horsepower (electric)	746	watts
inch	25.4*	millimeters
	2.54*	centimeters
	0.025 4*	meter
inch (head) of water @ 60F	248.84	pascals
inch (head) of mercury @ 60F	1.130 863 9	feet of water
	3 376.85	pascals
joule [J]	0.000 947 8	Btu
	0.737 562 18	foot-pound
	1	watt-second
kilogram [kg]	2.204 623	pounds
	1 000	grams
kgf/sq cm	98 066.5*	pascals
	14.223 344	lbf/sq in
kip	1 000	lbf
	4 448.221 615	newtons
ksi (kip per sq in)	6 894 757	pascals
	6.894 757	megapascals [MPa]

MULTIPLY	BY	TO OBTAIN
kilometer [km]	0.621 371 2	mile
kilowatt-hour	3 412.141 3	Btu
liter [L]	1 000	cubic centimeters
	61.023 744 1	cubic inches
	0.035 314 67	cubic feet
	0.264 172 1	gallon (liq. US)
	2.113 376 42	pints (liq. US)
	33.814 022 7	fluid ounces (US)
meter [m]	39.370 079	inches
	3.280 839 9	feet
	1.093 613 3	yards
	0.000 621 4	mile
	1 000	millimeters
	100	centimeters
	0.001*	kilometer
mil	0.001*	inch
	0.025 4*	millimeter
	0.000 025 4*	meter
micrometer (micron)	0.000 039 37	inch
	0.001*	millimeter
mile	1.609 344*	kilometers
	1 609.344*	meters
	5 280	feet
	1 760	yards
	8	furlongs
millimeter [mm]	0.1*	centimeter
	0.001*	meter
	0.039 370 79	inch
newton (N)	0.101 971 62	kilogram-force
	0.224 808 93	pound-force
newton/sq meter	1	pascal
ounce	28.349 523 12	grams
	0.028 349 5	kilogram
	0.278 013 85	newtons
pascal [Pa]	1	newton/sq meter
	0.000 145 04	pound/sq inch
pint	16	fluid ounces (US)
	20	fluid ounces (UK)
pound	16	ounces
	453.592 37*	grams
	0.453 592 37*	kilogram
	4.448 221 615	newtons
	7 000	grains
pound/sq in (psi)	6 894.757 2	pascals
	2.308 966	ft of water @ 60F
	2.041 772	inches of Hg @ 60F
pounds/cu in	27 679.905	kg/cu meter
pounds/sq ft	4.882 428	kg/sq meter
pounds/cu ft	16.018 463	kg/cubic meter
radian [rad]	57.295 779	degrees

CONTINUED FROM PREVIOUS PAGE -

MULTIPLY	BY	TO OBTAIN
rod (survey)	16.5*	feet
	5.029 2*	meters
square cm	0.155	square inch
square meter	0.000 247 1	acre
	1.195 99	square yards
	10.763 9	square feet
	10 000	square centimeters
sq kilometer	0.386 102 2	square mile
	247.105 383	acres
square inch	645.16*	square millimeters
square foot	0.092 903 04*	square meter
	144	square inches
square yard	0.836 127 36*	square meter
square mile	640	acres
	2.589 988	sq kilometers
	258.998 8	hectares
therm: Europe (EEC)	100 000	Btu
	105 506 000	joules
United States	100 000	Btu
	105 480 400	joules
ton (short-US, also net ton)	907.184 74*	kilograms
	2 000	pounds
	0.907 184 74*	metric ton
ton (long-UK, also gross ton)	1 016.046 91	kilograms
	2 240	pounds
	1.016 046 91	metric tons
	1.12*	short tons (US)

MULTIPLY	BY	TO OBTAIN
ton (metric) or tonne	1 000	kilograms
	2 204.623	pounds
	0.984 206 5	long ton (UK)
	1.102 311	short tons (US)
ton of refrigeration	12 000	Btu/hour
	200	Btu/minute
	3 517	watts
watt [W]	3.412 141 3	Btu/hour
	0.737 562 2	foot-pound/sec
	1	joule/second
watt-hour	3.412 141 3	Btu
yard [yd]	0.914 4*	meter

TEMPERATURE CONVERSION:

Fahrenheit to Celsius	$C = (F - 32) / 1.8$
Celsius to Fahrenheit	$F = (C \times 1.8) + 32$
Fahrenheit to kelvin	$K = (F + 459.67) / 1.8$
Celsius to kelvin	$K = C + 273.15$
Kelvin to Celsius	$C = K - 273.15$
Rankine to kelvin	$K = R / 1.8$

VISCOSITY:

centipoise (dynamic)	0.001	pascal second (Pa s)
centistokes (kinematic)	0.000 001	sq meter per second

Non-SI units: This table contains units combining kilogram in units of force and pressure. In SI, kilogram is the unit of mass, 'newton' is the unit of force, and 'pascal' is the unit of pressure

---

R U L E S   F O R   R O U N D I N G   V A L U E S

Reference: ASTM E 380

FIRST DISCARDED DIGIT	LAST RETAINED DIGIT
If less than 5	NO CHANGE
Equal to 5 and followed by at least one digit OTHER than 0	INCREASE BY ONE UNIT
If greater than 5	INCREASE BY ONE UNIT
Equal to 5 and followed ONLY by zeros	IF ODD: INCREASE BY ONE UNIT / IF EVEN: NO CHANGE

FROM 1866 TO 1959, THE METER WAS DEFINED AS 39.37-inches. IN 1959, THE U.S. YARD WAS REDEFINED, FROM 3600/3937m (0.914 401 828 037/m), TO 0.9144m EXACTLY. HOWEVER, DATA FROM GEODETIC SURVEYS WITHIN THE U.S. CONTINUED TO USE THE FOOT DERIVED FROM THE PRE-1959 STANDARD: THE U.S. SURVEY FOOT. THE FOOT DEFINED IN 1959, IS THE INTERNATIONAL FOOT, USED IN THIS TABLE, EXCEPT AS NOTED.

REFERENCES: US Department of Commerce/National Institute of Standards & Technology: National Aeronautics & Space Administration; American Society for Testing Materials: The American Society of Mechanical Engineers: National Physical Laboratory-UK

SPECIFIC HEATS - AVERAGE VALUES — TABLE M-8

Units: Btu/lb/°F = 4 186.8J/kg/K

Solids and Liquids

Material	Specific heat
Aluminum	0.214
Asbestos	0.20
Asphalt	0.40
Carbon	0.165
Carborundum	0.16
Cast iron	0.12 - 0.13
Cellulose	0.37
Cement, dry	0.37
Cement, powder	0.20
Chalk	0.215
Charcoal	0.20 - 0.24
Chromium	0.12
Coal	0.24 - 0.37
Coke	0.203
Concrete	0.19
Copper	0.092
Cork	0.48
Dowtherm A	0.50
Duralumin	0.23
Earth, dry	0.30
Fuel oil: sp gr 96	0.40
sp gr 91	0.44
sp gr 86	0.45
sp gr 81	0.51
Glass, plate	0.12
Glass, pyrex	0.20

Material	Specific heat
Glass, wool	0.16
Granite	0.19
Graphite	0.201
Ice: @ -112F	0.35
@ -40F	0.43
@ -4F	0.47
@ +32F	0.49 - 0.50
Kerosene	0.48 - 0.50
Lead	0.031
Limestone	0.217
Lucite	0.35
Magnesia	0.20 - 0.22
Malleable iron	0.12
Masonry, brick	0.20 - 0.22
Mineral wool	0.20
Mercury	0.033
Molybdenum	0.06
Nickel	0.109
Nylon	0.55
Olive oil	0.35 - 0.47
Paper	0.33
Plaster of Paris	1.14
Platinum	0.03 - 0.039
Polythene	0.53

Material	Specific heat
Quartz	0.17 - 0.21
Rocksalt	0.22
Rubber	0.27 - 0.48
Salt, granulated	0.21
Sand	0.195
Sandstone	0.22
Seawater, sp gr 1.023	0.94
Silica	0.191
Silicon	0.123
Soda	0.231
Sodium	0.295
Steel	0.117
Sucrose	0.30
Sugar, bulk	0.28
Stone	0.20
Sulfur	0.178
Tar, bituminous	0.35
Teflon	0.25
Tile	0.15
Tin	0.056
Tungsten	0.04
Water	1.00
Wood, fir	0.65
oak	0.57
pine	0.467
Wood shavings	0.52
Zinc	0.095

GASES

GASES	At Constant Pressure	At Constant Volume
Air	0.24	0.172
Ammonia	0.54	0.422
Argon	0.12	0.072
Carbon dioxide	0.20	0.150
Carbon monoxide	0.24	0.172
Carbon disulfide	0.16	0.132
Chlorine	0.11	0.082
Ethylene	0.40	0.332
Helium	1.25	0.75
Hydrogen	3.21	2.410
Hydrogen sulfide	0.25	0.189

GASES	At Constant Pressure	At Constant Volume
Iso-butane	0.39	0.355
Methane	0.59	0.446
Nitrogen	0.24	0.170
Nitrous oxide	0.21	0.166
Oxygen	0.22	0.157
Steam: @ 1.0 psia	0.46	0.349
@ 14.7 psia	0.47	0.359
@ 150.0 psia	0.54	0.421
Sulfur dioxide	0.15	0.119

Tables P-1 present calculated data as a guide only. Spans are for pipe arranged in pipeways with the following assumptions: Bare pipe - continuous straight run with welded joints and two or more straight spans at each end.

SPANS - calculated with lines full of water and a maximum bending stress of 4 000 PSI

SAG - (deflection) calculated with lines empty (drained condition)

The following factors were not considered in calculating spans for these tables:
Concentrated mechanical loads from flanges, valves, strainers, filters, and other inline equipment - weights of connecting branch lines - torsional loading from thermal movement - sudden reaction from lines(s) discharging contents - vibration - flattening effect of weight of contents in larger liquid filled lines - weight of insulation and pipe covering - weight of ice and snow - wind loads - seismic shock - reduction in wall thickness of pipe from threading or grooving.

DESIGN PRESSURE - calculated per ANSI B31.1 using allowable stress value of 9 000 PSI for seamless carbon steel pipe

BURSTING PRESSURE is approximate, calculated on yield strength of 30 000 PSI

PIPE DATA: DIMENSIONS & STRESS PARAMETERS — TABLES P-1

NPS (inch)	Piping Codes and Mfrs' Weights	O.D. (in.)	I.D. (in.)	Wall (in.)	Empty (lb/ft)	Waterfilled (lb/ft)	External (in²/ft)	Internal (in²/ft)	Flow (in²)	Metal (in²)	Moment of Inertia (in⁴)	Section Modulus (in³)	Radius of Gyration (in.)	Continuous Span (ft)	Sag (in.)	Design (kPSI)	Bursting (kPSI)
.375	SCH 40 STD API	.6750	.4930	.0910	.5690	.6516	25.45	18.59	.1909	.1670	.0073	.0216	.2090	11.5	.213	1.75	5.84
	SCH 80 XS API	.6750	.4230	.1260	.7406	.8015	25.45	15.95	.1405	.2173	.0086	.0255	.1991	11.3	.217	2.89	9.63
.500	SCH 40 STD API	.8400	.6220	.1090	.8531	.9847	31.67	23.45	.3039	.2503	.0171	.0407	.2613	12.9	.212	1.83	6.10
	SCH 80 XS API	.8400	.5460	.1470	1.091	1.192	31.67	20.58	.2341	.3200	.0201	.0478	.2505	12.7	.217	2.82	9.41
	SCH 160 API	.8400	.4660	.1880	1.312	1.386	31.67	17.49	.1691	.3851	.0222	.0528	.2399	12.3	.213	3.99	13.3
	XXS API	.8400	.2520	.2940	1.719	1.740	31.67	9.500	.0499	.5043	.0242	.0577	.2192	11.5	.194	7.56	25.2
.750	SCH 40 STD API	1.050	.8240	.1130	1.134	1.365	39.58	31.06	.5333	.3326	.0370	.0705	.3337	14.4	.203	1.33	4.42
	SCH 80 XS API	1.050	.7420	.1540	1.477	1.665	39.58	27.97	.4324	.4335	.0448	.0853	.3214	14.3	.215	2.14	7.13
	SCH 160 API	1.050	.6120	.2190	1.948	2.076	39.58	23.07	.2942	.5717	.0528	.1005	.3038	13.9	.213	3.55	11.8
	XXS API	1.050	.4340	.3080	2.447	2.511	39.58	16.36	.1479	.7180	.0579	.1103	.2840	13.3	.203	5.77	19.2
1.00	SCH 40 STD API	1.315	1.049	.1330	1.683	2.057	49.57	39.55	.8643	.4939	.0873	.1328	.4205	16.1	.199	1.27	4.24
	SCH 80 XS API	1.315	.9570	.1790	2.177	2.489	49.57	36.08	.7193	.6388	.1056	.1606	.4066	16.1	.213	2.00	6.66
	SCH 160 API	1.315	.8150	.2500	2.851	3.077	49.57	30.72	.5217	.8364	.1251	.1903	.3868	15.7	.216	3.21	10.7
	XXS API	1.315	.5990	.3580	3.668	3.790	49.57	22.58	.2818	1.076	.1405	.2136	.3613	15.0	.206	5.29	17.6
1.25	SCH 40 STD API	1.660	1.380	.1400	2.278	2.926	62.58	52.02	1.496	.6685	.1947	.2346	.5397	17.9	.187	1.02	3.40
	SCH 80 XS API	1.660	1.278	.1910	3.004	3.560	62.58	48.18	1.283	.8815	.2418	.2913	.5237	18.1	.207	1.64	5.47
	SCH 160 API	1.660	1.160	.2500	3.774	4.232	62.58	43.73	1.057	1.107	.2839	.3420	.5063	18.0	.216	2.40	8.00
	XXS API	1.660	.8960	.3820	5.227	5.500	62.58	33.78	.6305	1.534	.3411	.4110	.4716	17.3	.212	4.29	14.3
1.50	SCH 40 STD API	1.900	1.610	.1450	2.725	3.606	71.63	60.70	2.036	.7995	.3099	.3262	.6226	19.0	.179	.938	3.13
	SCH 80 XS API	1.900	1.500	.2000	3.640	4.405	71.63	56.55	1.767	1.068	.3912	.4118	.6052	19.3	.202	1.52	5.06
	SCH 160 API	1.900	1.338	.2810	4.871	5.480	71.63	50.44	1.406	1.429	.4824	.5078	.5810	19.3	.215	2.42	8.08
	XXS API	1.900	1.100	.4000	6.424	6.835	71.63	41.47	.9503	1.885	.5678	.5977	.5489	18.7	.215	3.89	13.0
2.00	SCH 40 STD API	2.375	2.067	.1540	3.662	5.115	89.54	77.92	3.356	1.075	.6657	.5606	.7871	20.9	.164	.736	2.45
	SCH 80 XS API	2.375	1.939	.2180	5.034	6.313	89.54	73.10	2.953	1.477	.8679	.7309	.7665	21.5	.193	1.26	4.22
	API	2.375	1.875	.2500	5.688	6.884	89.54	70.69	2.761	1.669	.9551	.8043	.7565	21.6	.202	1.54	5.13
	SCH 160 API	2.375	1.687	.3440	7.480	8.448	89.54	63.60	2.235	2.195	1.164	.9804	.7283	21.5	.215	2.38	7.93
	XXS API	2.375	1.503	.4360	9.051	9.820	89.54	56.66	1.774	2.656	1.311	1.104	.7027	21.2	.217	3.26	10.9

API = American Petroleum Institute's standard 5L, for 'Line pipe'. API pipe sizes: manufacturers' weights: Double-extra-strong (XXS), Extra-strong (XS), and Standard (STD), are included with schedule numbers in standard ANSI B36.10M. Also refer to 2.1.3

Thru NPS 10, wall thicknesses for SCH 40S and SCH 80S stainless steel pipes are the same as for SCH 40 and SCH 80 carbon steel pipes

PIPE DATA

TABLES P-1

NPS (inch)	PIPING CODES and MANUFACTURERS' WEIGHTS	DIMENSIONS O.D. (in.)	I.D. (in.)	Wall (in.)	WEIGHTS Empty (lb/ft)	Waterfilled (lb/ft)	AREAS External (in²/ft)	Internal (in²/ft)	Flow (in²)	Metal (in²)	Moment of Inertia (in⁴)	Section Modulus (in³)	Radius of Gyration (in.)	Continuous Spans Span (ft)	Sag (in.)	Code Pressures Design (kPSI)	Bursting (kPSI)
2.50	SCH 40 STD API	2.875	2.469	.2030	5.807	7.881	108.4	93.08	4.788	1.704	1.530	1.064	.9474	23.2	.172	.865	2.88
	SCH 80 XS API	2.875	2.323	.2760	7.680	9.515	108.4	87.58	4.238	2.254	1.924	1.339	.9241	23.7	.196	1.37	4.56
	SCH 160 API	2.875	2.125	.3750	10.04	11.57	108.4	80.11	3.547	2.945	2.353	1.637	.8938	23.8	.212	2.09	6.96
	XXS API	2.875	1.771	.5520	13.73	14.80	108.4	66.77	2.463	4.028	2.871	1.997	.8442	23.2	.216	3.49	11.6
3.00	SCH 40 STD API	3.500	3.068	.2160	7.595	10.80	131.9	115.7	7.393	2.228	3.017	1.724	1.164	25.3	.159	.777	2.59
	SCH 80 XS API	3.500	2.900	.3000	10.28	12.62	131.9	109.3	6.605	3.016	3.894	2.225	1.136	26.0	.183	1.14	3.80
	SCH 160 API	3.500	2.624	.4380	14.36	16.70	131.9	98.92	5.408	4.213	5.039	2.879	1.094	26.3	.210	2.07	6.89
	XXS API	3.500	2.300	.6000	18.63	20.43	131.9	86.71	4.155	5.466	5.993	3.424	1.047	25.9	.217	3.10	10.3
4	SCH 40 STD API	4.500	4.026	.2370	10.82	16.33	169.6	151.8	12.73	3.174	7.233	3.214	1.510	28.1	.144	.659	2.20
	SCH 80 XS API	4.500	3.826	.3370	14.99	20.28	169.6	144.2	11.50	4.407	9.610	4.271	1.477	29.2	.177	.924	3.08
	SCH 120 API	4.500	3.624	.4380	19.01	24.51	169.6	136.7	10.31	5.589	11.65	5.178	1.445	29.0	.170	1.32	4.41
	SCH 160 API	4.500	3.438	.5310	22.56	26.58	169.6	129.6	9.283	6.621	13.27	5.898	1.416	29.8	.208	1.54	5.14
	XXS API	4.500	3.152	.6740	27.61	30.99	169.6	118.8	7.803	8.101	15.28	6.793	1.374	29.6	.216	2.25	7.51
6	SCH 40 STD API	6.625	6.065	.2800	19.02	31.53	249.8	228.6	28.89	5.581	28.14	8.496	2.245	32.8	.122	.528	1.76
	SCH 80 XS API	6.625	5.761	.4320	28.64	39.93	249.8	217.2	26.07	8.405	40.49	12.22	2.195	34.0	.142	.791	2.64
	SCH 120 API	6.625	5.501	.5620	36.48	46.77	249.8	207.4	23.77	10.70	49.61	14.98	2.153	35.8	.187	1.03	3.45
	SCH 160 API	6.625	5.187	.7190	45.46	54.61	249.8	195.5	21.13	13.34	59.03	17.82	2.104	36.1	.204	1.30	4.33
	XXS API	6.625	4.897	.8640	53.29	61.45	249.8	184.6	18.83	15.64	66.33	20.02	2.060	36.1	.212	1.79	5.95
8	SCH 20 API	8.625	8.125	.2500	22.42	44.87	325.2	306.3	51.85	6.578	57.72	13.38	2.962	33.4	.074	.275	.915
	SCH 30 API	8.625	8.071	.2770	24.76	46.91	325.2	304.3	51.16	7.265	63.35	14.69	2.953	35.4	.095	.332	1.11
	SCH 40 STD API	8.625	7.981	.3220	28.62	50.29	325.2	300.9	50.03	8.399	72.49	16.81	2.938	36.6	.110	.430	1.43
	SCH 60 API	8.625	7.813	.4060	35.73	56.49	325.2	294.6	47.94	10.48	88.74	20.58	2.909	37.1	.125	.545	1.82
	SCH 80 XS API	8.625	7.625	.5000	43.50	63.27	325.2	287.5	45.66	12.76	105.7	24.51	2.878	39.4	.153	.684	2.28
	SCH 100 API	8.625	7.437	.5940	51.07	69.89	325.2	280.4	43.44	14.99	121.5	28.17	2.847	40.2	.170	.822	2.74
	SCH 120 API	8.625	7.187	.7190	60.86	78.43	325.2	270.9	40.57	17.86	140.7	32.62	2.807	40.8	.186	1.03	3.45
	SCH 140 API	8.625	7.001	.8120	67.92	84.59	325.2	263.9	38.50	19.93	153.7	35.65	2.777	41.1	.195	1.54	5.14
	XXS API	8.625	6.875	.8750	72.60	88.68	325.2	259.2	37.12	21.30	162.0	37.56	2.757	41.2	.200	1.69	5.65
	SCH 160 API	8.625	6.813	.9060	74.88	90.66	325.2	256.8	36.46	21.97	165.9	38.47	2.748	41.2	.202	1.77	5.90

Thru NPS 10, wall thicknesses for SCH 40S and SCH 80S stainless steel pipes are the same as for SCH 40 and SCH 80 carbon steel pipes

PIPE DATA

NPS (inch)	Piping Codes & Mfrs' Weights	O.D. (in.)	I.D. (in.)	Wall (in.)	Empty (lb/ft)	Waterfilled (lb/ft)	External (in^2/ft)	Internal (in^2/ft)	Flow (in^2)	Metal (in^2)	Moment of Inertia (in^4)	Section Modulus (in^3)	Radius of Gyration (in.)	Span (ft)	Sag (in.)	Design (kPSI)	Bursting (kPSI)
10	API	10.75	10.37	.1880	21.26	57.86	405.3	391.1	84.52	6.238	87.02	16.19	3.735	33.5	.047	.098	.325
	API	10.75	10.34	.2030	22.92	59.31	405.3	390.0	84.04	6.726	93.56	17.41	3.730	34.3	.052	.123	.410
	API	10.75	10.31	.2190	24.69	60.86	405.3	388.8	83.52	7.245	100.5	18.69	3.724	35.1	.058	.150	.500
	SCH 20 API	10.75	10.25	.2500	28.10	63.84	405.3	386.4	82.52	8.247	113.7	21.16	3.713	36.4	.067	.203	.676
	API	10.75	10.19	.2790	31.28	66.61	405.3	384.2	81.58	9.178	125.9	23.42	3.703	37.5	.076	.252	.841
	SCH 30 API	10.75	10.14	.3070	34.33	69.27	405.3	382.1	80.69	10.07	137.4	25.57	3.694	38.4	.084	.300	1.00
	API	10.75	10.06	.3440	38.33	72.76	405.3	379.3	79.52	11.25	152.4	28.35	3.681	39.5	.095	.364	1.21
	SCH 40 STD API	10.75	10.02	.3650	40.58	74.73	405.3	377.7	78.85	11.91	160.7	29.90	3.674	40.0	.100	.400	1.33
	API	10.75	9.874	.4380	48.36	81.52	405.3	372.2	76.57	14.19	188.9	35.15	3.649	41.5	.118	.528	1.76
	SCH 60 XS API	10.75	9.750	.5000	54.87	87.20	405.3	367.6	74.66	16.10	212.0	39.43	3.628	42.5	.131	.637	2.12
	SCH 80 API	10.75	9.562	.5940	64.59	95.68	405.3	360.5	71.81	18.95	245.2	45.62	3.597	43.7	.149	.805	2.68
	SCH 100 API	10.75	9.312	.7190	77.22	106.7	405.3	351.1	68.10	22.66	286.4	53.29	3.556	45.3	.167	1.03	3.44
	SCH 120 API	10.75	9.062	.8440	89.51	117.4	405.3	341.6	64.50	26.27	324.5	60.38	3.515	45.8	.181	1.26	4.21
	SCH 140 XXS API	10.75	8.750	1.000	104.4	130.4	405.3	329.9	60.13	30.63	367.8	68.43	3.465	46.0	.194	1.56	5.19
	SCH 160 API	10.75	8.500	1.125	115.9	140.5	405.3	320.4	56.75	34.02	399.3	74.29	3.426	46.0	.202	1.80	6.00
12	API	12.75	12.34	.2030	27.27	79.09	480.7	465.4	119.7	8.002	157.5	24.71	4.437	35.3	.042	.104	.345
	API	12.75	12.31	.2190	29.38	80.94	480.7	464.2	119.1	8.621	169.3	26.55	4.431	36.2	.046	.126	.421
	SCH 20 API	12.75	12.25	.2500	33.46	84.49	480.7	461.8	117.9	9.817	191.8	30.09	4.420	37.7	.055	.171	.569
	API	12.75	12.19	.2810	37.51	88.03	480.7	459.5	116.7	11.01	214.0	33.57	4.410	39.1	.063	.215	.717
	API	12.75	12.13	.3120	41.55	91.56	480.7	457.1	115.5	12.19	235.9	37.00	4.399	40.2	.071	.260	.866
	SCH 30 API	12.75	12.09	.3300	43.88	93.59	480.7	455.8	114.8	12.88	248.5	38.97	4.393	40.8	.076	.286	.953
	API	12.75	12.06	.3440	45.69	95.17	480.7	454.7	114.3	13.41	258.1	40.49	4.388	41.3	.080	.306	1.02
	STD API	12.75	12.00	.3750	49.69	98.66	480.7	452.4	113.1	14.58	279.3	43.82	4.377	42.1	.087	.351	1.17
	SCH 40 API	12.75	11.94	.4060	53.66	102.1	480.7	450.3	111.9	15.74	300.2	47.09	4.367	42.1	.094	.397	1.32
	API	12.75	11.87	.4380	57.74	105.7	480.7	447.6	110.7	16.94	321.4	50.42	4.356	43.7	.101	.443	1.48
	XS API	12.75	11.75	.5000	65.58	112.5	480.7	443.0	108.4	19.24	361.5	56.71	4.335	44.9	.114	.535	1.78
	SCH 60 API	12.75	11.63	.5620	73.34	119.3	480.7	438.3	106.2	21.52	400.4	62.81	4.314	45.9	.126	.627	2.09
	API	12.75	11.50	.6250	81.14	126.1	480.7	433.5	103.9	23.81	438.7	68.81	4.293	46.7	.137	.721	2.40
	SCH 80 API	12.75	11.37	.6880	88.85	132.8	480.7	428.8	101.6	26.07	475.7	74.62	4.271	47.4	.146	.816	2.72
	API	12.75	11.25	.7500	96.36	139.4	480.7	424.1	99.40	28.27	510.9	80.15	4.251	48.0	.155	.911	3.04
	SCH 100 API	12.75	11.06	.8440	107.6	149.2	480.7	417.0	96.11	31.57	562.2	88.19	4.220	48.6	.166	1.06	3.52
	SCH 120 XXS API	12.75	10.75	1.000	125.8	165.1	480.7	405.3	90.76	36.91	641.7	100.7	4.169	49.4	.181	1.30	4.33
	SCH 140 API	12.75	10.50	1.125	140.0	177.5	480.7	395.8	86.59	41.09	700.6	109.9	4.129	49.8	.190	1.50	4.99
	SCH 160 API	12.75	10.13	1.312	160.7	195.5	480.7	381.7	80.53	47.14	781.1	122.5	4.070	50.1	.201	1.80	6.01
14	API	14.00	13.58	.2100	31.01	93.72	527.8	512.0	144.8	9.098	216.3	30.90	4.876	36.3	.039	.103	.344
	API	14.00	13.56	.2190	32.31	94.87	527.8	511.3	144.5	9.481	225.1	32.16	4.873	36.8	.041	.115	.383
	SCH 10 API	14.00	13.50	.2500	36.80	98.79	527.8	508.9	143.1	10.80	255.3	36.47	4.862	38.4	.049	.155	.518
	API	14.00	13.44	.2810	41.27	102.7	527.8	506.6	141.8	12.11	285.0	40.72	4.851	39.8	.057	.196	.653
	SCH 20 API	14.00	13.38	.3120	45.72	106.6	527.8	504.3	140.5	13.42	314.4	44.91	4.841	41.1	.064	.236	.788
	API	14.00	13.31	.3440	50.30	110.6	527.8	501.9	139.2	14.76	344.2	49.18	4.830	42.2	.072	.279	.928
	SCH 30 STD API	14.00	13.25	.3750	54.70	114.4	527.8	499.5	137.9	16.05	372.8	53.25	4.819	43.1	.079	.319	1.06
	SCH 40 API	14.00	13.12	.4380	63.60	122.2	527.8	494.8	135.3	18.66	429.5	61.36	4.797	44.8	.093	.403	1.34
	API	14.00	13.06	.4690	67.94	126.0	527.8	492.4	134.0	19.94	456.8	65.26	4.787	45.5	.099	.444	1.48
	XS API	14.00	13.00	.5000	72.27	129.7	527.8	490.1	132.7	21.21	483.8	69.11	4.776	46.2	.105	.486	1.62
	SCH 60 API	14.00	12.81	.5940	85.26	141.1	527.8	483.0	128.9	25.02	563.1	80.44	4.744	47.8	.122	.613	2.04
	API	14.00	12.75	.6250	89.26	144.8	527.8	480.7	127.7	26.26	588.5	84.08	4.734	48.2	.127	.655	2.18
	SCH 80 API	14.00	12.50	.7500	106.4	159.5	527.8	471.2	122.7	31.22	687.3	98.19	4.692	49.6	.146	.826	2.75
	SCH 100 API	14.00	12.12	.9380	131.2	181.2	527.8	457.1	115.4	38.49	825.1	117.9	4.630	51.0	.167	1.09	3.63
	SCH 120 API	14.00	11.81	1.094	151.2	198.6	527.8	445.3	109.6	44.36	930.2	132.9	4.579	51.7	.181	1.31	4.37
	SCH 140 API	14.00	11.50	1.250	170.6	215.6	527.8	433.5	103.9	50.07	1027	146.7	4.529	52.2	.191	1.54	5.13
	SCH 160 API	14.00	11.19	1.406	189.6	232.2	527.8	421.8	98.31	55.63	1117	159.5	4.480	52.4	.199	1.77	5.90

Thru NPS 10, wall thicknesses for SCH 40S and SCH 80S stainless steel pipes are the same as for SCH 40 and SCH 80 carbon steel pipes

PIPE DATA

TABLES P-1

NPS (inch)	PIPING CODES and MANUFACTURERS' WEIGHTS	DIMENSIONS O.D. (in.)	I.D. (in.)	Wall (in.)	WEIGHTS Empty (lb/ft)	Waterfilled (lb/ft)	AREAS External (in^2/ft)	Internal (in^2/ft)	Flow (in^2)	Metal (in^2)	Moment of Inertia (in^4)	Section Modulus (in^3)	Radius of Gyration (in.)	Continuous Spans Span (ft)	Sag (in.)	Code Pressures Design (kPSI)	Bursting (kPSI)
16	API	16.00	15.56	.2190	37.00	119.4	603.2	586.7	190.2	10.86	338.1	42.26	5.580	37.6	.034	.101	.335
	SCH 10 API	16.00	15.50	.2500	42.16	123.9	603.2	584.3	188.7	12.37	383.7	47.96	5.569	39.4	.041	.136	.453
	API	16.00	15.44	.2810	47.29	128.3	603.2	582.0	187.2	13.88	428.7	53.59	5.558	40.9	.048	.171	.571
	SCH 20 API	16.00	15.38	.3120	52.40	132.8	603.2	579.7	185.7	15.38	473.2	59.16	5.548	42.2	.055	.207	.689
	API	16.00	15.31	.3440	57.66	137.4	603.2	577.2	184.1	16.92	518.6	64.83	5.537	43.4	.061	.243	.811
	SCH 30 STD API	16.00	15.25	.3750	62.73	141.8	603.2	574.9	182.7	18.41	562.1	70.26	5.526	44.5	.068	.279	.930
	API	16.00	15.12	.4380	72.98	150.8	603.2	570.2	179.6	21.41	648.7	81.09	5.504	46.4	.081	.352	1.17
	API	16.00	15.06	.4690	77.99	155.1	603.2	567.8	178.2	22.88	690.6	86.33	5.494	47.2	.087	.388	1.29
	SCH 40 XS API	16.00	15.00	.5000	82.98	159.5	603.2	565.5	176.7	24.35	731.9	91.49	5.483	47.9	.093	.424	1.41
	API	16.00	14.75	.6250	102.9	176.9	603.2	556.1	170.9	30.19	893.5	111.7	5.440	50.3	.114	.571	1.90
	SCH 60	16.00	14.69	.6560	107.8	181.1	603.2	553.7	169.4	31.62	932.3	116.5	5.430	50.7	.119	.608	2.03
	API	16.00	14.50	.7500	122.5	194.0	603.2	546.6	165.1	35.93	1047	130.9	5.398	52.0	.132	.720	2.40
	SCH 80	16.00	14.31	.8440	137.0	206.6	603.2	539.5	160.9	40.19	1157	144.7	5.367	52.9	.144	.833	2.78
	SCH 100	16.00	13.94	1.031	165.2	231.3	603.2	525.5	152.6	48.48	1364	170.6	5.305	54.3	.163	1.06	3.54
	SCH 120	16.00	13.56	1.219	192.9	255.5	603.2	511.3	144.5	56.61	1556	194.5	5.244	55.2	.178	1.30	4.32
	SCH 140	16.00	13.12	1.438	224.2	282.8	603.2	494.8	135.3	65.79	1761	220.1	5.173	55.8	.192	1.57	5.25
	SCH 160	16.00	12.81	1.594	245.9	301.7	603.2	483.0	128.9	72.14	1894	236.8	5.124	56.0	.199	1.78	5.92
18	SCH 10 API	18.00	17.50	.2500	47.51	151.7	678.6	659.7	240.5	13.94	549.1	61.02	6.276	40.1	.035	.121	.402
	API	18.00	17.44	.2810	53.31	156.7	678.6	657.4	238.8	15.64	614.0	68.23	6.265	41.7	.041	.152	.507
	SCH 20 API	18.00	17.38	.3120	59.09	161.8	678.6	655.1	237.1	17.34	678.2	75.36	6.255	43.2	.047	.183	.612
	API	18.00	17.31	.3440	65.03	167.0	678.6	652.6	235.4	19.08	743.8	82.65	6.244	44.5	.053	.216	.720
	STD API	18.00	17.25	.3750	70.76	172.0	678.6	650.3	233.7	20.76	806.6	89.63	6.233	45.7	.059	.248	.826
	API	18.00	17.19	.4060	76.48	177.0	678.6	648.0	232.0	22.44	868.8	96.53	6.222	46.7	.065	.279	.931
	SCH 30 API	18.00	17.12	.4380	82.36	182.1	678.6	645.6	230.3	24.17	932.2	103.6	6.211	47.7	.071	.312	1.04
	API	18.00	17.06	.4690	88.03	187.0	678.6	643.2	228.6	25.83	993.0	110.3	6.200	48.6	.077	.344	1.15
	XS API	18.00	17.00	.5000	93.68	192.0	678.6	640.9	227.0	27.49	1053	117.0	6.190	49.4	.082	.376	1.25
	SCH 40 API	18.00	16.88	.5620	104.9	201.8	678.6	636.2	223.7	30.79	1171	130.2	6.168	50.8	.093	.440	1.47
	API	18.00	16.75	.6250	116.3	211.7	678.6	631.5	220.4	34.12	1289	143.2	6.147	52.0	.103	.506	1.69
	SCH 60 API	18.00	16.50	.7500	138.5	231.1	678.6	622.0	213.8	40.64	1515	168.3	6.105	54.0	.120	.638	2.13
	SCH 80 API	18.00	16.12	.9380	171.3	259.8	678.6	607.9	204.2	50.28	1835	203.9	6.041	56.0	.143	.838	2.79
	SCH 100	18.00	15.69	1.156	208.5	292.2	678.6	591.4	193.3	61.17	2180	242.2	5.969	57.6	.163	1.08	3.58
	SCH 120	18.00	15.25	1.375	244.7	323.8	678.6	574.9	182.7	71.81	2498	277.6	5.898	58.6	.179	1.32	4.39
	SCH 140	18.00	14.88	1.562	274.9	350.2	678.6	560.8	173.8	80.66	2749	305.5	5.838	59.1	.189	1.53	5.10
	SCH 160	18.00	14.44	1.781	309.3	380.2	678.6	544.3	163.7	90.75	3020	335.6	5.769	59.4	.198	1.78	5.94
20	SCH 10 API	20.00	19.50	.2500	52.86	182.2	754.0	735.1	298.6	15.51	756.4	75.64	6.983	40.8	.030	.109	.362
	API	20.00	19.44	.2810	59.33	187.8	754.0	732.8	296.8	17.41	846.3	84.63	6.972	42.5	.035	.137	.456
	API	20.00	19.38	.3120	65.77	193.4	754.0	730.5	294.9	19.30	935.3	93.53	6.962	44.0	.041	.165	.550
	API	20.00	19.31	.3440	72.39	199.2	754.0	728.0	292.9	21.24	1026	102.6	6.951	45.4	.046	.194	.648
	SCH 20 STD API	20.00	19.25	.3750	78.79	204.8	754.0	725.7	291.0	23.12	1113	111.3	6.940	46.6	.052	.223	.742
	API	20.00	19.19	.4060	85.17	210.4	754.0	723.4	289.2	24.99	1200	120.0	6.929	47.8	.057	.251	.837
	API	20.00	19.12	.4380	91.74	216.1	754.0	721.0	287.2	26.92	1288	128.8	6.918	48.8	.063	.281	.936
	API	20.00	19.06	.4690	98.07	221.6	754.0	718.6	285.4	28.78	1373	137.3	6.907	49.8	.068	.309	1.03
	SCH 30 XS API	20.00	19.00	.5000	104.4	227.2	754.0	716.3	283.5	30.63	1457	145.7	6.897	50.6	.073	.338	1.13
	SCH 40	20.00	18.81	.5940	123.4	243.8	754.0	709.2	277.9	36.21	1706	170.6	6.864	52.9	.088	.425	1.42
	API	20.00	18.75	.6250	129.6	249.2	754.0	706.9	276.1	38.04	1787	178.7	6.854	53.6	.093	.454	1.51
	SCH 60 API	20.00	18.38	.8120	166.8	281.7	754.0	692.8	265.2	48.95	2257	225.7	6.790	56.6	.118	.631	2.10
	SCH 80	20.00	17.94	1.031	209.4	318.8	754.0	676.2	252.7	61.44	2772	277.2	6.716	59.0	.142	.841	2.80
	SCH 100	20.00	17.44	1.281	256.7	360.2	754.0	657.4	238.8	75.33	3315	331.5	6.634	60.7	.163	1.09	3.62
	SCH 120	20.00	17.00	1.500	297.1	395.4	754.0	640.9	227.0	87.18	3754	375.4	6.562	61.6	.177	1.30	4.35
	SCH 140	20.00	16.50	1.750	341.9	434.5	754.0	622.0	213.8	100.3	4216	421.6	6.482	62.3	.190	1.56	5.20
	SCH 160	20.00	16.06	1.969	380.1	467.9	754.0	605.5	202.6	111.5	4587	458.7	6.413	62.6	.198	1.79	5.96

PIPE DATA

TABLES P-1

NPS (inch)	Piping Codes / Sched.	Mfrs' Wt	O.D. (in.)	I.D. (in.)	Wall (in.)	Empty (lb/ft)	Waterfilled (lb/ft)	External (in^2/ft)	Internal (in^2/ft)	Flow (in^2)	Metal (in^2)	Moment of Inertia (in^4)	Section Modulus (in^3)	Radius of Gyration (in.)	Continuous Span (ft)	Sag (in.)	Design (kPSI)	Bursting (kPSI)
22	SCH 10	API	22.00	21.50	.2500	58.22	215.4	829.4	810.5	363.1	17.08	1010	91.84	7.690	41.3	.026	.099	.329
		API	22.00	21.44	.2810	65.34	221.6	829.4	808.2	361.0	19.17	1131	102.8	7.679	43.1	.031	.124	.414
		API	22.00	21.38	.3120	72.45	227.8	829.4	805.9	358.9	21.26	1250	113.7	7.669	44.7	.036	.150	.500
		API	22.00	21.31	.3440	79.76	234.2	829.4	803.4	356.7	23.40	1372	124.8	7.658	46.2	.041	.176	.588
	SCH 20 / STD	API	22.00	21.25	.3750	86.82	240.4	829.4	801.1	354.7	25.48	1490	135.4	7.647	47.5	.046	.202	.674
		API	22.00	21.19	.4060	93.87	246.5	829.4	798.8	352.6	27.54	1606	146.0	7.636	48.7	.051	.228	.760
		API	22.00	21.12	.4380	101.1	252.9	829.4	796.4	350.5	29.67	1725	156.8	7.625	49.8	.056	.255	.850
		API	22.00	21.06	.4690	108.1	259.0	829.4	794.0	348.4	31.72	1839	167.2	7.614	50.8	.061	.281	.936
	SCH 30 / XS	API	22.00	21.00	.5000	115.1	265.1	829.4	791.7	346.4	33.77	1952	177.5	7.603	51.8	.066	.307	1.02
		API	22.00	20.50	.7500	170.6	313.6	829.4	772.8	330.1	50.07	2830	257.2	7.518	57.3	.101	.519	1.73
	SCH 60	API	22.00	20.25	.8750	197.9	337.4	829.4	763.4	322.1	58.07	3245	295.0	7.475	59.1	.116	.627	2.09
	SCH 80	API	22.00	19.75	1.125	251.4	384.1	829.4	744.6	306.4	73.78	4030	366.4	7.391	61.8	.141	.845	2.82
	SCH 100	API	22.00	19.25	1.375	303.6	429.7	829.4	725.7	291.0	89.09	4759	432.6	7.308	63.5	.161	1.07	3.56
	SCH 120	API	22.00	18.75	1.625	354.5	474.1	829.4	706.9	276.1	104.0	5432	493.8	7.227	64.6	.176	1.29	4.31
	SCH 140	API	22.00	18.25	1.875	404.0	517.3	829.4	688.0	261.6	118.5	6054	550.3	7.146	65.2	.188	1.52	5.08
	SCH 160	API	22.00	17.75	2.125	452.2	559.3	829.4	669.2	247.4	132.7	6626	602.4	7.067	65.6	.197	1.76	5.87
24	SCH 10	API	24.00	23.50	.2500	63.57	251.4	904.8	885.9	433.7	18.65	1315	109.6	8.397	41.8	.023	.090	.301
		API	24.00	23.44	.2810	71.36	258.2	904.8	883.6	431.5	20.94	1473	122.7	8.387	43.6	.027	.114	.379
		API	24.00	23.38	.3120	79.13	265.0	904.8	881.3	429.2	23.22	1629	135.7	8.376	45.3	.032	.137	.458
		API	24.00	23.31	.3440	87.13	272.0	904.8	878.8	426.8	25.57	1789	149.1	8.365	46.8	.036	.162	.539
	SCH 20 / STD	API	24.00	23.25	.3750	94.85	278.7	904.8	876.5	424.6	27.83	1942	161.9	8.354	48.2	.041	.185	.618
		API	24.00	23.19	.4060	102.6	285.4	904.8	874.2	422.3	30.09	2095	174.6	8.343	49.5	.045	.209	.696
		API	24.00	23.12	.4380	110.5	292.3	904.8	871.8	420.0	32.42	2251	187.6	8.332	50.7	.050	.233	.778
		API	24.00	23.06	.4690	118.2	299.0	904.8	869.4	417.7	34.67	2401	200.1	8.321	51.7	.055	.257	.857
	XS	API	24.00	23.00	.5000	125.8	305.7	904.8	867.1	415.5	36.91	2549	212.4	8.310	52.7	.059	.281	.937
	SCH 30	API	24.00	22.88	.5620	141.0	319.0	904.8	862.4	411.0	41.38	2843	236.9	8.289	54.5	.068	.329	1.10
		API	24.00	22.75	.6250	156.4	332.4	904.8	857.7	406.5	45.90	3137	261.4	8.267	56.1	.077	.377	1.26
	SCH 40	API	24.00	22.62	.6880	171.7	345.8	904.8	852.9	402.0	50.39	3426	285.5	8.246	57.5	.085	.426	1.42
		API	24.00	22.50	.7500	186.7	358.9	904.8	848.2	397.6	54.78	3705	308.8	8.224	58.7	.093	.475	1.58
	SCH 60	API	24.00	22.06	.9690	238.9	404.5	904.8	831.7	382.3	70.11	4657	388.1	8.150	61.9	.117	.647	2.16
	SCH 80	API	24.00	21.56	1.219	297.3	455.4	904.8	812.9	365.1	87.24	5676	473.0	8.066	64.5	.140	.848	2.83
	SCH 100	API	24.00	20.94	1.531	368.3	517.4	904.8	789.3	344.3	108.1	6852	571.0	7.962	66.4	.163	1.10	3.67
	SCH 120	API	24.00	20.38	1.812	430.5	571.7	904.8	768.2	326.1	126.3	7825	652.0	7.871	67.5	.178	1.34	4.45
	SCH 140	API	24.00	19.88	2.062	484.4	618.7	904.8	749.3	310.3	142.1	8625	718.8	7.790	68.2	.188	1.55	5.16
	SCH 160	API	24.00	19.31	2.344	543.5	670.3	904.8	728.0	292.9	159.5	9458	788.2	7.701	68.6	.197	1.79	5.98
26	SCH 10	API	26.00	25.50	.2500	68.92	290.1	980.2	961.3	510.7	20.22	1676	129.0	9.104	42.2	.020	.083	.278
		API	26.00	25.44	.2810	77.38	297.5	980.2	959.0	508.2	22.70	1878	144.4	9.094	44.1	.024	.105	.350
		API	26.00	25.38	.3120	85.81	305.0	980.2	956.7	505.8	25.18	2077	159.8	9.083	45.8	.028	.127	.422
		API	26.00	25.31	.3440	94.49	312.4	980.2	954.2	503.2	27.73	2282	175.5	9.072	47.4	.032	.149	.497
	STD	API	26.00	25.25	.3750	102.9	319.7	980.2	951.9	500.7	30.19	2478	190.6	9.061	48.8	.037	.171	.570
		API	26.00	25.19	.4060	111.3	327.0	980.2	949.6	498.3	32.64	2674	205.7	9.050	50.2	.041	.193	.642
		API	26.00	25.12	.4380	119.9	334.8	980.2	947.2	495.8	35.17	2874	221.1	9.039	51.4	.045	.215	.718
		API	26.00	25.06	.4690	128.2	341.8	980.2	944.8	493.3	37.62	3066	235.9	9.028	52.5	.049	.237	.791
	XS	API	26.00	25.00	.5000	136.5	349.1	980.2	942.5	490.9	40.06	3257	250.5	9.017	53.6	.054	.259	.864
		API	26.00	24.88	.5620	153.1	363.5	980.2	937.8	486.0	44.91	3635	279.6	8.996	55.5	.062	.303	1.01
		API	26.00	24.75	.6250	169.8	378.1	980.2	933.1	481.1	49.82	4013	308.7	8.974	57.1	.070	.348	1.16
		API	26.00	24.50	.7500	202.8	406.9	980.2	923.6	471.4	59.49	4746	365.0	8.931	59.9	.085	.438	1.46
28	SCH 10	API	28.00	27.50	.2500	74.28	331.5	1056	1037	594.0	21.79	2098	149.9	9.812	42.5	.018	.077	.258
		API	28.00	27.44	.2810	83.39	339.4	1056	1034	591.3	24.47	2350	167.9	9.801	44.5	.022	.097	.325
		API	28.00	27.38	.3120	92.49	347.4	1056	1032	588.6	27.14	2601	185.8	9.790	46.3	.025	.118	.392
		API	28.00	27.31	.3440	101.9	355.6	1056	1030	585.9	29.89	2858	204.1	9.779	47.9	.029	.138	.461
	STD	API	28.00	27.25	.3750	110.9	363.5	1056	1027	583.2	32.54	3105	221.8	9.768	49.4	.033	.159	.529

NPS (inch)	Piping Codes	Mfr. Wt.	O.D. (in.)	I.D. (in.)	Wall (in.)	Empty (lb/ft)	Waterfilled (lb/ft)	External (in²/ft)	Internal (in²/ft)	Flow (in²)	Metal (in²)	Moment of Inertia (in⁴)	Section Modulus (in³)	Radius of Gyration (in.)	Span (ft)	Sag (in.)	Design (kPSI)	Bursting (kPSI)
	SCH 10	API	28.00	27.19	.4060	119.9	371.3	1056	1025	580.6	35.20	3351	239.3	9.757	50.8	.037	.179	.596
		API	28.00	27.12	.4380	129.3	379.5	1056	1023	577.8	37.93	3602	257.3	9.746	52.1	.041	.200	.666
		API	28.00	27.06	.4690	138.2	387.3	1056	1020	575.2	40.56	3844	274.6	9.735	53.3	.045	.220	.734
	SCH 20 XS	API	28.00	27.00	.5000	147.2	395.1	1056	1018	572.6	43.20	4085	291.8	9.724	54.3	.049	.240	.801
	SCH 30	API	28.00	26.75	.6250	183.2	426.5	1056	1008	562.0	53.75	5038	359.8	9.681	58.1	.064	.323	1.08
		API	28.00	26.50	.7500	218.8	457.6	1056	999.0	551.5	64.21	5964	426.0	9.638	61.0	.079	.406	1.35
30		API	30.00	29.44	.2810	89.41	384.1	1131	1110	680.6	26.24	2897	193.1	10.51	44.8	.019	.091	.303
	SCH 10	API	30.00	29.38	.3120	99.17	392.7	1131	1107	677.8	29.10	3206	213.8	10.50	46.7	.023	.110	.366
		API	30.00	29.31	.3440	109.2	401.4	1131	1105	675.2	32.05	3524	234.9	10.49	48.4	.026	.129	.430
	STD	API	30.00	29.25	.3750	118.9	409.9	1131	1103	674.4	34.90	3829	255.5	10.47	49.9	.030	.148	.493
		API	30.00	29.19	.4060	128.6	418.4	1131	1100	669.1	37.75	4133	275.5	10.46	51.3	.034	.167	.556
		API	30.00	29.12	.4380	138.6	427.1	1131	1098	666.2	40.68	4445	296.3	10.45	52.7	.037	.186	.621
		API	30.00	29.06	.4690	148.3	435.5	1131	1096	663.3	43.51	4744	316.3	10.44	53.9	.041	.205	.684
	XS	API	30.00	29.00	.5000	157.9	443.9	1131	1093	660.5	46.34	5042	336.1	10.43	55.0	.045	.224	.747
	SCH 20	API	30.00	28.75	.6250	196.6	477.7	1131	1084	649.2	57.68	6224	414.9	10.39	58.9	.059	.301	1.00
	SCH 30	API	30.00	28.50	.7500	234.9	511.1	1131	1074	637.9	68.92	7375	491.7	10.34	62.0	.073	.378	1.26
32	SCH 10	API	32.00	31.50	.2500	84.98	422.4	1206	1188	779.3	24.94	3142	196.4	11.23	43.1	.015	.068	.226
		API	32.00	31.44	.2810	95.43	431.6	1206	1185	776.2	28.00	3522	220.1	11.21	45.2	.018	.085	.284
		API	32.00	31.38	.3120	105.9	440.7	1206	1183	773.2	31.06	3899	243.7	11.20	47.0	.021	.103	.343
		API	32.00	31.31	.3440	116.6	450.0	1206	1180	770.0	34.21	4286	267.9	11.19	48.8	.024	.121	.403
	STD	API	32.00	31.25	.3750	127.0	459.1	1206	1178	767.0	37.26	4658	291.2	11.18	50.4	.027	.139	.462
		API	32.00	31.19	.4060	137.3	468.1	1206	1176	764.0	40.30	5029	314.3	11.17	51.8	.031	.156	.521
		API	32.00	31.12	.4380	148.0	477.5	1206	1173	760.8	43.43	5409	338.1	11.16	53.2	.034	.175	.582
		API	32.00	31.06	.4690	158.3	486.5	1206	1171	757.8	46.46	5775	360.9	11.15	54.5	.037	.192	.641
	XS	API	32.00	31.00	.5000	168.6	495.5	1206	1169	754.8	49.48	6139	383.7	11.14	55.7	.041	.210	.700
	SCH 20	API	32.00	30.75	.6250	209.9	531.5	1206	1159	742.6	61.60	7583	474.0	11.09	59.7	.055	.282	.940
	SCH 30	API	32.00	30.62	.6880	230.6	549.6	1206	1154	736.6	67.68	8298	518.6	11.07	61.4	.061	.318	1.06
	SCH 40	API	32.00	30.50	.7500	250.9	567.3	1206	1150	730.6	73.63	8993	562.1	11.05	63.0	.068	.354	1.18
34	SCH 10	API	34.00	33.50	.2500	90.34	472.0	1282	1263	881.4	26.51	3774	222.0	11.93	43.4	.013	.064	.212
		API	34.00	33.44	.2810	101.4	481.7	1282	1261	878.2	29.77	4231	248.9	11.92	45.5	.016	.080	.267
		API	34.00	33.38	.3120	112.5	491.4	1282	1258	874.9	33.02	4685	275.6	11.91	47.4	.019	.097	.323
		API	34.00	33.31	.3440	124.0	501.4	1282	1256	871.5	36.37	5151	303.0	11.90	49.2	.022	.114	.380
	STD	API	34.00	33.25	.3750	135.0	511.0	1282	1253	868.3	39.61	5599	329.4	11.89	50.8	.025	.130	.435
		API	34.00	33.19	.4060	146.0	520.6	1282	1251	865.1	42.85	6046	355.6	11.88	52.3	.028	.147	.490
		API	34.00	33.12	.4380	157.4	530.5	1282	1249	861.7	46.18	6504	382.6	11.87	53.7	.031	.164	.547
		API	34.00	33.06	.4690	168.4	540.1	1282	1246	858.5	49.40	6945	408.5	11.86	55.0	.034	.181	.603
	XS	API	34.00	33.00	.5000	179.3	549.7	1282	1244	855.3	52.62	7383	434.3	11.85	56.2	.038	.198	.659
	SCH 20	API	34.00	32.75	.6250	223.3	588.1	1282	1235	842.4	65.53	9128	536.9	11.80	60.4	.051	.265	.884
	SCH 30	API	34.00	32.62	.6880	245.4	607.4	1282	1230	835.9	72.00	9992	587.7	11.78	62.2	.057	.299	.998
	SCH 40	API	34.00	32.50	.7500	267.0	626.2	1282	1225	829.6	78.34	10832	637.2	11.76	63.8	.063	.333	1.11
36	SCH 10	API	36.00	35.50	.2500	95.69	524.3	1357	1338	989.8	28.08	4486	249.2	12.64	43.6	.012	.060	.201
		API	36.00	35.44	.2810	107.5	534.6	1357	1336	986.3	31.53	5029	279.4	12.63	45.5	.014	.076	.253
		API	36.00	35.38	.3120	119.2	544.8	1357	1334	982.9	34.98	5569	309.4	12.62	47.7	.017	.091	.305
		API	36.00	35.31	.3440	131.3	555.4	1357	1331	979.3	38.53	6124	340.2	12.61	49.5	.020	.108	.358
	STD	API	36.00	35.25	.3750	143.0	565.6	1357	1329	975.9	41.97	6659	369.9	12.60	51.1	.023	.123	.411
		API	36.00	35.19	.4060	154.7	575.8	1357	1327	972.5	45.40	7191	399.5	12.59	52.7	.026	.139	.463
		API	36.00	35.12	.4380	166.8	586.3	1357	1324	968.9	48.93	7737	429.8	12.57	54.1	.029	.155	.517
		API	36.00	35.06	.4690	178.4	596.5	1357	1322	965.5	52.35	8263	459.0	12.56	55.5	.032	.171	.569
	XS	API	36.00	35.00	.5000	190.0	606.7	1357	1319	962.1	55.76	8786	488.1	12.55	56.7	.035	.187	.622
	SCH 20	API	36.00	34.88	.5620	213.2	626.9	1357	1315	955.3	62.57	9825	545.8	12.53	59.0	.041	.218	.727
	SCH 30	API	36.00	34.75	.6250	236.7	647.4	1357	1310	948.4	69.46	10868	603.8	12.51	61.1	.047	.250	.834
	SCH 40	API	36.00	34.50	.7500	283.1	687.9	1357	1301	934.8	83.06	12906	717.0	12.47	64.6	.059	.314	1.05

PERSONNEL CLEARANCES

CLEARANCES TO MANUAL VALVES
AND SUGGESTED OPERATING HEIGHTS

OVERHEAD VALVES

FOR VALVE OPERATION ABOVE 6'-6" or 2 m, REFER TO 6.1.3, UNDER "OPERATING ACCESS TO VALVES

6'-6" or 2 m

MINIMUM ABOVE FLOOR or PLATFORM

INVERTED VALVES: REFER TO 6.1.3, UNDER 'ORIENTATION OF VALVE STEM'

VERTICAL VALVES

6'-6" or 2 m

4'-6" or 1.4 m
4'-3" or 1.3 m
3'-9" or 1140 mm

2'-0" or 610 mm

0'-0"

18" or 460 mm Max.

③
②

HORIZONTAL VALVES

6'-6" or 2 m

4'-6" or 1.4 m
3'-6" or 1070 mm
2'-0" or 610 mm

0'-0"

18" or 460 mm Max.

③
②
①

ZONES

☆ PREFERRED ELEVATIONS

▨ SECOND-CHOICE ELEVATIONS

▧ LEG OR HEAD HAZARD, UNLESS PROTECTION GIVEN BY RAILING, PIPING, EQUIPMENT, Etc.

NOTES

(1) TAKE CHAINS TO 3'-0" (OR 900 mm) FROM OPERATING FLOOR LEVEL. DO NOT HANG CHAINS IN A WALKWAY.

(2) DIMENSION APPLIES IF A RAILING IS PRESENT.

(3) IF A RAILING IS PRESENT, COMFORTABLE OPERATING ELEVATION IS 5'-0" TO 5'-6" (or 1.5 TO 1.7 m).

(4) GENERAL CLEARANCE FORMULAS ARE:
(a) 5.5 — (pitch angle/30) ft
(b) 1.68 — (pitch angle/100) m.

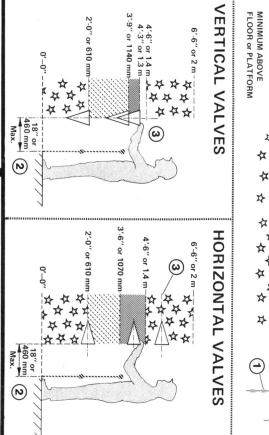

CLEARANCES AROUND STAIRWAYS & LADDERS
CHART P-2

DATA FROM THE CODE OF FEDERAL REGULATIONS 1910—"OCCUPATIONAL SAFETY AND HEALTH STANDARDS" (1984).

REFER TO CHART S-1.

7'-0" (or 2130 mm) minimum between tread and obstruction

2'-6" to 2'-10"
760 to 860 mm

30° to 50°

Minimum width of stairway is 22" or 560 mm

Minimum distances of rungs from obstruction at rear

WALL, Etc.

12"
12" Max.

4½" or 115 mm
1½" or 40 mm

Min. 2'-6" or 760 mm

WALL, Etc.

Minimum width of rungs = 16" or 400 mm

7" or 180 mm Min

15" or 380 mm
15" or 380 mm

Minimum lateral spacing for noncaged ladders

Preferred ladder pitch = 75° to 90°

④

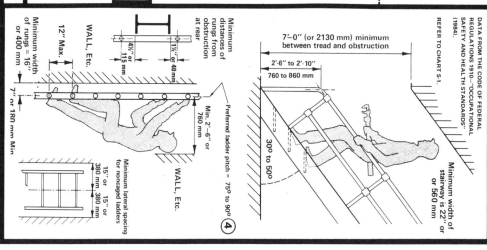

TABLE 6.1 GIVES ADDITIONAL DIMENSIONS

MATERIALS	FAHRENHEIT	CELSIUS
Aluminum	0.000 012 8	0.000 023 1
Carbon steel	0.000 006 5	0.000 011 7
Cast iron	0.000 005 9	0.000 010 62
Copper	0.000 009 3	0.000 016 8
Stainless steel	0.000 009 9	0.000 017 82

COEFFICIENTS OF EXPANSION OF DIFFERENT PIPING MATERIALS (in inches/degree/inch of length)

MATERIALS		FAHRENHEIT	CELSIUS
ABS:	Acrylonitrile-butadiene-styrene	0.000 035	0.000 063
HDPE:	High-density polyethylene	0.000 067	0.000 12
PE:	Polyethylene	0.000 083	0.000 15
CPVC:	Chlorinated polyvinyl chloride	0.000 044	0.000 079 2
PVC:	Polyvinyl chloride	0.000 028	0.000 050 4

SPANS OF HORIZONTAL PIPE

THESE TABLES GIVE SPANS SUITABLE FOR PIPE ARRANGED IN PIPEWAYS, AND APPLY WHEN THE SPAN IS PART OF A STRAIGHT PIPE, WITH TWO OR MORE SPANS AT EACH END.

TABLE S-1

FOR VALUES OF BENDING STRESS & MODULUS, REFER TO CHARTS S-2

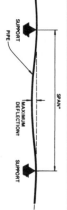

STEEL PIPE, SCHEDULE 160

NOMINAL PIPE SIZE	PIPE SPAN* Ft.	PIPE SPAN* In.	WEIGHT OF WATER-FILLED PIPE SPAN (Lb)	MAXIMUM DEFLECTION† (In.)
1.0-INCH	15	8.77	48	0.234
1.5-INCH	19	3.28	105	0.243
2.0-INCH	21	6.79	182	0.243
2.5-INCH	23	9.87	275	0.245
3.0-INCH	26	3.66	438	0.245
4.0-INCH	29	9.30	793	0.245
6.0-INCH	36	2.01	1,970	0.245
8.0-INCH	41	2.89	3,732	0.245
10.0-INCH	45	11.75	6,465	0.244
12.0-INCH	50	0.40	9,801	0.244
14.0-INCH	52	4.67	12,186	0.244
16.0-INCH	56	0.99	16,875	0.244
18.0-INCH	59	5.13	22,582	0.244
20.0-INCH	62	8.17	29,266	0.244
24.0-INCH	68	7.74	45,923	0.244

STEEL PIPE, SCHEDULE 80

NOMINAL PIPE SIZE	PIPE SPAN* Ft.	PIPE SPAN* In.	WEIGHT OF WATER-FILLED PIPE SPAN (Lb)	MAXIMUM DEFLECTION† (In.)
1.0-INCH	16	1.05	40	0.244
1.5-INCH	19	4.29	85	0.243
2.0-INCH	21	6.49	136	0.243
2.5-INCH	23	9.02	225	0.244
3.0-INCH	26	0.66	342	0.244
4.0-INCH	29	3.07	584	0.241
6.0-INCH	35	0.22	1,396	0.236
8.0-INCH	39	4.67	2,489	0.230
10.0-INCH	43	8.21	4,172	0.223
12.0-INCH	47	5.26	6,290	0.220
14.0-INCH	49	9.35	7,883	0.219
16.0-INCH	52	10.78	10,934	0.217
18.0-INCH	56	0.58	14,545	0.217
20.0-INCH	59	0.02	18,786	0.216
24.0-INCH	64	5.48	29,341	0.215

STEEL PIPE, SCHEDULE 40

NOMINAL PIPE SIZE	PIPE SPAN* Ft.	PIPE SPAN* In.	WEIGHT OF WATER-FILLED PIPE SPAN (Lb)	MAXIMUM DEFLECTION† (In.)
1.0-INCH	16	1.07	33	0.244
1.5-INCH	19	0.49	69	0.237
2.0-INCH	20	11.53	107	0.230
2.5-INCH	23	3.20	183	0.234
3.0-INCH	25	3.65	273	0.227
4.0-INCH	28	1.01	458	0.218
6.0-INCH	32	10.37	1,035	0.202
8.0-INCH	36	7.40	1,836	0.193
10.0-INCH	40	0.55	2,987	0.185
12.0-INCH	42	11.48	4,386	0.180
14.0-INCH	44	11.52	5,463	0.179
16.0-INCH	47	10.83	7,640	0.178
18.0-INCH	50	11.02	10,289	0.179
20.0-INCH	52	11.02	12,880	0.174
24.0-INCH	57	5.84	19,844	0.171

STEEL PIPE, SCHEDULE 20

NOMINAL PIPE SIZE	PIPE SPAN* Ft.	PIPE SPAN* In.	WEIGHT OF WATER-FILLED PIPE SPAN (Lb)	MAXIMUM DEFLECTION† (In.)
8.0-INCH	34	6.46	1,551	0.172
10.0-INCH	36	4.22	2,324	0.152
12.0-INCH	37	9.18	3,199	0.139
14.0-INCH	41	0.64	4,385	0.149
16.0-INCH	42	4.07	5,593	0.139
18.0-INCH	43	2.92	6,984	0.129
20.0-INCH	46	7.22	9,553	0.135
24.0-INCH	48	2.35	13,437	0.120
30.0-INCH	54	11.58	24,415	0.125

STEEL PIPE, SCHEDULE 10

NOMINAL PIPE SIZE	PIPE SPAN* Ft.	PIPE SPAN* In.	WEIGHT OF WATER-FILLED PIPE SPAN (Lb)	MAXIMUM DEFLECTION† (In.)
1.0-INCH	15	11.14	29	0.240
1.5-INCH	18	5.62	56	0.223
2.0-INCH	19	11.77	84	0.209
2.5-INCH	21	7.24	127	0.202
3.0-INCH	22	10.63	182	0.186
4.0-INCH	24	5.31	288	0.164
6.0-INCH	27	5.75	632	0.141
8.0-INCH	29	9.72	1,103	0.128
10.0-INCH	32	0.93	1,782	0.119
12.0-INCH	33	11.37	2,592	0.112
14.0-INCH	38	5.23	3,809	0.131
16.0-INCH	39	4.50	4,886	0.120
18.0-INCH	40	8.77	6,087	0.111
20.0-INCH	40	1.82	7,454	0.103
24.0-INCH	41	9.43	10,530	0.090

ALUMINUM PIPE, SCHEDULE 80

NOMINAL PIPE SIZE	PIPE SPAN* Ft.	PIPE SPAN* In.	WEIGHT OF WATER-FILLED PIPE SPAN (Lb)	MAXIMUM DEFLECTION† (In.)
1.0-INCH	17	4.67	18	0.414
1.5-INCH	20	2.26	41	0.386
2.0-INCH	22	0.19	66	0.367
2.5-INCH	24	5.26	110	0.374
3.0-INCH	26	4.25	169	0.357
4.0-INCH	28	11.94	295	0.336
6.0-INCH	33	11.69	719	0.314
8.0-INCH	37	6.31	1,306	0.294
10.0-INCH	39	8.42	1,935	0.264

ALUMINUM PIPE, SCHEDULE 40

NOMINAL PIPE SIZE	PIPE SPAN* Ft.	PIPE SPAN* In.	WEIGHT OF WATER-FILLED PIPE SPAN (Lb)	MAXIMUM DEFLECTION† (In.)
1.0-INCH	16	8.12	16	0.381
1.5-INCH	18	11.07	34	0.339
2.0-INCH	20	3.81	55	0.313
2.5-INCH	22	10.19	93	0.327
3.0-INCH	24	4.06	142	0.305
4.0-INCH	26	4.46	244	0.278
6.0-INCH	29	10.16	569	0.242
8.0-INCH	32	8.17	1,029	0.223
10.0-INCH	35	3.12	1,696	0.208

SPANS OF HORIZONTAL PIPE
WITH 3-FT. RISE OR FALL

CHARTS S-2

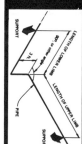

THESE CHARTS GIVE THE MAXIMUM LENGTH PERMISSIBLE FOR EITHER HORIZONTAL LIMB IN THE PIPING ARRANGEMENT SHOWN, AND APPLY WHEN THE SPAN INCLUDING THE RISE OR FALL IS CONTINUOUS WITH TWO OR MORE STRAIGHT SPANS AT EACH END.

LENGTH OF LOWER LIMB (FT.) *(upper chart top axis: 0, 5, 10, 15, 20, 25, 30, 35, 40)*

ALUMINUM PIPE SCHEDULE 40 (WATER FILLED)

1-INCH NPS
2-INCH NPS
3-INCH NPS
4-INCH NPS
6-INCH NPS
8-INCH NPS
10-INCH NPS

LENGTH OF UPPER LIMB (FT.) *(upper chart right axis: 45, 40, 35, 30, 25, 20, 15, 10, 5, 0)*

LENGTH OF UPPER LIMB (FT.) *(lower chart left axis: 0, 5, 10, 15, 20, 25, 30, 35, 40)*

ALUMINUM PIPE SCHEDULE 80 (WATER FILLED)

10-INCH NPS
8-INCH NPS
6-INCH NPS
4-INCH NPS
3-INCH NPS
2-INCH NPS
1-INCH NPS

LENGTH OF LOWER LIMB (FT.) *(lower chart bottom axis: 45, 40, 35, 30, 25, 20, 15, 10, 5, 0)*

LOWER CHART:
SCH 40, ALUMINUM

Data for water-filled steel pipe are based on a maximum bending stress of 4000 PSI, occurring at supports and due to bending by the weight of pipe plus water; applied stresses may increase the resultant tensile stress. These data apply to carbon-steel and stainless-steel pipe having a tensile modulus of elasticity of 29,000,000 PSI. For water-filled aluminum pipe, spans are similarly based on a stress of 2000 PSI and a modulus of 10,000,000 PSI.

UPPER CHART:
SCH 80, ALUMINUM

SPANS OF HORIZONTAL PIPE WITH 3-FT. RISE OR FALL

CHARTS S-2

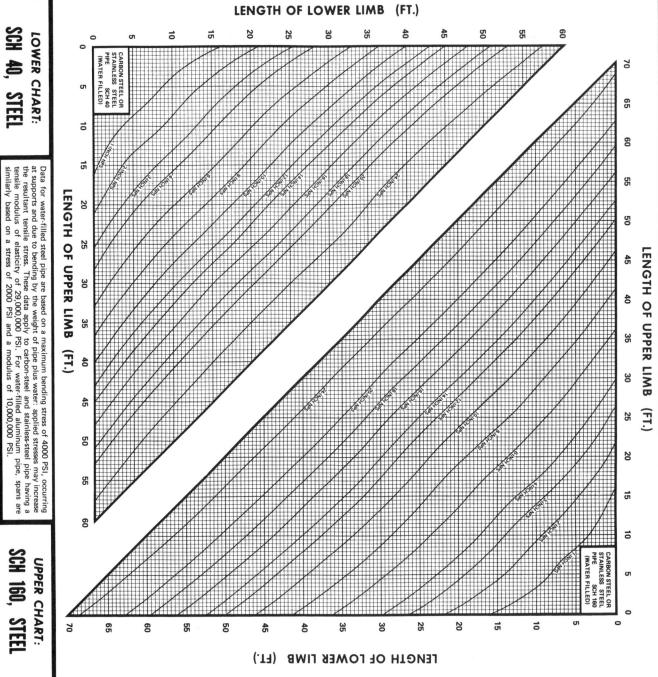

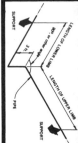

THESE CHARTS GIVE THE MAXIMUM LENGTH PERMISSIBLE FOR EITHER HORIZONTAL LIMB IN THE PIPING ARRANGEMENT SHOWN, AND APPLY WHEN THE SPAN INCLUDING THE RISE OR FALL IS CONTINUOUS WITH TWO OR MORE STRAIGHT SPANS AT EACH END.

LENGTH OF LOWER LIMB (FT.)

LENGTH OF UPPER LIMB (FT.)

CARBON STEEL OR STAINLESS STEEL PIPE SCH 40 (WATER FILLED)

CARBON STEEL OR STAINLESS STEEL PIPE SCH 160 (WATER FILLED)

LENGTH OF UPPER LIMB (FT.)

LENGTH OF LOWER LIMB (FT.)

LOWER CHART: SCH 40, STEEL

UPPER CHART: SCH 160, STEEL

Data for water-filled steel pipe are based on a maximum bending stress of 4000 PSI, occurring at supports and due to bending by the weight of pipe plus water; applied stresses may increase the resultant tensile stress. These data apply to carbon-steel and stainless-steel pipe having a tensile modulus of elasticity of 29,000,000 PSI. For water-filled aluminum pipe, spans are similarly based on a stress of 2000 PSI and a modulus of 10,000,000 PSI.

[57]

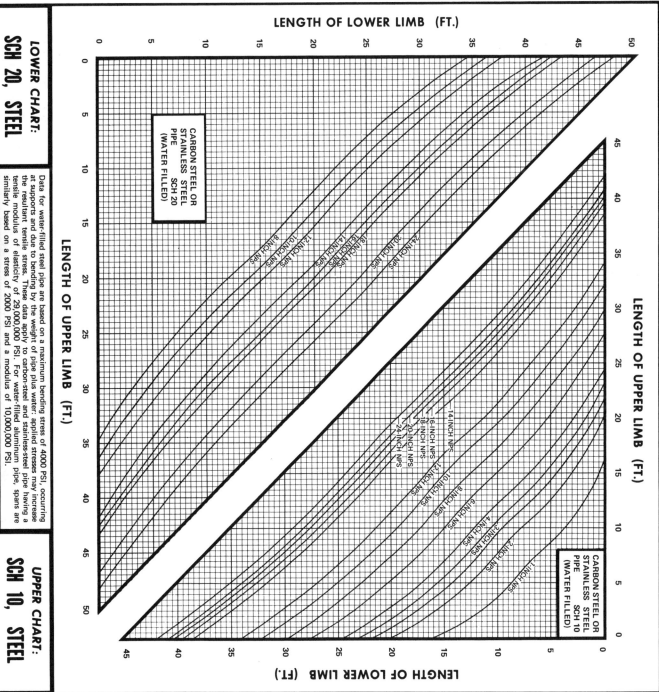

SPANS OF HORIZONTAL PIPE WITH 3-FT. RISE OR FALL

CHARTS S-2

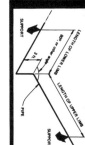

THESE CHARTS GIVE THE MAXIMUM LENGTH PERMISSIBLE FOR THE HORIZONTAL LIMB IN THE PIPING ARRANGEMENT SHOWN, AND APPLY WHEN THE SPAN INCLUDING THE RISE OR FALL IS CONTINUOUS WITH TWO OR MORE STRAIGHT SPANS AT EACH END.

LENGTH OF LOWER LIMB (FT.)

LENGTH OF UPPER LIMB (FT.)

CARBON STEEL OR STAINLESS STEEL PIPE SCH 20 (WATER FILLED)

8-INCH NPS
10-INCH NPS
12-INCH NPS
14-INCH NPS
16-INCH NPS
18-INCH NPS
20-INCH NPS
24-INCH NPS

CARBON STEEL OR STAINLESS STEEL PIPE SCH 10 (WATER FILLED)

14-INCH NPS
16-INCH NPS
18-INCH NPS
20-INCH NPS
24-INCH NPS
12-INCH NPS
10-INCH NPS
8-INCH NPS
6-INCH NPS
4-INCH NPS
3-INCH NPS
2-INCH NPS
1-INCH NPS

LENGTH OF LOWER LIMB (FT.)

LOWER CHART: SCH 20, STEEL

Data for water-filled steel pipe are based on a maximum bending stress of 4000 PSI, occurring at supports and due to bending by the weight of pipe plus water; applied stresses may increase the resultant tensile stress. These data apply to carbon-steel and stainless-steel pipe having a tensile modulus of elasticity of 29,000,000 PSI. For water-filled aluminum pipe, spans are similarly based on a stress of 2000 PSI and a modulus of 10,000,000 PSI.

UPPER CHART: SCH 10, STEEL

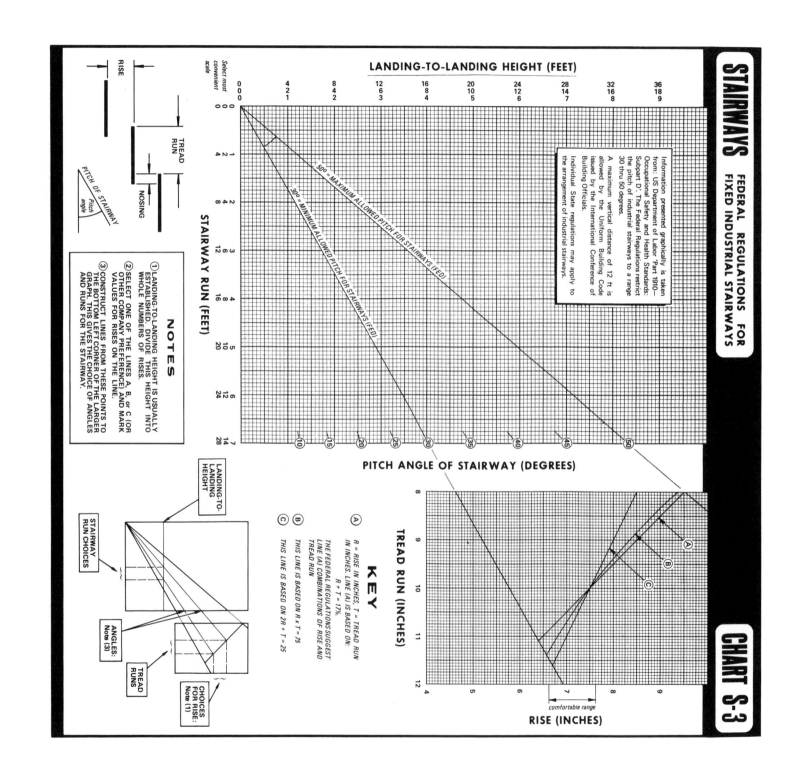

STRUCTURAL STEEL

TABLE S-4

W SHAPES

DECIMAL DIMENSIONS ARE NOMINAL – IN MULTIPLES OF 1/16"

I-beam cross-section labeled DEPTH, WIDTH, THICKNESS

> * INDICATES A DIMENSIONAL CHANGE OR SHAPE WAS DISCONTINUED (1978)

Column layout for each group:
DESIGNATION (NOM. SIZE × lb/ft) | DEPTH | WIDTH | THICK — DIMENSIONS: inches

W 36
DESIGNATION	DEPTH	WIDTH	THICK
W 36 × 300	36.75	16.62	1.69 *
× 280	36.50	16.62	1.56 *
× 260	36.25	16.50	1.44 *
× 245	36.12	16.50	1.38 *
× 245	36.12	16.50	1.38 *
× 230	35.88	16.50	1.26 *
× 210	36.75	12.12	1.38 *
× 194	36.50	12.12	1.26 *
× 182	36.38	12.12	1.18 *
× 170	36.12	12.12	1.12 *
× 160	36.00	12.00	1.02 *
× 150	35.88	12.00	0.94 *
× 135	35.50	12.00	0.81 *

W 33
DESIGNATION	DEPTH	WIDTH	THICK
W 33 × 241	34.12	15.88	1.38 *
× 240	34.50	15.88	1.38 *
× 221	33.88	15.75	1.25 *
× 201	33.62	15.75	1.12 *
× 152	33.50	11.62	1.06 *
× 141	33.25	11.50	0.94 *
× 130	33.12	11.50	0.88 *
× 118	32.88	11.50	0.75 *

W 30
DESIGNATION	DEPTH	WIDTH	THICK
W 30 × 211	31.00	15.12	1.31 *
× 210	30.38	15.12	1.31 *
× 191	30.62	15.12	1.19 *
× 173	30.50	15.00	1.06 *
× 172	30.12	15.00	1.06 *
× 132	30.25	10.50	1.00 *
× 124	30.12	10.50	0.94 *
× 116	30.00	10.50	0.88 *
× 108	29.88	10.50	0.75 *
× 99	29.62	10.50	0.62 *

W 27
DESIGNATION	DEPTH	WIDTH	THICK
W 27 × 178	27.75	14.12	1.19 *
× 177	27.25	14.12	1.19 *
× 161	27.62	14.00	1.06 *
× 160	27.12	14.00	1.06 *
× 146	27.38	14.00	1.00 *
× 145	27.25	14.00	1.00 *
× 114	27.12	10.12	0.94 *
× 102	27.12	10.00	0.81 *
× 94	26.88	10.00	0.75 *
× 84	26.75	10.00	0.62 *

W 24
DESIGNATION	DEPTH	WIDTH	THICK
W 24 × 162	25.00	13.00	1.25 *
× 160	24.75	14.12	1.12 *
× 146	24.75	12.88	1.06 *
× 145	24.50	12.88	1.00 *
× 131	24.50	12.88	0.94 *
× 130	24.25	14.00	0.88 *
× 120	24.12	12.88	0.94 *
× 117	24.25	12.75	0.88 *
× 104	24.00	12.75	0.75 *
× 100	24.00	12.75	0.75 *
× 94	24.25	9.12	0.88 *
× 84	24.12	9.00	0.75 *
× 76	24.00	9.00	0.69 *
× 68	23.88	9.00	0.59 *
× 62	23.75	7.00	0.59 *
× 61	23.75	7.00	0.56 *
× 55	23.62	7.00	0.50 *
× 55	23.50	7.00	0.50 *

W 21
DESIGNATION	DEPTH	WIDTH	THICK
W 21 × 147	22.00	12.50	1.12 *
× 142	21.50	13.12	1.06 *
× 132	21.88	12.50	1.06 *
× 127	21.62	13.00	1.00 *
× 122	21.62	12.38	1.00 *
× 111	21.50	12.38	0.88 *
× 101	21.38	12.25	0.81 *
× 93	21.62	8.38	0.94 *
× 83	21.38	8.38	0.81 *
× 73	21.25	8.25	0.75 *
× 68	21.12	8.25	0.69 *
× 62	21.00	8.25	0.62 *
× 57	21.12	6.50	0.69 *
× 55	20.75	8.25	0.56 *
× 50	20.88	6.50	0.56 *
× 49	20.88	6.50	0.56 *
× 44	20.62	6.50	0.44 *

W 18
DESIGNATION	DEPTH	WIDTH	THICK
W 18 × 119	19.00	11.25	1.06 *
× 114	18.50	11.88	0.94 *
× 106	18.75	11.25	0.94 *
× 105	18.38	11.75	0.88 *
× 97	18.62	11.12	0.88 *
× 96	18.12	11.75	0.81 *
× 86	18.38	11.12	0.75 *
× 85	18.38	8.75	0.81 *
× 77	18.12	8.75	0.75 *
× 76	18.25	11.00	0.69 *
× 71	18.50	7.62	0.81 *
× 70	18.00	8.75	0.69 *
× 65	18.38	7.62	0.75 *
× 64	17.88	8.50	0.62 *
× 60	18.25	7.50	0.69 *
× 55	18.12	7.50	0.62 *
× 50	18.00	7.50	0.56 *
× 46	18.06	6.00	0.62 *
× 45	17.88	7.50	0.50 *
× 40	17.88	6.00	0.56 *
× 35	17.75	6.00	0.44 *

W 16
DESIGNATION	DEPTH	WIDTH	THICK
W 16 × 100	17.00	10.38	0.88 *
× 96	16.38	11.50	0.88 *
× 89	16.75	10.38	0.88 *
× 88	16.12	11.38	0.81 *
× 78	16.50	10.25	0.69 *
× 77	16.50	10.25	0.62 *
× 71	16.38	10.12	0.75 *
× 67	16.38	10.25	0.62 *
× 64	16.62	8.50	0.62 *
× 58	16.38	8.50	0.62 *
× 57	16.50	7.12	0.62 *
× 50	16.25	7.12	0.62 *
× 45	16.12	7.00	0.56 *
× 40	16.00	7.00	0.50 *
× 36	15.88	7.00	0.44 *
× 31	15.88	5.50	0.44 *
× 26	15.62	5.50	0.38 *

W 14
DESIGNATION	DEPTH	WIDTH	THICK
W 14 × 730	22.38	17.88	4.94 *
× 730	22.50	17.88	4.94 *
× 665	21.62	17.62	4.50 *
× 605	20.88	17.38	4.16 *
× 550	20.25	17.25	3.81 *
× 500	19.62	17.00	3.50 *
× 455	19.00	16.88	3.19 *
× 426	18.62	16.75	3.06 *
× 398	18.25	16.62	2.88 *
× 370	17.88	16.50	2.69 *

W 12
DESIGNATION	DEPTH	WIDTH	THICK
W 12 × 336	16.88	13.38	2.94 *
× 305	16.38	13.25	2.69 *
× 279	15.88	13.12	2.50 *
× 252	15.38	13.00	2.25 *
× 230	15.00	12.88	1.88 *
× 210	14.75	12.75	1.75 *
× 190	14.38	12.62	1.50 *
× 170	14.00	12.50	1.38 *
× 161	13.88	12.50	1.25 *
× 152	13.75	12.38	1.25 *
× 136	13.38	12.38	1.12 *
× 133	13.75	12.12	1.25 *
× 120	13.12	12.38	1.12 *
× 106	12.88	12.25	0.99 *
× 96	12.75	12.12	0.88 *
× 92	12.62	12.12	0.88 *
× 87	12.50	12.12	0.81 *
× 85	12.50	12.12	0.81 *
× 79	12.38	12.12	0.75 *
× 72	12.25	12.00	0.69 *
× 65	12.12	12.00	0.62 *
× 58	12.25	10.00	0.62 *
× 53	12.00	10.00	0.56 *

W 14
DESIGNATION	DEPTH	WIDTH	THICK
W 14 × 50	12.00	8.12	0.62 *
× 45	12.25	8.00	0.56 *
× 40	12.00	8.00	0.50 *
× 35	12.50	6.50	0.44 *
× 30	12.25	6.50	0.44 *
× 27	12.00	6.50	0.38 *
× 22	12.25	4.00	0.44 *
× 16.5	12.12	4.00	0.38 *
× 14	11.88	4.00	0.25 *

W 10
DESIGNATION	DEPTH	WIDTH	THICK
W 10 × 112	11.38	10.38	1.25 *
× 100	11.12	10.38	1.12 *
× 89	10.88	10.25	1.00 *
× 77	10.62	10.25	0.88 *
× 72	10.50	10.12	0.81 *
× 66	10.38	10.12	0.75 *
× 60	10.25	10.12	0.69 *
× 54	10.12	10.00	0.62 *
× 49	10.00	10.00	0.56 *
× 45	10.12	8.00	0.62 *
× 39	9.88	8.00	0.50 *
× 33	9.75	8.00	0.44 *
× 30	10.50	5.75	0.50 *
× 29	10.25	5.75	0.50 *
× 25	10.12	5.75	0.44 *
× 22	10.12	5.75	0.38 *
× 21	9.88	5.75	0.38 *
× 19	10.25	4.00	0.44 *
× 17	10.12	4.00	0.38 *
× 15	10.00	4.00	0.31 *
× 11.5	9.88	4.00	0.19 *

W 8
DESIGNATION	DEPTH	WIDTH	THICK
W 8 × 67	9.00	8.25	0.94 *
× 58	8.75	8.25	0.81 *
× 48	8.50	8.12	0.69 *
× 40	8.25	8.12	0.56 *
× 35	8.12	8.00	0.50 *
× 31	8.00	8.00	0.44 *
× 28	8.06	6.50	0.46 *
× 24	7.88	6.50	0.40 *
× 21	8.25	5.25	0.38 *
× 20	8.12	5.25	0.38 *
× 18	8.12	5.25	0.31 *
× 17	8.00	5.25	0.31 *
× 15	8.12	4.00	0.31 *
× 13	8.00	4.00	0.25 *
× 10	7.88	4.00	0.19 *

W 6
DESIGNATION	DEPTH	WIDTH	THICK
W 6 × 25	6.38	6.12	0.44 *
× 20	6.25	6.00	0.38 *
× 16	6.25	6.00	0.25 *
× 15.5	6.00	6.00	0.25 *
× 12	6.00	6.00	0.19 *
× 9	5.88	4.00	0.19 *
× 8.5	5.88	4.00	0.19 *

W 5
DESIGNATION	DEPTH	WIDTH	THICK
W 5 × 19	5.12	5.00	0.44 *
× 18.5	5.12	5.00	0.44 *
× 16	5.00	5.00	0.38 *

W 4
DESIGNATION	DEPTH	WIDTH	THICK
W 4 × 13	4.12	4.00	0.38

CHANNEL DATA
AMERICAN STANDARD

AMERICAN STANDARD CHANNELS (diagram showing DEPTH, WIDTH, AVERAGE THICKNESS)

DESIGNATION Depth (nom) x wgt lb/ft	DIMENSIONS IN INCHES DEPTH	WIDTH	THICK
C 15x50	15.00	3.75	0.62
x40	15.00	3.50	0.62
x33.9	15.00	3.38	0.62
C 12x30	12.00	3.12	0.50
x25	12.00	3.00	0.50
x20.7	12.00	3.00	0.50
C 10x30	10.00	3.00	0.44
x25	10.00	2.88	0.44
x20	10.00	2.75	0.44
x15.3	10.00	2.62	0.44
C 9x20	9.00	2.62	0.44
x15	9.00	2.50	0.44
x13.4	9.00	2.38	0.44
C 8x18.75	8.00	2.50	0.38
x13.75	8.00	2.38	0.38
x11.5	8.00	2.25	0.38
C 7x14.75	7.00	2.25	0.38
x12.25	7.00	2.25	0.38
x9.8	7.00	2.12	0.38
C 6x13	6.00	2.12	0.38
x10.5	6.00	2.00	0.38
x8.2	6.00	1.88	0.38
C 5x9	5.00	1.88	0.31
x6.7	5.00	1.75	0.31
C 4x7.25	4.00	1.75	0.31
x5.4	4.00	1.62	0.31
C 3x6	3.00	1.62	0.25
x5	3.00	1.50	0.25
x4.1	3.00	1.38	0.25

ANGLE DATA
WEIGHTS IN POUNDS PER LINEAR FOOT

TABLES S-5

UNEQUAL LEGS

EXAMPLE DESIGNATION: L 2 x 1 1/2 x 1/4

SIZE	1	7/8	3/4	5/8	9/16	1/2	7/16	3/8	5/16	1/4	3/16	1/8
1 3/4 x 1 1/4 x										2.34	1.8	1.23
2 x 1 1/4 x										2.55	1.96	
2 x 1 1/2 x										2.77	2.12	1.44
2 1/2 x 1 1/2 x									3.92	3.19	2.44	
2 1/2 x 2 x								5.3	4.5	3.62	2.75	
3 x 2 x						7.7	6.8	5.9	5.0	4.1	3.07	
3 x 2 1/2 x						8.5	7.6	6.6	5.6	4.5	3.39	
3 1/2 x 2 1/2 x						9.4	8.3	7.2	6.1	4.9		
3 1/2 x 3 x						10.2	9.1	7.9	6.6	5.3		
4 x 3 x				13.6		11.1	9.8	8.5	7.2	5.8		
4 x 3 1/2 x				14.7		11.9	10.6	9.1	7.7	6.2		
5 x 3 x				15.7		12.8	11.3	9.8	8.2	6.6		
5 x 3 1/2 x			19.8	16.8		13.6		10.4	8.7	7.0		
6 x 3 1/2 x						15.3		11.7	9.8			
6 x 4 x			23.6	20.0	18.1	16.2	14.3	12.3	10.3			
7 x 4 x			26.2	22.1		17.9	15.7	13.6				
8 x 4 x	37.4	33.1	28.7	24.2	21.9	19.6	17.2					
8 x 6 x	44.2	39.1	33.8	28.5	25.7	23.0	20.2					
9 x 4 x	40.8	36.1	31.3	26.3	23.8	21.3						

EQUAL LEGS

EXAMPLE DESIGNATION: L 3 x 3 x 3/8

SIZE	1 1/8	1	7/8	3/4	5/8	9/16	1/2	7/16	3/8	5/16	1/4	3/16	1/8
1 x 1 x											1.49	1.16	0.80
1 1/4 x 1 1/4 x											1.92	1.48	1.01
1 1/2 x 1 1/2 x											2.34	1.80	1.23
1 3/4 x 1 3/4 x											2.77	2.12	1.44
2 x 2 x									4.7	3.92	3.19	2.44	1.65
2 1/2 x 2 1/2 x							7.7		5.9	5.0	4.1	3.07	
3 x 3 x							9.4	8.3	7.2	6.1	4.9	3.71	
3 1/2 x 3 1/2 x							11.1	9.8	8.5	7.2	5.8		
4 x 4 x				18.5	15.7		12.8	11.3	9.8	8.2	6.6		
5 x 5 x			27.2	23.6	20.0		16.2	14.3	12.3	10.3			
6 x 6 x		37.4	33.1	28.7	24.2	21.9	19.6	17.2	14.9				
8 x 8 x	56.9	51.0	45.0	38.9	32.7	29.6	26.4						

REPRODUCED BY COURTESY OF STOCKHAM VALVES AND FITTINGS

The following dimensional data for copper tube conform to ASTM B-88, which specifies general requirements for Wrought Seamless Copper Alloy Pipe and Tube.

TYPE K TUBE

Heavy wall thickness, hard or soft, is furnished for interior plumbing and underground service; steam and hot water heating systems; fuel oil lines; industrial process applications carrying liquids, air and gases; air conditioning, refrigeration, and low pressure hydraulic lines. Hard copper tube is used for gas service lines because its rigidity eliminates traps caused by sagging lines.

Nominal Size	NOMINAL DIMENSIONS			THEORETICAL AREAS BASED ON NOMINAL DIMENSIONS			
	Outside Diameter (Inches)	Inside Diameter (Inches)	Wall Thickness (Inches)	Cross Sectional Area of Bore (Sq. Inches)	External Surface (Sq. Ft. Per Lin. Ft.)	Internal Surface (Sq. Ft. Per Lin. Ft.)	Theoretical Weight (Pounds Per Foot)
1/4	.375	.305	.035	.073	.098	.080	0.145
3/8	.500	.402	.049	.127	.131	.105	0.269
1/2	.625	.527	.049	.218	.164	.138	0.344
3/4	.875	.745	.065	.436	.229	.195	0.641
1	1.125	.995	.065	.778	.294	.261	0.839
1 1/4	1.375	1.245	.065	1.22	.360	.326	1.04
1 1/2	1.625	1.481	.072	1.72	.425	.388	1.36
2	2.125	1.959	.083	3.01	.556	.513	2.06
2 1/2	2.625	2.435	.095	4.66	.687	.638	2.93
3	3.125	2.907	.109	6.64	.818	.761	4.00

TYPE L TUBE

Medium wall thickness, hard or soft, is used for medium pressure interior plumbing and for steam and hot water house-heating systems, panel heating, plumbing vent systems, industrial and process applications.

Nominal Size	NOMINAL DIMENSIONS			THEORETICAL AREAS BASED ON NOMINAL DIMENSIONS			
	Outside Diameter (Inches)	Inside Diameter (Inches)	Wall Thickness (Inches)	Cross Sectional Area of Bore (Sq. Inches)	External Surface (Sq. Ft. Per Lin. Ft.)	Internal Surface (Sq. Ft. Per Lin. Ft.)	Theoretical Weight (Pounds Per Foot)
1/4	.375	.315	.030	.078	.098	.082	0.126
3/8	.500	.430	.035	.145	.131	.113	0.198
1/2	.625	.545	.040	.233	.164	.143	0.285
3/4	.875	.785	.045	.484	.229	.206	0.455
1	1.125	1.025	.050	.825	.294	.268	0.655
1 1/4	1.375	1.265	.055	1.26	.360	.331	0.884
1 1/2	1.625	1.505	.060	1.78	.425	.394	1.14
2	2.125	1.985	.070	3.09	.556	.520	1.75
2 1/2	2.625	2.465	.080	4.77	.687	.645	2.48
3	3.125	2.945	.090	6.81	.818	.771	3.33

TYPE M TUBE

Light wall thickness, hard only, furnished for applications requiring little or no pressure or tensions on the lines.

Nominal Size	NOMINAL DIMENSIONS			THEORETICAL AREAS BASED ON NOMINAL DIMENSIONS			
	Outside Diameter (Inches)	Inside Diameter (Inches)	Wall Thickness (Inches)	Cross Sectional Area of Bore (Sq. Inches)	External Surface (Sq. Ft. Per Lin. Ft.)	Internal Surface (Sq. Ft. Per Lin. Ft.)	Theoretical Weight (Pounds Per Foot)
1 1/4	1.375	1.291	.042	1.31	.360	.338	0.682
1 1/2	1.625	1.527	.049	1.83	.425	.400	0.940
2	2.125	2.009	.058	3.17	.556	.526	1.460
2 1/2	2.625	2.495	.065	4.89	.687	.653	2.030
3	3.125	2.981	.072	6.98	.818	.780	2.680

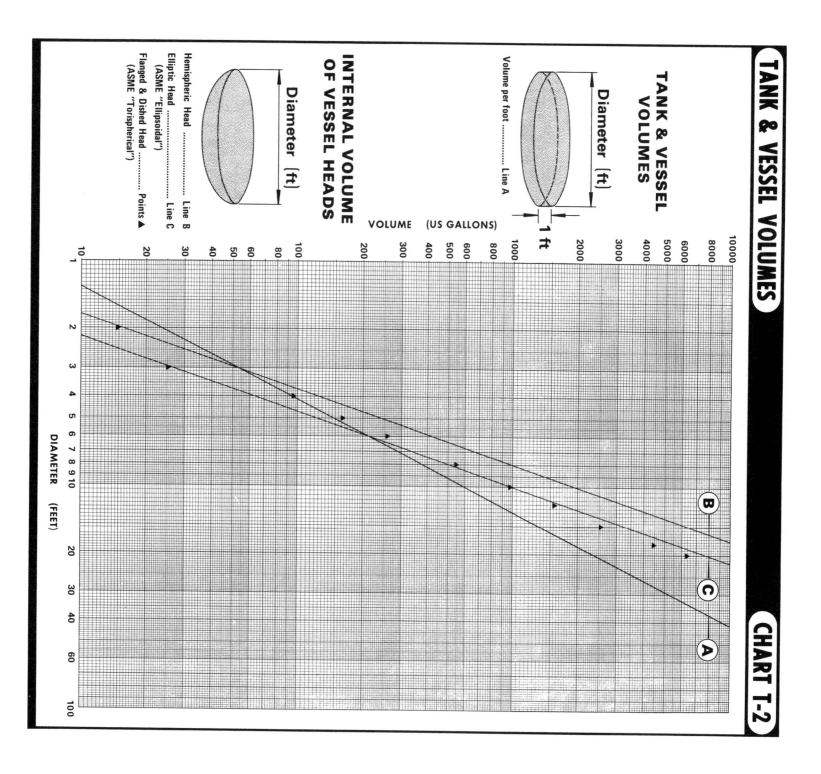

TANK & VESSEL
VOLUMES

Diameter (ft)

Volume per foot .... Line A

INTERNAL VOLUME
OF VESSEL HEADS

Diameter (ft)

Hemispheric Head .... Line B
Elliptic Head .... Line C
(ASME "Ellipsoidal")
Flanged & Dished Head .... Points ▲
(ASME "Torispherical")

VOLUME   (US GALLONS)

DIAMETER   (FEET)

VALVE DATA – RUN LENGTHS

DIMENSIONS IN INCHES — **TABLE V-1**

NOMINAL PIPE SIZE [NPS]

STEEL GATE VALVES — SOLID WEDGE & DOUBLE-DISC (SPLIT-WEDGE)

FLANGE CLASS	2	2 1/2	3	4	6	8	10	12	14	16	18	20	24
FLANGED 150	7	7.5	8	9	10.50	11.50	13	14	15	16	17	18	20
BEVELED 150	8.5	9.5	11.12	12	15.88	19.50	24.50	27.50					
300	8.5	9.5	11.12	12	15.88	16.50	18	19.75	30	33	36	39	45
600	11.50	13	14	17	22	26	31	33	35	39	43	47	55
900	14.50	16.50	15	18	24	29	33	38	40.50	44.50	48	52	61
1500	14.50	16.50	18.50	21.50	27.75	32.75	39	44.50	49.50	54.50	60.50	65.50	76.50
2500	17.75	20	22.75	26.50	36	40.25	50	56					

STEEL GLOBE VALVES / LIFT CHECK VALVES

FLANGE CLASS	2	2 1/2	3	4	6	8	10	12
150	8	8.5	9.5	11.50	16	19.50	24.50	28
300	10.50	11.50	12.50	14	17.50	22	24.50	28
600	11.50	13	14	17	22	26	31	33
900	14.50	16.50	15	18	24	29	33	38
1500	14.50	16.50	18.50	21.50	27.75	32.75	39	44.50
2500	17.75	20	22.75	26.50	36	40.25	50	56

SWING CHECK VALVES / TILTING DISC CHECK VALVES

(cells show two stacked values — swing check / tilting disc (T-D); "—" = not listed)

FLANGE CLASS	2	2 1/2	3	4	6	8	10	12
2500	17.75	20	22.75	26.50	36	40.25	50	56
T-D 1500	14.50 / 14.50	16.50 / 16.50	18.50 / 18.50	21.50 / 21.50	27.75 / 27.75	32.75 / 32.75	39 / 39	— / 44.50
T-D 900	14.50 / —	16.50 / —	15 / 15	18 / 18	24 / 24	29 / 29	33 / 33	38 / 38
T-D 600	11.50 / 11.50	13 / 13	14 / 14	17 / 17	22 / 22	26 / 26	31 / 31	33 / 33
T-D 300	10.50 / 10.50	11.50 / 11.50	12.50 / 12.50	14 / 14	17.50 / 17.50	21 / 21	24.50 / 24.50	28 / 28
T-D 150	8 / 8	8.5 / 8.5	9.5 / 9.5	11.50 / 11.50	14 / 14	19.50 / 19.50	24.50 / 24.50	27.50 / 27.50

NOTES

DIMENSIONS IN THIS TABLE CONFORM TO ANSI B16.10 AND APPLY TO FLANGED VALVES AND VALVES WITH ENDS BEVELLED FOR WELDING AS SHOWN:

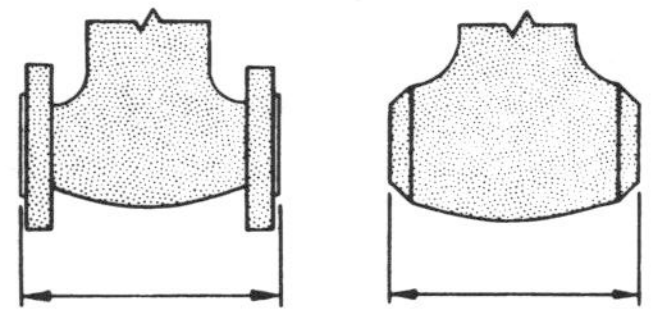

Tabled Dimension

FOR FLANGED VALVES THE TABLED DIMENSION INCLUDES ALLOWANCE FOR BOTH RAISED FACES OF THE VALVE. FOR CLASSES 150 AND 300 VALVES, 0.06-inch HAS BEEN INCLUDED FOR EACH RAISED FACE AND FOR VALVES OF CLASS 600 AND ABOVE, 0.25-inch HAS BEEN INCLUDED FOR EACH RAISED FACE.

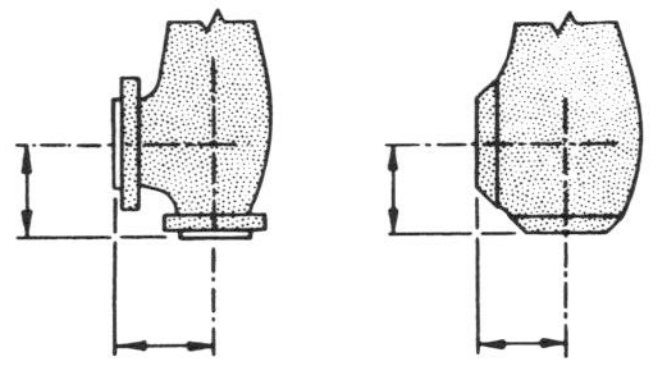

Half Tabled Dimension

FOR ANGLE GLOBE & ANGLE LIFT-CHECK VALVES, HALVE THE TABLED DIMENSION TO OBTAIN CENTER-TO-FACE DIMENSIONS.

NOTES

A factor in the design of piping supports is the weight of the piping to be supported. Calculation of the loadings involve the weights of pipe, fittings, flanges, valves, insulation, the conveyed fluid, and other related items that are also to be supported as part of the piping system.

Tables show weights of piping components. Data are subject to variation from manufacturing tolerances.

PIPE

For Schedule numbers, Manufacturers' weights (traditional designations: STD, XS, etc.), weight per unit length, weight filled with water, thickness of wall – refer to Tables P-1.

VALVES

Weights for valves do not include weights of powered operators or other devices specified for particular valves. Weights shown for valves in these tables are from data available as indicated from the Henry Vogt Machine Co. and from the Crane Company. Information herein is not intended to indicate the complete range of valves available from either manufacturer. Weights shown are for valves having conventional ports.

As valve features vary between manufacturers, actual weights of valves should be obtained from the specified manufacturer or supplier.

INSULATION

Weights of insulation are shown for both calcium silicate and for conventional 85% magnesia (alone or in combination with diatomaceous silica). The assumed densities are 11 pounds per cubic foot for calcium silicate and 85% magnesia, and 21 pounds per cubic foot for diatomaceous silica.

Insulation weights assumed include estimated weights of canvas, cement, paint, wire and bands, but not weatherproofing or other special protection. Pipe coverings of other compositions will have different densities. Data for insulation are based on conventional thickness recommendations and may not correspond with insulation specifications for a particular project.

UNITS OF WEIGHT

Weights in the following tables are in pounds – avoirdupois

TABLES W-1 — WEIGHTS OF PIPING

NOMINAL PIPE SIZE: 1"

FITTINGS

FORGED STEEL SOCKET WELD:

	PRESSURE CLASS 3000	6000	9000
90 ELBOW	1.00	2.35	3.19
45 ELBOW	0.94	1.91	2.50
TEE	1.31	3.31	3.75
COUP/RED ***	0.56	1.00	1.69
SOCKOLET **	0.60	1.30	1.30

FORGED STEEL THREADED:

	PRESSURE CLASS 2000	3000	6000
90 ELBOW	1.13	2.27	3.50
45 ELBOW	1.06	1.99	2.79
TEE	1.36	3.03	4.63
COUP/RED ***	--	0.63	2.13
THREDOLET **	--	0.62	1.23

MALL. IRON THREADED:

	PRESSURE CLASS 150	300
90 ELBOW	0.67	1.15
45 ELBOW	0.59	1.07
TEE	0.93	1.62
COUPLING	0.46	1.03
REDUCER ***	0.44	0.82

FLANGES

FORGED STEEL:

	PRESSURE CLASS 150	300	600	1500	2500
WELDING NECK	2.5	4	4	8.5	13
SLIP-ON	2	3	3.5	7.5	12
THREADED	2	3	3.5	7.5	12
LAP JOINT	2	3	4	8	
SOCKET	2	3	4	8	

VALVES

FORGED & CAST STEEL:

	PRESSURE CLASS 150	300	600	1500	2500
GATE-FLGD	12.1	15.4	17.2	41.3	
GLOBE-FLGD	11.9	15.6	17	43.6	
CHECK-FLGD	9	13.7	16.3	30	
GATE-THRD/SW			24.7		
GLOBE-THRD/SW			26.9		
CHECK-THRD/SW			15		
GATE PSB-SW			22		
GATE PSB-BW			21		
GLOBE PSB-SW			20		
VOGT VALVES					
CRANE VALVES			12.7		

NOMINAL PIPE SIZE: 1 1/2"

FITTINGS

FORGED STEEL SOCKET WELD:

	PRESSURE CLASS 3000	6000	9000
90 ELBOW	2.13	5.25	6.69
45 ELBOW	1.63	4.31	4.81
TEE	2.64	7.48	7.88
COUP/RED ***	1.00	2.00	2.19
SOCKOLET **	1.04	2.00	2.00

FORGED STEEL THREADED:

	PRESSURE CLASS 2000	3000	6000
90 ELBOW	2.18	3.50	7.50
45 ELBOW	1.74	3.00	5.75
TEE	2.80	7.04	9.63
COUP/RED ***	--	2.19	4.38
THREDOLET **	--	1.00	1.96

MALL. IRON THREADED:

	PRESSURE CLASS 150	300
90 ELBOW	1.36	2.57
45 ELBOW	1.17	2.30
TEE	1.85	3.46
COUPLING	0.93	2.10
REDUCER ***	0.85	1.69

FLANGES

FORGED STEEL:

	PRESSURE CLASS 150	300	600	1500	2500
WELDING NECK	4	7	8	14	28
SLIP-ON	3	6.5	6.5	14	25
THREADED	3	6.5	6.5	14	25
LAP JOINT	3	6.5	7	15	24
SOCKET	3	6	6.5		

VALVES

FORGED & CAST STEEL:

	PRESSURE CLASS 150	300	600	1500	2500
GATE-FLGD	21.5	29.2	30.0	80	
GLOBE-FLGD	25	29.9	33.5	80	
CHECK-FLGD	20.7	27.9	33	57	
GATE-THRD/SW			58.4		
GLOBE-THRD/SW			49.1		
CHECK-THRD/SW			59		
GATE PSB-SW			39		
GATE PSB-BW			37		
GLOBE PSB-SW			45		
VOGT VALVES					
CRANE VALVES			12.7		

INSULATION

TEMPERATURE RANGE deg F

	100-199	200-299	300-399	400-499	500-599	600-699	700-799	800-899	900-999	1000-1199

NOMINAL PIPE SIZE: 1"

	100-199	200-299	300-399	400-499	500-599	600-699	700-799	800-899	900-999	1000-1199
Cal Sil. in.	0.7	1	1	1.5	1.9	1.9	2.5	2.5	3	3
Weight lb/ft	0.7	1	1.5	1.9	2.5	2.5	2.8	3.3	3.7	3.7
85% Mag in.	0.7	1	1	1.5	1.9	1.9	2.8	2.8	3.7	3.7
Weight lb/ft	0.7	1	1.5	1.9	2.5	2.5	2.8	3	4.7	4.7
H.T.C. in.	0.7	1	1.2	1.9	2.5	2.5	3.3	3.3	4.7	4.7
Weight lb/ft	0.7	1.5	2	2.8	3.3	3.3	4.7	4.7	5.6	5.6

NOMINAL PIPE SIZE: 1 1/2"

	100-199	200-299	300-399	400-499	500-599	600-699	700-799	800-899	900-999	1000-1199
Cal Sil. in.	0.8	1	1.2	1.9	2.5	2.5	3.5	3.5	4.5	4.5
Weight lb/ft	0.8	1	1.4	1.9	2.5	2.5	3.5	4.2	4.5	4.5
85% Mag in.	0.8	1	1.2	1.9	2.5	2.5	3.5	3.5	4.5	4.5
Weight lb/ft	0.8	1.5	1.4	2.5	2.5	2.5	3.5	4.2	5.6	5.6
H.T.C. in.	0.8	1.5	2	2.5	3.5	3.5	4.2	5.6	5.6	5.6

BOLTS *

NOMINAL PIPE SIZE 1"	1	2	2	6	6	6
NOMINAL PIPE SIZE 1 1/2"	1	3.5	3.5	9	9	12

---

*weights for bolts are for one complete flange set. **weights are for reducing Sockolets and Thredolets. ***weights for reducers are for one pipe size reduction. PSB indicates valves having pressure seal bonnets. Other weights for valves are for valves having flanged bonnets.

TABLES W-1

WEIGHTS OF PIPING

BUTT WELDING FITTINGS — MFR'S WEIGHT

SCHEDULE NO.	2" 40 STD	2" 80 XS	2" 160 –	2" XXS –	3" 40 STD	3" 80 XS	3" 160 –	3" XXS –
LR 90 ELBOW	1.60	2.20	3.25	3.50	5.00	6.50	8.50	11.0
SR 90 ELBOW	1.00	1.50	–	3.50	3.00	4.25	–	–
LR 45 ELBOW	0.81	1.19	1.56	2.00	2.63	3.50	4.38	5.75
TEE	3.50	4.00	5.00	6.25	7.00	8.50	10.0	13.5
REDUCER ***	0.90	1.20	1.60	2.38	1.80	2.60	3.40	5.00
WELDOLET **	1.75	1.75	2.13	2.13	4.00	4.00	6.32	6.32

FORGED STEEL SOCKET WELD — PRESSURE CLASS

Fitting	2" 3000	2" 6000	2" 9000	3" 3000	3" 6000
90 ELBOW	3.13	6.66	6.69	10.9	19.3
45 ELBOW	2.71	4.81	9.62	10.5	14.3
TEE	4.07	8.24	8.75	12.5	23.5
COUP/RED ***	2.00	3.88	4.66	3.88	6.63
SOCKOLET **	1.60	5.13	5.13	3.80	–

FORGED STEEL THREADED — PRESSURE CLASS

Fitting	2" 2000	2" 3000	2" 6000	3" 2000	3" 3000	3" 6000
90 ELBOW	3.14	5.92	13.4	10.9	14.4	39.1
45 ELBOW	2.88	4.93	13.6	11.3	13.6	30.6
TEE	4.46	7.55	18.5	12.9	23.1	47.5
COUP/RED ***	–	3.13	5.35	–	6.75	13.5
THREDOLET **	–	1.75	5.13	–	4.35	–

MALL. IRON THREADED — PRESSURE CLASS

Fitting	2" 150	2" 300	3" 150	3" 300
90 ELBOW	2.16	4.00	5.37	9.46
45 ELBOW	1.82	3.70	4.75	8.54
TEE	2.81	5.35	7.77	13.2
COUPLING	1.48	3.60	3.72	8.00
REDUCER ***	1.47	2.88	3.87	6.60

FORGED STEEL FLANGES — CLASS

Flange	2" 150	2" 300	2" 600	2" 1500	2" 2500	3" 150	3" 300	3" 600	3" 1500	3" 2500
WELDING NECK	6	8	10	24	42	11.5	18	18	48	94
SLIP-ON	5	7	8	22	–	9	13	13	–	–
THREADED	5	7	8	22	38	10	14	15	48	83
LAP JOINT	5	7	9	21	37	9	14.5	14	38	80
SOCKET	5	7	9	24	–	8	13	16	–	–

CAST STEEL VALVES — CLASS

Valve	2" 150	2" 300	2" 600	2" 1500	2" 2500	3" 150	3" 300	3" 600	3" 1500	3" 2500
GATE-FLGD	46	74	84	180		76	108	160	370	
GLOBE-FLGD	47	83	90			91	135	160		
CHECK-FLGD	35	60	70	160		70	115	135	280	
GATE-BW	45	49	72	155		62	85	140	300	
GLOBE-BW	34	72	78			75	105	140		
CHECK-BW	25	47	55			50	87	105		
GATE PSB-			130	53 SW			100	210	70 FLGD	105
GATE PSB-BW				47 SW					110	130
GLOBE PSB-				55 SW					220 BW	155 BW / 170 BW / 670 BW

INSULATION — TEMPERATURE RANGE deg F

	100-199	200-299	300-399	400-499	500-599	600-699	700-799	800-899	900-999	1000-1199
2" Cal Sil. in.	1.0	1.0	1.5	2	2.5	2.5	3.5	4.2	4.2	
2" Weight lb/ft	1.0	1.0	1.7	2.5	2.5	2.5	3.5	4.2	4.2	
2" H.T.C. in. / 85% Mag in.	1	1	1.5	2	2.5	3	3	3	3	3
2" Weight lb/ft	1.0	1.0	1.7	2.5	3.0	3.0	5.9	5.9	6.9	6.9
3" Cal Sil. in.	1	1	1.3	1.3	1.5	2	3.0	5.1	5.2	6.7
3" Weight lb/ft	1	1	1.3	1.3	1.5	2.1	3.0	5.2	5.2	3.5
3" H.T.C. in. / 85% Mag in.	1	1	1.3	1.3	1.5	3.0	3.0	3	3	3.5
3" Weight lb/ft	1	1.3	2.1	3.0	3.0	5.1	6.9	6.9	6.9	9.2

BOLTS*

	2"	3"
BOLTS*	1.5	1.5

*Weights for bolts are for one complete flange set. **Weights are for reducing Sockolets, Weldolets and Thredolets. ***Weights for reducers are for one pipe size reduction. PBS indicates valves having pressure seal bonnets. Other weights for valves are for valves having flanged bonnets.

TABLES W-1

WEIGHTS OF PIPING

BUTT-WELDING FITTINGS

NOMINAL PIPE SIZE: 4"

SCHEDULE No.: / MFR'S WEIGHT:	40 STD	80 XS	160	–	XXS
LR 90 ELBOW	9.00	13.5	18.0	–	20.0
SR 90 ELBOW	6.25	8.50	–	–	–
LR 45 ELBOW	4.50	6.10	8.75	–	10.8
TEE	12.0	15.8	25.0	25.0	25.0
REDUCER ***	3.38	4.50	6.40	6.40	9.00
WELDOLET **	6.30	6.40	10.5	10.5	10.5

NOMINAL PIPE SIZE: 6"

SCHEDULE No.: / MFR'S WEIGHT:	40 STD	80 XS	160	–	XXS
LR 90 ELBOW	24.5	35.0	57.0	–	65.0
SR 90 ELBOW	18.0	23.0	–	–	–
LR 45 ELBOW	12.0	17.5	30.0	–	32.0
TEE	34.0	40.0	62.0	–	68.0
REDUCER ***	8.25	11.5	16.5	–	22.0
WELDOLET **	12.0	23.0	28.0	28.0	28.0

FLANGES — FORGED STEEL

NOMINAL PIPE SIZE: 4"

CLASS	150	300	600	1500	2500
WELDING NECK	16.5	26.5	37	69	146
SLIP-ON	13	23.5	33	–	–
THREADED	13	24	33	73	127
LAP JOINT	12	24	31	75	122

NOMINAL PIPE SIZE: 6"

CLASS	150	300	600	1500	2500
WELDING NECK	26	45	73	164	378
SLIP-ON	17	36	80	–	–
THREADED	19.5	36	78	164	323
LAP JOINT	18	38	78	170	314

VALVES — CAST STEEL

NOMINAL PIPE SIZE: 4"

CLASS	150	300	600	1500	2500
GATE-FLGD	110	165	300	610	
GLOBE-FLGD	143	220	320		
CHECK-FLGD	115	185	255	630	
GATE-BW	95	120	270	520	
GLOBE-BW	122	180	230		
CHECK BW	92	140	170	390	
GATE PSB-FLGD			190		
GATE PSB-BW			110	190	335
GLOBE PSB-BW			230	530	750

NOMINAL PIPE SIZE: 6"

CLASS	150	300	600	1500	2500
GATE-FLGD	175	320	640	1410	
GLOBE-FLGD	250	390	640		
CHECK-FLGD	200	330	530	1360	
GATE-BW	165	245	520	1250	
GLOBE-BW	230	350	560		
CHECK BW	165	280	420	790	
GATE PSB-FLGD			425		
GATE PSB-BW			285	490	840
GLOBE PSB-BW			600	880	1440

INSULATION

NOMINAL PIPE SIZE: 4"

TEMPERATURE RANGE deg F	100 199	200 299	300 399	400 499	500 599	600 699	700 799	800 899	900 999	1000 1199
Cal Sil. in. / Weight lb/ft	1 / 1.6	1 / 1.6	1.5 / 2.6	2 / 3.6	2.5 / 4.7	2.5 / 4.7	3 / 6.1	3.5 / 7.5	3.5 / 7.5	3.5 / 7.5
H.T.C. in. / 85% Mag in. / Weight lb/ft	1 / 1.6	1 / 1.6	1.5 / 2.6	2 / 3.6	2.5 / 4.7	2.5 / 4.7	3 / 6.1	3.5 / 7.5	3.5 / 7.5	3.5 / 7.5

NOMINAL PIPE SIZE: 6"

TEMPERATURE RANGE deg F	100 199	200 299	300 399	400 499	500 599	600 699	700 799	800 899	900 999	1000 1199
Cal Sil. in. / Weight lb/ft	1 / 2.1	1 / 2.1	1.5 / 3.3	2 / 4.6	2.5 / 6.1	2.5 / 6.1	3 / 7.6	3.5 / 9.8	3.5 / 9.8	3.5 / 9.8
H.T.C. in. / 85% Mag in. / Weight lb/ft	1 / 2.1	1 / 2.1	1.5 / 3.3	2 / 4.6	2.5 / 6.1	2.5 / 6.1	3 / 7.6	3.5 / 9.8	3.5 / 9.8	3.5 / 9.8

BOLTS

Weights*					
NOMINAL PIPE SIZE: 4"	4	7.5	12.5	34	61
NOMINAL PIPE SIZE: 6"	6	11.5	30	76	145

*Weights for bolts are for one complete flange set. **Weights are for reducing weldolets. ***Weights for reducers are for one pipe size reduction. PSB indicates valves having pressure seal bonnets. All other weights are for valves having flanged bonnets.

TABLES W-1

WEIGHTS OF PIPING

BUTT-WELDING FITTINGS — MFR'S WEIGHT

	\ 8" \ 40 / STD	80 / XS	160 / —	XXS	\ 10" \ 40 / STD	60 / XS	160 / —	XXS
SCHEDULE No.:	40 / STD	80 / XS	160 / —	XXS	40 / STD	60 / XS	160 / —	XXS
LR 90 ELBOW	50	71	120	118	88	107	260	1068
SR 90 ELBOW	34	47.5	—	—	58	70	—	—
LR 45 ELBOW	23	35	62	60	43	53	130	—
TEE	55	75	110	120	85	105	260	925
REDUCER ***	13.3	18.8	31	36	22	29.5	57.5	—
WELDOLET **	23	37	(refer to Mfr)		36	46	(refer to Mfr)	

FLANGES — FORGED STEEL

	\ 8" CLASS \ 150	300	600	1500	2500	\ 10" CLASS \ 150	300	600	1500	2500
WELDING NECK	42	69	112	273	576	54	100	189	454	1068
SLIP-ON	28	56	97	—	—	40	77	177	—	—
THREADED	30	56	97	258	485	41	80	177	436	925
LAP JOINT	28	55	112	286	471	36	88	195	485	897

VALVES — CAST STEEL

	\ 8" CLASS \ 150	300	600	1500	2500	\ 10" CLASS \ 150	300	600	1500	2500
GATE-FLGD	310	500	1080	2600	—	455	760	1790	4910	—
GLOBE-FLGD	420	740	800	—	—	570	1010	—	—	—
CHECK-FLGD	390	620	900	2100	—	470	640	—	—	—
GATE-BW	260	410	940	1900	—	410	625	1250	3690	—
GLOBE-BW	390	640	670	—	—	480	850	1580	—	—
CHECK-BW	350	510	740	1320	—	370	590	1030	—	—
GATE PSB-FLGD	—	—	855	—	—	—	—	1300	—	—
CHECK PSB-FLGD	—	—	615	—	—	—	—	915	—	—
GATE PSB-BW	—	—	800	900	1500	—	—	1620	1540	2490
GLOBE PSB-BW	—	—	800	—	1440	—	—	1620	—	2500
GLOBE PSB-BW	—	—	—	—	1700	—	—	—	—	3500

INSULATION

TEMPERATURE RANGE deg F	100/199	200/299	300/399	400/499	500/599	600/699	700/799	800/899	900/999	1000/1199
8" Cal Sil. in.	1.5	1.5	2	2	2.5	3	3.5	4	4	4
Weight lb/ft	4.1	4.1	5.6	5.6	7.9	9.5	12	14	14	14
H.T.C. in.	1.5	1.5	2	2.5	3	3.5	4	4	4	
85% Mag in.	1.5	2	2.5	3	3.5	4	4	4		
Weight lb/ft	4.1	4.1	5.6	8	13	16	20	20	20	
10" Cal Sil. in.	1.5	1.5	2	2.5	2.5	2.5	3.5	4	4	4
Weight lb/ft	5.2	5.2	7.1	8.9	8.9	8.9	11	13	16	16
H.T.C. in.	1.5	1.5	2	2.5	2.5	2.5	3.5	4	4	4
85% Mag in.	1.5	2	2.5	2.5	2.5	3.5	4	4		
Weight lb/ft	5.2	5.2	7.1	8.9	8.9	8.9	15	19	23	23

BOLTS*

	150	300	600	1500	2500
8"	6.5	18	40	121	232
10"	15	38	52	184	445

*weights for bolts are for one complete flange set. **weights are for reducing Weldolets.
***weights for reducers are for one pipe size reduction. PSB indicates valves having
pressure seal bonnets. All other weights for valves are for valves having flanged bonnets.

TABLES W-1 — WEIGHTS OF PIPING

BUTT-WELDING FITTINGS

	NOMINAL PIPE SIZE: 12"				NOMINAL PIPE SIZE: 14"			
SCHEDULE No.:	—	—	160	—	30	—	160	—
MFR'S WEIGHT:	STD	XS	—	XXS	STD	XS	—	XXS
LR 90 ELBOW	125	160	450	—	160	205	572	—
SR 90 ELBOW	80	104	---	—	---	140	---	—
LR 45 ELBOW	62	84	225	—	80	100	286	—
TEE	120	160	480	—	165	240	---	—
REDUCER ***	34	43.5	96	—	60	80	---	—
WELDOLET **	59	61	(refer to Mfr)		66	70	(refer to Mfr)	

FLANGES — FORGED STEEL

	NOMINAL PIPE SIZE: 12" — CLASS					NOMINAL PIPE SIZE: 14" — CLASS				
	150	300	600	1500	2500	150	300	600	1500	2500
WELDING NECK	88	142	226	690	1608	114	206	347	(refer to Mfr)	
SLIP-ON	61	113	215	---	---	83	159	259	(refer to Mfr)	
THREADED	65	110	215	667	1300	85	164	259	(refer to Mfr)	
LAP JOINT	60	139	240	749	1262	77	184	290	(refer to Mfr)	

VALVES — CAST STEEL

	NOMINAL PIPE SIZE: 12" — CLASS					NOMINAL PIPE SIZE: 14" — CLASS				
	150	300	600	1500	2500	150	300	600	1500	2500
GATE-FLGD	650	1020	2570	7150		860	1380	3455	8580	
GLOBE-FLGD	1431	1675	1830			1525				
CHECK-FLGD	635	950	2160			1200	1340			
GATE-BW	580	890				730	1220	2960	6420	
GLOBE-BW	1310	1455				1360				
CHECK-BW	560	720	4650			1010	1150			
GATE PSB-FLGD			1410	1750	2250			1900		
GATE PSB-BW			1405	2400	3850				2710	4410
GLOBE PSB-BW			1405	2780	5000				3510	

INSULATION

	NOMINAL PIPE SIZE: 12"	NOMINAL PIPE SIZE: 14"
TEMPERATURE RANGE deg F	100-199, 200-299, 300-399, 400-499, 500-599, 600-699, 700-799, 800-899, 900-999, 1000-1199	100-199, 200-299, 300-399, 400-499, 500-599, 600-699, 700-799, 800-899, 900-999, 1000-1199
Cal Sil. in.	1.5, 2, 2.5, 3, 3, 3.5, 4, 4, 5	1.5, 2, 2.5, 3, 3, 3.5, 4, 4, 5
Weight lb/ft	6, 8, 11, 13, 13, 15, 18, 18, 24	6.2, 8.4, 11, 13, 16, 19, 19, 26
H.T.C. in. 85% Mag in.	1.5, 1.5, 2, 2.5, 3, 3, 3.5, 4, 4, 5	1.5, 1.5, 2, 2.5, 3, 3.5, 4, 4, 5
Weight lb/ft	6, 6, 8, 11, 13, 13, 15, 18, 18, 24	6.2, 6.2, 8.4, 11, 13, 16, 19, 19, 26

BOLTS*

	NOMINAL PIPE SIZE: 12"	NOMINAL PIPE SIZE: 14"
BOLTS*	15, 49, 91, 306, 622, 118	22, 62, 118

*weights for bolts are for one complete flange set. **weights are for reducing weldolets.
***weights for reducers are for one pipe size reduction. PSB indicates valves having
pressure seal bonnets. All other weights are for valves having flanged bonnets.

TABLES W-1 — WEIGHTS OF PIPING

BUTT-WELDING FITTINGS

	NOMINAL PIPE SIZE: 16"		NOMINAL PIPE SIZE: 18"	
SCHEDULE No.:	30	40	–	–
MFR'S WEIGHT:	STD	XS	STD	XS
LR 90 ELBOW	206	276	260	340
SR 90 ELBOW	132	174	167	219
LR 45 ELBOW	100	135	126	167
TEE	195	280	249	332
REDUCER ***	71	91	85	115
WELDOLET **	75	102	97	130

FLANGES — FORGED STEEL

	16" CLASS 150	300	600	1500	18" CLASS 150	300	600	1500
WELDING NECK	142	249	481	(refer	165	306	555	(refer
SLIP-ON	106	210	366	to	109	253	476	to
THREADED	93	220	366	Mfr)	120	280	476	Mfr)
LAP JOINT	104	234	400		146	305	469	

VALVES — CAST STEEL

	16" CLASS 150	300	600	1500	2500	18" CLASS 150	300	600	1500	2500
GATE-FLGD	1120	1960	4375			1400	2450	6020		
GLOBE-FLGD										
CHECK-FLGD	1450	1650								
GATE-BW	960	1620	3675			1250	2000	4460		
GLOBE-BW										
CHECK-BW	1250	1220								
GATE PSB-FLGD										
GATE PSB-BW			2575					3400		
GLOBE PSB										

INSULATION

TEMPERATURE RANGE deg F	100/199	200/299	300/399	400/499	500/599	600/699	700/799	800/899	900/999	1000/1199
16" Cal Sil. in.	1.5	1.5	2	2.5	3	3.5	4	4	4	5
16" Cal Sil. Weight lb/ft	6.9	6.9	9.3	12	15	18	21	21	21	28
16" H.T.C. in. / 85% Mag in.	1.5	1.5	2	2.5	3	3.5	4	4	4	5
16" H.T.C. Weight lb/ft	6.9	6.9	9.3	12	15	20	25	31	31	42
18" Cal Sil. in.	1.5	1.5	2	2.5	3	3.5	4	4	4	5
18" Cal Sil. Weight lb/ft	7.7	7.7	10	13	16	19	23	23	23	31
18" H.T.C. in. / 85% Mag in.	1.5	1.5	2	2.5	3	3.5	4	4	4	5
18" H.T.C. Weight lb/ft	7.7	7.7	10	13	16	23	28	34	34	46

BOLTS*

	16"			18"		
BOLTS*	31	83	152	41	101	193

*Weights for bolts are for one complete flange set. **Weights are for reducing Weldolets. ***Weights for reducers are for one pipe size reduction. PSB indicates valves having pressure seal bonnets. All other weights for valves are for valves having flanged bonnets.

TABLES W-1 — WEIGHTS OF PIPING

NOMINAL PIPE SIZE: 20"

BUTT-WELDING FITTINGS
(SCHEDULE No. / MFR'S WEIGHT)

	20 STD	30 XS
LR 90 ELBOW	320	420
SR 90 ELBOW	210	275
LR 45 ELBOW	160	206
TEE	342	480
REDUCER ***	125	170
WELDOLET **	118	158

FLANGES — FORGED STEEL (CLASS)

	150	300	600	1500
WELDING NECK	197	369	690	
SLIP-ON	148	307	612	
THREADED	155	325	612	
LAP JOINT	159	375	604	(refer to Mfr)

VALVES — CAST STEEL (CLASS)

	150	300	600	1500	2500
GATE-FLGD	2125	3890	7015		
GLOBE-FLGD					
CHECK-FLGD					
GATE-BW	1855	3370	5755		
GLOBE-BW					
CHECK-BW					
GATE PSB-FLGD				5200	
GATE PSB-BW					
GLOBE PSB-BW					

INSULATION

TEMPERATURE RANGE deg F	100-199	200-299	300-399	400-499	500-599	600-699	700-799	800-899	900-999
Cal Sil. in.	1.5	2	2.5	3	3	3	4	4	5
Weight lb/ft	8.5	12	15	18	18	21	25	25	34
H.T.C. in. / 85% Mag in.	1.5	2	2.5	3	3.5	3.5	4	4	5
Weight lb/ft	8.5	12	15	18	25	31	37	37	50

BOLTS*

150	300	600
52	105	242

NOMINAL PIPE SIZE: 24"

BUTT-WELDING FITTINGS
(SCHEDULE No. / MFR'S WEIGHT)

	20 STD	-- XS
LR 90 ELBOW	460	600
SR 90 ELBOW	298	392
LR 45 ELBOW	238	300
TEE	528	610
REDUCER ***	150	200
WELDOLET **	220	290

FLANGES — FORGED STEEL (CLASS)

	150	300	600	1500
WELDING NECK	268	579	977	
SLIP-ON	204	490	876	
THREADED	210	490	876	
LAP JOINT	195	530	866	(refer to Mfr)

VALVES — CAST STEEL (CLASS)

	150	300	600	1500	2500
GATE-FLGD	3120	5955	9360		
GLOBE-FLGD					
CHECK-FLGD					
GATE-BW	2500	4675	8020		
GLOBE-BW					
CHECK-BW					
GATE PSB-FLGD				6800	
GATE PSB-BW					
GLOBE PSB-BW					

INSULATION

TEMPERATURE RANGE deg F	100-199	200-299	300-399	400-499	500-599	600-699	700-799	800-899	900-999	1000-1199
Cal Sil. in.	1.5	1.5	2	2.5	3	3	3.5	4	4	5
Weight lb/ft	10	10	13	17	21	21	25	29	29	39
H.T.C. in. / 85% Mag in.	1.5	1.5	2	2.5	3	3	3.5	4	4	5
Weight lb/ft	10	10	13	17	21	21	25	29	29	58

BOLTS*

150	300	600
71	174	360

---

*Weights for bolts are for one complete flange set. **Weights are for reducing Weldolets.
***Weights for reducers are for one pipe size reduction. PSB indicates valves having
pressure seal bonnets. All other weights are for valves having flanged bonnets.

WEIGHTS OF MATERIALS — TABLE W-2

MATERIAL	specific gravity	lb/in³	lb/ft³	lb/ft²·in	Kg/m³	lb/US gal	lb/Imp gal
METALS & ALLOYS							
Aluminum (2S)	2.71	0.0978	169	14.1	2710		
Aluminum bronze	7.70	0.278	481	40.1	7700		
Brasses: %Cu %Zn							
Red brass 85 15	8.75	0.316	546	45.5	8750		
Low brass 80 20	8.67	0.313	541	45.1	8670		
Cartridge brass 70 30	8.52	0.308	532	44.3	8520		
Muntz metal 60 40	8.39	0.303	524	43.7	8390		
Bronze, %Cu=80-95,%Sn=20-5	8.84	0.319	552	46.0	8840		
Copper	8.91	0.322	556	46.3	8900		
Iron, gray-cast	7.21	0.260	450	37.5	7210		
malleable	7.34	0.267	461	38.4	7380		
wrought	7.69	0.278	480	40.0	7690		
Lead	11.37	0.411	710	59.2	11370		
Monel	8.83	0.319	551	45.9	8830		
Nickel	8.87	0.321	554	46.2	8870		
Steel, carbon	7.85	0.284	490	40.8	7850		
stainless, %Cr=18,%Ni=8	7.93	0.286	495	41.3	7930		
LIQUIDS							
Fuel oil	0.95	0.034	59		950	7.9	9.5
Gasoline	0.67 thru 0.75	0.024 thru 0.027	42 thru 47		670 thru 750	5.6 thru 6.3	6.7 thru 7.5
Lube oil	0.90	0.032	56		900	7.5	9.0
Jet fuel	0.82	0.030	51		820	6.8	8.2
Water, fresh	1.00	0.036	62.3		1000	8.33	10.0
salt (seawater)	1.03	0.037	64		1030	8.6	10.3
INSULATING MATERIALS							
Abestos	2.45	0.0885	153	12.8	2450		
Cork	0.24	0.0087	15.0	1.25	240		
Fiberglas (Owens/Corning "Kaylo")	0.176	0.0064	11.0	0.92	176		
Magnesia (85%)	0.18	0.0064	11.0	0.92	176		
Plastic foam	0.08 thru 0.10	0.0029 thru 0.0038	5.0 thru 6.5	0.42 thru 0.54	80 thru 104		
MATERIALS OF CONSTRUCTION							
Brick, common	1.92	0.069	120	10.0	1920		
Concrete, plain	2.31	0.083	144	12.0	2310		
reinforced	2.40	0.088	150	12.5	2400		
Earth, dry, loose	1.22	0.044	76	6.3	1220		
dry, packed	1.52	0.055	95	7.9	1520		
moist, loose	1.25	0.045	78	6.5	1250		
moist, packed	1.54	0.056	96	8.0	1540		
Glass	2.50	0.090	156	13.0	2500		
Gravel, dry	1.60	0.058	100	8.3	1600		
wet	1.92	0.069	120	10.0	1920		
Sand, dry	1.60	0.058	100	8.3	1600		
wet	1.92	0.069	120	10.0	1920		
Snow, loose	0.13	0.0046	8	0.7	130		

FRACTIONAL EQUIVALENTS

0.06	0.12	0.19	0.25	0.31	0.38	0.44	0.50	0.56	0.62	0.69	0.75	0.81	0.88	0.94
1/16	1/8	3/16	1/4	5/16	3/8	7/16	1/2	9/16	5/8	11/16	3/4	13/16	7/8	15/16

The USA uses two systems of weight and measures: the United States system of English origin, and the metric system of French origin.

The English or Imperial system was a customary system with origins in Babylonian, Egyptian, Greek, Roman, Anglo-Saxon, French (Norman) and other civilizations and cultures. The English system evolved over centuries from simple measures and practices, eventually attaining precision through legislation and standardization. Although some standardization resulted from reform (sometimes a royal decree), the overwhelming pressure came from expansion in industry and commerce.

Imperial Rome established a system of weights and measures used from England to Asia. But, with the decline of the Roman Empire, what was once an almost universal system degenerated into local customary systems in continental Europe and England.

By the 17th and 18th centuries, through colonization and dominance in commerce, the English system had developed to a point where it was in use in many parts of the world, including the American colonies. The French, however, decided to abandon the confusion of European customary units (which varied not only from country to country, but from province to province and sometimes, from city to city), and to create an entirely new system to rationalize weights and measures - the Metric System.

The metric system was the result of years of scientific investigation and recommendations for reform. It was adopted in the late 18th century by the post-revolutionary government of France and, subsequently, by other nations. The standardized units and decimal base were particularly well suited for science and engineering.

In 1960, at the General Conference of Weights and Measures (Conference Generale des Poids et Mesures [CGPM]), the modern version of the metric system was designated the International System of Units (Le Systeme International d'Unites), and endorsed by the International Organization for Standardization (ISO) - a federation of national standardization bodies representing most countries of the world. The international symbol for this system is SI.

SI, now the primary world system of units of measurement, is a rationalized selection of units from the metric system with which ISO seeks to establish international standards, especially those for universal interchangeability of components. SI simplifies measurement by logically coordinating unique units for length, mass, temperature, time, etc., in a decimal system in which the magnitude of a unit is changed by moving the decimal point (or, for example, by using a prefix such as 'milli' with meter for the factor 0.001).

The customary system is more complicated as it uses three types of subdivision: duodecimal (twelfths), decimal (tenths), and binary (halves), and requires conversion, for example, between different units of length (such as inches, feet and yards), or of mass (such as ounces, pounds and tons).

Changing from customary units to SI units is straightforward, but changing from traditional metric units to SI units is more difficult in countries already using the metric system. Because of this difficulty, although not in keeping with the goals of ISO, a limited number of traditional metric units are temporarily being used with SI; one such unit is bar, the unit for pressure, referred to below under 'Flange Classes and Pressure Numbers'.

Without a legislative mandate, full implementation of SI in the United States is unlikely; however, technical and economic requirements of American companies operating internationally are encouraging voluntary transition; for example, manufacturers of equipment and components are now presenting dimensional and other data in SI units (and temporary metric units in use with SI) in addition to U.S. customary units.

The USA uses two systems of weight and measures: the United States system of English origin, and the metric system of French origin.

By the middle of the 20th century, principal manufacturing countries not using the metric system were Britain, the British Commonwealth countries and the United States. Although in 1866 the U.S. Congress legalized the metric system for use throughout the United States and, in 1975 passed the Metric Conversion Act, the United States is the only major industrial nation today, neither to have adopted nor mandated use of the metric system as its primary system of measurement.

NOMINAL LINES SIZES

Sizes of pipe, fittings, flanges, and valves are given in nominal diameters - in inch units as NPS (Nominal Pipe Size) and in metric units as DN (Diametre Nominale [Nominal Diameter]). The following table gives equivalent diameters in nominal inch units and nominal millimeter units:

CUSTOMARY NPS (inch)	METRIC DN (mm)	CUSTOMARY NPS (inch)	METRIC DN (mm)	CUSTOMARY NPS (inch)	METRIC DN (mm)
1/8	6	6	150	30	750
1/4	8	8	200	32	800
3/8	10	10	250	36	900
1/2	15	12	300	40	1000
3/4	20	14	350	42	1100
1	25	16	400	48	1200
1 1/4*	32	18	450	54	1400
1 1/2	40	20	500	60	1500
2	50	22	550	64	1600
2 1/2*	65	24	600	72	1800
3	80	26	650	80	2000
4	100	28	700	88	2200

* These sizes may be used in special applications; they are not normally used in new industrial construction.

FLANGE CLASSES and PRESSURE NUMBERS

Earlier classifications of flanges for steel pipe (and flanged fittings): 150-lb, 300-lb, 400-lb, 600-lb, etc., referred to 'Primary Service Pressure Ratings in pounds (pounds-force) per square-inch'. (Flanges, however, are suitable for service over a range of pressure, with actual pressures depending on operating temperatures, and materials of construction.) These classifications have been supplanted by pressure rating class designations: Class 150, Class 300, etc., in which each class identifies a group of flanges conforming to established dimensions, for a range of pipe sizes.

Standards publish 'Pressure-Temperature Ratings' for each class of flange. These ratings are maximum allowable non-shock (gage), working (or service) pressures over a range of temperature for different materials of construction, including bolts and gaskets.

In addition to class designations, flange tables in this section of the 'PIPING GUIDE' also show 'PN' designations according to ANSI B16.5-1981 (until re-issued 1988), and MSS-SP-86-1981 (re-issued 1987), which states "....the recommendation for metric pressure designations is the use of the prefix PN, which may be thought of as 'Pressure Number'."

Pressure Numbers (PN), similar to class designations, identify groups of flanges conforming to established dimensions, and for each class of flange express the pressure rating within the temperature range -20 to +100F (refer to Table F-9), as a nominal bar* value.

Class and corresponding PN designations are shown in the following table:

CLASS	150	300	400	600	900	1500	2500
PN	20	50	68	100	150	250	420

[* Bar is not an SI unit; pascal (Pa) is the SI unit for pressure (and stress). The pascal is a small unit. For stating process or service pressure it is used with a prefix such as kPa for kilopascal (1000 pascals), or MPa for megapascal (1 000 000 pascals), although megapascal is more suitable for the greater values of stress. Bar, equal to 100 000 pascals, is a traditional metric unit in widespread use internationally in industry and technology. Until it is displaced, bar is in temporary use with SI units. (Temporary units are specific, widely used, traditional metric units whose use in future work is discouraged.)]

Contemporary references and suppliers' literature refer to bar values and PN designations. Flange tables in this section of the 'PIPING GUIDE' include PN references for information only.

In the following pages, selected data from PT II of the 'PIPING GUIDE' are presented in SI units. For identification, these tables and charts are given the suffix 'M'.

TABLES GIVE THE MINIMUM SPACING. INCREASE DIMENSIONS:
1) FOR INSULATION
2) IF THERMAL MOVEMENT WOULD REDUCE CLEARANCE

LINES WITHOUT FLANGES — DIMENSION 'C'

NOMINAL DIAMETER (DN)

DN	50	80	100	150	200	250	300	350	400	450	500	600
50	115	125	140	165	190	220	245	260	285	310	335	385
80	125	140	155	180	205	235	260	275	300	325	350	400
100	140	155	165	195	220	245	270	285	315	340	365	415
150	165	180	195	220	245	275	300	315	340	365	390	440
200	190	205	220	245	270	300	325	350	375	390	415	465
250	220	235	245	275	300	325	350	365	390	415	445	495
300	245	260	270	300	325	350	375	390	420	445	470	520
350	260	275	285	315	340	365	390	410	435	460	485	535
400	285	300	315	340	365	390	420	435	460	485	510	560
450	310	325	340	365	390	415	445	460	485	510	535	585
500	335	350	365	390	415	445	470	485	510	535	560	610
600	385	400	415	440	465	495	520	535	560	585	610	660

SURFACE-TO-CENTER OF PIPE DIMENSION

DN	'S' (WITHOUT FLANGES)	'SF' FLANGE CLASS 150	'SF' FLANGE CLASS 300	'SF' FLANGE CLASS 600
50	85	105	110	110
80	95	125	130	130
100	110	140	155	165
150	135	165	185	205
200	160	200	220	235
250	190	230	250	280
300	215	270	290	305
350	230	295	320	330
400	255	325	350	370
450	280	345	385	400
500	305	375	415	435
600	355	435	485	495

LINES WITH FLANGES — DIMENSION 'CF'

CLASS 150 & CLASS 150 FLANGES

NOMINAL DIAMETER (DN) OF FLANGED PIPE

DN	50	80	100	150	200	250	300	350	400	450	500	600
50	135	155	170	200	230	260	300	315	355	375	405	465
80	155	170	185	210	245	275	315	340	370	390	420	480
100	170	185	200	225	255	285	325	350	380	405	435	490
150	200	210	225	250	285	310	340	370	405	430	460	515
200	230	245	255	285	315	340	380	405	430	455	485	540
250	260	275	290	315	350	380	405	430	455	480	505	555
300	300	315	325	350	380	405	435	455	490	505	530	580
350	315	325	340	365	390	420	455	480	505	525	555	605
400	355	370	385	410	435	465	490	505	530	555	580	630
450	375	390	405	430	455	480	505	525	550	575	605	665
500	405	420	435	460	485	515	540	555	580	605	630	690
600	465	480	490	520	545	570	595	610	640	665	690	740

CLASS 300 & CLASS 300 FLANGES

NOMINAL DIAMETER (DN) OF FLANGED PIPE

DN	50	80	100	150	200	250	300	350	400	450	500	600
50	140	165	185	215	250	280	320	350	380	415	445	515
80	165	175	200	230	265	295	335	365	395	430	460	530
100	185	200	210	245	275	305	345	375	405	440	470	540
150	215	230	245	270	305	335	375	405	435	470	500	570
200	250	265	275	305	330	360	400	430	460	495	525	595
250	280	295	305	335	360	385	425	455	490	520	550	620
300	320	335	345	375	400	425	450	480	515	545	575	645
350	350	365	375	405	430	455	480	500	530	560	595	665
400	380	395	405	435	460	490	515	530	560	585	620	690
450	415	430	440	470	495	520	545	560	585	610	645	715
500	445	460	470	500	525	550	575	595	620	645	670	740
600	515	530	540	570	595	620	645	665	690	715	740	790

INSULATION

DIMENSIONS IN THESE TABLES ARE SPACINGS FOR BARE PIPE. FOR INSULATED LINES, ADD THE THICKNESS OF INSULATION AND COVERING TO THESE FIGURES

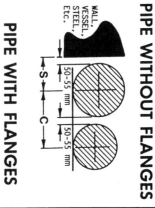

PIPE WITHOUT FLANGES

WALL, VESSEL, STEEL, Etc.
50-55 mm — S — C — 50-55 mm

PIPE WITH FLANGES

WALL, VESSEL, STEEL, Etc.
CF — 25-30 mm — SF — CF
25-30 mm / 25-30 mm

CLASS 150 & CLASS 300 FLANGES

NOMINAL DIAMETER (DN) OF FLANGED PIPE

DN	50	80	100	150	200	250	300	350	400	450	500	600
50	140	165	185	215	250	280	320	350	380	415	445	515
80	155	175	200	230	265	295	335	365	395	430	460	530
100	170	190	210	245	275	305	345	375	405	440	470	540
150	200	215	240	270	305	335	370	405	435	470	500	570
200	230	245	265	295	330	360	400	430	460	495	525	595
250	260	275	290	325	355	385	425	455	490	520	550	620
300	300	315	325	355	380	410	450	480	515	545	575	645
350	325	340	350	380	405	430	470	500	530	560	595	665
400	355	370	385	410	435	465	505	535	570	595	620	690
450	375	390	405	430	455	480	515	550	580	610	645	715
500	405	420	435	460	485	515	540	575	605	640	670	740
600	465	480	490	520	545	570	595	625	655	690	720	790

CLASS 300 & CLASS 500 FLANGES

NOMINAL DIAMETER (DN) OF FLANGED PIPE

DN	50	80	100	150	200	250	300	350	400	450	500	600
50	140	165	195	235	270	310	340	360	400	430	465	530
80	155	175	210	250	280	325	350	375	415	445	480	540
100	185	200	220	265	295	340	365	390	430	455	490	555
150	215	230	250	290	320	365	390	415	455	485	520	580
200	250	265	275	315	345	390	415	440	480	510	545	605
250	280	295	305	345	370	420	445	465	505	535	570	635
300	320	335	345	370	400	445	470	490	535	560	595	660
350	350	365	375	405	430	460	485	505	550	575	610	675
400	380	395	410	435	460	490	515	535	575	600	630	700
450	415	430	440	470	495	520	545	570	610	635	665	725
500	445	460	470	500	525	550	575	595	640	665	690	750
600	515	530	540	570	595	620	645	665	715	740	805	

CLASS 150 & CLASS 600 FLANGES

NOMINAL DIAMETER (DN) OF FLANGED PIPE

DN	50	80	100	150	200	250	300	350	400	450	500	600
50	140	165	195	235	270	310	340	360	400	430	465	530
80	155	175	210	250	280	325	350	375	415	445	480	540
100	170	190	220	265	295	340	365	390	430	455	490	555
150	200	215	235	290	320	365	390	415	455	485	520	580
200	230	245	275	295	345	390	415	440	480	510	545	605
250	260	275	300	340	375	420	445	465	505	535	570	635
300	300	315	325	365	390	415	470	490	535	560	595	660
350	325	340	350	385	415	440	465	505	550	575	610	675
400	355	370	385	415	430	445	490	535	575	605	640	700
450	375	390	405	435	455	485	515	550	605	630	665	725
500	405	420	435	465	490	510	535	575	610	640	690	750
600	465	480	490	530	555	580	605	635	660	675	805	

CLASS 600 & CLASS 600 FLANGES

NOMINAL DIAMETER (DN) OF FLANGED PIPE

DN	50	80	100	150	200	250	300	350	400	450	500	600
50	140	165	195	235	270	310	340	360	400	430	465	530
80	165	175	210	250	280	325	350	375	415	445	480	540
100	185	200	220	265	295	340	365	390	430	455	490	555
150	215	230	250	290	320	365	390	415	455	485	520	580
200	250	265	275	315	345	390	415	440	480	510	545	605
250	280	295	305	345	370	420	445	465	505	535	570	635
300	320	335	345	370	400	445	470	490	535	560	595	660
350	350	365	375	405	430	460	485	505	550	575	610	675
400	380	395	410	435	460	490	515	535	575	605	640	700
450	415	430	440	470	495	520	545	570	610	635	665	725
500	445	460	470	500	525	550	575	595	640	665	690	750
600	515	530	540	570	595	620	645	665	715	740	805	

PIPEWAY WIDTH

When the order of lines, line sizes, flange classes (for lines with flanges), and insulation thicknesses for insulated lines have been decided, determine pipeway width from Tables A-1M, A-2M and A-3M, adding 25% so that the final design includes 20% (distributed) space for future piping. Additional space will usually be required for electrical and instrument trays/raceways.

For a **tentative** estimate of the pipeway width required for a selection of lines without flanges, of nominal sizes in the range DN 50 thru DN 200, either of the following factors may be used - the first is preferable:

(1) If all pipe sizes are known, add their nominal sizes in millimeters together and multiply by 4.1 to estimate the width in millimeters

(2) If only the number of lines is known, multiply number of lines by 436 to estimate the width in millimeters

Either factor gives a pipeway width which includes insulation for 25% of lines, allows 20% of the width for the addition and re-sizing of lines, and allocates a further 20% of the width for future piping.

45° JUMPOVERS

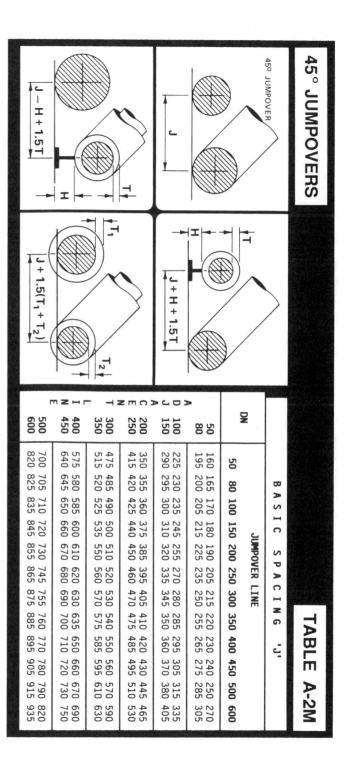

45° JUMPOVER

Diagram labels: J, J − H + 1.5T, H, T, T₁, J + 1.5(T₁ + T₂), J + H + 1.5T, T₂, H, T

TABLE A-2M

BASIC SPACING 'J'

DN	\ JUMPOVER LINE	50	80	100	150	200	250	300	350	400	450	500	600
50	L	160	165	170	180	190	205	215	220	230	240	250	270
80	I	195	200	205	215	225	235	250	255	265	275	285	305
100	N	225	230	235	245	255	270	280	285	295	305	315	335
150	E	290	295	300	310	320	335	345	350	360	370	380	405
200		350	355	360	375	385	395	405	410	420	430	445	465
250		415	420	425	440	450	460	470	475	485	495	510	530
300		475	485	490	500	510	520	530	540	550	560	570	590
350		515	520	525	535	550	560	570	575	585	595	610	630
400		575	580	585	600	610	620	630	635	650	660	670	690
450		640	645	650	660	670	680	690	700	710	720	730	750
500		700	705	710	720	730	745	755	760	770	780	790	820
600		820	825	835	845	855	865	875	885	895	905	915	935

45° RUNUNDERS

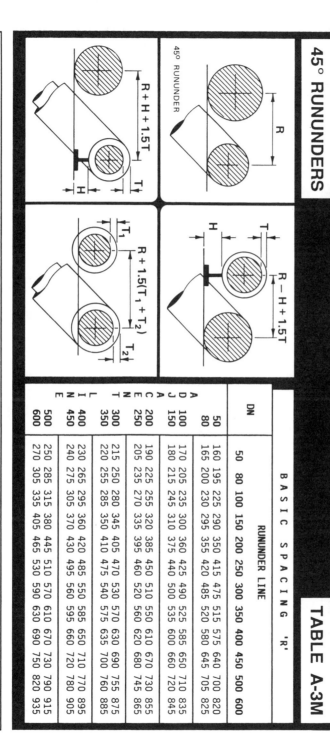

45° RUNUNDER

Diagram labels: R, R + H + 1.5T, H, T, T₁, R + 1.5(T₁ + T₂), R − H + 1.5T, T₂, H, T

TABLE A-3M

BASIC SPACING 'R'

DN	\ RUNUNDER LINE	50	80	100	150	200	250	300	350	400	450	500	600
50	L	160	195	225	290	350	415	475	515	575	640	700	820
80	I	165	200	230	295	355	420	485	520	580	645	705	825
100	N	170	205	235	300	360	425	490	525	585	650	710	835
150	E	180	215	245	310	375	440	500	535	600	660	720	845
200		190	225	255	320	385	450	510	550	610	670	730	855
250		205	235	270	335	395	460	520	560	620	680	745	865
300		215	250	280	345	405	470	530	570	630	690	755	875
350		220	255	285	350	410	475	540	575	635	700	760	885
400		230	265	295	360	420	485	550	585	650	710	770	895
450		240	275	305	370	430	495	560	595	660	720	780	905
500		250	285	315	380	445	510	570	610	670	730	790	915
600		270	305	335	405	465	530	590	630	690	750	820	935

NOTES FOR TABLES A-2M & A-3M

(1) SPACING SHOWN IN THE DIAGRAMS ALLOWS A MINIMUM CLEARANCE OF 50mm. COMPARE BASIC SPACING 'J' or 'R' WITH APPROPRIATE 'C' or 'CF' SPACING IN TABLE A-1M AND USE THE LARGER DIMENSION

(2) 'H' IS THE EFFECTIVE SHOE HEIGHT AND 'T' IS THE THICKNESS OF INSULATION (WITH COVERING)

(3) FOR SIMPLICITY, THE VALUE 1.5 HAS BEEN SUBSTITUTED FOR THE COEFFICIENT $1/\sin 45$ (1.414.....)

CLASS 600 — BUTT-WELDED PIPING DIMENSIONS — TABLE D-1M

DIMENSIONS IN THIS TABLE INCLUDE 6.4 mm RAISED FACE ON FLANGES

FITTINGS
DIMENSIONS FROM ANSI B16.5, B16.9, B16.28 AND MANUFACTURERS DATA

VALVES
DIMENSIONS FROM ANSI B16.10 AND MANUFACTURERS DATA

- **STRAIGHT TEE** — TABLE D-6M FOR REDUCING TEES (dimension T)
- **WELDOLET** — STANDARD AND EXTRA-STRONG (branch diameters 100 / 80 / 50)
- **REDUCERS** — CONCENTRIC & ECCENTRIC / REGULAR & REDUCING
- **90° LR ELLS** — REGULAR & REDUCING
- **90° SR ELL**
- **45° ELL (LR)** — dimension A
- **OFFSET (TWO 45° ELLS)** — dimensions A, B, C
- **ROLLED-ELL (45° ELL + 90° LR ELL)** — dimensions C, D
- **90° LR ELL + WELDING-NECK RAISED-FACE FLANGE** — dimensions E, F, G
- **PLUG** — VENTURI PATTERN: DN 50–600 / REGULAR PATTERN: DN 50–400 (dimensions H, I)
- **GATE** — DIMENSIONS ALSO APPLY TO GATE VALVES WITH BUTT-WELDING ENDS (dimension J)
- **BALL** — LONG PATTERN
- **GLOBE** — DIMENSIONS ALSO APPLY TO GLOBE VALVES WITH BUTT-WELDING ENDS (dimensions K, L, M)
- **CHECK** — SWING: DN 50–600 / TILTING DISC: DN 50–600 / LIFT: DN 50–300 (FLANGED & BUTT-WELDING)

Dimensions (mm) by Nominal Diameter (DN)

Item	Dim	50	80	100	150	200	250	300	350	400	450	500	600
STRAIGHT TEE	T	64	86	105	143	178	216	254	279	305	343	381	432
WELDOLET (branch 50)		68	83	95	122	148	181	200	216	241	267	292	343
WELDOLET (branch 80)		—	89	102	129	154	187	206	222	248	273	298	349
WELDOLET (branch 100)		—	—	108	135	160	187	203	213	229	254	279	305
REDUCERS		Swage–Table D-4M	76	89	102	127	152	178	203	229	254	279	305
90° LR ELLS		76	114	152	229	305	381	457	533	610	686	762	914
90° SR ELL		51	76	102	152	203	254	305	356	406	457	508	610
45° ELL (LR)	A	35	51	64	95	127	159	190	222	254	286	318	381
OFFSET (two 45° ells)	B	119	173	217	325	434	542	650	759	867	976	1092	1297
ROLLED-ELL	C	114	168	216	324	432	540	648	757	865	973	1081	1297
(45°ELL + 90°LR ELL)	D	156	203	260	352	444	540	619	705	794	876	959	1124
90°LR ELL + WN RF flange	E	79	117	153	229	305	382	458	534	611	686	763	916
	F	165	210	273	356	419	508	603	650	686	743	813	940
	G	79	89	108	124	140	159	162	171	184	191	197	210
PLUG (venturi)	H	229	305	406	559	610	711	762	914	965	965	1067	1067
PLUG (regular R / venturi V)	I	292 R/V	356	432	559	660	787	838	889	991 V	1092 V	1194 V	1397 V
GATE	J	292	356	432	559	660	787	838	889	991	1092	1194	1397
BALL (long pattern)		533	660	838	1194	1346	1676	1854	2057	2362	2515	2718	3200
GLOBE	K	305	356	457	610	914							
GLOBE	L	533	686	838	1118	1194							
GLOBE	M	292	356	432	559	660	787	838	889	991	1092	1194	1397
CHECK (flanged / butt-welding)		292 / 356 L	356 / 432 L	432 / 559 L	559 / 660 L	660 / 787 L	787 / 838 S	838 / 889 S	889 / 991 S	965 / 991 V	1067 / 1092 V	1194 V	1397 T
OFFSET (open, G)		49	72	90	135	180	225	269	314	359	404	449	539

DIMENSIONS ALSO APPLY TO GATE VALVES WITH BUTT-WELDING ENDS.
DIMENSIONS ALSO APPLY TO GLOBE VALVES WITH BUTT-WELDING ENDS.

Notes
- DIMENSIONS FOR COMBINATIONS OF FITTINGS AND INSTALLED WELDOLETS DO NOT INCLUDE THE 'WELD GAP' — REFER TO TEXT: SECTION 5.3.5
- DIMENSIONS IN THIS TABLE ARE NOMINAL AND FOR COMBINATIONS OF FITTINGS ARE ROUNDED TO 1 mm
- 'H', 'I', AND 'L' ARE THE LARGEST DIMENSIONS FOR MANUALLY-OPERATED CAST-STEEL VALVES FROM A SELECTION OF MANUFACTURERS
- GUIDELINES FOR THE USE OF GEAR AND POWERED OPERATORS WITH VALVES ARE GIVEN IN SECTION 3.1.2. OF THE TEXT

FITTINGS
DIMENSIONS FROM ANSI B16.5, B16.9, B16.28 AND MANUFACTURERS DATA

VALVES
DIMENSIONS FROM ANSI B16.10 AND MANUFACTURERS DATA

WELDOLET — STANDARD AND EXTRA-STRONG — TABLE D-6M FOR REDUCING TEES

NOMINAL DIAMETER (DN)	50	80	100	150	200	250	300	350	400	450	500	600
STRAIGHT TEE (branch = run)	64	86	105	143	178	216	254	279	305	343	381	432
Straight tee – branch 50	—	68	89	122	154	181	206	222	248	273	298	349
Straight tee – branch 80	—	—	102	129	148	175	200	216	241	267	292	343
Straight tee – branch 100	—	—	—	135	160	187	213	229	254	279	305	356
REDUCERS (concentric & eccentric; Swage Table D-4M)	—	89	102	140	152	178	203	330	356	381	508	610
90° LR ELLS (regular & reducing)	76	114	152	229	305	381	457	533	610	686	762	914
90° SR ELL	51	76	102	152	203	254	305	356	406	457	508	610
45° ELL (LR)	35	51	64	95	127	159	190	222	254	286	318	381
A — OFFSET (two 45° ells)	119	173	217	325	434	542	650	759	867	976	1084	1301
B — OFFSET (two 45° ells)	49	72	90	135	180	225	269	314	359	404	449	539
C — ROLLED-ELL (45° ell + 90° LR ell)	114	168	216	324	432	540	648	757	865	973	1081	1297
D — ROLLED-ELL (45° ell + 90° LR ell)	79	117	153	229	305	382	458	534	611	687	763	916
E — RAISED-FACE FLANGE / WELDING-NECK	146	194	238	327	416	498	587	676	756	845	924	1083
F — RAISED-FACE FLANGE / WELDING-NECK	165	210	254	318	381	444	521	584	648	711	711	914
G — RAISED-FACE FLANGE / WELDING-NECK	70	79	86	98	111	117	130	143	146	159	162	168
H — 90° LR ELL + WELDING-NECK	203	254	305	406	508	610	610	711	711	813	914	914
I — PLUG	533	635	737	991	1245	1499	1702	1930	2057	2337	2591	3124
J — GATE	216 (S/V)	283 (S/V)	305 (S/V)	403 (S/V)	419 (S/V)	457 (S/V)	502 (S/V)	762 (V/R)	838 (V/R)	914 (V/R)	991 (V/R)	1143 (V/R)
K — BALL	254	305	356	406	559	660	914	—	—	—	—	—
L — GLOBE	508 (T/L/S)	610	686	813	1041	1245	1321	—	—	—	—	—
M — CHECK	267 (S/T/L)	318	356	444	533	622	711	762	838	914	991	1143

Fitting / valve notes

- **REDUCERS:** CONCENTRIC & ECCENTRIC.
- **90° LR ELLS:** REGULAR & REDUCING.
- **OFFSET:** TWO 45° ELLS.
- **ROLLED-ELL:** 45° ELL + 90° LR ELL.
- **RAISED-FACE FLANGE / WELDING-NECK.**
- **90° LR ELL + WELDING-NECK.**
- **PLUG:** VENTURI PATTERN: DN 50–600; SHORT PATTERN: DN 50–300; REGULAR PATTERN: DN 350–600.
- **GATE:** DIMENSIONS ALSO APPLY TO GATE VALVES WITH BUTT-WELDING ENDS.
- **BALL:** SHORT PATTERN: DN 50–150.
- **GLOBE:** DIMENSIONS ALSO APPLY TO GLOBE VALVES WITH BUTT-WELDING ENDS. LONG PATTERN: DN 50–600; SHORT PATTERN: DN 50–150.
- **CHECK:** SWING: DN 50–600; TILTING DISC: DN 50–300; LIFT: DN 50–150, 250–300.

BRANCH DIAMETER: 100 / 80 / 50

General notes

- DIMENSIONS FOR COMBINATIONS OF FITTINGS AND INSTALLED WELDOLETS DO NOT INCLUDE THE 'WELD GAP' — REFER TO TEXT: SECTION 5.3.5
- DIMENSIONS IN THIS TABLE ARE NOMINAL AND FOR COMBINATIONS OF FITTINGS ARE ROUNDED TO 1 mm
- 'H', 'I', AND 'L' ARE THE LARGEST DIMENSIONS FOR MANUALLY-OPERATED CAST-STEEL VALVES FROM A SELECTION OF MANUFACTURERS
- GUIDELINES FOR THE USE OF GEAR AND POWERED OPERATORS WITH VALVES ARE GIVEN IN SECTION 3.1.2. OF THE TEXT

CLASS 150 — BUTT-WELDED PIPING DIMENSIONS — TABLE D-3M

DIMENSIONS IN THIS TABLE INCLUDE 1.6 mm RAISED FACE ON FLANGES

FITTINGS
DIMENSIONS FROM ANSI B16.5, B16.9, B16.28 AND MANUFACTURERS DATA

VALVES
DIMENSIONS FROM ANSI B16.10 AND MANUFACTURERS DATA

Fitting / valve descriptions

- **STRAIGHT TEE** — TABLE D-6M FOR REDUCING TEES
- **WELDOLET** — STANDARD AND EXTRA-STRONG (BRANCH DIAMETER 100 / 80 / 50)
- **REDUCERS** — CONCENTRIC & ECCENTRIC / REGULAR & REDUCING
- **90° LR ELLS** — REGULAR & REDUCING
- **90° SR ELL**
- **45° ELL (LR)**
- **OFFSET (TWO 45° ELLS)**
- **ROLLED-ELL** (45° ELL + 90° LR ELL)
- **90° LR ELL + WELDING-NECK RAISED-FACE FLANGE**
- **PLUG** — SHORT PATTERN: DN 50-300, VENTURI PATTERN: DN 50-100, 350-600, REGULAR PATTERN: DN 350-600
- **GATE** — REFER TO TABLE V-1M FOR END-TO-END DIMENSIONS OF GATE VALVES WITH BUTT-WELDING ENDS. DIMENSIONS ALSO APPLY TO GATE VALVES WITH BUTT-WELDING ENDS.
- **BALL** — LONG PATTERN: DN 50-600, SHORT PATTERN: DN 50-400, USE 'J'; ABOVE FOR GATE VA
- **GLOBE** — DIMENSIONS ALSO APPLY TO GLOBE VALVES WITH BUTT-WELDING ENDS
- **CHECK** — SWING: DN 50-600, TILTING DISC: DN 50-350, LIFT: DN 50-100, 200-350. FLANGED & BUTT-WELDING

Dimension table (values in mm)

Item / Ref	50	80	100	150	200	250	300	350	400	450	500	600
STRAIGHT TEE	64	86	105	143	178	216	254	279	305	343	381	432
WELDOLET (branch 100)	—	—	89	102	108	129	148	160	181	187	200	222
WELDOLET (branch 80)	—	89	95	122	148	154	175	181	200	206	216	216
WELDOLET (branch 50)	68	83	102	108	135	160	187	213	229	254	279	356
REDUCERS (Swage – Table D-4M)	Swage D-4M	—	108	127	152	178	203	229	254	279	305	356
90° LR ELL (A)	76	114	152	229	305	381	457	533	610	686	762	914
90° SR ELL (B)	51	76	102	152	203	254	305	356	406	457	508	610
45° ELL (LR) (C)	35	51	64	95	127	159	190	222	254	286	318	381
OFFSET — A	49	72	90	135	180	225	269	314	359	404	449	539
OFFSET — B	119	173	217	325	434	542	650	759	867	976	1084	1301
ROLLED-ELL — C	140	184	216	318	406	483	572	660	737	826	906	1067
ROLLED-ELL — D	114	168	216	324	432	540	648	757	865	973	1081	1297
ROLLED-ELL — E	79	117	153	229	305	382	458	534	611	687	763	916
LR ELL + FLANGE — F	152	190	229	279	343	406	483	533	597	635	660	660
LR ELL + FLANGE — G	64	70	76	89	102	114	114	127	127	140	144	152
PLUG (H) S/V/R	178 (S/V)	203 (S/V)	229 (S/V)	267 (S/V)	292 (S/V)	330 (S/V)	356 (S/V)	686 (R/V)	762 (R/V)	864 (R/V)	914 (R/V)	1067 (R/V)
GATE (H) face-to-face	203	254	305	356	406	457	508	610	660	762	762	914
GATE (I) open	483	584	711	940	1194	1346	1549	1803	2032	2261	2489	2870
BALL (J)	178	203	229	394	457	533	610	686	762	864	914	1067
GLOBE (K)	381	483	533	660	838	813	1067	1245		914		
GLOBE (L)	203	241	292	356	406	495	610	686	762	762	762	1067
GLOBE (M)	203	241	292	406	457	508	508	610	406	432	457	508
CHECK (flanged L / butt-welding T)	203 (L/T)	241 (L/T)	292 (L/T)	356 (L/T)	406/495	495/622	698/698	787/787	864/914	978/914	978	1295

Notes

- DIMENSIONS FOR COMBINATIONS OF FITTINGS AND INSTALLED WELDOLETS DO NOT INCLUDE THE 'WELD GAP' — REFER TO TEXT: SECTION 5.3.5
- DIMENSIONS IN THIS TABLE ARE NOMINAL AND FOR COMBINATIONS OF FITTINGS ARE ROUNDED TO 1 mm
- 'H', 'I', AND 'L' ARE THE LARGEST DIMENSIONS FOR MANUALLY-OPERATED CAST-STEEL VALVES FROM A SELECTION OF MANUFACTURERS
- GUIDELINES FOR THE USE OF GEAR AND POWERED OPERATORS WITH VALVES ARE GIVEN IN SECTION 3.1.2. OF THE TEXT

SWAGES

TABLE D-4M

DN (mm) LARGE END	SMALL END	LENGTHS
50	8-40	165
65	8-50	178
80	15-65	203
90	50-80	203
100	25-90	229
125	50-100	279
150	40-125	305
200	50-150	330
250	100-200	381

Dimensions in this table are for Mills Iron Works swages, available with ends plain, threaded, bevelled, Victaulic grooved, and in any combination of these terminations.

ELBOLETS: THREADED/SOCKET & BUTT-WELDING

DIMENSIONS IN MILLIMETERS — TABLE D-5M

NOMINAL DIAMETER OF MAIN RUN [DN]

CLASS 3000 THREADED & SOCKET-WELDING* – STD AND XS BUTT-WELDING

LR ELL

DN OF BRANCH	50	80	100	150	200	250	300	350	400	450	500	600
15	90	151	184	254	321	391	458	511	578	645	713	847
20	122	158	191	261	329	398	465	518	585	652	720	854
25	130	166	199	269	337	406	473	526	593	660	728	862
40	141	177	210	280	348	417	484	537	604	672	739	873
50	156	191	225	294	362	431	498	552	618	686	761	887
65		207	241	310	378	447	514	568	634	702	769	903
80			258	328	395	464	532	585	652	719	787	921
100				371	438	507	575	628	695	762	829	964
150					464	533	600	653	720	787	855	989
200						579	646	699	766	833	901	1035
250							672	725	791	859	938	1060
300												

Dimensions converted from BONNEY FORGE data. Dimensions for Elbolets are nominal. Size DN 50 Elbolets are designed to fit the different sizes of run pipe; in sizes larger than DN 50, each size of Elbolet is designed to fit a range of run pipe sizes.

* Threaded and socket-welding Elbolets are not available in sizes DN 150 and larger.

REDUCING BUTT-WELDING TEES

TABLE D-6M

WEIGHTS: STD and XS. SCH 160 thru DN 300. XXS thru DN 200.

NOMINAL DIAMETER OF MAIN RUN [DN]

DIMENSION 'A'

DN ▶	80	100	150	200	250	300	350	400	450	500	600
50	76	89	105	143	178	216	254				
80		98	124	156	194	219	248				
100			130	168	184	229	238	257			
150					203	216	254	279	305		
200						229	264	295	330		
250							273	283	321	356	
300							298	308	333	397	432
350								324	346	356	
400									321	406	406
450										419	419
500											432

DIMENSION 'A'

DN	A
80	86
100	105
150	143
200	178
250	216
300	254
350	305
400	356
450	400
500	450

CHECK VALVES - WAFER-TYPE

TABLE D-7M

FACE-TO-FACE DIMENSIONS BY CLASS FOR VALVES CONFORMING TO API 594

DN	FLANGE CLASSES					
	150	300	600	900	1500	2500
50	60	60	60	70	70	70
80	73	73	73	83	83	86
100	73	73	79	102	102	105
150	98	98	136	159	159	159
200	127	127	165	206	206	206
250	146	146	213	241	248	254
300	181	181	229	292	305	305
350	184	222	273	356	356	
400	190	232	305	384	384	
450	203	264	362	451	468	
500	219	292	368	451	533	
600	222	318	438	495	559	

SINGLE AND DUAL PLATES

SOCKET WELDING FITTINGS - FORGED STEEL

'R' DIMENSIONS INCLUDE EXPANSION GAP - NOTE 1

TABLE D-8M

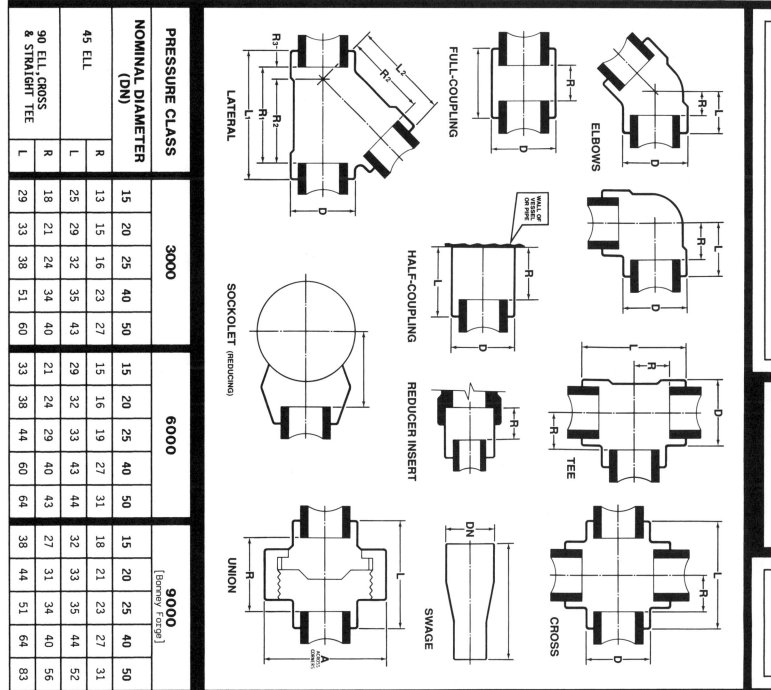

LATERAL · FULL-COUPLING · ELBOWS · SOCKOLET (REDUCING) · HALF-COUPLING · REDUCER INSERT · TEE · UNION · SWAGE · CROSS

WALL OF VESSEL OR PIPE

PRESSURE CLASS	NOMINAL DIAMETER (DN)	45 ELL		90 ELL, CROSS & STRAIGHT TEE	
		R	L	R	L
3000	15	13	25	18	29
	20	15	29	21	33
	25	16	32	24	38
	40	23	35	34	51
	50	27	43	40	60
6000	15	15	29	21	33
	20	16	32	24	38
	25	19	33	29	44
	40	27	43	40	60
	50	31	44	43	64
9000 [Bonney Forge]	15	18	32	27	38
	20	21	33	31	44
	25	23	35	34	51
	40	27	44	40	64
	50	31	52	56	83

Fitting		Dim																
FULL-COUPLING		R	14	14	17	17	23	14	14	17	17	23	14	14	17	17	23
		L	35	38	44	51	64	35	38	44	51	64	35	38	44	51	64
HALF-COUPLING [Bonney Forge]	15	R	14	14	17	17	23	14	14	17	17	23	14	14	17	17	23
	20																
	40																
		L	35	38	44	51	64	35	38	44	51	64	35	38	44	51	64
REDUCER INSERT [Bonney Forge]	40	R		24	26	18	23	27		29	31	44	51	64		35	24	26	31
	25				31	21	26	29		31	27	29	31	27	29	32	38	44	51
	20					44	34	27		32	27	29	31	34	27	34	31	27	27
	15							50				31		50		54		51	64
LATERAL [Bonney Forge & Ladish]		L2	54	65	76	100	121	65	76	89	121	175	76	89	100	127	
		L1	76	90	105	137	164	90	105	122	164	229	105	122	137	168	
		R3	12	13	15	19	23	13	15	18	23	34	16	19	21	21	
		R2	43	53	62	83	100	53	62	73	100	154	64	75	85	107	
		R1	55	66	77	102	123	66	77	91	123	188	80	94	106	128	
DIAMETER		D	33	40	48	64	78	40	47	57	78	92	47				
SOCKOLET (REDUCING) [Bonney Forge]	40	B R A N C H	15					51					54				62
	25			30	33	40	46	52	38	41	48	54	56	37	40	48	54	58
	20				33	40	46			41	48				43	50	56	
	15					46	52				50	56				52	58	
UNION [Bonney Forge]		R	32	33	39	52	56	33	39	47	56	64	33	39	47	56	64
		L	51	59	62	76	89	59	62	73	89	105	59	62	73	89	105
		A	49	60	71	94	112	60	71	85	112	133					
SWAGE [Bonney Forge]		A	70	76	89	114	165	70	76	89	114	165	70	76	89	114	165

(1) 'R' DIMENSIONS ('REMOVED RUN' OF PIPE) HAVE BEEN ROUNDED TO 1.0 mm AND INCLUDE 2 mm EXPANSION GAP(S) FOR WELDING. REFER TO 'SOCKET-WELDING PIPING' - CHART 2.2

(2) DIMENSIONS ARE FROM THE FOLLOWING SUPPLIERS' DATA: BONNEY FORGE, ITT GRINNEL, LADISH AND VOGT

(3) UNLESS THE SUPPLIER IS STATED, 'L' & 'D' DIMENSIONS ARE THE LARGEST QUOTED BY BONNEY FORGE, ITT GRINNEL, LADISH AND VOGT

(4) FITTINGS CONFORM TO ANSI B16.11, EXCEPT LATERALS AND REDUCER INSERTS, WHICH ARE MADE TO MANUFACTURERS' STANDARDS

(5) FOR INFORMATION ON THE BORE DIAMETER AND RATING OF FITTINGS, REFER TO 'SOCKET-WELDED PIPING' - CHART 2.2

(6) UNIONS CONFORM TO MSS-SP-83

(7) DIMENSIONS FOR INSTALLED SOCKOLETS EXCLUDE THE 'ROOT GAP' - REFER TO 'DIMENSIONING SPOOLS (WELDED ASSEMBLIES)' - 5.3.5

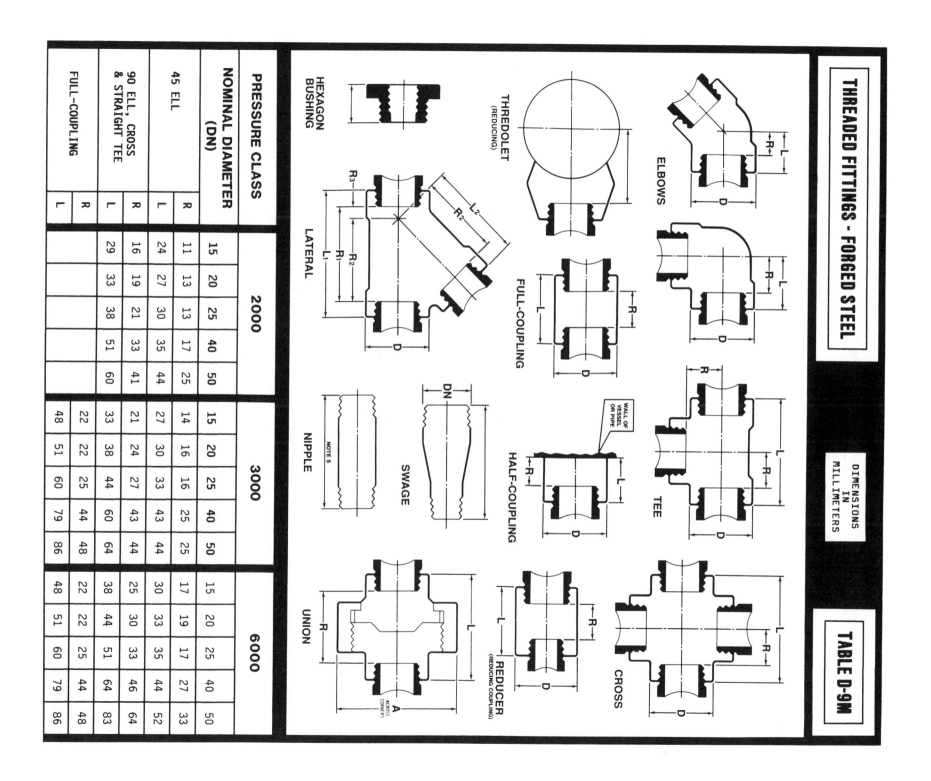

THREADED FITTINGS - FORGED STEEL

DIMENSIONS IN MILLIMETERS

TABLE D-9M

HEXAGON BUSHING

THREDOLET (REDUCING)

ELBOWS

LATERAL

FULL-COUPLING

TEE

NIPPLE — NOTE 5

SWAGE

HALF-COUPLING — WALL OF VESSEL OR PIPE

CROSS

UNION — ACROSS CORNERS — A

REDUCER (REDUCING COUPLING)

PRESSURE CLASS	NOMINAL DIAMETER (DN)	45 ELL R	45 ELL L	90 ELL, CROSS & STRAIGHT TEE R	90 ELL, CROSS & STRAIGHT TEE L	FULL-COUPLING R	FULL-COUPLING L
2000	15	11	24	16	29		
	20	13	27	19	33		
	25	13	30	21	38		
	40	17	35	33	51		
	50	25	44	41	60		
3000	15	14	27	21	33	22	48
	20	16	30	24	38	22	51
	25	16	33	27	44	25	60
	40	25	43	43	60	44	79
	50	25	44	44	64	48	86
6000	15	17	30	25	38	22	48
	20	19	33	30	44	22	51
	25	17	35	33	51	25	60
	40	27	44	46	64	44	79
	50	33	52	64	83	48	86

Dimensional data table for threaded pipe fittings (dimensions in mm). The table is grouped into three column sets (each covering nominal sizes 15, 20, 25, 40, 50).

Fitting	Dim	15	20	25	40	50	15	20	25	40	50	15	20	25	40	50
HALF-COUPLING	R	11	11	13	22	24	11	11	13	22	24	11	11	13	22	24
	L															
REDUCER (15 / 20 / 25 / 40)	R	24	25	30	40	43	24	25	30	40	43	24	25	30	40	43
			29	48	52			29	48	52			29	48	52	
				44	49				44	49				44	49	
					49					49					49	
LATERAL [Bonney Forge & Ladish]	L							51	60	79	86		51	60	79	86
	R1	53	63	72	105	129	67	78	90	132	191	81	97	105	194	194
	R2	42	51	59	84	103	53	63	73	105	156	64	76	84	159	157
	R3	10	12	12	21	25	13	15	17	27	35	17	21	21	37	35
	L1	78	92	106	140	167	92	106	125	167	229	106	125	140	229	232
	L2	55	66	77	102	122	66	77	90	122	175	77	90	102	175	178
DIAMETER	D	33	40	47	66	78	40	47	57	78	92	47	57	64	92	110
THREDOLET (REDUCING) [Bonney Forge] (40 / 25 A / 20 / 15)	BRANCH H C N A R B						26	29	37	40	48	32	33	39	46	56
								29	37	43	46		36	43	49	52
									40	46	52		39	46	52	56
UNION [Bonney Forge]	R	25	30	27	42	50				40	46	32	33	38	52	56
	L	51	59	62	76	88					48	58	62	73	87	104
	A	49	61	71	94	112	49	61	70	85	112	61	70	85	112	133
HEX BUSH		24	25	27	33	37	24	25	27	33	37	24	25	27	33	37
SWAGE		70	76	89	114	165	70	76	89	114	165	70	76	89	114	165
THREAD ENGAGEMENT		13	14	17	17	19	13	14	17	17	19	13	14	17	17	19

(1) 'R' DIMENSIONS ('REMOVED RUN' OF PIPE) ARE BASED ON NORMAL THREAD ENGAGEMENT BETWEEN MALE AND FEMALE THREADS TO MAKE TIGHT JOINTS — ROUNDED TO 1.00 mm

(2) DIMENSIONS FOR FITTINGS ARE FROM THE FOLLOWING SUPPLIERS' DATA: BONNEY FORGE, ITT GRINNEL, LADISH AND VGT

(3) UNLESS THE SUPPLIER IS STATED, 'L' & 'D' DIMENSIONS ARE THE LARGEST QUOTED BY BONNEY FORGE, ITT GRINNEL, LADISH AND VGT

(4) FITTINGS CONFORM TO ANSI B16.11, EXCEPT LATERALS, WHICH ARE MADE TO MANUFACTURERS' STANDARDS. UNIONS CONFORM TO MSS-SP-83

(5) FOR SIZES AND AVAILABILITIES OF PIPE NIPPLES, REFER TO 'MALLEABLE-IRON PIPE FITTINGS' - TABLE D-11m

(6) DIMENSIONS FOR INSTALLED THREDOLETS EXCLUDE THE 'ROOT GAP' - REFER TO 'DIMENSIONING SPOOLS (WELDED ASSEMBLIES)' - 5.3.5

CLASS 800 VALVES

API CLASS 800 FORGED-STEEL GATE, GLOBE & CHECK VALVES

TABLE D-10M

'R' is the 'REMOVED RUN' of pipe occupied by the valve

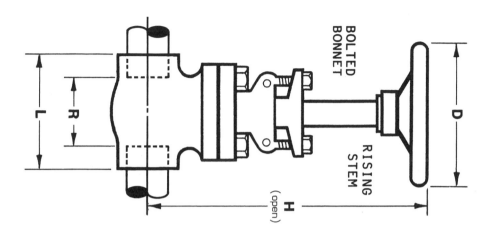

BOLTED BONNET

RISING STEM

H (open)

VALVES WITH THREADED ENDS

	DN	15	20	25	40	50
GATE	D	102	102	140	168	168
	H	162	184	217	279	317
	L	89	98	108	140	144
	R	64	70	73	105	106
GLOBE	D	102	102	102	117	168
	H	162	167	173	206	257
	L*	83	89	114	159	184
	R*	57	60	79	124	146

* These dimensions also apply to horizontal lift-check valves

'R' dimensions are based on normal thread engagement for tight joints

VALVES WITH SOCKET ENDS

	DN	15	20	25	40	50
GATE	D	102	102	140	168	168
	H	162	184	217	279	317
	L	89	98	108	140	144
	R	52	58	71	80	96
GLOBE	D	102	102	102	117	168
	H	162	167	173	206	257
	L*	83	89	114	159	184
	R*	61	64	87	125	144

'R' dimensions include 2 mm expansion gaps for welding. Refer to text: Chart 2.2

THREADED FITTINGS - MALLEABLE-IRON

TABLE D-11M

DIMENSIONS ROUNDED TO 1.00 mm

Fitting diagrams (left column):

- **45° ELL**
- **90° ELL**
- **90° STREET ELL** (dimensions A, B)
- **RETURN BEND** (OPEN, MEDIUM, CLOSE; dimensions A, B)
- **STRAIGHT TEE** — (These data also apply to the center-to-end dimension for straight cross.)
- **LATERAL** (dimensions A, C)
- **UNION** — OCTAGONAL
  - S — GRINNELL: COPPER ALLOY-TO-IRON
  - S — STOCKHAM: BRASS-TO-IRON or ALL-IRON (E, A, T, S)
- **COUPLING** — CLOSE NIPPLE
- **NIPPLE** — CARBON-STEEL (TANK NIPPLES ARE 150mm LONG)
  - AVAILABILITIES OF SHORT AND LONG NIPPLES: AVAILABLE IN 50, 65, 75, 90, 100, 115, 125, 140, 150, 180, 200, 230, 255, 280 & 305 mm LENGTHS (DN 15 and DN 20 nipples are also available 40 mm long)
- **SWAGE** — MILLS IRON WORKS, CARBON-STEEL (DN)
- **REDUCER** — CARBON-STEEL (DN)
- **THREAD ENGAGEMENT** — TAPER/TAPER (Engagement)

PRESSURE CLASS		150						300					
NOMINAL DIAMETER [DN]		**15**	**20**	**25**	**40**	**50**	**80**	**15**	**20**	**25**	**40**	**50**	**80**
45° ELL		22	25	29	37	43	56	25	29	33	43	51	64
90° ELL		29	33	38	49	57	78	32	37	41	54	64	86
90° STREET ELL	B	41	48	54	68	83	114	51	56	65	79	94	130
90° STREET ELL	A	29	33	38	49	57	78	32	37	41	54	64	86
RETURN BEND	A OPEN	38	51	64	89	102							
RETURN BEND	A MEDIUM	32	38	48	64	76							
RETURN BEND	A CLOSE	25	32	38	56	67							
RETURN BEND	B	29	33	38	49	57	78	32	37	41	54	64	86
STRAIGHT TEE	A	29	33	38	49	57	78	32	37	41	54	64	86
LATERAL	A	59	71	84	111	132	184						
LATERAL	C	43	52	62	83	100	141						
UNION	A	46	51	56	67	78	98	49	57	62	76	86	125
COUPLING	A	33	38	43	54	64	81	48	54	60	73	92	105
NIPPLE	CLOSE NIPPLE	30	35	40	45	50	65	30	35	40	45	50	65
SWAGE	DN	70	76	89	114	165	203	70	76	89	114	165	203
REDUCER	DN	32	37	43	59	71	94	43	44	51	68	81	103
THREAD ENGAGEMENT		13	14	17	17	19	25	13	14	17	17	19	25

DIMENSIONS IN THIS TABLE ARE FOR BANDED FITTINGS AND CONFORM TO ANSI STANDARD B16.3, AND FEDERAL SPECIFICATION WW-P-521. UNIONS CONFORM TO ANSI B16.39. DATA FROM ITT GRINNELL CORPORATION AND STOCKHAM VALVES AND FITTINGS

FORGED-STEEL FLANGES & LAP-JOINT STUB-ENDS

FLANGE CLASSES 150-2500 (PN20-PN420)

TABLES FM

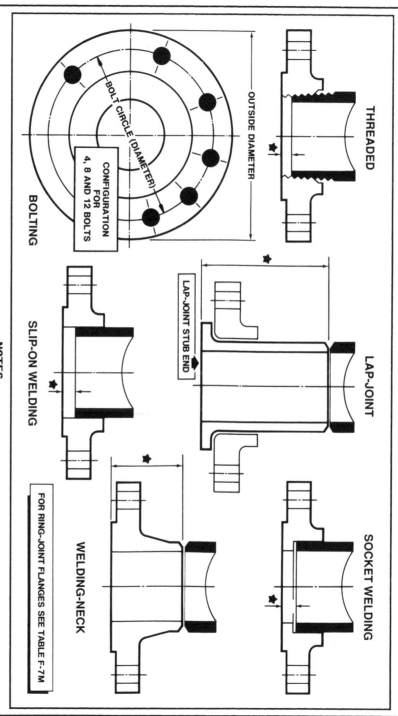

THREADED

BOLTING

BOLT CIRCLE (DIAMETER)

OUTSIDE DIAMETER

CONFIGURATION FOR 4, 8 AND 12 BOLTS

SLIP-ON WELDING

LAP-JOINT

LAP-JOINT STUB END

WELDING-NECK

SOCKET WELDING

FOR RING-JOINT FLANGES SEE TABLE F-7M

NOTES

[1] FLANGE DIMENSIONS: INTERNATIONAL STANDARD ISO 2229, ANSI STANDARD B16.5 AND MANUFACTURERS' DATA

[2] BLIND FLANGES: DATA FOR FLANGE DIAMETERS AND BOLTING IN THESE TABLES ALSO APPLIES TO BLIND FLANGES

[3] REDUCING FLANGES: AVAILABLE IN SLIP-ON, THREADED AND WELDING-NECK TYPES

[4] LAP-JOINT STUB-ENDS: ANSI B16.9 (Long Pattern) & MSS SP-43 (Short Pattern)

[5] STUDBOLT THREAD LENGTHS FOR LAP-JOINTS

FLANGE COMBINATION	FLANGE CLASS	STUDBOLT THREAD LENGTHS FOR LAP-JOINTS
Lapped to non-lapped	150 or 300	Thickness of lap
	Over 300	Thickness of lap
Lapped to lapped	150 - 2500	Thickness of lap minus 6.4 mm
		Thickness of two laps

INCREASE IN STUDBOLT LENGTH OVER LENGTHS IN TABLES F-1M thru F-6M

Thickness of lap = Thickness of pipe wall + 0 mm + 1.6 mm

[90]

CLASS 150 FLANGE DATA — PN20 — TABLE F-1M

* DIMENSIONS INCLUDE 1.6 mm RAISED FACE ON FLANGES (except lap-joint)
** DIMENSIONS INCLUDE 2 mm GAP FOR WELDING – REFER TO CHART 2.2

NOMINAL DIAMETER: DN	15	20	25	40	50	80	100	150	200	250	300	350	400	450	500	600
OUTSIDE DIAMETER	89	98	108	127	152	190	229	279	343	406	483	533	597	635	698	813
END OF PIPE TO FACE OF FLANGE or LAP JOINT ★ — WELD-NECK	48	52	56	62	64	76	89	102	114	127	127	140	140	152	144	152
SLIP-ON	9	7	7	8	10	12										
SOCKET **	2	2	0	5	7	9										
THREADED	2	2	5	7	9											
L-J STUB END • — ANSI	76	76	102	102	152	152	203	203	254	254	254	305	305	305	305	305
L-J STUB END • — MSS	51	51	51	51	64	76	76	89	102	127	152	152	152	152	152	152
BORE: WELD-NECK & SOCKET	15.8	21	26.2	40.9	52.5	77.9	102.3	154.1	202.7	254.5	304.9	[Order to match pipe ID]				
L-J STUB END: RF	57	57	64	70	76	89	95	102	114	127	127	140	146	152	159	165
L-J STUB END: RJ	-	-	76	83	89	102	108	114	127	140	146	152	159	165	171	184
STUDBOLT THREAD length – except lap-joint: Note 5 — RF	57	57	64	70	76	89	95	102	108	114	127	127	140	146	152	171
STUDBOLT THREAD length – except lap-joint: Note 5 — RJ	-	-	76	83	89	102	108	114	127	140	146	152	159	165	171	184
BOLTS PER FLANGE	4	4	4	4	4	4	8	8	8	12	12	12	16	16	20	20
BOLT CIRCLE DIAMETER	60.3	69.8	79.4	98.4	120.6	152.4	190.5	241.3	298.4	362	431.8	476.2	539.8	577.8	635	749.3
DIAMETER OF BOLT (IN)	1/2	1/2	1/2	1/2	5/8	5/8	5/8	3/4	3/4	7/8	7/8	7/8	1	1 1/8	1 1/8	1 1/4

(Bore of weld-neck & socket for larger sizes: Wall thickness of pipe + 2 mm)

CLASS 300 FLANGE DATA — PN50 — TABLE F-2M

* DIMENSIONS INCLUDE 1.6 mm RAISED FACE ON FLANGES (except lap-joint)
** DIMENSIONS INCLUDE 2 mm GAP FOR WELDING – REFER TO CHART 2.2

NOMINAL DIAMETER: DN	15	20	25	40	50	80	100	150	200	250	300	350	400	450	500	600
OUTSIDE DIAMETER	95	117	124	156	165	210	254	318	381	444	521	584	648	711	775	914
END OF PIPE TO FACE OF FLANGE or LAP JOINT — WELD-NECK	52	57	62	68	70	79	86	98	111	117	130	143	146	159	162	168
SLIP-ON	15	16	17	16	18	25										
SOCKET **	2	2	0	5	6	11	9	15	18	19	19					
THREADED	2	2	5	6	9	15	18	19	19							
L-J STUB END — ANSI	76	76	102	102	152	152	203	203	254	254	254	305	305	305	305	305
L-J STUB END — MSS	51	51	51	51	64	76	76	89	102	127	152	152	152	152	152	152
BORE: WELD-NECK & SOCKET	15.8	21	26.2	40.9	52.5	77.9	102.3	154.1	202.7	254.5	304.9	[Order to match pipe ID]				
BOLTS PER FLANGE	4	4	4	4	8	8	8	12	12	16	16	20	24	24	24	24
BOLT CIRCLE DIAMETER	66.7	82.6	88.9	114.3	127	168.3	200	269.9	330.2	387.4	450.8	514.4	571.5	628.6	685.8	812.8
DIAMETER OF BOLT (IN)	1/2	5/8	5/8	3/4	5/8	3/4	3/4	3/4	7/8	7/8	7/8	1	1 1/8	1 1/8	1 1/4	1 1/4

(Bore of weld-neck & socket for larger sizes: Wall thickness of pipe + 2 mm)

PN references are discussed under 'FLANGE CLASSES and PRESSURE NUMBERS' – page 75 (Part II)

* DIMENSIONS INCLUDE 6.4 mm RAISED FACE ON FLANGES (except lap-joint)
** DIMENSIONS INCLUDE 2 mm GAP FOR WELDING - REFER TO CHART 2.2

NOMINAL DIAMETER: DN	15	20	25	40	50	80	100	150	200	250	300	350	400	450	500	600
OUTSIDE DIAMETER	95	117	124	156	165	210	273	356	419	508	559	603	686	743	813	940
END OF PIPE TO FACE OF FLANGE or LAP JOINT — WELD-NECK	58	63	68	76	79	89	108	123	139	158	162	171	184	190	196	209
SLIP-ON																
SOCKET **	21	22	23	24	28	34										
THREADED	9	8	7	11	12	15	15	22	24	25	26					
L-J STUB END ANSI	76	76	102	102	152	152	152	203	203	254	254	305	305	305	305	305
L-J STUB END MSS	51	51	51	51	64	76	89	102	102	152	152	152	152	152	152	152
BORE: WELD-NECK & SOCKET	Wall thickness of pipe + 2 mm — Order to match Internal Diameter of Pipe															
BOLTS PER FLANGE	4	4	4	4	4	8	8	8	12	12	12	16	16	20	20	24
BOLT CIRCLE DIAMETER	66.7	82.6	88.9	114.3	127	168.3	215.9	292.1	349.2	431.8	489	527	603.2	654	723.9	838.2
DIAMETER OF BOLT (IN)	1/2	5/8	5/8	3/4	3/4	3/4	7/8	1	1 1/8	1 1/4	1 1/4	1 3/8	1 1/2	1 5/8	1 5/8	1 7/8
STUDBOLT THREAD length - except lap-joint: Note 5 — RF	76	83	89	89	102	121	140	165	190	210	216	229	248	267	286	324
STUDBOLT THREAD length - except lap-joint: Note 5 — RJ	76	83	89	102	102	127	146	171	197	216	222	235	254	273	292	337

PN references are discussed under 'FLANGE CLASSES and PRESSURE NUMBERS' - page 75 (Part II)

* DIMENSIONS INCLUDE 6.4 mm RAISED FACE ON FLANGES (except lap-joint)

NOMINAL DIAMETER: DN	15	20	25	40	50	80	100	150	200	250	300	350	400	450	500	600
OUTSIDE DIAMETER — WELD-NECK	121	130	149	178	216	241	292	381	470	546	610	641	705	787	857	1041
END OF PIPE TO FACE OF FLANGE or LAP JOINT — WELD-NECK	66	76	79	89	108	108	120	146	168	190	206	219	222	235	254	298
SLIP-ON / SOCKET	Not available in this class															
THREADED	15	17	19	21	27	12	16	22	24	24	26					
L-J STUB END ANSI	76	76	102	102	152	152	152	203	203	254	254	305	305	305	305	305
L-J STUB END MSS	51	51	51	51	64	64	76	89	102	127	152	152	152	152	152	152
BORE: WELD-NECK	Wall thickness of pipe + 2 mm — Order to match Internal Diameter of Pipe															
BOLTS PER FLANGE	4	4	4	4	8	8	8	12	12	16	20	20	20	20	20	20
BOLT CIRCLE DIAMETER	82.6	88.9	101.6	123.8	165.1	190.5	235	317.5	393.7	469.9	533.4	558.8	616	685.8	749.3	901.7
DIAMETER OF BOLT (IN)	3/4	3/4	7/8	7/8	7/8	1	1 1/8	1 1/8	1 3/8	1 3/8	1 1/2	1 5/8	1 7/8	2	2 1/2	
STUDBOLT THREAD length - except lap-joint: Note 5 — RF	102	108	108	121	140	165	190	216	229	248	267	279	324	343	432	
STUDBOLT THREAD length - except lap-joint: Note 5 — RJ	102	108	121	133	146	171	190	222	216	229	235	248	254	267	356	457

CLASS 1500 FLANGE DATA — PN250 — TABLE F-5M

★ DIMENSIONS INCLUDE 6.4 mm RAISED FACE ON FLANGES (except lap-joint)
★ DIMENSIONS INCLUDE 2 mm GAP FOR WELDING - REFER TO CHART 2.2

NOMINAL DIAMETER: DN	15	20	25	40	50	80	100	150	200	250	300	350	400	450	500	600
OUTSIDE DIAMETER — WELD-NECK	121	130	149	178	216	267	311	394	483	584	673	749	826	914	984	1168
END OF PIPE TO FACE OF FLANGE or LAP JOINT ★ — WELD-NECK	Wall thickness of pipe + 2 mm															
SLIP-ON ★★	66	76	79	89	108	123	130	177	219	260	289	304	317	333	362	412
SOCKET ★★	31	32	37	36	48											
THREADED	8	8	8	11	16	13	15	22	24	25	26					
L-J STUB END — ANSI	76	76	76	102	152	152	152	203	254	254	305	305	305	305	305	
STUB END ★ — MSS	51	51	51	51	51	64	76	89	102	127	152	152	152	152	152	
BOLT CIRCLE DIAMETER	82.6	88.9	101.6	123.8	165.1	203.2	241.3	317.5	393.7	482.6	571.5	635	704.8	774.7	831.8	990.6
DIAMETER OF BOLT (IN)	3/4	3/4	7/8	1	7/8	1 1/8	1 1/4	1 3/8	1 5/8	1 7/8	2	2 1/4	2 1/2	2 3/4	3	3 1/2
BOLTS PER FLANGE	4	4	4	4	8	8	8	12	12	12	16	16	16	16	16	16
STUDBOLT THREAD length - except lap-joint: Note 5 — RF	102	108	121	133	140	171	190	254	286	337	375	406	444	489	533	610
RJ	102	108	121	133	146	178	197	260	298	343	387	425	470	514	565	648
BORE: WELD-NECK & SOCKET	Order to match Internal Diameter of Pipe															

PN references are discussed under 'FLANGE CLASSES and PRESSURE NUMBERS' - page 75 (Part II)

CLASS 2500 FLANGE DATA — PN420 — TABLE F-6M

★ DIMENSIONS INCLUDE 6.4 mm RAISED FACE ON FLANGES (except lap-joint)

NOMINAL DIAMETER: DN	15	20	25	40	50	80	100	150	200	250	300
OUTSIDE DIAMETER — WELD-NECK	133	140	159	203	235	305	356	483	552	673	762
END OF PIPE TO FACE OF FLANGE or LAP JOINT ★ — WELD-NECK	Wall thickness of pipe + 2 mm										
SLIP-ON	79	85	95	117	133	174	196	279	324	425	470
SOCKET	Not available in this class										
THREADED	9	12	17	23	13	15	22	24	27	26	
L-J STUB END — ANSI	76	76	102	102	152	152	203	203	254	254	254
STUB END ★ — MSS	51	51	51	64	64	76	89	102	102	127	152
BOLT CIRCLE DIAMETER	88.9	95.2	107.9	146	171.4	228.6	273	368.3	438.1	539.7	619.1
DIAMETER OF BOLT (IN)	3/4	3/4	7/8	1 1/8	1 1/4	1 1/2	1 3/4	2	2	2 1/2	2 3/4
BOLTS PER FLANGE	4	4	4	4	8	8	8	8	12	12	12
STUDBOLT THREAD length - except lap-joint: Note 5 — RF	121	121	133	165	216	248	279	343	381	483	533
RJ	121	121	133	171	216	222	273	356	425	508	559
BORE: WELD-NECK	Order to match Internal Diameter of Pipe										

ISO STANDARD 2229 IDENTIFIES FLANGES IN CLASSES 150 THRU 2500.

DIMENSIONAL DATA ARE SIMILAR TO FLANGES SPECIFIED BY ANSI STANDARD B16.5 EXCEPT FOR BOLT LENGTHS.

ANSI B16.5 SPECIFIES LONGER BOLTS. SHORTER BOLTS ARE ACCEPTABLE PROVIDING FULL THREAD ENGAGEMENT IS OBTAINED WHEN FLANGES ARE ASSEMBLED.

ISO 2229 SPECIFIES BOLT DIAMETERS IN INCHES.

RING-JOINT GASKET DATA

TABLE F-7M

DIMENSIONS IN MILLIMETERS

DATA FOR WELDING-NECK FLANGES

L = LENGTH THRU HUB OF WELDING-NECK FLANGE WITH RING JOINT

G = GAP BETWEEN FLANGE FACES UNDER NORMAL COMPRESSION

◆ FOR OUTSIDE DIAMETERS OF FLANGES AND BOLTING REFER TO TABLES F-1M THRU F-6M

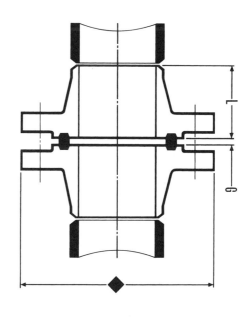

FLANGE CLASSES

DN	150 L	150 G	150 RING No	300 L	300 G	300 RING No	600 L	600 G	600 RING No	900 L	900 G	900 RING No	1500 L	1500 G	1500 RING No	2500 L	2500 G	2500 RING No
600	158	3	R 76	179	6.3	R 77	214	5.6	R 77	308	5.6	R 78	427	11.1	R 79			
500	150	3	R 72	172	5.6	R 73	200	4.8	R 73	261	4.8	R 74	374	9.5	R 75			
450	146	3	R 68	167	5.6	R 69	192	4.8	R 69	242	4.8	R 70	345	7.9	R 71			
400	133	3	R 64	154	5.6	R 65	186	4.8	R 65	227	4	R 66	329	7.9	R 67			
350	133	3	R 59	151	5.6	R 61	173	4.8	R 61	224	4	R 62	314	5.6	R 63			
300	120	4	R 56	138	5.6	R 57	164	4.8	R 57	208	4	R 57	297	4	R 58	482	7.9	R 60
250	108	4	R 52	125	5.6	R 53	160	4.8	R 53	192	4	R 53	265	4	R 54	437	6.3	R 55
200	108	4	R 48	119	5.6	R 49	141	4.8	R 49	170	4	R 49	224	4	R 50	332	4.8	R 51
150	95	4	R 43	106	5.6	R 45	125	4.8	R 45	148	4	R 45	181	3.2	R 46	286	4	R 47
100	82	4	R 36	94	5.6	R 37	110	4.8	R 37	122	4	R 37	132	3.2	R 39	201	4	R 38
80	76	4	R 29	87	5.6	R 31	91	4.8	R 31	110	4	R 31	125	3.2	R 35	178	3.2	R 32
50	70	4	R 22	78	5.6	R 23	81	4.8	R 23	110	3.2	R 24	110	3.2	R 24	135	3.2	R 26
40	68	4	R 19	74	4	R 20	76	4	R 20	89	4	R 20	89	4	R 20	119	4	R 23
25	62	4	R 15	68	4	R 16	68	4	R 16	79	4	R 16	79	4	R 16	95	4	R 18
20	–	–	–	63	4	R 13	63	4	R 13	76	4	R 14	76	4	R 14	85	4	R 16
15	–	–	–	58	3.2	R 11	58	3.2	R 11	66	4	R 12	66	4	R 12	79	4	R 13

SLIP-ON FLANGES ON BUTT-WELDING ELBOWS — TABLE F-8M

FOR USE ON BUTT-WELDING ELBOWS AS PERMITTED BY THE PIPING SPECIFICATION FOR THE PROJECT

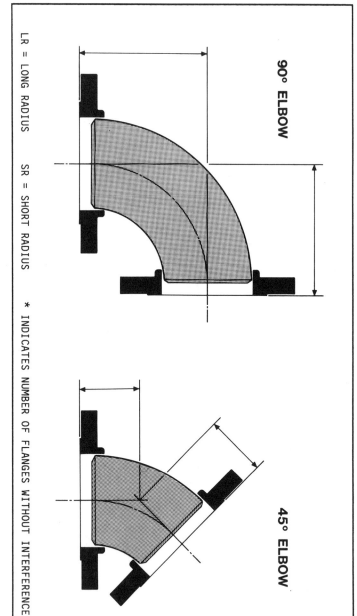

90° ELBOW

45° ELBOW

LR = LONG RADIUS   SR = SHORT RADIUS   * INDICATES NUMBER OF FLANGES WITHOUT INTERFERENCE

DN	CLASS 150 FLANGES						CLASS 300 FLANGES					
	90 LR	*	90 SR	*	45 LR	*	90 LR	*	90 SR	*	45 LR	*
50	89	1	68	1	48	1	97	1	76	1	56	1
80	130	2	97	1	67	1	143	1	110	1	79	1
100	168	2	124	1	79	1	183	2	138	1	94	1
150	243	2	175	1	110	1	256	2	187	1	122	2
200	319	2	227	2	141	2	337	2	244	2	159	2
250	397	2	276	2	175	2	408	2	294	2	186	2
300	473	2	332	2	206	2	487	2	349	2	221	2
350	549	2	376	2	238	2	559	2	395	2	248	2
400	625	2	432	2	270	2	632	2	451	2	276	2
450	702	2	484	2	302	2	711	2	505	2	311	2
500	778	2	533	2	333	2	794	2	556	2	349	2
600	930	2	645	2	395	2	951	2	668	2	418	2

DIMENSIONS IN MILLIMETERS

PRESSURE / TEMPERATURE RATINGS FOR CARBON STEEL FLANGES

TABLE F-9M

Maximum ratings for flanges conforming to ISO Standard 2229 dimensions and material specification ASTM A-105

GAGE PRESSURE IN kilopascals (kPa) FOR FLANGE CLASSES 150 - 2500

TEMPERATURE CELSIUS	FLANGE CLASSES						
	150	300	400	600	900	1500	2500
-29 to 38	1 900	4 960	6 610	9 920	14 900	24 830	41 380
50	1 830	4 940	6 560	9 850	14 840	24 670	41 130
100	1 630	4 800	6 400	9 600	14 450	24 020	40 060
150	1 450	4 680	6 260	9 400	14 130	23 500	39 200
200	1 260	4 600	6 150	9 190	13 850	23 060	38 380
250	1 070	4 370	5 850	8 780	13 170	21 920	36 600
300	940	3 960	5 300	7 900	11 900	19 810	33 060
350	810	3 480	4 670	6 930	10 490	17 440	29 000
375	750	3 210	4 340	6 380	9 660	15 800	26 770
400	690	2 910	3 940	5 820	8 780	14 580	24 360
425	640	2 550	3 440	5 100	7 730	12 820	21 250
450	580	2 140	2 880	4 270	6 460	10 760	17 820
475	510	1 680	2 240	3 340	5 070	8 420	14 030
500	400	1 240	1 620	2 460	3 720	6 170	10 330
525	340	810	1 080	1 600	2 410	4 030	6 720
538	260	550	760	1 100	1 650	2 820	4 590

ISO 2229 flange dimensions are similar to those of standard ANSI B16.5. Both standards limit the prolonged use of flanges manufactured from carbon steels made to material specification ASTM A-105 at elevated temperatures. ANSI B16.5 also makes recommendations regarding the use of threaded and socket-welding flanges. Refer to footnote: Table F-9.

Ratings are for non-shock conditions. Values in this table do not prevail over limitations imposed by codes, standards, regulations or other obligations which may pertain to projects.

Tables P-1M present calculated data as a guide only. Spans are for pipe arranged in pipeways with the following assumptions: Bare pipe - continuous straight run with welded joints and two or more straight spans at each end.

SPANS - calculated with lines full of water and a maximum bending stress of 4 000 PSI

SAG - (deflection) calculated with lines empty (drained condition)

The following factors were not considered in calculating spans for these tables:
Concentrated mechanical loads from flanges, valves, strainers, filters, and other inline equipment - weights of connecting branch lines - torsional loading from thermal movement - sudden reaction from lines(s) discharging contents - vibration - flattening effect of weight of contents in larger liquid filled lines - weight of insulation and pipe covering - weight of ice and snow - wind loads - seismic shock - reduction in wall thickness of pipe from threading or grooving.

DESIGN PRESSURE - calculated per ANSI B31.1 using allowable stress value of 9 000 PSI for seamless carbon steel pipe

BURSTING PRESSURE is approximate, calculated on yield strength of 30 000 PSI

[E in these tables is for 'Exponent', the power of 10 to which the number must be raised. Example: 1.0E5 = 100 000]

API = American Petroleum Institute's standard 5L, for 'Line pipe'. API pipe sizes; manufacturers' weights: Double-extra-strong (XXS), Extra-strong (XS), and Standard (STD), are included with schedule numbers in standard ANSI B36.10M. Also refer to 2.1.3

PIPE DATA: DIMENSIONS & STRESS PARAMETERS — TABLES P-1M

DN (mm) [NPS]	PIPING CODES and MANUFACTURERS' WEIGHTS	DIMENSIONS O.D. (mm)	I.D. (mm)	Wall (mm)	WEIGHTS Empty (kg/m)	Waterfilled (kg/m)	AREAS External (mm²/mm)	Internal (mm²/mm)	Flow (mm²)	Metal (mm²)	Moment of Inertia (10⁴ mm⁴)	Section Modulus (10³ mm³)	Radius of Gyration (mm)	Continuous Spans Span (m)	Sag (mm)	Code Pressures Design (MPa)	Bursting (MPa)
10 / .375	SCH 40 STD API	17.15	12.52	2.311	.8434	.9666	53.86	39.34	123.2	107.7	3.035	.3540	5.308	3.52	5.42	12.1	40.3
	SCH 80 XS API	17.15	10.74	3.200	1.098	1.188	53.86	33.75	90.66	140.2	3.587	.4185	5.058	3.45	5.52	19.9	66.4
15 / .500	SCH 40 STD API	21.34	15.80	2.769	1.265	1.461	67.03	49.63	196.0	161.5	7.114	.6669	6.637	3.93	5.39	12.6	42.1
	SCH 80 XS API	21.34	13.87	3.734	1.617	1.768	67.03	43.57	151.1	206.5	8.357	.7833	6.362	3.87	5.53	19.5	64.8
	SCH 160 API	21.34	11.79	4.775	1.945	2.054	67.03	37.03	109.1	248.4	9.225	.8648	6.094	3.77	5.63	27.5	91.6
	XXS API	21.34	6.401	7.468	2.548	2.580	67.03	20.11	32.18	325.4	10.09	.9458	5.569	3.52	4.94	52.1	174
20 / .750	SCH 40 STD API	26.67	20.93	2.870	1.680	2.024	83.79	65.75	344.0	214.6	15.42	1.156	8.475	4.39	5.17	9.14	30.5
	SCH 80 XS API	26.67	18.85	3.912	2.190	2.469	83.79	59.21	279.7	279.7	18.64	1.398	8.164	4.37	5.48	14.8	49.2
	SCH 160 API	26.67	15.54	5.563	2.888	3.078	83.79	48.84	189.8	368.9	21.97	1.647	7.717	4.25	5.48	24.5	81.7
	XXS API	26.67	11.02	7.823	3.627	3.722	83.79	34.63	95.44	463.2	24.11	1.808	7.215	4.05	5.16	39.8	133
25 / 1.00	SCH 40 STD API	33.40	26.64	3.378	2.495	3.052	104.9	83.71	557.6	318.6	36.35	2.177	10.68	4.91	5.08	8.77	29.2
	SCH 80 XS API	33.40	24.31	4.547	3.227	3.691	104.9	76.37	464.1	412.1	43.96	2.632	10.33	4.91	5.43	13.8	45.9
	SCH 160 API	33.40	20.70	6.350	4.225	4.562	104.9	65.03	336.6	539.6	52.08	3.119	9.824	4.80	5.51	24.1	73.7
	XXS API	33.40	15.21	9.093	5.437	5.619	104.9	47.80	181.8	694.4	58.46	3.501	9.176	4.59	5.25	36.5	122
32 / 1.25	SCH 40 STD API	42.16	35.05	3.556	3.377	4.342	132.5	110.1	965.0	431.3	81.04	3.844	13.71	5.47	4.75	7.03	23.4
	SCH 80 XS API	42.16	32.46	4.851	4.453	5.280	132.5	102.0	827.6	568.7	100.6	4.774	13.30	5.52	5.26	11.3	37.7
	SCH 160 API	42.16	29.46	6.350	5.594	6.276	132.5	92.56	681.8	714.5	118.2	5.604	12.86	5.49	5.49	16.5	55.2
	XXS API	42.16	22.76	9.703	7.748	8.155	132.5	71.50	406.8	989.5	142.0	6.734	11.98	5.28	5.41	29.5	98.5
40 / 1.50	SCH 40 STD API	48.26	40.89	3.683	4.039	5.352	151.6	128.5	1313	515.8	129.0	5.346	15.81	5.81	4.54	6.46	21.5
	SCH 80 XS API	48.26	38.10	5.080	5.396	6.536	151.6	119.7	1140	689.1	162.8	6.748	15.37	5.90	5.14	10.5	34.9
	SCH 160 API	48.26	33.99	7.137	7.220	8.127	151.6	106.8	907.1	922.1	200.8	8.321	14.76	5.88	5.48	16.7	55.7
	XXS API	48.26	27.94	10.16	9.522	10.14	151.6	87.78	613.1	1216	236.4	9.795	13.94	5.71	5.47	26.8	89.4
50 / 2.00	SCH 40 STD API	60.32	52.50	3.912	5.428	7.593	189.5	164.9	2165	693.2	277.1	9.187	19.99	6.39	4.17	5.07	16.9
	SCH 80 XS API	60.32	49.25	5.537	7.463	9.368	189.5	154.7	1905	953.1	361.3	11.98	19.47	6.57	4.91	8.72	29.1
	SCH 160 API	60.32	42.85	8.738	11.09	12.53	189.5	134.6	1442	1416	484.6	16.07	18.50	6.58	5.47	16.4	54.7
	XXS API	60.32	38.18	11.07	13.42	14.56	189.5	119.9	1145	1713	545.8	18.10	17.85	6.48	5.52	22.5	75.0

Thru DN 250, wall thicknesses for SCH 40S and SCH 80S stainless steel pipes are the same as for SCH 40 and SCH 80 carbon steel pipes

PIPE DATA   TABLES P-1M

 Dimensions — O.D., I.D., Wall (mm). **Weights** — Empty, Waterfilled (kg/m). **Areas** — External, Internal (mm²/mm); Flow, Metal (mm²). **Moment of Inertia** (10⁴ mm⁴). **Section Modulus** (10³ mm³). **Radius of Gyration** (mm). **Continuous Spans** — Span (m), Sag (mm). **Code Pressures** — Design (MPa), Bursting (MPa).

DN (mm) / [NPS]	Piping Codes / Mfrs' Wts	O.D.	I.D.	Wall	Empty	Waterfilled	External	Internal	Flow	Metal	Mom. of Inertia	Section Modulus	Radius of Gyration	Span	Sag	Design	Bursting
65 / 2.50	SCH 40 STD API	73.03	62.71	5.156	8.608	11.70	229.4	197.0	3089	1099	636.6	17.44	24.06	4.37	1.98	5.96	19.9
	SCH 80 XS API	73.03	59.00	7.010	11.38	14.12	229.4	185.4	2734	1454	800.9	21.94	23.47	4.99	3.11	9.43	31.4
	SCH 160 API	73.03	53.98	9.525	14.88	17.17	229.4	169.6	2288	1900	979.3	26.82	22.70	5.39	4.40	14.4	48.0
	XXS API	73.03	44.98	14.02	20.35	21.94	229.4	141.3	1589	2599	1195	32.73	21.44	5.51	5.53	24.0	80.1
80 / 3.00	API	88.90	82.55	3.175	6.695	12.05	279.3	259.4	5352	855.1	786.5	17.69	30.33				
	API	88.90	80.98	3.962	8.279	13.43	279.3	254.4	5150	1057	955.6	21.50	30.06				
	API	88.90	79.35	4.775	9.882	14.83	279.3	249.3	4945	1262	1120	25.20	29.79				
	SCH 40 STD API	88.90	77.93	5.486	11.26	16.03	279.3	244.8	4769	1438	1256	28.25	29.55				
	API	88.90	76.20	6.350	12.89	17.45	279.3	239.4	4560	1647	1411	31.75	29.27				
	API	88.90	74.63	7.137	14.36	18.73	279.3	234.4	4374	1833	1544	34.73	28.86				
	SCH 80 XS API	88.90	73.66	7.620	15.27	19.50	279.3	231.4	4261	1946	1621	36.47	27.78				
	SCH 160 API	88.90	66.65	11.13	21.28	24.77	279.3	209.4	3489	2718	2097	47.19	27.78				
	XXS API	88.90	58.42	15.24	27.61	30.29	279.3	183.5	2680	3527	2494	56.11	26.59				
100 / 4	API	114.3	108.0	3.175	8.679	17.83	359.1	339.1	9152	1108	1712	29.96	39.30				
	API	114.3	106.4	3.962	10.75	19.64	359.1	334.2	8887	1374	2093	36.62	39.04				
	API	114.3	104.7	4.775	12.87	21.48	359.1	329.1	8618	1643	2468	43.19	38.76				
	API	114.3	103.2	5.563	14.88	23.24	359.1	324.1	8361	1900	2816	49.27	38.49				
	SCH 40 STD API	114.3	102.3	6.020	16.03	24.25	359.1	321.3	8213	2048	3010	52.68	38.34				
	API	114.3	101.6	6.350	16.86	24.97	359.1	319.2	8107	2154	3148	55.08	38.23				
	API	114.3	100.0	7.137	18.81	26.67	359.1	314.2	7858	2403	3465	60.62	37.97				
	SCH 80 XS API	114.3	98.45	7.925	20.74	28.35	359.1	309.3	7612	2648	3767	65.91	37.71				
	API	114.3	97.18	8.560	22.26	29.68	359.1	305.3	7417	2844	4000	69.99	37.51				
	SCH 120 API	114.3	92.04	11.13	28.?	33.?	359.1	289.2	6653	3607	4856	84.97	36.69				
	SCH 160 API	114.3	87.33	13.49	33.45	39.44	359.1	274.3	5989	4272	5524	96.65	35.96				
	XXS API	114.3	80.06	17.12	40.92	45.96	359.1	251.5	5034	5227	6362	111.3	34.89				
150 / 6	API	168.3	158.7	4.775	19.21	38.99	528.7	498.6	19787	2453	8203	97.50	57.83				
	API	168.3	157.1	5.563	22.26	41.66	528.7	493.5	19396	2843	9421	112.0	57.56				
	API	168.3	155.6	6.350	25.29	44.30	528.7	488.8	19009	3230	10603	126.0	57.29				
	SCH 40 STD API	168.3	154.1	7.112	28.19	46.83	528.7	484.0	18639	3601	11714	139.2	57.04				
	API	168.3	152.4	7.925	31.26	49.51	528.7	478.9	18248	3992	12862	152.9	56.76				
	API	168.3	150.8	8.738	34.29	52.15	528.7	473.8	17860	4379	13975	166.1	56.49				
	SCH 80 XS API	168.3	146.3	10.97	42.46	59.28	528.7	459.7	16817	5423	16853	200.3	55.75				
	SCH 120 API	168.3	139.7	14.27	54.08	69.41	528.7	439.0	15333	6906	20649	245.4	54.68				
	SCH 160 API	168.3	131.7	18.26	67.39	81.02	528.7	413.9	13633	8607	24569	292.0	53.43				
	XXS API	168.3	124.4	21.95	78.99	91.14	528.7	390.8	12151	10089	27610	328.2	52.31				
200 / 8	API	219.1	209.5	4.775	25.17	59.65	688.2	658.2	34479	3215	18464	168.6	75.79				
	API	219.1	208.8	5.156	27.13	61.36	688.2	655.8	34229	3465	19833	181.1	75.65				
	API	219.1	207.9	5.563	29.22	63.18	688.2	653.3	33963	3731	21277	194.2	75.51				
	SCH 20 API	219.1	206.4	6.350	33.23	66.68	688.2	648.4	33451	4244	24026	219.2	75.24				
	SCH 30 API	219.1	205.0	7.036	36.70	69.71	688.2	644.0	33007	4687	26369	240.7	75.01				
	API	219.1	203.2	7.925	41.16	73.60	688.2	638.5	32437	5257	28025	267.2	74.72				
	SCH 40 STD API	219.1	202.7	8.179	42.43	74.71	688.2	636.9	32275	5419	30172	275.5	74.62				
	API	219.1	201.6	8.738	45.21	77.13	688.2	633.3	31921	5774	31985	292.0	74.43				
	API	219.1	200.0	9.525	49.10	80.52	688.2	628.4	31424	6271	34489	314.9	74.16				
	SCH 60 API	219.1	198.5	10.31	52.96	83.89	688.2	623.4	30931	6763	36935	337.2	73.90				
	API	219.1	196.8	11.13	56.91	87.33	688.2	618.3	30426	7268	39399	359.7	73.63				
	SCH 80 XS API	219.1	193.7	12.70	64.47	93.93	688.2	608.4	29460	8234	44002	401.7	73.10				
	SCH 100 API	219.1	188.9	15.09	75.71	103.7	688.2	593.4	28025	9669	50566	461.6	72.32				
	SCH 120 API	219.1	182.5	18.26	90.21	116.3	688.2	573.4	26173	11521	58556	534.6	71.29				
	SCH 140 API	219.1	177.6	20.62	100.7	125.5	688.2	558.7	24836	12859	63984	584.1	70.54				
	XXS API	219.1	174.6	22.23	107.6	131.6	688.2	548.6	23950	13744	67423	615.5	70.04				
	SCH 160 API	219.1	173.1	23.01	111.0	134.5	688.2	543.7	23520	14174	69048	630.4	69.79				

Thru DN 250, wall thicknesses for SCH 40S and SCH 80S stainless steel pipes are the same as for SCH 40 and SCH 80 carbon steel pipes.

PIPE DATA — TABLES P-1 M

DN 250 / NPS 10 (O.D. 273.1 mm)

Piping Codes & Mfrs' Weights	I.D. (mm)	Wall (mm)	Empty (kg/m)	Waterfilled (kg/m)	External (mm²/mm)	Internal (mm²/mm)	Flow (mm²)	Metal (mm²)	Moment of Inertia (10⁴ mm⁴)	Section Modulus (10³ mm³)	Radius of Gyration (mm)	Span (m)	Sag (mm)	Design (MPa)	Bursting (MPa)
API	263.5	4.775	31.51	86.04	857.8	827.8	54532	4025	36218	265.3	94.86	10.2	1.20	.672	2.24
API	262.7	5.156	33.98	88.20	857.8	825.4	54217	4340	38944	285.3	94.73	10.4	1.33	.847	2.82
API	261.9	5.563	36.60	90.48	857.8	822.9	53882	4674	41825	306.4	94.59	10.7	1.46	1.03	3.45
SCH 20 API	260.3	6.350	41.66	94.89	857.8	817.9	53236	5320	47331	346.7	94.32	11.1	1.71	1.40	4.66
API	258.9	7.087	46.36	99.00	857.8	813.3	52635	5921	52393	383.8	94.07	11.4	1.93	1.74	5.80
SCH 30 API	257.5	7.798	50.88	102.9	857.8	808.8	52058	6498	57198	419.0	93.82	11.7	2.07	2.07	6.90
API	255.6	8.738	56.81	108.1	857.8	802.9	51301	7255	63428	464.6	93.50	12.0	2.41	2.51	8.37
STD SCH 40 API	254.5	9.271	60.16	111.0	857.8	799.6	50874	7683	66903	490.0	93.32	12.2	2.55	2.76	9.20
API	250.8	11.13	71.68	121.1	857.8	787.9	49402	9154	78647	576.1	92.69	12.7	3.00	3.45	11.5
XS SCH 60 API	247.7	12.70	81.33	129.5	857.8	778.0	48169	10388	88220	646.2	92.16	13.0	3.34	4.39	14.6
SCH 80 API	242.9	15.09	95.74	142.1	857.8	763.0	46329	12227	1.0E5	747.5	91.36	13.3	3.78	5.55	18.5
SCH 100 API	236.5	18.26	114.5	158.4	857.8	743.1	43938	14618	1.2E5	873.3	90.31	13.6	4.24	7.11	23.7
SCH 120 API	230.2	21.44	132.7	174.3	857.8	723.1	41611	16946	1.4E5	989.4	89.28	13.8	4.60	8.71	29.0
XXS SCH 140 API	222.2	25.40	154.7	193.5	857.8	698.2	38795	19762	1.5E5	1121	88.02	14.0	4.94	10.7	35.8
SCH 160 API	215.9	28.58	171.8	208.5	857.8	678.3	36610	21947	1.7E5	1217	87.02	14.0	5.13	12.4	41.4

DN 300 / NPS 12 (O.D. 323.9 mm)

Piping Codes & Mfrs' Weights	I.D. (mm)	Wall (mm)	Empty (kg/m)	Waterfilled (kg/m)	External (mm²/mm)	Internal (mm²/mm)	Flow (mm²)	Metal (mm²)	Moment of Inertia (10⁴ mm⁴)	Section Modulus (10³ mm³)	Radius of Gyration (mm)	Span (m)	Sag (mm)	Design (MPa)	Bursting (MPa)
API	313.5	5.156	40.42	117.6	1017	985.0	77209	5162	65558	404.9	112.7	10.8	1.06	.714	2.38
API	312.7	5.563	43.55	120.4	1017	982.5	76809	5562	70458	435.1	112.5	11.1	1.18	.871	2.90
SCH 20 API	311.2	6.350	49.59	125.6	1017	977.5	76038	6334	79843	493.1	112.3	11.5	1.39	1.18	3.92
API	309.6	7.137	55.61	130.9	1017	972.6	75270	7102	89088	550.2	112.0	11.9	1.60	1.48	4.95
API	308.0	7.925	61.59	136.1	1017	967.6	74506	7865	98192	606.4	111.7	12.3	1.81	1.79	5.97
SCH 30 API	307.1	8.382	65.05	139.1	1017	964.7	74064	8307	1.0E5	638.7	111.6	12.4	1.93	1.97	6.57
API	306.4	8.738	67.73	141.4	1017	962.5	73722	8650	1.1E5	663.5	111.2	12.6	2.02	2.11	7.04
STD API	304.8	9.525	73.65	146.6	1017	957.6	72966	9406	1.2E5	718.0	111.1	12.9	2.21	2.42	8.07
SCH 40 API	303.2	10.31	79.54	151.7	1017	952.6	72214	10158	1.3E5	771.7	110.9	13.1	2.39	2.73	9.11
API	301.6	11.13	85.58	157.0	1017	947.5	71442	10930	1.3E5	826.2	110.6	13.3	2.58	3.06	10.2
XS API	298.5	12.70	97.20	167.2	1017	937.6	69957	12414	1.5E5	929.4	110.1	13.7	2.90	3.69	12.3
SCH 60 API	295.3	14.27	108.7	177.2	1017	927.7	68489	13883	1.7E5	1029	109.6	14.0	3.20	4.32	14.4
API	292.1	15.88	120.3	187.3	1017	917.7	67012	15360	1.8E5	1128	109.0	14.3	3.47	4.97	16.6
SCH 80 API	288.9	17.48	131.7	197.3	1017	907.6	65552	16820	2.0E5	1223	108.5	14.5	3.72	5.63	18.8
API	285.8	19.05	142.8	207.0	1017	897.7	64130	18241	2.1E5	1313	108.0	14.6	3.93	6.28	20.9
SCH 100 API	281.0	21.44	159.5	221.5	1017	882.7	62005	20367	2.3E5	1445	107.2	14.8	4.22	7.28	24.3
SCH 120 API	273.1	25.40	186.5	245.0	1017	857.8	58556	23815	2.7E5	1649	105.9	15.1	4.60	8.96	29.9
SCH 140 API	266.7	28.58	207.6	263.4	1017	837.9	55865	26507	2.9E5	1801	104.9	15.2	4.84	10.3	34.4
SCH 160 API	257.2	33.32	238.2	290.1	1017	808.0	51956	30416	3.3E5	2008	103.4	15.3	5.10	12.4	41.4

DN 350 / NPS 14 (O.D. 355.6 mm)

Piping Codes & Mfrs' Weights	I.D. (mm)	Wall (mm)	Empty (kg/m)	Waterfilled (kg/m)	External (mm²/mm)	Internal (mm²/mm)	Flow (mm²)	Metal (mm²)	Moment of Inertia (10⁴ mm⁴)	Section Modulus (10³ mm³)	Radius of Gyration (mm)	Span (m)	Sag (mm)	Design (MPa)	Bursting (MPa)
API	344.9	5.334	45.96	139.4	1117	1084	93445	5870	90034	506.4	123.9	11.1	1.04	.712	2.37
API	344.5	5.563	47.90	141.1	1117	1082	93198	6117	93711	527.1	123.8	11.2	1.14	.793	2.64
SCH 10 API	342.9	6.350	54.55	146.9	1117	1077	92347	6967	1.1E5	597.7	123.5	11.7	1.35	1.07	3.57
API	341.3	7.137	61.18	152.7	1117	1072	91501	7814	1.2E5	667.3	123.2	12.1	1.56	1.35	4.50
SCH 20 API	339.8	7.925	67.78	158.4	1117	1067	90659	8656	1.3E5	736.0	123.0	12.5	1.77	1.63	5.43
API	338.1	8.738	74.55	164.3	1117	1062	89793	9521	1.4E5	805.9	122.7	12.9	1.99	1.92	6.40
STD SCH 30 API	336.6	9.525	81.09	170.0	1117	1057	88959	10356	1.6E5	872.6	122.4	13.2	2.17	2.20	7.34
SCH 40 API	333.3	11.13	94.27	181.5	1117	1047	87275	12040	1.8E5	1005	121.9	13.7	2.56	2.78	9.26
XS API	330.2	12.70	107.1	192.8	1117	1037	85634	13681	2.0E5	1132	121.6	13.9	2.82	3.35	11.2
SCH 60 API	325.4	15.09	126.4	209.6	1117	1022	83175	16140	2.3E5	1318	120.5	14.1	3.27	4.23	14.1
API	323.9	15.88	132.7	215.0	1117	1017	82372	16943	2.4E5	1378	120.2	14.1	3.40	4.52	15.1
SCH 80 API	317.5	19.05	157.7	236.9	1117	997.5	79173	20142	2.9E5	1609	119.2	14.7	3.90	5.70	19.0
SCH 100 API	307.9	23.83	194.4	268.9	1117	967.5	74482	24833	3.4E5	1932	117.6	15.6	4.48	7.51	25.0
SCH 120 API	300.0	27.79	224.1	294.8	1117	942.6	70698	28617	3.9E5	2178	116.3	15.8	4.84	9.04	30.1
SCH 140 API	292.1	31.75	252.9	319.9	1117	917.7	67012	32303	4.3E5	2405	115.0	15.9	4.86	10.6	35.4
SCH 160 API	284.2	35.71	281.0	344.4	1117	892.8	63425	35889	4.6E5	2614	113.8	16.0	5.07	12.2	40.7

[99]

Thru DN 250, wall thicknesses for SCH 40S and SCH 80S stainless steel pipes are the same as for SCH 40 and SCH 80 carbon steel pipes

PIPE DATA

TABLES P-1 M

DN (mm) / NPS	PIPING CODES and MANUFACTURERS' WEIGHTS	O.D. (mm)	I.D. (mm)	Wall (mm)	Empty (kg/m)	Water-filled (kg/m)	External mm^2/mm	Internal mm^2/mm	Flow mm^2	Metal mm^2	Moment of Inertia ($10^4\ mm^4$)	Section Modulus ($10^3\ mm^3$)	Radius of Gyration (mm)	Continuous Span (m)	Sag (mm)	Design (MPa)	Bursting (MPa)
400 / 16	API	406.4	395.3	5.563	54.85	177.6	1277	1242	1.2E5	7005	1.4E5	692.5	141.7	11.5	.862	.693	2.31
	SCH 10 API	406.4	393.7	6.350	62.49	184.2	1277	1237	1.2E5	7981	1.6E5	785.9	141.5	12.5	1.04	.936	3.12
	API	406.4	392.1	7.137	70.10	190.9	1277	1232	1.2E5	8953	1.8E5	878.2	141.2	12.5	1.21	1.18	3.93
	SCH 20 API	406.4	390.6	7.925	77.68	197.5	1277	1227	1.2E5	9921	2.0E5	969.4	140.9	12.9	1.38	1.42	4.75
	API	406.4	388.9	8.738	85.47	204.3	1277	1222	1.2E5	10916	2.2E5	1062	140.6	13.2	1.56	1.68	5.59
	SCH 30 STD API	406.4	387.4	9.525	92.99	210.8	1277	1217	1.2E5	11876	2.3E5	1151	140.4	13.6	1.72	1.92	6.41
	API	406.4	384.1	11.13	108.2	224.1	1277	1207	1.1E5	13815	2.7E5	1329	139.8	14.1	2.05	2.43	8.09
	API	406.4	382.6	11.91	115.6	230.6	1277	1202	1.1E5	14764	2.9E5	1415	139.5	14.4	2.20	2.67	8.92
	SCH 40 XS API	406.4	381.0	12.70	123.0	237.0	1277	1197	1.1E5	15708	3.0E5	1499	139.3	14.6	2.35	2.92	9.75
	API	406.4	374.7	15.88	152.2	262.7	1277	1177	1.1E5	19477	3.7E5	1830	138.2	15.3	2.89	3.94	13.1
	SCH 60 API	406.4	373.1	16.66	159.7	269.1	1277	1172	1.1E5	20401	3.9E5	1910	137.9	15.5	3.02	4.19	14.0
	API	406.4	368.3	19.05	181.5	288.0	1277	1157	1.1E5	23182	4.4E5	2145	137.1	15.9	3.36	4.96	16.5
	SCH 80 API	406.4	363.5	21.44	203.0	306.8	1277	1142	1.0E5	25927	4.8E5	2371	136.3	16.1	3.66	5.74	19.1
	SCH 100 API	406.4	354.0	26.19	244.9	343.4	1277	1112	98437	31280	5.7E5	2795	134.7	16.6	4.16	7.32	24.4
	SCH 120 API	406.4	344.5	30.96	285.9	379.1	1277	1082	93198	36520	6.5E5	3188	133.2	16.8	4.54	8.93	29.8
	SCH 140 API	406.4	333.3	36.53	332.3	419.6	1277	1047	87275	42442	7.3E5	3607	131.4	17.0	4.87	10.9	36.2
	SCH 160 API	406.4	325.4	40.49	364.4	447.6	1277	1022	83175	46542	7.9E5	3880	130.2	17.1	5.05	12.3	40.8
450 / 18	SCH 10 API	457.2	444.5	6.350	70.42	225.6	1436	1396	1.6E5	8994	2.3E5	999.9	159.4	12.2	.880	.832	2.77
	API	457.2	442.9	7.137	79.02	233.1	1436	1391	1.5E5	10092	2.6E5	1118	159.1	12.7	1.03	1.05	3.49
	SCH 20 API	457.2	441.4	7.925	87.58	240.6	1436	1387	1.5E5	11185	2.8E5	1235	158.9	13.2	1.19	1.27	4.22
	API	457.2	439.7	8.738	96.39	248.3	1436	1381	1.5E5	12310	3.1E5	1354	158.6	13.6	1.35	1.49	4.97
	STD API	457.2	438.2	9.525	104.9	255.7	1436	1376	1.5E5	13396	3.4E5	1469	158.3	13.9	1.50	1.71	5.69
	API	457.2	436.6	10.31	113.4	263.1	1436	1372	1.5E5	14478	3.6E5	1582	158.0	14.2	1.65	1.93	6.42
	SCH 30 API	457.2	434.9	11.13	122.1	270.7	1436	1366	1.5E5	15591	3.9E5	1697	157.8	14.5	1.80	2.15	7.18
	API	457.2	433.4	11.91	130.5	278.0	1436	1361	1.5E5	16665	4.1E5	1808	157.5	14.8	1.94	2.37	7.91
	XS API	457.2	431.8	12.70	138.9	285.3	1436	1357	1.5E5	17735	4.4E5	1918	157.2	15.1	2.08	2.59	8.65
	SCH 40 API	457.2	428.7	14.27	155.5	299.8	1436	1347	1.4E5	19863	4.9E5	2133	156.7	15.5	2.35	3.04	10.1
	API	457.2	425.5	15.88	172.3	314.5	1436	1337	1.4E5	22010	5.4E5	2347	156.1	15.9	2.60	3.49	11.6
	SCH 60 API	457.2	419.1	19.05	205.3	343.3	1436	1317	1.4E5	26222	6.3E5	2758	155.1	16.5	3.06	4.40	14.7
	SCH 80 API	457.2	409.5	23.83	254.0	385.7	1436	1287	1.3E5	32438	7.6E5	3341	153.5	17.1	3.63	5.78	19.3
	SCH 100 API	457.2	398.5	29.36	309.0	433.7	1436	1252	1.2E5	39466	9.1E5	3969	151.6	17.6	4.15	7.41	24.7
	SCH 120 API	457.2	387.3	34.93	362.8	480.6	1436	1217	1.2E5	46332	1.0E6	4548	149.8	17.9	4.51	9.09	30.3
	SCH 140 API	457.2	377.9	39.67	407.5	519.6	1436	1187	1.1E5	52041	1.1E6	5006	148.3	18.0	4.82	10.5	35.1
	SCH 160 API	457.2	366.7	45.24	458.4	564.0	1436	1152	1.1E5	58547	1.3E6	5499	146.5	18.1	5.04	12.3	40.9
500 / 20	SCH 10 API	508.0	495.3	6.350	78.36	271.0	1596	1556	1.9E5	10007	3.1E5	1240	177.4	12.4	.757	.748	2.49
	API	508.0	493.7	7.137	87.94	279.4	1596	1551	1.9E5	11231	3.5E5	1387	177.1	12.9	.895	.943	3.14
	API	508.0	492.2	7.925	97.48	287.7	1596	1546	1.9E5	12450	3.9E5	1533	176.8	13.4	1.03	1.14	3.79
	API	508.0	490.5	8.738	107.3	296.3	1596	1541	1.9E5	13705	4.3E5	1682	176.5	13.8	1.18	1.34	4.46
	SCH 20 STD API	508.0	489.0	9.525	116.8	304.6	1596	1536	1.9E5	14916	4.6E5	1825	176.3	14.2	1.32	1.54	5.12
	API	508.0	487.4	10.31	126.2	312.8	1596	1531	1.9E5	16124	5.0E5	1966	176.0	14.6	1.45	1.73	5.77
	API	508.0	485.7	11.13	136.0	321.3	1596	1526	1.9E5	17366	5.4E5	2111	175.7	14.9	1.59	1.94	6.45
	API	508.0	484.2	11.91	145.4	329.5	1596	1521	1.8E5	18566	5.7E5	2250	175.4	15.2	1.73	2.13	7.11
	SCH 30 XS API	508.0	482.6	12.70	154.7	337.7	1596	1516	1.8E5	19762	6.1E5	2387	175.2	15.4	1.86	2.33	7.77
	SCH 40 API	508.0	477.8	15.09	182.9	362.3	1596	1501	1.8E5	23364	7.1E5	2796	174.4	16.1	2.23	2.93	9.78
	API	508.0	476.3	15.88	192.2	370.3	1596	1496	1.8E5	24544	7.4E5	2928	174.1	16.3	2.35	3.13	10.4
	SCH 60 API	508.0	466.7	20.62	247.3	418.4	1596	1466	1.7E5	31579	9.4E5	3698	172.5	17.3	2.99	4.35	14.5
	SCH 80 API	508.0	455.6	26.19	310.4	473.4	1596	1431	1.6E5	39639	1.2E6	4542	170.6	18.5	3.60	5.80	19.3
	SCH 100 API	508.0	442.9	32.54	380.5	534.6	1596	1391	1.5E5	48601	1.4E6	5432	168.5	18.8	4.14	7.49	25.0
	SCH 120 API	508.0	431.8	38.10	440.4	586.8	1596	1357	1.5E5	56245	1.6E6	6152	166.7	19.0	4.51	8.99	30.0
	SCH 140 API	508.0	419.1	44.45	506.9	644.8	1596	1317	1.4E5	64732	1.8E6	6908	164.6	19.0	4.82	10.7	35.8
	SCH 160 API	508.0	408.0	50.01	563.4	696.2	1596	1282	1.3E5	71959	1.9E6	7516	162.9	19.1	5.03	12.3	41.1

PIPE DATA · TABLES P-1M

DN (mm) [NPS]	CODES	MFR	O.D. (mm)	I.D. (mm)	Wall (mm)	Empty (kg/m)	Waterfilled (kg/m)	External (mm²/mm)	Internal (mm²/mm)	Flow (mm²)	Metal (mm²)	Moment of Inertia (10⁴ mm⁴)	Section Modulus (10³ mm³)	Radius of Gyration (mm)	Cont. Span (m)	Sag (mm)	Design (MPa)	Bursting (MPa)
550 / 22	SCH 10	API	558.8	546.1	6.350	86.29	320.5	1756	1716	$2.3\text{E}5$	11021	$4.2\text{E}5$	1505	195.3	12.6	.658	.680	2.27
		API	558.8	544.5	7.137	96.86	329.7	1756	1711	$2.3\text{E}5$	12370	$4.7\text{E}5$	1684	195.1	13.1	.781	.857	2.86
		API	558.8	543.0	7.925	107.4	338.9	1756	1706	$2.3\text{E}5$	13715	$5.2\text{E}5$	1862	194.8	13.6	.906	1.03	3.45
		API	558.8	541.3	8.738	118.2	348.4	1756	1701	$2.3\text{E}5$	15099	$5.7\text{E}5$	2044	194.5	14.1	1.04	1.22	4.06
	SCH 20 STD	API	558.8	539.8	9.525	128.7	357.5	1756	1696	$2.3\text{E}5$	16436	$6.2\text{E}5$	2219	194.2	14.5	1.16	1.39	4.65
		API	558.8	538.2	10.31	139.1	366.6	1756	1691	$2.3\text{E}5$	17770	$6.7\text{E}5$	2392	194.0	14.8	1.29	1.57	5.24
		API	558.8	536.5	11.13	149.9	376.0	1756	1686	$2.3\text{E}5$	19142	$7.2\text{E}5$	2570	193.7	15.2	1.42	1.76	5.86
		API	558.8	535.0	11.91	160.3	385.0	1756	1681	$2.2\text{E}5$	20467	$7.7\text{E}5$	2740	193.4	15.5	1.54	1.94	6.45
	SCH 30 XS	API	558.8	533.4	12.70	170.6	394.1	1756	1676	$2.2\text{E}5$	21788	$8.1\text{E}5$	2909	193.1	15.8	1.66	2.12	7.05
		API	558.8	520.7	19.05	252.9	465.9	1756	1636	$2.1\text{E}5$	32303	$1.2\text{E}6$	4215	190.9	17.5	2.56	3.58	11.9
	SCH 60	API	558.8	514.3	22.23	293.3	501.1	1756	1616	$2.1\text{E}5$	37465	$1.4\text{E}6$	4834	189.7	18.0	2.94	4.32	14.4
	SCH 80	API	558.8	501.7	28.58	372.7	570.3	1756	1576	$2.0\text{E}5$	47599	$1.7\text{E}6$	6004	187.7	18.8	3.58	5.82	19.4
	SCH 100	API	558.8	489.0	34.93	450.1	637.8	1756	1536	$1.9\text{E}5$	57480	$2.0\text{E}6$	7089	185.6	19.4	4.08	7.36	24.5
	SCH 120	API	558.8	476.3	41.27	525.4	703.6	1756	1496	$1.8\text{E}5$	67107	$2.3\text{E}6$	8092	183.6	19.7	4.47	8.92	29.7
	SCH 140	API	558.8	463.6	47.63	598.8	767.6	1756	1456	$1.7\text{E}5$	76481	$2.5\text{E}6$	9018	181.5	19.9	4.77	10.5	35.0
	SCH 160	API	558.8	450.9	53.98	670.3	829.9	1756	1416	$1.6\text{E}5$	85602	$2.8\text{E}6$	9872	179.5	20.0	5.00	12.1	40.4
600 / 24	SCH 10	API	609.6	596.9	6.350	94.23	374.1	1915	1875	$2.8\text{E}5$	12034	$5.5\text{E}5$	1796	213.3	12.7	.577	.623	2.08
		API	609.6	595.3	7.137	105.8	384.1	1915	1870	$2.8\text{E}5$	13509	$6.1\text{E}5$	2011	213.0	13.3	.688	.785	2.62
		API	609.6	593.8	7.925	117.3	394.2	1915	1865	$2.8\text{E}5$	14980	$6.8\text{E}5$	2224	212.7	13.8	.801	.947	3.16
		API	609.6	592.1	8.738	129.1	404.5	1915	1860	$2.8\text{E}5$	16494	$7.4\text{E}5$	2443	212.5	14.3	.920	1.11	3.72
	SCH 20 STD	API	609.6	590.6	9.525	140.6	414.5	1915	1855	$2.7\text{E}5$	17956	$8.1\text{E}5$	2652	212.2	14.7	1.04	1.28	4.26
		API	609.6	589.0	10.31	152.0	424.5	1915	1850	$2.7\text{E}5$	19415	$8.7\text{E}5$	2860	211.9	15.1	1.15	1.44	4.80
		API	609.6	587.3	11.13	163.8	434.7	1915	1845	$2.7\text{E}5$	20917	$9.4\text{E}5$	3074	211.6	15.4	1.27	1.61	5.36
		API	609.6	585.8	11.91	175.1	444.6	1915	1840	$2.7\text{E}5$	22368	$9.9\text{E}5$	3278	211.4	15.8	1.39	1.77	5.91
	XS	API	609.6	584.2	12.70	186.5	454.5	1915	1835	$2.7\text{E}5$	23815	$1.1\text{E}6$	3481	211.1	16.1	1.50	1.94	6.46
	SCH 30	API	609.6	581.1	14.27	209.0	474.2	1915	1825	$2.7\text{E}5$	26698	$1.2\text{E}6$	3883	210.5	16.6	1.72	2.27	7.56
		API	609.6	577.9	15.88	231.9	494.1	1915	1815	$2.7\text{E}5$	29611	$1.3\text{E}6$	4284	210.0	17.1	1.94	2.60	8.68
	SCH 40	API	609.6	574.6	17.48	254.5	513.9	1915	1805	$2.6\text{E}5$	32508	$1.4\text{E}6$	4678	209.4	17.5	2.15	2.94	9.80
		API	609.6	571.5	19.05	276.7	533.3	1915	1795	$2.6\text{E}5$	35343	$1.5\text{E}6$	5060	208.9	17.9	2.35	3.27	10.9
	SCH 60	API	609.6	560.4	24.61	354.2	600.8	1915	1760	$2.4\text{E}5$	45233	$1.9\text{E}6$	6359	207.0	18.9	2.98	4.46	14.9
	SCH 80	API	609.6	547.7	30.96	440.7	676.3	1915	1721	$2.4\text{E}5$	56285	$2.4\text{E}6$	7751	204.9	19.7	3.56	5.84	19.5
	SCH 100	API	609.6	531.8	38.89	545.9	768.1	1915	1671	$2.2\text{E}5$	69723	$2.9\text{E}6$	9357	202.2	20.3	4.13	7.60	25.3
	SCH 120	API	609.6	517.6	46.02	638.1	848.4	1915	1626	$2.1\text{E}5$	81488	$3.3\text{E}6$	10685	199.9	20.6	4.52	9.21	30.7
	SCH 140	API	609.6	504.9	52.37	717.9	918.1	1915	1586	$2.0\text{E}5$	91686	$3.6\text{E}6$	11778	197.9	20.8	4.79	10.7	35.6
	SCH 160	API	609.6	490.5	59.54	805.6	994.6	1915	1541	$1.9\text{E}5$	$1.0\text{E}5$	$3.9\text{E}6$	12916	195.6	20.9	5.02	12.4	41.2
650 / 26		API	660.4	647.7	6.350	102.2	431.7	2075	2035	$3.3\text{E}5$	13048	$7.0\text{E}5$	2113	231.3	12.9	.510	.575	1.92
		API	660.4	646.1	7.137	114.7	442.6	2075	2030	$3.3\text{E}5$	14648	$7.8\text{E}5$	2367	231.0	13.4	.610	.724	2.41
	SCH 10	API	660.4	644.6	7.925	127.2	453.5	2075	2025	$3.3\text{E}5$	16244	$8.6\text{E}5$	2618	230.7	14.0	.713	.874	2.91
		API	660.4	642.9	8.738	140.1	464.7	2075	2020	$3.3\text{E}5$	17888	$9.5\text{E}5$	2876	230.4	14.5	.822	1.03	3.43
	STD	API	660.4	641.4	9.525	152.5	475.6	2075	2015	$3.2\text{E}5$	19477	$1.0\text{E}6$	3124	230.1	14.9	.928	1.18	3.93
		API	660.4	639.8	10.31	164.9	486.4	2075	2010	$3.2\text{E}5$	21061	$1.1\text{E}6$	3370	229.9	15.3	1.03	1.33	4.43
		API	660.4	638.1	11.13	177.7	497.5	2075	2005	$3.2\text{E}5$	22693	$1.2\text{E}6$	3622	229.6	15.7	1.15	1.48	4.95
		API	660.4	636.6	11.91	190.0	508.3	2075	2000	$3.2\text{E}5$	24269	$1.3\text{E}6$	3865	229.3	16.0	1.25	1.64	5.45
	SCH 20 XS	API	660.4	635.0	12.70	202.3	519.0	2075	1995	$3.2\text{E}5$	25842	$1.4\text{E}6$	4106	229.0	16.3	1.36	1.79	5.95
		API	660.4	631.9	14.27	226.9	540.4	2075	1985	$3.1\text{E}5$	28976	$1.5\text{E}6$	4582	228.5	16.9	1.57	2.09	6.97
		API	660.4	628.6	15.88	251.7	562.1	2075	1975	$3.1\text{E}5$	32144	$1.7\text{E}6$	5058	227.9	17.4	1.78	2.40	8.00
		API	660.4	622.3	19.05	300.5	604.7	2075	1955	$3.0\text{E}5$	38383	$2.0\text{E}6$	5982	226.9	18.3	2.17	3.02	10.1
700 / 28		API	711.2	698.5	6.350	110.1	493.3	2234	2194	$3.8\text{E}5$	14061	$8.7\text{E}5$	2456	249.2	13.0	.455	.534	1.78
		API	711.2	696.9	7.137	123.6	505.1	2234	2189	$3.8\text{E}5$	15787	$9.8\text{E}5$	2751	248.9	13.6	.545	.672	2.24
	SCH 10	API	711.2	695.4	7.925	137.1	516.8	2234	2185	$3.8\text{E}5$	17509	$1.1\text{E}6$	3045	248.7	14.1	.639	.811	2.70
		API	711.2	693.7	8.738	151.0	529.0	2234	2179	$3.8\text{E}5$	19283	$1.2\text{E}6$	3345	248.4	14.6	.738	.954	3.18
	STD	API	711.2	692.2	9.525	164.4	540.7	2234	2174	$3.8\text{E}5$	20997	$1.3\text{E}6$	3635	248.1	15.1	.836	1.09	3.65

PIPE DATA

TABLES P-1.M

DN (mm) / [NPS]	PIPING CODES / MANUF. WEIGHTS	O.D. (mm)	I.D. (mm)	Wall (mm)	Empty (kg/m)	Waterfilled (kg/m)	External (mm²/mm)	Internal (mm²/mm)	Flow (mm²)	Metal (mm²)	Moment of Inertia (10⁴ mm⁴)	Section Modulus (10³ mm³)	Radius of Gyration (mm)	Continuous Span (m)	Sag (mm)	Design (MPa)	Bursting (MPa)
700 / 28	API	711.2	690.6	10.31	177.8	552.3	2234	2170	3.7E5	22707	1.4E6	3922	247.8	15.5	.935	1.23	4.11
	API	711.2	688.9	11.13	191.6	564.4	2234	2164	3.7E5	24468	1.5E6	4216	247.5	15.9	1.04	1.38	4.59
	SCH 20 XS API	711.2	687.4	11.91	204.9	576.0	2234	2159	3.7E5	26171	1.6E6	4500	247.3	16.2	1.14	1.52	5.06
	SCH 30 API	711.2	685.8	12.70	218.2	587.6	2234	2155	3.7E5	27869	1.7E6	4781	247.0	16.6	1.24	1.66	5.52
	API	711.2	679.5	15.88	271.5	634.1	2234	2135	3.6E5	34678	2.1E6	5897	245.9	17.7	1.63	2.23	7.42
	API	711.2	673.1	19.05	324.3	680.2	2234	2115	3.6E5	41423	2.5E6	6981	244.8	18.6	2.00	2.80	9.33
750 / 30	API	762.0	747.7	7.137	132.5	571.6	2394	2349	4.4E5	16926	1.2E6	3165	266.9	13.7	.490	.627	2.09
	API	762.0	746.2	7.925	147.0	584.3	2394	2344	4.4E5	18774	1.3E6	3503	266.6	14.2	.576	.757	2.52
	SCH 10 API	762.0	744.5	8.738	161.9	597.3	2394	2339	4.4E5	20677	1.5E6	3850	266.3	14.7	.667	.890	2.97
	STD API	762.0	743.0	9.525	176.3	609.8	2394	2334	4.4E5	22517	1.6E6	4184	266.1	15.2	.757	1.02	3.40
	API	762.0	741.4	10.31	190.7	622.4	2394	2329	4.3E5	24353	1.8E6	4515	265.8	15.6	.849	1.15	3.83
	API	762.0	739.7	11.13	205.5	635.3	2394	2324	4.3E5	26244	1.8E6	4856	265.5	16.1	.944	1.28	4.28
	API	762.0	738.2	11.91	219.8	647.8	2394	2319	4.3E5	28072	2.0E6	5183	265.2	16.4	1.04	1.42	4.72
	SCH 20 API	762.0	736.6	12.70	234.1	660.2	2394	2314	4.3E5	29896	2.1E6	5508	265.0	16.8	1.13	1.55	5.15
	SCH 30 API	762.0	730.3	15.88	291.4	710.2	2394	2294	4.2E5	37211	2.6E6	6800	263.9	18.0	1.50	2.08	6.92
	XS API	762.0	723.9	19.05	348.1	759.7	2394	2274	4.1E5	44464	3.1E6	8057	262.8	18.9	1.86	2.61	8.69
800 / 32	SCH 10 API	812.8	800.1	6.350	126.0	628.7	2553	2514	5.0E5	16088	1.3E6	3218	285.1	13.1	.367	.467	1.56
	API	812.8	798.5	7.137	141.5	642.3	2553	2509	5.0E5	18065	1.5E6	3607	284.9	13.8	.443	.588	1.96
	API	812.8	797.0	7.925	156.9	655.7	2553	2504	5.0E5	20039	1.6E6	3993	284.6	14.3	.522	.709	2.36
	STD API	812.8	795.3	8.738	172.8	669.6	2553	2499	5.0E5	22072	1.8E6	4390	284.3	14.9	.606	.834	2.78
	API	812.8	793.8	9.525	188.2	683.0	2553	2494	4.9E5	24037	1.9E6	4771	284.0	15.4	.689	.956	3.19
	API	812.8	792.2	10.31	203.6	696.4	2553	2489	4.9E5	25999	2.1E6	5151	283.7	15.8	.774	1.08	3.59
	API	812.8	790.5	11.13	219.4	710.2	2553	2484	4.9E5	28019	2.3E6	5540	283.5	16.2	.862	1.20	4.01
	API	812.8	789.0	11.91	234.7	723.6	2553	2479	4.9E5	29973	2.4E6	5915	283.2	16.6	.949	1.33	4.42
	SCH 20 API	812.8	787.4	12.70	250.0	736.9	2553	2474	4.9E5	31923	2.6E6	6287	282.9	17.0	1.04	1.45	4.83
	SCH 30 API	812.8	781.1	15.88	311.2	790.3	2553	2454	4.8E5	39745	3.2E6	7767	281.8	18.2	1.39	1.94	6.48
	XS API	812.8	777.8	17.48	341.9	817.1	2553	2444	4.8E5	43663	3.5E6	8499	281.3	18.7	1.56	2.19	7.32
	SCH 40 API	812.8	774.7	19.05	372.0	843.3	2553	2434	4.7E5	47504	3.7E6	9211	280.7	19.2	1.72	2.44	8.14
850 / 34	SCH 10 API	863.6	850.9	6.350	133.9	702.6	2713	2673	5.7E5	17101	1.6E6	3638	303.1	13.2	.333	.439	1.46
	API	863.6	849.3	7.137	150.4	716.9	2713	2668	5.7E5	19204	1.8E6	4078	302.8	13.9	.402	.553	1.84
	API	863.6	847.8	7.925	166.8	731.3	2713	2663	5.6E5	21303	1.9E6	4516	302.5	14.4	.475	.667	2.22
	STD API	863.6	846.1	8.738	183.7	746.0	2713	2658	5.6E5	23466	2.1E6	4965	302.3	15.0	.552	.785	2.62
	API	863.6	844.6	9.525	200.1	760.3	2713	2653	5.6E5	25557	2.3E6	5397	302.0	15.5	.629	.899	3.00
	API	863.6	843.0	10.31	216.5	774.6	2713	2648	5.6E5	27644	2.5E6	5828	301.7	15.9	.708	1.01	3.38
	API	863.6	841.3	11.13	233.3	789.3	2713	2643	5.6E5	29795	2.7E6	6269	301.4	16.4	.791	1.13	3.77
	API	863.6	839.8	11.91	249.6	803.5	2713	2638	5.5E5	31874	2.9E6	6694	301.1	16.8	.872	1.25	4.16
	SCH 20 API	863.6	838.2	12.70	265.8	817.6	2713	2633	5.5E5	33950	3.1E6	7117	300.9	17.1	.953	1.36	4.54
	SCH 30 API	863.6	831.9	15.88	331.0	874.5	2713	2613	5.5E5	42278	3.8E6	8798	299.8	18.4	1.28	1.83	6.09
	XS API	863.6	828.6	17.48	363.7	903.0	2713	2603	5.4E5	46452	4.2E6	9631	299.2	19.0	1.45	2.06	6.88
	SCH 40 API	863.6	825.5	19.05	395.8	931.0	2713	2593	5.4E5	50544	4.5E6	10442	298.7	19.5	1.61	2.30	7.66
900 / 36	SCH 10 API	914.4	901.7	6.350	141.8	780.4	2873	2833	6.4E5	18115	1.9E6	4084	321.1	13.3	.303	.415	1.38
	API	914.4	900.1	7.137	159.3	795.6	2873	2828	6.4E5	20343	2.1E6	4578	320.8	13.9	.367	.522	1.74
	API	914.4	898.6	7.925	176.7	810.8	2873	2823	6.3E5	22568	2.3E6	5070	320.5	14.5	.434	.630	2.10
	STD API	914.4	896.9	8.738	194.7	826.5	2873	2818	6.3E5	24861	2.5E6	5576	320.2	15.1	.506	.741	2.47
	API	914.4	895.4	9.525	212.0	841.6	2873	2813	6.3E5	27077	2.8E6	6062	319.9	15.6	.577	.849	2.83
	API	914.4	893.8	10.31	229.3	856.7	2873	2808	6.3E5	29290	3.0E6	6546	319.7	16.1	.651	.957	3.19
	API	914.4	892.1	11.13	247.2	872.3	2873	2803	6.3E5	31570	3.2E6	7043	319.4	16.5	.728	1.07	3.56
	API	914.4	890.6	11.91	264.5	887.4	2873	2798	6.2E5	33775	3.4E6	7522	319.1	16.9	.804	1.18	3.93
	SCH 20 API	914.4	889.0	12.70	281.7	902.4	2873	2793	6.2E5	35976	3.7E6	7999	318.9	17.3	.880	1.28	4.29
	SCH 30 API	914.4	885.9	14.27	316.1	932.4	2873	2783	6.1E5	40367	4.1E6	8944	318.3	18.0	1.03	1.50	5.01
	XS API	914.4	882.6	15.88	350.9	962.8	2873	2773	6.1E5	44812	4.5E6	9895	317.7	18.6	1.19	1.73	5.75
	SCH 40 API	914.4	876.3	19.05	419.6	1023	2873	2753	6.0E5	53584	5.4E6	11750	316.6	19.7	1.50	2.17	7.22

STRUCTURAL STEEL

W SHAPES

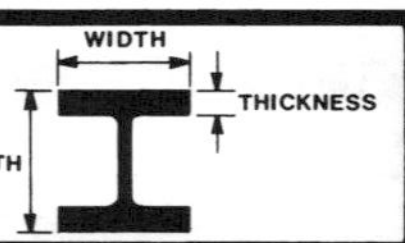

TABLE S-4M

Diagram: I-shape with labels DEPTH, WIDTH, THICKNESS.

Column headings for every group:
DESIGNATION — NOM. SIZE × lb/ft | DIMENSIONS: mm — DEPTH | WIDTH | THICK

W 36

NOM. SIZE × lb/ft	DEPTH	WIDTH	THICK
W 36 × 300	933	423	42.7
× 280	928	422	39.9
× 260	921	420	36.6
× 245	916	419	34.3
× 230	912	418	32.0
× 194	927	308	34.5
× 182	923	307	32.0
× 170	919	306	29.2
× 160	915	305	26.8
× 150	911	304	24.8
× 135	903	304	20.1

W 33

NOM. SIZE × lb/ft	DEPTH	WIDTH	THICK
W 33 × 241	868	403	35.6
× 221	851	403	32.4
× 201	838	401	29.2
× 152	851	294	24.4
× 141	846	294	26.8
× 130	840	292	21.7
× 118	835	292	18.8

W 30

NOM. SIZE × lb/ft	DEPTH	WIDTH	THICK
W 30 × 211	786	384	33.4
× 191	772	382	30.2
× 173	773	381	27.1
× 132	770	268	25.4
× 124	759	267	23.6
× 116	762	267	21.6
× 108	758	266	19.3
× 99	753	265	17.0

W 27

NOM. SIZE × lb/ft	DEPTH	WIDTH	THICK
W 27 × 178	706	358	30.2
× 161	701	356	27.4
× 146	695	355	24.8
× 114	683	256	21.1
× 102	688	254	21.6
× 94	684	254	18.9
× 84	678	253	16.3

W 24

NOM. SIZE × lb/ft	DEPTH	WIDTH	THICK
W 24 × 162	635	329	31.0
× 146	628	328	27.7
× 131	622	327	25.1
× 117	616	325	22.1
× 104	611	324	19.0
× 94	617	230	22.2
× 84	612	229	19.6
× 76	608	228	17.3
× 68	603	228	14.9
× 62	603	179	15.0
× 55	599	178	12.8

W 21

NOM. SIZE × lb/ft	DEPTH	WIDTH	THICK
W 21 × 147	560	318	29.2
× 142	545	334	27.8
× 132	554	316	26.3
× 122	551	315	24.4
× 111	546	312	22.1
× 101	543	312	20.3
× 96	537	310	22.2
× 83	544	212	23.7
× 73	539	211	21.6
× 68	537	210	20.3
× 62	535	209	16.5
× 57	535	166	16.5
× 50	529	166	13.3
× 49	529	166	13.5
× 44	525	165	11.4

W 18

NOM. SIZE × lb/ft	DEPTH	WIDTH	THICK
W 18 × 119	482	286	26.9
× 114	469	301	21.8
× 105	476	284	23.9
× 97	465	283	22.1
× 86	467	282	19.9
× 85	472	298	21.1
× 77	461	300	21.6
× 76	463	280	17.3
× 71	469	194	20.6
× 70	457	223	19.1
× 65	457	222	17.7
× 64	457	222	17.7
× 60	454	221	16.0
× 55	460	191	14.5
× 50	457	190	14.5
× 46	460	154	15.4
× 45	457	191	12.7
× 40	454	190	10.7
× 35	450	152	10.8

W 16

NOM. SIZE × lb/ft	DEPTH	WIDTH	THICK
W 16 × 100	431	265	25.0
× 89	425	263	22.2
× 77	420	261	19.2
× 67	415	260	16.9
× 57	417	181	18.2
× 50	413	180	16.0
× 45	410	179	14.4
× 40	407	178	12.8
× 36	403	177	10.9
× 31	403	140	11.2
× 26	399	140	8.8

W 14

NOM. SIZE × lb/ft	DEPTH	WIDTH	THICK
W 14 × 730	569	454	124.7
× 665	550	448	114.8
× 605	531	442	105.7
× 550	514	437	97.0
× 500	498	432	88.9
× 455	483	428	81.5
× 426	474	424	77.1
× 398	455	421	72.3
× 370	455	418	67.6
× 342	446	416	62.7
× 320	427	424	53.2

W 12

NOM. SIZE × lb/ft	DEPTH	WIDTH	THICK
W 12 × 336	427	340	75.1
× 305	415	336	68.7
× 279	403	334	62.7
× 252	391	330	57.2
× 230	382	328	52.6
× 210	365	325	44.1
× 190	374	322	44.4
× 170	353	319	37.7
× 161	348	317	35.6
× 152	340	315	35.7
× 136	341	314	31.8
× 133	333	313	28.1
× 120	327	310	28.1
× 106	340	306	23.1
× 99	324	309	22.9
× 92	323	308	20.7
× 87	318	308	18.7
× 79	314	307	18.7
× 72	311	306	17.0
× 65	308	305	15.4
× 58	310	254	16.3
× 53	306	254	14.6
× 50	310	205	16.3
× 45	306	204	14.6
× 40	303	203	13.1

W 14 (light)

NOM. SIZE × lb/ft	DEPTH	WIDTH	THICK
W 14 × 311	437	412	58.0
× 283	435	409	57.4
× 257	425	406	53.2
× 233	419	404	48.0
× 211	416	401	49.2
× 193	409	399	42.9
× 176	413	398	41.4
× 159	406	393	39.3
× 145	403	398	36.6
× 132	407	400	33.3
× 120	399	400	30.7
× 109	393	399	30.2
× 99	397	400	26.2
× 90	391	400	25.9
× 84	375	394	27.7
× 82	375	395	21.8
× 78	375	396	20.7
× 74	372	374	21.8
× 68	371	373	19.8
× 61	368	373	17.7
× 53	362	371	16.0
× 48	360	370	14.5
× 43	357	372	13.5
× 38	358	267	13.1
× 34	354	256	11.6
× 30	352	305	9.8
× 26	353	305	10.7
× 22	349	127	8.5

W 10

NOM. SIZE × lb/ft	DEPTH	WIDTH	THICK
W 10 × 112	289	265	31.7
× 100	282	263	28.4
× 89	276	261	25.1
× 77	269	259	22.1
× 72	264	258	20.6
× 68	264	257	19.6
× 60	260	256	17.3
× 54	256	254	15.6
× 49	253	254	14.2
× 45	257	204	15.7
× 39	252	202	13.1
× 33	247	202	11.0
× 30	266	148	11.2
× 26	262	147	9.4
× 22	258	146	8.6
× 19	260	102	8.4
× 17	257	102	7.5
× 16.5	254	102	6.9
× 15	254	101	6.9
× 14	251	101	5.2
× 11.5	251	100	5.2

W 12 (light)

NOM. SIZE × lb/ft	DEPTH	WIDTH	THICK
W 12 × 167	311	324	13.7
× 133	317	317	13.2
× 106	307	306	11.2
× 96	304	305	10.2
× 79	310	305	10.9
× 65	309	305	8.9
× 16.5	305	102	6.7
× 14	303	101	5.7

W 8

NOM. SIZE × lb/ft	DEPTH	WIDTH	THICK
W 8 × 67	229	210	23.7
× 58	222	209	21.6
× 48	216	206	17.4
× 40	210	205	14.2
× 31	203	203	11.0
× 28	207	166	11.8
× 24	201	165	10.2
× 21	210	134	10.2
× 18	207	133	8.4
× 17	203	102	8.4
× 15	206	100	8.0
× 13	203	102	6.5
× 11.5	200	100	5.2

W 6

NOM. SIZE × lb/ft	DEPTH	WIDTH	THICK
W 6 × 25	162	154	11.6
× 20	157	153	9.3
× 16	160	152	10.3
× 15.5	153	152	6.6
× 12	152	152	6.6
× 8.5	148	100	4.9

W 5

NOM. SIZE × lb/ft	DEPTH	WIDTH	THICK
W 5 × 19	131	128	10.9
× 18.5	130	128	10.7
× 16	127	127	9.1

W 4

NOM. SIZE × lb/ft	DEPTH	WIDTH	THICK
W 4 × 13	106	103	8.8

* INDICATES A DIMENSIONAL CHANGE OR SHAPE WAS DISCONTINUED (1978)

References:
- The Rolling Program for American Wide Flange Structural Shapes - Arbed S.A., Luxembourg
- The Rolling Schedule for Wide Flange Shapes - Nippon Steel Corporation, Japan
- The American Institute of Steel Construction

CHANNEL DATA
AMERICAN STANDARD

AMERICAN STANDARD CHANNELS (diagram showing DEPTH, WIDTH, AVERAGE THICKNESS)

DESIGNATION Depth (ins) x wgt lb/ft	DIMENSIONS IN mm DEPTH	WIDTH	Av Th
C 15x50 / x40 / x33.9	381 / 381 / 381	94 / 89 / 86	16.5 / 16.5 / 16.5
C 12x30 / x25 / x20.7	305 / 305 / 305	81 / 77 / 75	12.7 / 12.7 / 12.7
C 10x30 / x25 / x20 / x15.3	254 / 254 / 254 / 254	77 / 73 / 70 / 66	11.1 / 11.1 / 11.1 / 11.1
C 9X20 / x15 / x13.4	229 / 229 / 229	67 / 63 / 62	10.5 / 10.5 / 10.5
C 8x18.75 / x13.75 / x11.5	203 / 203 / 203	64 / 60 / 57	9.9 / 9.9 / 9.9
C 7x14.75 / x12.25 / x9.8	178 / 178 / 178	58 / 56 / 53	9.3 / 9.3 / 9.3
C 6x13 / x10.5 / x8.2	152 / 152 / 152	55 / 52 / 49	8.7 / 8.7 / 8.7
C 5x9 / x6.7	127 / 127	48 / 48	8.1 / 8.1
C 4x7.25 / x5.4	102 / 102	44 / 40	7.5 / 7.5
C 3x6 / x5 / x4.1	76 / 76 / 76	41 / 38 / 36	6.9 / 6.9 / 6.9

ANGLE DATA
WEIGHTS IN KILOGRAMS PER LINEAR METER

TABLES S-5M

UNEQUAL LEGS

SIZES & THICKNESSES (INCHES)	(MILLIMETERS)	1	7/8	3/4	5/8	9/16	1/2	7/16	3/8	5/16	1/4	3/16	1/8
(mm)		25.4	22.2	19	15.9	14.3	12.7	11.1	9.5	7.9	6.4	4.8	3.2
9 x 4 x	229 x 102 x				39.1	35.4	31.7		24.1	20.2			
8 x 6 x	203 x 152 x	65.8	58.2	50.3	42.4	38.2	34.2	30.1					
8 x 4 x	203 x 102 x	55.6	49.3	42.7		32.6	29.2	25.6	22.2				
7 x 4 x	178 x 102 x			39	32.9		26.6	23.5	20.2	16.8			
6 x 4 x	152 x 102 x		40.5	35.1	29.8	26.9	24.1	21.3	18.3	15.3			
6 x 3 1/2 x	152 x 89 x						22.8		17.4	14.6			
5 x 3 1/2 x	127 x 89 x			29.5	25.0		20.2		15.5	12.9	10.4		
5 x 3 x	127 x 76 x						19.0	16.8	14.6	12.2	9.8		
4 x 3 1/2 x	102 x 89 x						17.7	15.8	13.5	11.5	9.2		
4 x 3 x	102 x 76 x				20.2		16.5	14.6	12.6	10.7	8.6		
3 1/2 x 3 x	89 x 76 x						15.2	13.5	11.8	9.8	8.0		
3 1/2 x 2 1/2 x	89 x 64 x						14.0	12.4	10.7	9.1	7.3		
3 x 2 1/2 x	76 x 64 x						12.6	11.3	9.8	8.3	6.7	5.0	
3 x 2 x	76 x 51 x						11.5	10.1	8.8	7.4	6.1	4.6	
*2 1/2 x 2 x	64 x 51 x								7.9	6.7	5.4	4.1	
*2 1/2 x 1 1/2 x	64 x 38 x								5.8		4.7	3.6	
*2 x 1 1/2 x	51 x 38 x										4.1	3.2	2.1
*2 x 1 1/4 x	51 x 32 x										3.8	2.9	
*1 3/4 x 1 1/4 x	44 x 32 x										3.5	2.7	1.8

EQUAL LEGS

SIZES & THICKNESSES (INCHES)	(MILLIMETERS)	1 1/8	1	7/8	3/4	5/8	9/16	1/2	7/16	3/8	5/16	1/4	3/16	1/8
(mm)		28.6	25.4	22.2	19	15.9	14.3	12.7	11.1	9.5	7.9	6.4	4.8	3.2
8 x 8 x	203 x 203 x	84.7	75.9	67	57.9	48.7	44	39.3						
6 x 6 x	152 x 152 x		55.7	49.3	42.7	36	32.6	29.2	25.6	22.2	18.5			
5 x 5 x	127 x 127 x			40.5	35.1	29.8		24.1	21.3	18.3	15.3			
4 x 4 x	102 x 102 x				27.5	23.4		19	16.8	14.6	12.2	9.8		
3 1/2 x 3 1/2 x	89 x 89 x							16.5	14.6	12.6	10.7	8.6		
3 x 3 x	76 x 76 x							14	12.4	10.7	9.1	7.3	5.5	
2 1/2 x 2 1/2 x	64 x 64 x							11.5		8.8	7.4	6.1	4.6	
2 x 2 x	51 x 51 x									7	5.8	4.8	3.6	
*1 3/4 x 1 3/4 x	44 x 44 x									5	4.1	3.2	2.5	
*1 1/2 x 1 1/2 x	38 x 38 x											3.5	2.7	1.8
*1 1/4 x 1 1/4 x	32 x 32 x											2.9	2.2	1.5
*1 x 1 x	25 x 25 x											2.2	1.7	1.2

THE MOST COMMON SIZES OF PAPERS FOR GENERAL USE ARE THE ISO "A" SERIES DESIGNATED: A0, A1, A2, A3, ETC., WITH LENGTH TO WIDTH RATIO OF: LENGTH = WIDTH x SQUARE ROOT 2 (1.414). THE AREA OF THE LARGEST SHEET, A0, IS EQUAL TO ONE SQUARE METER

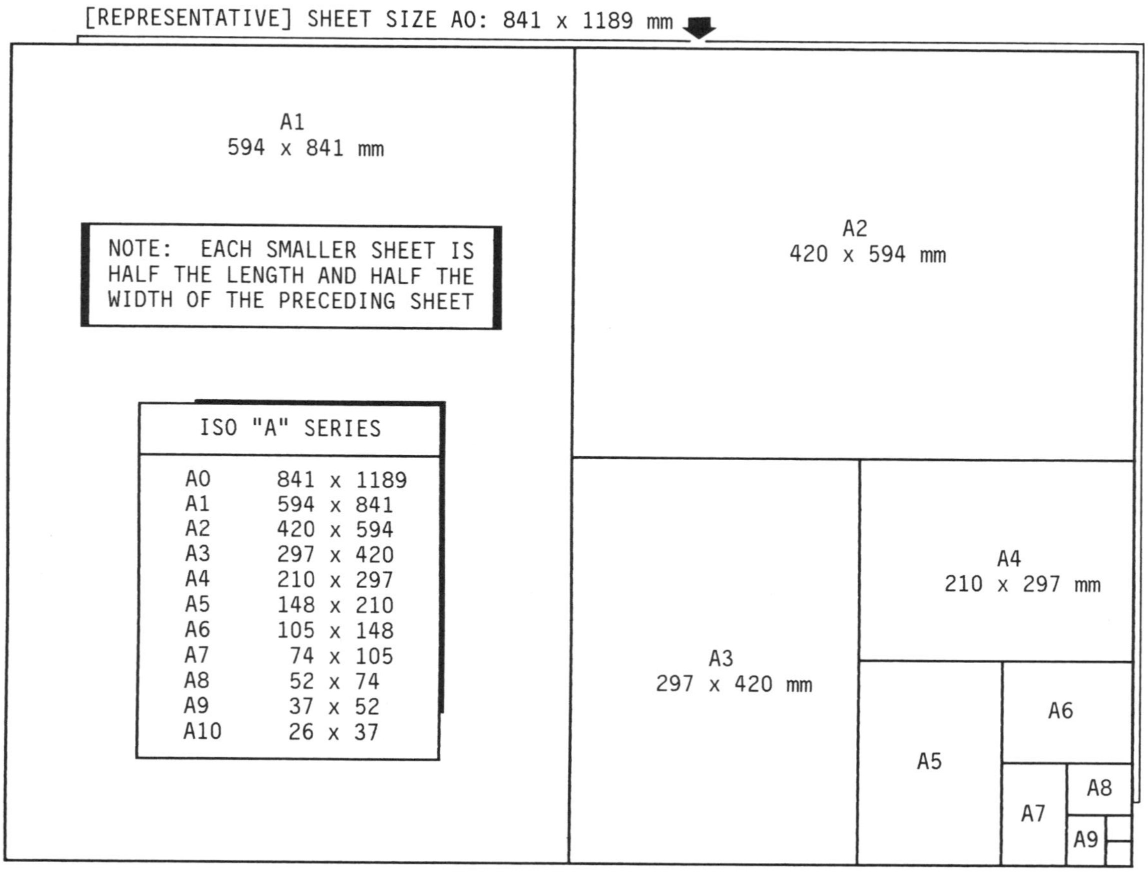

[REPRESENTATIVE] SHEET SIZE A0: 841 x 1189 mm ◄

NOTE: EACH SMALLER SHEET IS HALF THE LENGTH AND HALF THE WIDTH OF THE PRECEDING SHEET

ISO "A" SERIES

A0	841 × 1189
A1	594 × 841
A2	420 × 594
A3	297 × 420
A4	210 × 297
A5	148 × 210
A6	105 × 148
A7	74 × 105
A8	52 × 74
A9	37 × 52
A10	26 × 37

A1
594 × 841 mm

A3
297 × 420 mm

A2
420 × 594 mm

A4
210 × 297 mm

A5

A6

A7

A8

A9

COMPARISON OF ISO SHEET SIZES WITH USA SHEET SIZES, FOR DRAWINGS

ISO DESIGNATION	WIDTH		LENGTH		US SIZES	
	mm	inches	mm	inches	LETTER	inches
--	--	--	--	--	F	28.0 × 40.0
A0	841	33.11	1189	46.81	E	34.0 × 44.0
A1	594	23.39	841	33.11	D	22.0 × 34.0
A2	420	16.54	594	23.39	C	17.0 × 22.0
A3	297	11.69	420	16.54	B	11.0 × 17.0
A4	210	8.27	297	11.69	A	8.5 × 11.0

VALVE DATA - RUN LENGTHS

DIMENSIONS IN MILLIMETERS

TABLE V-1M

	FLANGE CLASS	50	65	80	100	150	200	250	300	350	400	450	500	600
STEEL GATE VALVES SOLID WEDGE & DOUBLE-DISC (SPLIT-WEDGE)	FLANGED 150	178	190	203	229	267	292	330	356	381	406	432	457	508
	BEVELED 150	216	241	283	305	403	419	457	502	572	610	660	711	813
	300	216	241	283	305	403	419	457	502	762	838	914	991	1143
	600	292	330	356	432	559	660	787	838	889	991	1092	1194	1397
	900	368	419	381	457	610	737	838	965	1029	1130	1219	1321	1549
	1500	368	419	470	546	705	832	991	1130	1257	1384	1537	1664	1943
	2500	451	508	578	673	914	1022	1270	1422					
STEEL GLOBE VALVES LIFT CHECK VALVES	150	203	216	241	292	406	495	622	698					
	300	267	292	318	356	444	559	622	711					
	600	292	330	356	432	559	660	787	838					
	900	368	419	381	457	610	737	838	965					
	1500	368	419	470	546	705	832	991	1130					
	2500	451	508	578	673	914	1022	1270	1422					
SWING CHECK VALVES TILTING DISC CHECK VALVES	T-D 150	203 203	216 216	241 241	292 292	356 356	495 495	622 622	698 698					
	T-D 300	267 267	292 292	318 318	356 356	444 444	533 533	622 622	711 711					
	T-D 600	292 292	330 330	356 356	432 432	559 559	660 660	787 787	838 838					
	T-D 900	- 368	- 419	381 381	457 457	610 610	737 737	838 838	965 965					
	T-D 1500	368 368	419 419	470 470	546 546	705 705	832 832	991 991	- 1130					
	2500	451	508	578	673	914	1022	1270	1422					

(Column headings 50–600: NOMINAL DIAMETER [DN] OF PIPE)

NOTES

DIMENSIONS IN THIS TABLE CONFORM TO ANSI B16.10 AND APPLY TO FLANGED VALVES AND VALVES WITH ENDS BEVELLED FOR WELDING AS SHOWN:

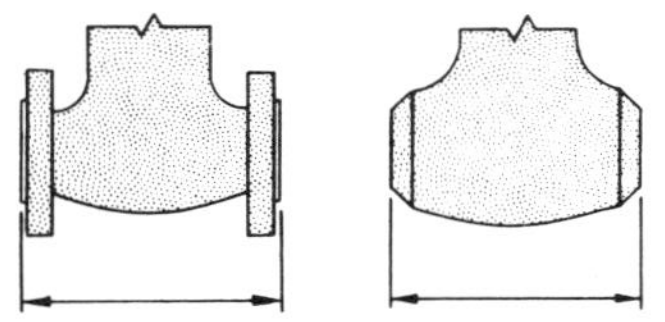

Tabled Dimension

FOR FLANGED VALVES THE TABLED DIMENSION INCLUDES ALLOWANCE FOR BOTH RAISED FACES OF THE VALVE. FOR CLASSES 150 AND 300 VALVES, 1.6mm HAS BEEN INCLUDED FOR EACH RAISED FACE AND FOR VALVES OF CLASS 600 AND ABOVE, 6.4mm HAS BEEN INCLUDED FOR EACH RAISED FACE.

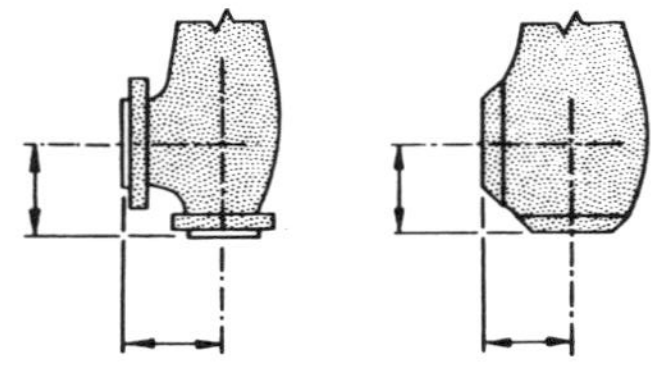

Half Tabled Dimension

FOR ANGLE GLOBE & ANGLE LIFT-CHECK VALVES, HALVE THE TABLED DIMENSION TO OBTAIN CENTER-TO-FACE DIMENSIONS.

ACKNOWLEDGMENTS

Photographs and illustrations reproduced courtesy of the following companies. [Page numbers are bracketed]:

[6]   BACKING RING - Tube Turns (Div of Chemtron Inc)

[7]   ELBOWS & RETURNS - Taylor Forge Inc
      REDUCERS - Tube Turns (Div of Chemtron Inc)

[8]   FLANGES: WELDING NECK, SLIP-ON, REDUCING SLIP-ON - Taylor Forge Inc
      EXPANDER FLANGE - Tube Turns (Div of Chemtron Inc)

[9]   LAP-JOINT FLANGE - Ladish Company
      TEES - Tube Turns (Div of Chemtron Inc)
      WELDOLET - Bonney Forge

[10] SWEEPOLET - Bonney Forge
      BUTT-WELDING CROSS, LATERAL, NIPPLE - Tube Turns (Div of Chemtron Inc)

[11] BUTT-WELDING CAP - Crane Company

[12] FULL-COUPLING, REDUCER - Crane Company
      SOCKET-WELDING REDUCER INSERTS - Ladish Company

[13] SOCKET-WELDING FLANGE - Taylor Forge Inc
      SOCKET WELDING: ELBOWS, TEE, LATERAL and CROSS - Crane Company

[14] SOCKET-WELDING HALF-COUPLING - Crane Company
      SOCKOLET - Bonney Forge
      SOCKET-WELDING CAP - Henry Vogt Machine Co

[15] FULL-COUPLING - Crane Company

[16] REDUCING COUPLING - Crane Company
      UNION - Stanley G. Flagg & Co Inc
      HEXAGON BUSHING - Crane Company

[17] THREADED ELBOWS, 45 and 90 DEGREE - Crane Company
      THREADED FLANGE - Taylor Forge Inc

[18] THREADED LATERAL, THREADED CROSS - Crane Company
      THREDOLET, THREADED ELBOLET, THREADED LATROLET - Bonney Forge

[19] THREADED CAP - Henry Vogt Machine Co
      THREADED BARSTOCK PLUG - Ladish Company

[21] MACHINE BOLT & NUT, and STUDBOLT & NUTS - Crane Company

[23] VICTAULIC COMPRESSION SLEEVE COUPLING - Victaulic Company

[25] REINFORCING SADDLES - Crane Company

[31] GATE VALVE (OS&Y, bolted bonnet, rising stem), GLOBE VALVE (OS&Y, bolted bonnet, rising stem), GATE VALVE (IS, bolted bonnet, non-rising stem) - Jenkins Bros. Valve Manufacturers

[32] LANTERN RING - Wm. Powell Co
      PACKLESS VALVE - Crane Co
      BELLOWS-SEAL VALVE - Henry Vogt Machine Co
      COCKS - Wm. Powell Co
      HAMMER-BLOW HANDWHEEL - Wm. Powell Co

[33] SPUR-GEAR OPERATOR and BEVEL-GEAR OPERATOR - Crane Company

[33] ELECTRIC MOTOR OPERATOR, PNEUMATIC OPERATOR - Wm. Powell Co
      QUICK-ACTING VALVES:
        ROTATING STEM ON GLOBE VALVE - Jenkins Bros. Valve Manufacturers
        SLIDING STEM ON GATE VALVE - Lunkenheimer Company

[35] SOLID WEDGE GATE VALVE - Wm. Powell Co
      SINGLE-DISC PARALLEL-SEATS GATE VALVE - Henry Vogt Machine Co
      PLUG GATE VALVE - Crane Company

[36] GLOBE VALVES - Henry Vogt Machine Co, WYE-BODY GLOBE VALVE (incorporating composition disc) - Jenkins Bros. Valve Manufacturers
      NEEDLE VALVE, ROTARY-BALL VALVES - Lunkenheimer Company

[37] BUTTERFLY VALVE (WAFER TYPE) - Lunkenheimer Company
      SWING CHECK VALVES - Jenkins Bros. Valve Manufacturers, Walworth Co,
      PISTON-CHECK VALVE & STOP CHECK VALVE - Rockwell Mfg Co

[38] SAFETY VALVE, RELIEF VALVE, BALL FLOAT VALVE, BLOWOFF VALVE - Crane Co
      FLUSH-BOTTOM TANK VALVE (GLOBE TYPE) - Wm. Powell Co

[39] INVERTED-BUCKET TRAP Armstrong Machine Works

[93] DRIPSHIELD - Wm. Powell Co

[110] SWARTWOUT HEAD - Crane Co

[116] SHELL-AND-TUBE HEAT EXCHANGER WITH REMOVABLE TUBE BUNDLE - Bell & Gosset and California Hydronics Corporations

[119] LEVEL GAGE ASSEMBLY - Wm. Powell Co

[120] ROTAMETER - Instruments Division of Schutte & Koerting Company

[123] JACKETED PIPE & HOSE - Parkes-Cramer Company